The Management of Industrial Forest Plantations

Managing Forest Ecosystems

Volume 33

Series Editors:

Aims & Scope:

Well-managed forests and woodlands are a renewable resource, producing essential raw material with minimum waste and energy use. Rich in habitat and species diversity, forests may contribute to increased ecosystem stability. They can absorb the effects of unwanted deposition and other disturbances and protect neighbouring ecosystems by maintaining stable nutrient and energy cycles and by preventing soil degradation and erosion. They provide much-needed recreation and their continued existence contributes to stabilizing rural communities.

Forests are managed for timber production and species, habitat and process conservation. A subtle shift from *multiple-use management to ecosystems management* is being observed and the new ecological perspective of *multi-functional forest management* is based on the principles of ecosystem diversity, stability and elasticity, and the dynamic equilibrium of primary and secondary production.

Making full use of new technology is one of the challenges facing forest management today. Resource information must be obtained with a limited budget. This requires better timing of resource assessment activities and improved use of multiple data sources. Sound ecosystems management, like any other management activity, relies on effective forecasting and operational control.

The aim of the book series ***Managing Forest Ecosystems*** is to present state-of-the-art research results relating to the practice of forest management. Contributions are solicited from prominent authors. Each reference book, monograph or proceedings volume will be focused to deal with a specific context. Typical issues of the series are: resource assessment techniques, evaluating sustainability for even-aged and uneven-aged forests, multi-objective management, predicting forest development, optimizing forest management, biodiversity management and monitoring, risk assessment and economic analysis.

For further volumes:
http://www.springer.com/series/6247

José G. Borges • Luis Diaz-Balteiro
Marc E. McDill • Luiz C.E. Rodriguez
Editors

The Management of Industrial Forest Plantations

Theoretical Foundations and Applications

Springer

Editors
José G. Borges
Forest Research Centre
School of Agriculture
University of Lisbon
Lisbon, Portugal

Marc E. McDill
Department of Ecosystem Science
and Management
Pennsylvania State University
University Park, PA, USA

Luis Diaz-Balteiro
Departamento de Economía y Gestión
Forestal
ETS Ingenieros de Montes
Technical University of Madrid
Madrid, Spain

Luiz C.E. Rodriguez
Department of Forest Sciences
University of São Paulo
São Paulo, Brazil

ISSN 1568-1319 ISSN 2352-3956 (electronic)
ISBN 978-94-017-8898-4 ISBN 978-94-017-8899-1 (eBook)
DOI 10.1007/978-94-017-8899-1
Springer Dordrecht Heidelberg New York London

Library of Congress Control Number: 2014947533

Printed on acid-free paper

Springer is part of Springer Science+Business Media (www.springer.com)

Preface

The Management of Industrial Forest Plantations: Theoretical Foundations and Applications answers calls from both the research and the professional communities for a synthesis of current knowledge about industrial forestry management planning processes. The emphasis of the book is thus on covering components of the forest supply chain ranging from modelling techniques to management planning approaches and information and communication technology support. We think that this synthesis may provide effective support to research and outreach activities that focus on forest management and specifically on industrial plantations management. It may contribute further to support forest managers when developing industrial plantations management plans. Moreover, the structure of the book as well the standardized design of individual chapters – that address a wide range of topics relevant to industrial plantations management scheduling – targets the support of both undergraduate and graduate forest management classes. For that purpose, the introduction to each chapter includes a *What you will learn in this chapter* section to highlight and convey the main objectives of the chapter and thus help structure the reading by learners and trainees. All chapters include also a section with exercises, problems and review questions to further support the structuring of the knowledge acquired by students.

The book addresses thoroughly the theoretical foundations for the integration of information and knowledge to support industrial plantations management scheduling. It presents the whole range of models and methods that are useful to support all industrial forest management planning levels – e.g. strategic, tactical and operational as well as transportation and routing. It further presents approaches to enhance the efficiency and the effectiveness of industrial plantations management planning in a context of global change – e.g. the development of decision support systems to integrate management planning levels, the dynamic design of management units taking advantage of recent remote sensing techniques or the use of advanced techniques for developing integrated sustainability indexes. The book further addresses emergent topics that frame the forestry business – e.g. carbon sequestration and certification – as well as approaches to enhance its competitiveness, e.g. supply chain management.

Nevertheless the presentation is structured such that it may be followed by learners and trainees with diverse backgrounds and education. No specific preparation is assumed in any scientific field. Chapters may be read independently of each other and yet there is a logical sequence and a read thread linking chapters in each part and parts in the book. In fact, the presentation of concepts, models, methods and management planning approaches is complemented by the discussion of applications in 27 Management Planning in Actions boxes distributed all over the book. These applications highlight the linkage between theory and practice and the contribution of models, methods and management planning approaches to the efficiency and the effectiveness of industrial plantations management planning.

The first part of *The Management of Industrial Forest Plantations: Theoretical Foundations and Applications* addresses the context as well as the models and methods to develop industrial plantations management plans. It starts by presenting the objectives and the importance of industrial forest plantations and by highlighting the need of tools to address its economic goals as well as its social and environmental contexts (Chap. 1). Afterwards, it provides an overview of the planning process, ranging from data acquisition to monitoring and re-planning sub-processes, as well as of the management of information needed for it to unfold (Chap. 2). The ensuing discussion of planning levels, of data requirements, acquisition, and management, and of planning models sets the stage for the remaining chapters in the first part of the book. The book presents and explores models to project the growth of industrial plantations (Chap. 3) under key silvicultural activities (Chap. 4). Silviculture and growth and yield models provide the information needed to enumerate adequate technical plans (prescriptions) to manage each stand in an industrial forest plantation. The economic concepts and methods needed for the forest manager to select, from within this set of feasible technical prescriptions, the plan to be implemented are presented and discussed in Chap. 5. Chapters 3, 4 and 5 thus present in a logical sequence the information that must be fed into management scheduling models to address strategic (Chap. 6), tactical and operational (Chap. 7) and transportation and routing (Chap. 8) planning processes. Building from the Chap. 2 overview of planning models, a wide range of management planning techniques and approaches is presented in detail in Chap. 6, starting with a simple and yet comprehensive numerical example and going up to real-world applications. The emphasis is in building competences to understand and develop methods to address strategic management scheduling and to interpret its solutions and assess its relative merits. The potential of these techniques to address tactical and operational and transportation and routing planning processes is discussed further in Chaps. 7 and 8, respectively.

The second part of the book addresses advanced management planning topics that are relevant to industrial plantations. The role of information and communication technology support is discussed in Chap. 9. The focus is on the design and architecture of decision support systems that may help integrate management planning levels and manage the forest products supply chain. The impact of risk and uncertainty in industrial plantations management scheduling and the development of methods to address it are discussed in Chap. 10. Methods and criteria to determine clone recommendations for each industrial forest plantation management unit are presented in Chap. 11. The dynamic design of management units to improve the efficiency of industrial forest management scheduling as compared with permanent stand compartments is discussed in Chap. 12. The last chapter in this part of the book (Chap. 13) presents a methodology to aggregate sustainability indicators in order to select the optimal sustainable management plan for an industrial plantation.

The third part of the book starts by addressing emergent topics that frame the forestry business. Chapter 14 focusses on the measurement and modelling of carbon stock changes as well as on the economic optimisation of plantations with joint production of logs and carbon. It further addresses the trading of plantation-derived carbon units. Chapter 15 examines the differences in the certification standards for industrial forest plantations around the world, and compares these to the standards used for the harvesting of natural and semi-natural forests. The forestry raw material supply chain is characterized in Chap. 16 while the problems of the pulpwood supply chain – procurement, transportation, inventory, production and demand planning – are discussed and modelled in Chap. 17. Finally Chap. 18 discusses the use of a Life Cycle Assessment (LCA) methodology to assess the environmental impacts of forest operations and pulp manufacture.

Addressing successfully the wide range of topics relevant to industrial plantations management scheduling was only possible because of the voluntary effort and dedication of all authors over a 3-year period. We acknowledge and thank the valuable contribution by all authors involved in this project. We also thank the reviewers of our book proposal for their valuable comments as well as the editors of this Springer Book Series for their encouragement to the development of this project, and we acknowledge the support from the Erasmus Mundus Master Course Mediterranean Forestry and Natural Resources Management (MEDfOR) and from the European Union's Seventh Programme for research, technological development and demonstration under grant agreement n° PIRSES-GA-2010-269257 (ForEAdapt - Knowledge exchange between Europe and America on forest growth models and optimization for adaptive forestry). We hope that the book structure and contents are able to meet the challenge of providing a learned and effective synthesis of the current knowledge about industrial forestry management planning processes and may thus become a useful learning and reference tool to students, researchers and forest practitioners interested in this field.

Lisbon, Portugal	José G. Borges
Madrid, Spain	Luis Diaz-Balteiro
University Park, PA, USA	Marc E. McDill
Piracicaba, Brazil	Luiz C.E. Rodriguez

Contents

Part II Advanced Management Planning Topics

Part III The Industrial Forestry Supply Chain

Contributors

Editors and Authors:

José G. Borges Forest Research Centre, School of Agriculture, University of Lisbon, Lisbon, Portugal

Luis Diaz-Balteiro Departamento de Economía y Gestión Forestal, ETS Ingenieros de Montes, Technical University of Madrid, Madrid, Spain

Marc E. McDill Department of Ecosystem Science and Management, Pennsylvania State University, University Park, PA, USA

Luiz C.E. Rodriguez Department of Forest Sciences, University of São Paulo, São Paulo, Brazil

Other Authors:

Roland Agnemo Domsjö Fabriker AB, Örnsköldsvik, Sweden

Clayton Alcarde Alvares Instituto de Pesquisas e Estudos Florestais (IPEF), Forest Productivity Cooperative (FPC), Piracicaba, SP, Brazil

Jean François Audy Université Laval, Québec, QC, Canada

Maurel Behling Embrapa Agrossilvipastoril – CPAMT, Sinop, MT, Brazil

Staffan Berg The Forestry Research Institute of Sweden (Skogforsk), Uppsala Science Park, Uppsala, Sweden

Mercedes Bertomeu Grado en Ingeniería Forestal y del Medio Natural, Centro Universitario de Plasencia, University of Extremadura, Spain

Kevin Boston Department of Forest Engineering, Resources and Management, Oregon State University, Corvallis, OR, USA

Marc-André Carle Department of Operations and Decision Systems, Laval University, Laval, Canada

Sophie D'Amours Department of Mechanical Engineering, Laval University, Laval, Canada

Rafael Epstein Department of Industrial Engineering, University of Chile, Santiago, Chile

Ljusk Ola Eriksson Department of Forest Management and Geomatics, Swedish University of Agricultural Science, Uppsala, Sweden

André O. Falcão Faculdade de Ciências, University of Lisbon, Lisbon, Portugal

Gumersindo Feijoo Department of Chemical Engineering, University of Santiago de Compostela, A Coruña, Spain

Liliana Ferreira School of Technology and Management, Polytechnic Institute of Leiria, Leiria, Portugal

Jordi Garcia-Gonzalo Forest Research Centre, School of Agriculture, University of Lisbon, Lisbon, Portugal

Juan Carlos Giménez Grado en Ingeniería Forestal y del Medio Natural, Centro Universitario de Plasencia, University of Extremadura, Spain

José Leonardo de Moraes Gonçalves ESALQ, Departamento de Ciências Florestais, USP, Universidade de São Paulo, Piracicaba, SP, Brazil

Sara González-García Department of Chemical Engineering, University of Santiago de Compostela, A Coruña, Spain

Tero Heinonen University of Eastern Finland, Joensuu, Finland

John L. Innes Department of Forest Resources, University of British Columbia, Vancouver, BC, Canada

Bruce Manley School of Forestry, University of Canterbury, Christchurch, New Zealand

Manuel F. Marey Department of Agroforestry Engineering, Universidade de Santiago de Compostela, Lugo, Spain

Alexandra S. Marques INESC Tec, Porto, Portugal

Susete Marques Forest Research Centre, School of Agriculture, University of Lisbon, Lisbon, Portugal

John Paul McTague Warnell School of Forestry, University of Georgia, Athens, GA, USA

María Teresa Moreira Department of Chemical Engineering, University of Santiago de Compostela, A Coruña, Spain

Silvana Ribeiro Nobre Department of Forest Management, Atrium Forest Consulting, São Paulo, Brazil

Petteri Packalén University of Eastern Finland, Joensuu, Finland

Maria Pasalodos-Tato Instituto Nacional de Investigación Agraria, Madrid, Spain

Timo Pukkala University of Eastern Finland, Joensuu, Finland

Carlos Romero ETS Ingenieros de Montes, Technical University of Madrid, Madrid, Spain

Mikael Rönnqvist FORAC, Université Laval, Québec, QC, Canada

Luciana Duque Silva ESALQ, Departamento de Ciências Florestais, USP, Universidade de São Paulo, Piracicaba, SP, Brazil

Anna Tikina Department of Forest Resources, University of British Columbia, Vancouver, BC, Canada

Victor A. Valdebenito School of Forest Resources, Pennsylvania State University, University Park, PA, USA

Andres Weintraub Department of Industrial Engineering, University of Chile, Santiago, Chile

Aaron R. Weiskittel School of Forest Resources, University of Maine, Orono, ME, USA

Part I
Industrial Forest Plantations Management Planning

Chapter 1
The Importance of Industrial Forest Plantations

Luiz C.E. Rodriguez, Maria Pasalodos-Tato, Luis Diaz-Balteiro, and John Paul McTague

1.1 Introduction

The tools used by planners to manage trees planted for industrial purposes are the main subject of this book. And why should we care about these tools? Well, they are probably being used to help us on how we manage approximately 264 million hectares of forest plantations (FAO 2010) currently cultivated to satisfy a significant part of our current levels of demand for wood, fibers and other forest products. The area in which these forest plantations grow, when compared to agricultural standards, represent 17.4 % of the total arable land and land under permanent crops in the world (FAO 2013a), and is equivalent in extent to the area globally harvested in 2010 to feed the world with rice and soybeans.[1]

[1] According to FAO the area of soybeans and rice paddy (rice grain after threshing and winnowing, also known as rice in the husk and rough rice, used mainly for human food) harvested in 2012 was approximately 105 and 163,2 million hectares, respectively. (FAO 2014).

L.C.E. Rodriguez (✉)
Department of Forest Sciences, University of São Paulo, São Paulo, Brazil
e-mail: lcer@usp.br

M. Pasalodos-Tato
Instituto Nacional de Investigación Agraria, Madrid, Spain
e-mail: pasalodos.maria@inia.es

L. Diaz-Balteiro
Departamento de Economía y Gestión Forestal, ETS Ingenieros de Montes, Technical University of Madrid, Madrid, Spain
e-mail: luis.diaz.balteiro@upm.es

J.P. McTague
Warnell School of Forestry, University of Georgia, Athens, GA, USA
e-mail: forestbiometrician@yahoo.com

J.G. Borges et al. (eds.), *The Management of Industrial Forest Plantations: Theoretical Foundations and Applications*, Managing Forest Ecosystems 33,
DOI 10.1007/978-94-017-8899-1_1, © Springer Science+Business Media Dordrecht 2014

We plant trees to respond to increasing levels of demand in many different ways. However, and essentially, we manage planted forests to function as main sources of raw materials for vital industrial activities like fuel wood for drying grains, fibers for the production of pulp, resins for the chemical industry, sawn wood for building houses and even cork for the winery industry.

Planted in large scales, these forest plantations require skilled professionals to plan and accommodate technical requirements and conflicting economic, social and environmental interests. Most of the techniques presented in this book have been developed as a response to challenges most forest managers face on a daily basis.

The decision-making processes handled by forest managers demand analytical systems and computational tools to deal with scheduling, logistical and assignment problems. Moreover, this book provides a comprehensive overview of these and many other forest planning techniques, authored by contributors with many years of experience in their areas of expertise.

Initially, it might be necessary to summarize the importance of industrial forests for the current level of industrial development on a global scale. Although regularly challenged by new technological advances, it is important to realize that trees are still one of the most efficient ways to produce highly demanded materials, with minimal collateral negative impacts. The argument stands in the principle that, being a renewable organism that synthesizes carbon fueled by sun light, the sustainable use of planted forests reduces the impact over other natural resources and fossil fuels. Furthermore, the antagonism between planted forests and natural forests has diminished (Paquette and Messier 2010), as well as its complementary and compatible coexistence with wild land native forests in many different landscapes and regional contexts (Boyle 1999).

1.1.1 Definition of Industrial Forest Plantations

Planted forests, *forest plantations* or *industrial forest plantations* are terms used in this book interchangeably, and refer to areas dedicated to the cultivation of trees that end up being harvested or treated to satisfy direct or indirect human consumption needs. Our main interest here involves the concept of planning for the efficient and effective cultivation of forests planted to attend human goals.

Despite this clarification, it is noteworthy to comment what these terms mean in different literature and sources of information. The issue of using a universally accepted and clear set of terms to address the result of humans planting trees is surprisingly controversial and has been addressed by different experts (Evans et al. 2009; Helms et al. 2003).

Moreover, the meaning of these terms has been changing along the years, implying that the differentiation among them is not as straightforward as one could imagine. It could be said that the concept of forest plantations is the one that suffered many modifications in its definition over the years.

Focusing initially on reports delivered by FAO, a first definition of man-made planted forests was stated as "a forest crop raised artificially, either by sowing or planting. This could be interpreted to include all forms of artificial regeneration but no natural regeneration" (FAO 1967). Some years later, the Tropical Forest Resources Assessment Project (FAO 1981; Varmola and Del Lungo 2003) consolidates the definition a bit more by excluding those stands that, although being established by artificial regeneration, were essentially similar to the ones they were replacing. This publication was the first one that introduced the differences between industrial and non-industrial forest plantations.

Ten years later, FAO's FRA 1990 modified again the term, and plantation forests becomes "forests established artificially by afforestation on lands which previously did not carry forest within living memory; or forests established artificially by reforestation of land which carried forest before and involves the replacement of indigenous species by a and essentially different species or genetic variety".

CIFOR (2002) contributed with their own definition of forest plantation and industrial plantation. A forest plantation, for CIFOR, is a forest established by planting or/and seeding introduced species or, in some cases, indigenous species, in the process of afforestation or reforestation. Industrial plantations are intensively managed forest stands, usually even aged with regular spacing, established to provide material for sale locally or outside the immediate region, by planting or/and seeding in the process of afforestation or reforestation. Moreover, CIFOR refers to industrial plantations as constituted with introduced species (all planted stands) and/or of one or two indigenous species, usually either large scale or integrated to one of a few large-scale industrial enterprises in the landscape.

Simultaneously, a joint effort in 2002 to harmonize forest definitions involved FAO, ITTO, CIFOR, IPCC, UNEP and IUFRO. The meeting used the following interim working definition of planted forest: forest stand in which trees have predominantly been established by planting, deliberate seeding or coppicing, where the coppicing is of previously planted trees (includes all stands established by planting or seeding of both native and nonnative species). Planted forest were divided into forest plantation and other planted forest. There was also agreement on a working definition for forest plantation: forest stand in which trees have been established by planting or/and deliberate seeding or coppicing (where the coppicing is of previously planted trees) with either native species or non-native species that meet all the following criteria: one or two or a few species; even-aged; regular spacing. The joint effort also recommended that the term plantation forest, which in essence denotes the same as forest plantation, should preferably be replaced by the term forest plantation.

Another decade after and the FRA 2010 (FAO 2010) recognizes the existence of a continuum from primary forests with undetected human activity to intensively managed planted forests of introduced species, primarily managed for a single product, often on a relatively short rotation, and frequently consisting of only one species – in some cases a single clone. Three groups were accordingly created: primary forests, other naturally regenerated forests and planted forests, in which a sub-group of planted forests composed primarily of introduced species is considered.

Similar to FAO's definitions for planted forest (FAO 2002a, b), the Forest Stewardship Council (FSC 2003) refers to plantations as "forest areas lacking most of the principal characteristics and key elements of native ecosystems, which result from the human activities of planting, sowing or intensive silvicultural treatments". FSC's definition includes also naturally regenerated forests that have been simplified or degraded by silvicultural treatments, resulting in the loss of the key elements of the native forest ecosystem. In both FSC (2003) and FAO (2012a) definitions of planted forests and forest plantations, both categories comprise productive and protective forests.

Currently, the term planted forest, when composed of trees established through planting and/or through deliberate seeding of native or introduced species, extends the concept of forest plantations used in previous global FAO assessments, merging the existing categories until FRA 2005 (FAO 2006), namely plantation forests and semi-natural forests. The basis for the merge was the argument that semi-natural forests have more similarities with plantation forest than with natural forest.

Consistently with all above definitions, Overbeek et al. (2012) refers to industrial plantations as large-scale, intensively-managed, even-aged monocultures, mostly exotic trees like fast-growing eucalyptus, pine and acacia species, destined for industrial processes that produce pulp and paper as well as rubber and oil palm products. Complementarily and synthetically stated, it is also noteworthy recall Evans' definition of industrial forest plantations: a type of forest plantation that have, as their main objective, to grow a product efficiently (Evans 2004).

Increasingly, the world plants more trees in intensively managed small areas to produce enough wood to meet the world's requirements, mitigating the pressure of timber harvests over large areas of natural forests. It seems like we are getting close to the desirable scenario proposed by Botkin and Sedjo (1997) where more than 90 % of the world's demand for wood to be used as pulp or construction material could be satisfied with what is known as commodity-grade wood, which can be produced in managed, high-yield, intensive forest plantations.

What You Will Learn in This Chapter

The importance of industrial forest plantations as

- the source of raw material for many essential products;
- an economic agent that creates employment, economic growth, human development and climate change mitigation;
- a mechanism that promotes environmental restoration and provides environmental services.

1.2 Essential Roles

1.2.1 Historical Role as a Driving Force Behind Economic Growth

Comparative advantage: Virtually all countries can point some parts of their country that are maintained in forest coverage. Perhaps the forested area sustains a vibrant regional economy and is sustainable under active forest management. Likewise some forest areas retain their ecosystem integrity because of either their remoteness, or they are under government protection. What then governs the transformation of a forested area to a plantation? If the progression of a natural forest to a plantation is viewed as a type of product or sector specialization, then the explanation can be partially attributed to the economic concept of comparative advantage. Although comparative advantage is generally described in terms of international trade, it is clear that there is increased efficiency and utility in some forest-based sectors when management activity is concentrated, and in many cases simplified, by growing plantations. Increased globalization of economies only accentuates the comparative advantage and role that modern day forest plantations can fulfill. In several cases, and especially true in the southern hemisphere, the plantations are established with fast-growing introduced or exotic species. While some trial and error has been evident in matching the best species to the specific land-base, several key natural and policy factors are key to transforming a forested area to a plantation.

1. Land availability. The United States, Canada, BRIC (Brazil, Russia, India, and China) nations, and Australia would appear to have a distinct advantage here, however temperate, sub-tropical, and tropical climates are essential. Unless the plantations are managed by government concessions, bare land prices for plantation forests still remain typically below that of agricultural land.
2. High productivity in terms of forest growth. Growth rates of 70 m^3/ha/year from operational forest plantations are currently experienced in some locations. Favorable innate factors such as temperature, rainfall, and edaphic characteristics are important, but modern forest management practices are also of key relevance. Forest research and science is behind much of the success in improved forest genetics and intensive silvicultural practices that promote higher productivity.
3. Gains in management efficiency. Uniform plantations are easier to manage and less expensive to harvest. Utilization is typically far higher with monocultures and less forest residuals are evident on the forest floor after an intermediate or final harvest. Sustainable harvest plans are easier to formulate, have less risk associated with uncertainty, and are easier to valuate.
4. Astute government policies. Raw material scarcity and free market prices governed by supply and demand are essential forces behind the creation of plantations in developed nations; however, they alone are not sufficient in developing nations. Population increase and income elasticity for forest products

certainly drive up the demand, however this can easily be met by increased imports unless the government maintains favorable macroeconomic policies. Several countries in South America promoted the creation of forest plantations with the use of tax incentives for planting and the use of forest zone planning to direct the location of planting activity. A controversial aspect to government planning is whether the concepts of import substitution should be used to create demand for plantation raw material following the establishment of forest products industries. Overvalued exchange rates or tariff barriers are common instruments used to assist nascent industries get established. Typically, however, the domestic sawmills, veneer mills, or panel producers are of moderate capacity and they fail to capture economies of scale. Once erected, tariff barriers are difficult to phase out.

Beneficial results derived from plantations: Several authors of this book have first-hand experience of working in developing countries and witnessing the benefits derived from the start-up of a new forest products mill and the afforestation of tens of thousands of hectares of understocked forest land of low productivity. New industries in the rural sector assist in alleviating underemployment, abating the migration of an untrained workforce to urban areas, and they foster a dramatic improvement in health and education standards. The early phases of forest plantation establishment have heavy requirements for nursery production and early plantation tending (competing vegetation and insect control). These two activities offer an immediate boost to employment of the untrained rural workforce. Balanced growth is viewed positively by the authors, and measures that diminish the spread in wealth between the traditional rural sector and the modern urban enclave are considered imperative for equitable and sustainable development. Since plantations are often established in areas that are considered quite remote, the linkages are quite strong for establishing new utilities, commerce, municipal government and private services. Plantations often replace degraded lands, and they also assist in furnishing forest protection in terms of fire control, erosion control on steep slopes, and replenishment of local fauna abundance. Plantations can also be used as a Clean Development Mechanism for generating income with programs for sequestering carbon. Finally, many foresters view plantations as a vehicle for saving native or natural forests from overcutting or deforestation. Essentially, the raw material derived from the fast-growing plantations are viewed as a substitute for the round wood extracted from slow-growing, extensively or unmanaged natural forests. For more "pros and cons", see Cossalter and Pye-Smith (2003).

Obstacles to successful establishment and sustainable management of forest plantations: The acquisition of land and the initial expenditure associated with furnishing a modern mill with a guaranteed wood supply, requires considerable capital. In many nations outside the European Union or North America, the cost of capital can be 3–6 % higher than within the developed nations. The premium attached to the cost of capital is linked to perceived risk and it makes the hurdle rate for plantation-based investment far more restrictive and difficult to attain. Capital inflows from foreign-based multinationals are often accompanied with some restrictions, which are frequently related to the acquisition of land in concentrated sub-regions. While the cost of capital is lower in the developed nations, the growth rate and productivity of the plantations are typically lower also. Since the cost of land acquisition is a major component of plantation development, many forest-based firms seek to diminish their cost with land owner assistance programs and guaranteed fiber supply agreements with land owners. These programs are helpful in reducing the demand for capital, but they are difficult to manage over the extended forest plantation rotation. While it is relatively easy in developed nations to outsource activities, such as forest planting, it is difficult to purchase expensive harvesting equipment, where few have the capability to secure loans at permissible interest rates. Several other socio-economic and institutional obstacles to the financial success of plantation management are listed below. Most, but not all, are related to the problems of plantation management in less developed regions.

1. Large-scale plantations are in many occasions located in remote and under-developed areas. The local workforce is easily absorbed for activities such as nursery management and plantation tending, but it is difficult to attract and retain qualified workers for operating the industrial project. Over time, this problem attenuates as the local standard of living improves with improved housing, health standards, and education.
2. Land tenure problems can be vexing, especially related to disputed land titles and native or landless groups with land use grievances. Clearly, the national and local governments have a role in resolving these issues, but many times, there is a lack of political will to take on these thorny and emotional issues.
3. Lack of infrastructure. This can require expensive projects to meet basic energy requirements, and furnish sanitation, housing, education, and other services.
4. Inadequate or expensive transportation. Forest-based industries are located close to remote plantations and the mills are situated far from export ports or centers of consumption. The delay and fuel cost associated with inefficient road transportation can represent a substantial setback.

Drawbacks to modern forest plantation management: One only needs to read the historical accounts of the rubber plantations established in the 1930s by Henry Ford in the Amazon to know the perils of pest and disease in a monoculture plantation. Large-scale failures can be responsible for generating desperation and feelings of ill will among the local population and environmentalists that linger for decades. Plantations that are managed exclusively for the export market possess their share of unique risks. For instance, the consumption of biofuels derived from plantation wood chips is currently incentivized in the European Union with energy legislation that is focused on reducing the production of tons of CO_2 equivalent with carbon neutral sources. Should the legislation, and hence subsidy, for the consumption of plantation wood pellets be altered or dropped, it is quite likely that the wood pellet industry in southern United States would completely cease production since this energy source is not economically competitive in the United States. Countries without export controls frequently find themselves in the position of exporting plantation logs. The Far East, comprising of Japan, S. Korea, China, and Vietnam are large importers of logs from the Pacific Rim nations (United States, Canada, and Chile) and South Pacific nations such as New Zealand and Australia. The value-added to log exports from plantations is low, and it appears anachronistic to utilize fast-growing plantations for the transportation and delivery of products that are approximately 50 % water. The export of plantation-based products are especially susceptible to the often times unpredictable changes in exchange rates and maritime shipping rates.

Vertically integrated forest product industries can sometimes absorb costs for cultural treatments and research and development activities that are not reflected in market economics with higher product prices. For instance, the pruning and the growing of plantation wood with higher specific gravity, represents two practices that enhance the intrinsic value of wood. Yet in the United States, there does not exist any market premium for clear wood or wood of higher specific gravity. Vertically integrated firms can engage in pruning or enhancement of specific gravity with improved genetics and realize the economic benefit. Since year 2000 however, the North American forest products industry has either sold or converted their forest timberlands and plantations to forest investment firms that pay considerably less in taxes. While tax efficient, these new forest investment management firms are reluctant to engage in plantation treatments and improvements unless they are compensated by higher prices when selling the plantation round wood to processing industries.

1.2.2 Current Role as the Source of Raw Material

Initially, it is necessary to specify what forest raw materials are the main providers. Usually referred to as forest products, these raw materials comprise a substantially large list. Led by FAO, efforts to standardize the classification and definitions of forest products do exist since the 1970s. In the 1980s, important revisions started to produce harmonized forest commodity descriptions and coding systems, taking into consideration changes in technology, industry and trade practice and the appearance of new products. As a result, the effort built the basis for an internationally accepted and harmonized classification system,[2] which groups forest products into ten large groups (FAO 1982):

1. Wood in the rough
2. Residues of wood processing; recoverable wood products
3. Wood chips and particles
4. Wood simply worked or processed

---

[2]The goal was to conciliate terms used by organizations like the UN Standard Industrial Classification of All Economic Activity (ISIC); UN Classification by Broad Economic Categories (BEC); UN Standard International Trade Classification (SITC); Customs Cooperation Council Nomenclature for the Classification of Goods in Customs Tariffs (CCCN); and Customs Cooperation Council Harmonized Commodity Description and Coding Systems (HS).

Table 1.1 Industrialized forest products categories

Group	Industrialized forest product
1	Roundwood; wood fuel, including wood for charcoal; industrial roundwood (wood in the rough); sawlogs and veneer logs; pulpwood, round and split; other industrial roundwood
2	Wood charcoal, wood chips and particles, wood residues, wood pellets and other agglomerates
3	Sawnwood
4	Wood-based panels; veneer sheets; plywood; particle board; oriented strand board; fiber board; hardboard; medium density fiber board (MDF); insulating board
5	Wood pulp; mechanical wood pulp; semi-chemical wood pulp; chemical wood pulp; unbleached sulphite pulp; bleached sulphite pulp; unbleached sulphate pulp; bleached sulphate pulp; dissolving wood pulp; other fiber pulp; recovered paper
6	Paper and paperboard; newsprint; printing and writing paper; other paper and paperboard; household and sanitary paper; wrapping and packaging paper and paperboard

5. Wood sawn lengthwise; veneer sheets
6. Wood-based panels (including panels from ligno-cellulosic materials)
7. Pulp of wood, other fibrous ligno-cellulosic materials and waste paper
8. Paper and paperboard
9. Waste paper
10. Raw, semi-processed and worked cork

Even contemplating a large amount of forest products, and intended to cover wood and wood-based products for which FAO collect statistics on a regular basis, some products were intentionally dismissed. Examples are small ornamental trees; nuts, berries, seeds, roots, or other parts of plants gathered in forest areas; gums, balsams, lacs, etc.; wood derivatives such as turpentine, tall oil, sulphite dye and other chemicals. Important functions were also excluded, such as reforestation for fire protection, protection and management of watersheds, of forest wildlife and of forest recreational areas.

The classification and definition system published by FAO in 1982 has suffered minor changes and is still used to report annual data on the production and trade of forest products. The most important categories mentioned in these reports are summarized in Table 1.1.

Management Planning in Action 1.1: Industrial Forest Plantations and the Job Market in the USA and Brazil

In 2012, the United States Forest Service revealed that the US share of world wood products was declining. It also stated that the swift and dramatic changes since 2005, implied that forestry and the wood-processing manufacturing of the United States were at the major decision point.

(continued)

(continued)

The future of the industries, employment levels and wages, and even the conditions of forest plantations were dependent upon market forces and competition. The downward trend could be altered with the introduction of new government policies to incentivize the sector. In other words, either the forest products industry could become more efficient with the unappealing 'survival of the fittest', or actions could be taken to stimulate construction of 'green' structures using energy efficient forest products. Indeed, the following graph reveals the trend of employment during the last two decades in the United States (Fig. 1.1).

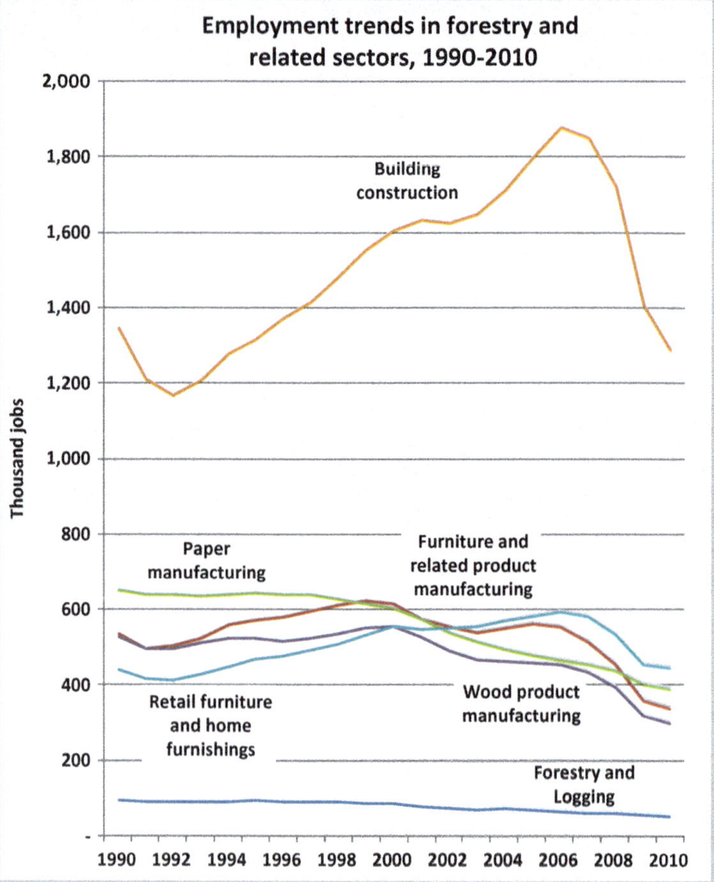

Fig. 1.1 Absolute full and part time job losses in the United States from 2005 to 2009 in all primary wood sectors (294,000 jobs lost)

(continued)

(continued)

On a brighter note, the pulp and paper industry of the United States has remained relatively stable in terms of capacity during the same period. This is especially true in the US South, where the pulp mills are furnished primarily with wood from plantations. The US Forest Service has stated that during 2002–2012, the US pulp and paper sector has positioned itself to compete in global markets and thus achieved far greater resiliency during periods of economic down-turns (Fig. 1.2).

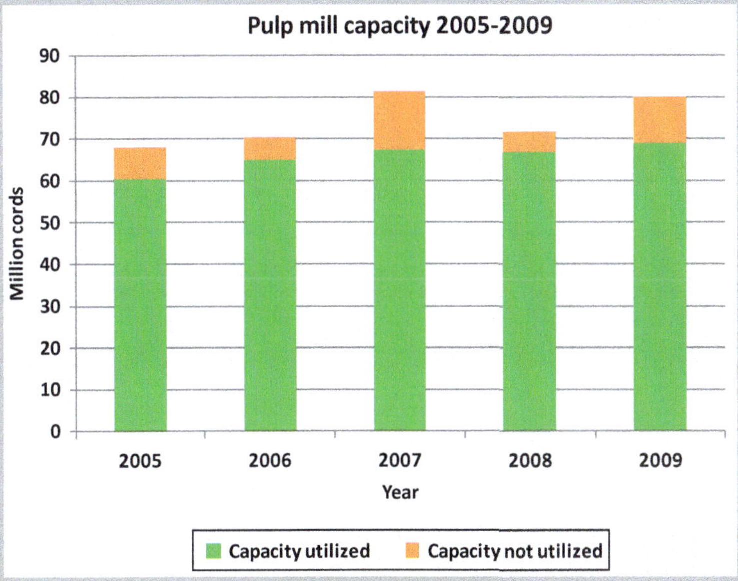

Fig. 1.2 US pulp mill capacity has remained steady at greater than 90 % overall capacity

The increase in pulpwood production during the last decade is largely attributed to the increase in productivity rather than an increase in plantation acreage. The higher productivity has been achieved with better site preparation, control of competing vegetation with the use of herbicides, fertilization of plantations, and improved genetic plantation stock (Fig. 1.3).

Brazil, in contrast to the United States is enjoying a period of rapid employment expansion in the forest products labor market. In 2011, Brazil employed 634, 217 in the forest sector and was experiencing a growth of 3.6 % a year in job creation. The distribution of jobs is displayed in the figure below (Fig. 1.4).

(continued)

(continued)

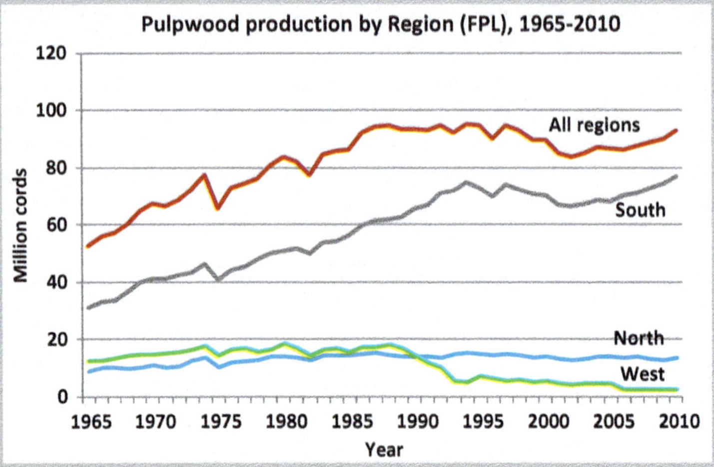

Fig. 1.3 2010 US pulp production represented a 7.4 % increase from 2005 levels. The gains were predominantly in the US South, where the bulk of the pulpwood is furnished from plantations

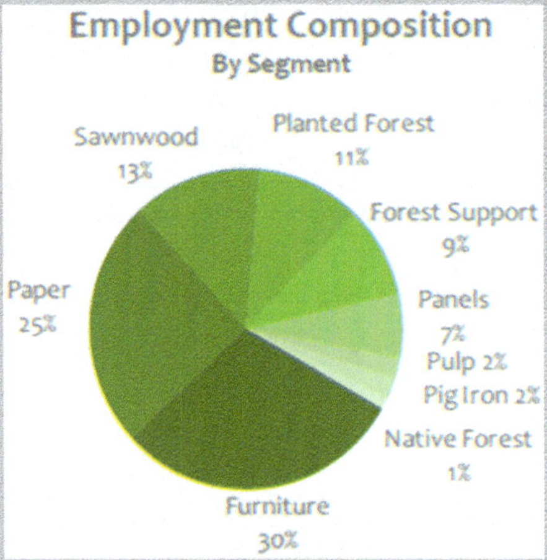

Fig. 1.4 Distribution of forest products labor (formal) by industry in Brazil (2011) (Source: Consufor, Advisory & Research)

Although Brazil faces the hurdle of capital creation and overcoming challenges posed by antiquated infrastructure and transportation, it is clear that Brazil enjoys a long-term comparative advantage in the forest products

(continued)

(continued)

sector. A large portion of the recent increase in plantation-based employment has been for areas designated for energy forests. Wood from these forests is converted into charcoal and are used to power pig-iron plants or steel mills. The graph below depicts the growth of employment used for the establishment of plantations (Fig. 1.5).

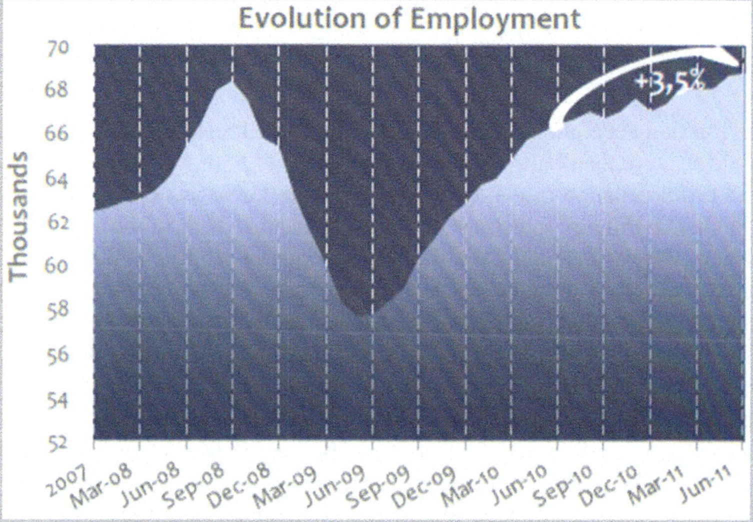

Fig. 1.5 Brazil employs in 2011, 68,831 forest-based plantation workers. Employment grew at a rate 3.5 % during 2010–2011 (Source: Consufor Advisory & Research)

1.3 Main Producers and International Players

1.3.1 Main Statistics

More precise and reliable estimates for the quantity and quality of different forest types based on regular and consistent large-scale assessments are still in the wish list of most forest managers. A global effort supported by satellite imagery and coordinated by FAO and the European Commission Joint Research Centre looks promising (FAO and JRC 2012), but it is still not fully implemented. Meanwhile, the most comprehensive source of information has been the Global Forest Resources Assessment, coordinated by FAO (2010), which states that the world's total forest area in 2010 is estimated to be over 4 billion hectares (31 % of the earth surface), corresponding to an average of 0.6 ha of forest per capita.

Table 1.2 Area expansion of planted forests (regions and sub-regions) from 1990 to 2010

	Area of planted forests (1,000 ha)				Annual increment (%)		
Region	1990	2000	2005	2010	1990–2000	2000–2005	2005–2010
Africa	11,580	12,873	14,032	15,326	1.1	1.7	1.8
Asia	70,873	92,871	109,670	122,777	2.7	3.4	2.3
Europe	58,166	65,309	68,500	69,318	1.2	1.0	0.2
North-Central America	20,056	30,186	35,786	38,659	4.2	3.5	1.6
Oceania	2,542	3,322	3,849	4,100	2.7	3.0	1.3
South America	8,115	10,058	11,123	13,821	2.2	2.0	4.4

The Russian Federation, Brazil, Canada, the United States of America and China account for more than half of the total forest area (53 %), while 64 countries with a combined population of two billion people have forest on no more than 10 % of their land area. Primary forests are estimated to occupy 36 % of the total forest area. Other naturally regenerated forests make up some 57 %. The total area of planted forest (composed of trees established through planting and/or through deliberate seeding of native or introduced species) is estimated to be 264 million hectares, corresponding to 6.6 % of the forest area. Since 1990 the planted forest area has steadily increased in all regions and sub regions. It increased by more than 3.6 million ha per year from 1990 to 2000, by 5.6 million ha per year from 2000 to 2005, and by 4.2 million ha per year from 2005 to 2010. Around 76 % of planted forests have production as their primary function (FAO 2005).

The area covered by planted forests is increasing all over the world. Although the major increase took place during the 1990–2000 decade, the trend is still clear. Regarding the proportion of area covered by planted forests, Asia is the region that presents the higher proportion of this type of forests with 47 % of the total area, Europe occupies the second place with 26 %, followed by North and Central America in the third place (15 %) (Table 1.2).

Most of the planted forest was established through afforestation particularly in China. Three-quarters of all planted forests consist of native species while 25 % comprises introduced species. East Asia, Europe and North America reported the greatest area of planted forests, together accounting for about 75 % of global planted forest area.

Among the ten countries with the highest annual increase in planted forest areas, Brazil tops the rank in the 2005–2010 period, followed by China, Canada and India (Table 1.3). These four countries together account for an average annual increase in planted forests of 2.6 million ha over this period. Given this trend, a further rise in the planted forest area up to 260 million ha by 2020 can be anticipated.

Planted forest expansion helped reduce the net loss of forest area to 5.2 million ha per year in the last decade, compared with the gross rate of loss through deforestation and natural causes, estimated at 13 million ha per year in the last decade (largely converted to agriculture or lost through natural causes) (FAO 2010).

Table 1.3 The ten most important countries in terms of area planted from 1990 to 2010

	Area of planted forests (1,000 ha)				Annual change rate (%)		
Country	1990	2000	2005	2010	1990–2000	2000–2005	2005–2010
China	41,950	54,394	67,219	77,157	2.63	4.32	2.80
U.S. of America	17,938	22,560	24,425	25,363	2.32	1.60	0.76
Russian Federation	12,651	15,360	16,963	16,991	1.96	2.01	0.03
Japan	10,287	10,331	10,324	10,326	0.04	−0.01	0.00
India	5,716	7,167	9,486	10,211	2.29	5.77	1.48
Canada	1,357	5,820	8,048	8,963	15.67	6.70	2.18
Poland	8,511	8,645	8,767	8,889	0.16	0.28	0.28
Brazil	4,984	5,176	5,765	7,418	0.38	2.18	5.17
Sudan	5,424	5,639	5,854	6,068	0.39	0.75	0.72
Finland	4,393	4,956	5,904	5,904	1.21	3.56	0.00

Moreover, planted forests are established for different purposes and not all of them are designated for production of wood or non-wood forest products. The protective function of the forest implies mainly the provision of services or non-industrial purposes, such as environmental, recreation, fuel wood on a small scale, etc. Planted forests established for protective purposes account for 25 % of total area; however, if the proportions are analyzed at the interregional level, the pattern varies significantly. In America and Oceania, almost 100 % of the planted forests are devoted to production activities, in Europe and Africa the proportion is not extreme but still is very unbalanced towards the productive side, around 81 % and 79 % respectively. Asia is the only region whose numbers are above the media with 35 % of protective planted forests. Since the area of planted forests provided by Asia is almost half of the world area covered by planted forests, global numbers are highly affected by this region. The performance of these trends is rather regular over the different periods analyzed (FAO 2010).

Disaggregating the area covered by planted forests into those ones belonging to plantations and to semi-natural forests, the numbers implied that the proportion of area included on each category is almost the same at the world-wide level, though a bit lower for semi-natural forests. Nevertheless, these numbers only reflect the pattern at the global level. At the regional level, the picture looks quite different. Again, South America and Oceania follow a rather similar trend. These two regions do not present semi-natural forests almost at all. In North and Central America, only 25 % of the planted forests corresponds to semi-natural forests. However, in Asia the proportion of both types of forests is quite similar, although slightly deviated towards semi-natural forests. In Europe, this trend appears as an extreme with more than 65 % of the planted forests consisting on semi-natural forests.

Moreover, although not that impressive anymore, one thing that attracts the attention is the huge increase in plantation area in North and Central America and in Oceania from 1990 until 2000. It would be interesting to compare these rates with the ones from year 2000 until 2010, but in FRA 2010 plantation forests are not considered as an independent category anymore. Their areas appear merged with semi-natural forests creating a new category called planted forests. In South America the increase in area covered by plantations is also important, as well as in Europe and Asia. Asia is the region that presents the biggest amount of plantation forest. It is interesting to note that this area is increasing as an important proportion through the years.

Regarding the objective of both plantations and semi-natural forests, it is clear that the majority of the plantations have been established for productive purposes. In America and Oceania, the objective of plantations is mainly productive. In Africa and Europe plantations with protective purposes account for approximately 20 % of the total planted area, while in Europe the proportion has declined over the last years.

1.3.2 Global Trends

Demand for forest products will continue to grow as world population and incomes grow. The most recent FAO projections estimate that by 2030 global consumption of industrial round wood will rise by 60 % over current levels, to around 2,400 million m^3. Substantial rises are also likely in the consumption of paper and paperboard products. Will the world's forest resources be able to cope?

Until the early 1990s, expert assessments were pessimistic, but most global analysts today foresee positive signs of recovery from the 2011 crisis in terms of wood production and supply (Bruinsma 2003). Whiteman (2014), for example, after analyzing nine scenarios built by three different organizations, plotted 2050 projections for forest area. For some of these projections, the forest area increases up to 6 %, for the rest of these projections the forest area can decrease as low as 16 %. The same author and others like Carle and Holmgren (2008) debate the role played by forest plantations in countries where the area of planted forests increases like Brazil, India, China and in a small group of temperate countries.

Clearly, China stands out as the leader in reforestation and planting of new forests (afforestation). In 2005, it was estimated that China was establishing plantations at a rate of 5.2 million hectares/year. Barr and Sayer (2012) have enumerated several reasons why the centrally planned economy of China has reached this level of prominence. Over the planned period of 2001–2015, US$ 8.6 billion of subsidized funding is planned to be directed to 99 priority plantation projects for the establishment of 13.3 million acres of commercial tree plantations. The Ministry of Finance and state policy banks fund 90 % of the plantation costs with reduced interest rates and extended repayment periods. While local government bodies in China cover 3 % of the cost, entities such as state forest farms, cooperatives, and private companies are expected to contribute 7 % of the program costs with commercial loans or equity contributions.

FAO figures have shown that the economic recession that preceded and followed the 2011 financial crisis clearly affected trade statistics. Globally, the production of industrial roundwood, sawnwood, wood-based panels, pulp and paper declined in 2009 and gradually recovered in 2010–2012. Anyhow, and although exceeding the 2008 levels, amounts recorded for 2012 still indicated values below the pre-recession levels. The good news were the fastest paces of recovery observed in Asia, North America and Latin America and Caribbean (FAO 2012a).

Specifically regarding tropical wood products, ITTO statistics show some gradual recovery (ITTO 2012), with non-added value products like tropical logs facing an increasingly supply constrained trade environment in the international market and a rising domestic consumption in the most important supplying countries. Large tropical log consumers like China have tried diversification strategies, which lead to a more heavily reliance on countries with questionable sustainability issues, like Papua New Guinea, Solomon Islands, Myanmar, Republic of Congo, Malaysia, Cameroon, Equatorial Guinea, Mozambique and Benin. Value-added wood processing continued to expand in the African region. Tropical sawnwood production remained relatively stable, with significant export downs in 2011 and 2012 and countries like Brazil maintaining high domestic consumption levels. Asia-Pacific continues to dominate the tropical sawnwood trade. EU tropical sawnwood imports slump to very low levels, with imports declining significantly between 2007 and 2009, down to nearly half the peak level of 2007. Tropical plywood production continued to shift towards cost competitive China, away from Malaysia and Indonesia, slowly recovering from the lowest export demand prices in 2011 and diverting to domestic markets. For secondary processed wood product trade, the levels have not reached pre-crisis values yet, but a slow rate of growth in wooden furniture is still observed in China and the imports of builders' woodwork and joinery is still dominated by the US although the wood's share in products like windows and doors remain challenged by substitution.

On a large-scale market analysis, Buongiorno et al. (2012) make extensive use of the Global Forest Products Model to forecast land use and forest products consumption by continent until 2060. Their econometric model is a function of population growth, change in gross domestic product, energy use, land-use changes, resource availability, and pace of technological change. These authors provide sensitivity analysis with various scenarios of the predicted changes to the economic and resource variables. These scenarios were created by the Intergovernmental Panel on Climate Change. While useful at a macro level for forest products and forest cover, the Global Forest Products Model does not address the role of forest plantations.

Outside of government policies and forest development plans, there is a role for the use of stand-level forest economics in inspecting the potential for private equity investment in forest plantations. Sohngen et al. (2008) partially address this and observe that fast-growing plantations of semi-tropical regions of South America, mainly in Brazil, enjoy a clear advantage almost tripling the net present value observed in some US forest plantations.

Lambin and Meyfroidt (2011) also point to several developing countries that have managed a land use transition over recent decades that promote an increase in forest cover and agricultural production. These authors embrace globalization as a methodology for increasing land use efficiency and curtailing uncontrolled land use expansion. Government policies, such as land use zoning and forest protection, and an increasing reliance on imported food are key elements to deterring deforestation. At a global scale, land is scarce, demanding increased productivity as a means to achieving more efficient land use allocation. Lambin and Meyfroidt (2011) predict that during the period of 2000–2030, the additional land demand for forest plantations will be in the range of 126–215 million hectares, of which 56–109 million hectares will be destined for industrial forest use.

While not necessarily suggesting causal effect, it is interesting to detect from the work of Brockerhoff et al. (2008), that developed countries of smaller land area tend to have a larger percentage of their forest cover in plantations. According to their assessment the percentage of forest cover with plantation forests in countries where exotic species play important roles vary more significantly (1.1 % in Brazil, 3.8 % in Indonesia, 22.3 % in New Zealand and 67.6 % in UK) than countries where native plantations are dominant (15.9 % in China, 12.7 % in France and 5.6 % in USA).

Brockerhoff et al. (2008) and Paquette and Messier (2010) refute the notion that plantation forests are synonymous with in biological deserts. These authors advocate a new trend toward the creation of multi-resource plantations that can assist in mitigating climate change by carbon sequestration, while protecting native primary forests from deforestation. Clearly however, there is an increasing demand upon government policy makers that primary native forests should remain intact while the new industrial plantations are allocated to degraded lands or secondary forests.

Barr and Sayer (2012) indicate that for several countries in the Asia-Pacific region, the transition from traditional forest practices to modern forest plantations is tarnished with mixed success. There have been losses in biodiversity and cases of marginalization or displacement of rural communities. Large capital subsidies to political elites are fraught with allegations of corruption. Gerber (2011) has documented many cases of unrest between rural communities and the administrative management of industrial plantations. Solutions to these complex confrontations between the rural poor and the management of modern industrial forests require both ingenuity and good governance.

Forest managers will remain pressed to resolve conflicting objectives, and it is rational to expect that the currently observed lower world population growth rate and wide-spreading substitution processes will affect future wood consumption. Large-scale industrial forest operations and fast growing forest plantations can significantly benefit from current new computational technologies and systems modeling approaches specially developed to plan large-scale landscape interventions, from new plantation establishments to thinning and clear-cut harvests. All this will certainly press the forest management community to be more efficient, more adaptable and more effective.

1.4 Planning for the Future of Industrial Forests

Starting in the mid nineteenth century, and on a large scale, investments in planted forests, for both the production of wood and protection against desertification, resulted in the establishment of roughly 80 million hectares, according to the 2012 FAO State of the World's Forests (FAO 2012a). Currently, planted forests span approximately 7 % of all forest area covering the planet (FAO 2010). From this proportionally small area, though, a significant amount of forest services and goods has flowed to the market raising the attention of the world to a level where the environmental and economic importance of these plantations is unanimously taken into consideration.

Until 2030 the world volume produced in planted areas will increase from 1.4 billion m^3 in 2005 to 1.7 billion m^3 or 2.1 billion m^3 in 2030, depending on how technology and genetic improves over the years. Considering the most optimistic technological scenario, the highest yield improvements will happen in South America (Carle and Holmgren 2008). According to FAO (2010), in 2030 commercial plantations will probably supply at least one third of the round wood industrial demand (Table 1.4).

Estimates published by Carle and Holmgren (2008), and presented in Table 1.4, show that more than 90 % of the wood produced in planted forests in 2030, considering the pessimistic scenario, will be consumed by the processing industry and pulp mills. The other 10 % will be converted to some sort of biofuel. Assuming no substitute product replaces wood in the future, the pressure for higher productivity from these forests will build up, once the growth in income per capita and better distributed wealth in most emerging economies generates higher of consumption on products like panels, furniture, papers and cardboards, timber for housing etc.

Table 1.4 Wood volume produced in planted forests by regions and use at 2005 and 2030 for the scenario that the area changes have been based on past trends and are assumed to continue at the same rate until 2030 (millions m^3. $year^{-1}$)

	Fuel/bioenergy		Pulp/fiber		Wood products	
Region	2005	2030	2005	2030	2005	2030
Africa	11	10	9	15	55	56
Asia	79	88	141	146	264	321
Europe	17	18	123	129	166	185
Southern Europe	3	6	26	55	26	56
America	7	8	98	117	24	31
South America	19	23	133	173	91	115
Oceania	1	1	11	13	31	36
Total	136	155	540	647	659	800

1.4.1 Conflicting Goals

Although the wealth provided by industrial forest plantations to the regions where they are established is undeniable, the success of this type of plantations is subject to the success in overcome the conflicting outcomes resulting from this type of practices. In several countries, industrial plantations have a prominent role in the economy. Moreover, some of these plantations are promoted to mitigate climate change or produce agro-fuels. However, the conflicts regarding industrial plantations is an issue that needs to be taken into account.

Gerber (2011) defines industrial plantations conflicts as "physical mobilizations coming from neighboring populations and targeted at the perceived negative effects of the plantation. These effects may be economic, socio-cultural or environmental". The conflicts mainly arise in the southern hemisphere, where industrial plantations occupy large areas, are intensively managed, and consist mainly of even-aged monocultures of exotic species with a clear productive vocation (Overbeek et al. 2012). Generally, these types of plantations are owned and/or promoted by corporate actors with a low degree of local community's involvement. In the northern hemisphere, forest plantations are on a smaller-scale and, most of the times, they are not perceived as a threat to the environment or society.

Conflicts faced by industrial forest plantations usually have an environmental nature or deflagrate social and economic impacts (Sawyer 1993). Among the environmental impacts, the use of exotic instead of native species plays an important role. On one side, indigenous or native species are more adapted to the habitat, are also habitat-providers for other species, and have the capacity to regenerate naturally. On the other hand, most often they grow at slower rates when compared to exotic species and are difficult to establish for productive purposes. Related also with the introduction of exotic tree species is the invasive nature of some introduced species. That affects negatively the provision of non-wood products and environmental services. Most of the times, industrial plantations based on intensive genetic improvement programs produce tree populations with low levels of genetic variation, decreasing their genetic diversity and their capacity for adapting to changes.

Tree plantations, depending on how they are cultivated, can cause severe damage to the soil, water flows and ecosystems (Jackson et al. 2005), but this trade-offs also depend on the characteristics of the area where the plantation has been established. Tree plantations reduce average stream flow and groundwater recharge, but may also have some minor environmental planning benefits in some environments (van Dijk and Keenan 2007). Something similar occurs with the loss of biodiversity, which is a function of the ecosystem they are replacing (Evans and Turnbull 2004). When industrial plantations are established by afforestation, the loss of diversity becomes less relevant when compared to reforestation activities that replace natural forests with forest plantations that dramatically reduce local biodiversity.

The occasional negative social and economic impacts of industrial forest plantations are usually the result of disregarding local community perceptions and its participation in the introduction of these plantations. In these instances, locals perceive the plantations as a threat to their right to use the land, as a loss of habitat and ecosystem degradation and therefore as a menace to their daily livelihood. Although potentially guided towards the creation of job opportunities, these industrial plantations can give rise to social conflicts that continue to be an important issue closely related to the future sustainability of this type of production activity (Gerber 2011).

1.4.2 Management Tools to Balance Future Demands

Bruinsma (2003) summarized projections suggesting that since early 1960s the largest source of increase in crop production observed in developing and developed countries has yield improvements as its main driver. In developing countries, the expansion of the area cultivated accounted for a quarter of production growth for the whole group, with larger contributing factors being observed in regions with more abundant land like sub-Saharan Africa and in Latin America. These projections also show that the same trends for the developing countries will continue, at least until 2030, and land expansion will account for approximately 20 % of production growth, yield improvements for about 70 % and increased cropping intensity for the remainder. The same author states that for sub-Saharan Africa and Latin America, land expansion will still be important, although increasingly outweighed by yield increases.

Simultaneously, it is important to consider some issues constraining the world's effort to increase yield and more intense cultivation (Bruinsma 2003):

- countrywide, unequal development may create wealthier consumers in some places while in other resource-less places households may not be able to afford even basic foods;
- processes not reflecting environmental costs of expanding and intensifying agriculture and the failure to internalize resource costs may curb investment in research, holding back the potential for future growth in yields; and
- land or water scarcities and other problems will continue to arise at country and local levels, with serious consequences for poverty and food security.

The development of land use and forest management support systems have been challenged to simultaneously consider all these factors. Capacity building to overcome these challenges have to reach countries where the use of land and forest management support systems are not widespread. This is especially important when considering the fact that the pool of unused suitable cropland is very unevenly distributed. More than half of the remaining global available cropland is concentrated in just seven countries (Angola, Argentina, Bolivia, Brazil, Colombia, Democratic Republic of Congo and the Sudan), where most of the times the suitable cropland is in practice unavailable, or already assigned to other valuable uses (Bruinsma 2003).

Demand for forest products will continue to grow as world population and incomes grow. Consequently, the major challenge will be to improve the sustainable management of forests and to ensure equitable distribution of the benefits of forest use.

In addition, by 2030 annual world consumption of industrial round wood is expected to rise by 60 % over current levels, to around 2,400 million m^3. Production of industrial round wood from plantations is expected to double by 2030, from 400 million m^3 today, to around 800 million m^3.

1.5 Summary

The chapter initially addresses the definition of planted forests, industrial forests and industrial forest plantations. It also contextualizes historically the evolution of the term forest plantations, citing the several international organizations involved and different attempts to satisfy many actors and opinions. Facts and figures are also presented to illustrate the evolving roles that forest plantations have undertaken over the years. Some comprehensive statistics are included to emphasize the importance and benefits of its expansion. The chapter also presents some important future trends that guide the need for the forest planning and forest management techniques discussed in this book. These tools will support and overcome the challenges brought by the more complex and conflicting social, economic and environmental goals confronted by the forest sector.

1.6 Problems

Download the global tables from the FAO FRA 2010 interactive database (Excel format: http://countrystat.org/home.aspx?c=FOR) and answer the following questions:

1. What countries present the largest percentage of introduced species in planted forest?
2. Present a list of the top five countries in terms of annual change rate in the area of planted forest for the period 2005–2010.
3. List the top10 countries in terms of percentage of forested area assigned to production.

Acknowledgements The work of Luis Diaz-Balteiro was funded by the Spanish Ministry of Economy and Competitiveness under project AGL2011-25825.

References

Barr CM, Sayer JA (2012) The political economy of reforestation and forest restoration in Asia-Pacific: critical issues for REDD+. Biol Conserv 154:9–19

Botkin D, Sedjo RA (1997) Using forest plantations to spare natural forests. Environment 39(10):14+. Academic OneFile. Web. 24 Aug 2013

Boyle JM (1999) Planted forests: views and viewpoints. New Forests 17:5–9

Brockerhoff EG, Jactel H, Parrotta JA, Quine CP, Sayer J (2008) Plantation forests and biodiversity: oxymoron or opportunity? Biodivers Conserv 17:925–951

Bruinsma J (2003) World agriculture: towards 2015/2030 – an FAO perspective. Earthscan, London, p 432

Buongiorno J, Zhu S, Raunikar R, Prestemon JP (2012) Outlook to 2060 for world forest and forest industries: a technical document supporting the Forest Service 2010 RPA assessment. USDA Forest Service, General Technical Report SRS-151

Carle J, Holmgren P (2008) Wood from planted forests: a global outlook 2005–2030. Forest Prod J 58(12):6–18

CIFOR (2002) Typology of planted forests. InfoBrief, Bogor, Indonesia. http://www.cifor.org/publications/pdf_files/Typology/John-typology.pdf

Cossalter C, Pye-Smith C (2003) Fast-wood forestry – myths and realities. CIFOR – Center for International Forestry Research, Bogor, 60 p

Evans J (2004) Forest plantations. In: Burley J, Evans J, Youngquist J (eds) Encyclopedia of forest sciences, vol 2. Elsevier, Oxford, pp 822–828

Evans J, Turnbull JW (2004) Plantation forestry in the tropics: the role, silviculture, and use of planted forests for industrial, social, environmental, and agroforestry purposes, 3rd edn. Oxford University Press, Oxford, UK

Evans J, Carle JB, Del Lungo A (2009) The question of definitions. In: Evans J (ed) Planted forests: uses, impacts and sustainability. CABI, Cambridge, p 213

FAO (1967) World symposium on man-made forests and their industrial importance. Unasylva 21(86–87):116

FAO (1981) Forest resources of tropical Africa, Tropical forest resources assessment project (in the framework of GEMS). Series forest resources of tropical Africa, tropical forest resources assessment project (in the framework of GEMS) Technical Report 2, UN 32/6.1301-78-04. FAO, Rome, Italy, 586 pp

FAO (1982) Classification and definitions of forest products. FAO forestry paper 32. Food and Agriculture Organisation of the United Nations (FAO), Rome, Italy, 258 p

FAO (2002a) Second expert meeting on harmonizing forest-related definitions for use by various stakeholders. Proceedings. Rome, Italy

FAO (2002b) World agriculture: towards 2015/2030: summary report. Food and Agriculture Organization, Rome

FAO (2005) Third expert meeting on harmonizing forest-related definitions for use by various stakeholders. Proceedings, Rome, 17–19 Jan 2005, 137 p

FAO (2006) Global forest resources assessment 2005: progress towards sustainable forest management (FRA 2005). FAO forestry paper 147, Rome, Italy

FAO (2010) Global forest resources assessment 2010: main report. FAO forestry paper 163. Rome, Italy

FAO (2012a) State of the world's forests. Food and Agriculture Organization of the United Nations, Rome, 60 p

FAO (2012b) Global forest products facts and figures. Food and Agriculture Organization of the United Nations, Rome, 17 p

FAO (2013a) Statistical yearbook – world food and agriculture. FAO, Rome, 289 p

FAO (2013b) FAO yearbook of forest products 2011, FAO forestry series 46//FAO statistics series 202. Food and Agriculture Organisation of the United Nations (FAO), Rome, 358 p

FAO (2014) FAOSTAT. Food and Agriculture Organization, Statistics Division online Database. Rome, Italy. http://faostat.fao.org

FAO, JRC (2012) Global forest land-use change 1990–2005. In: Lindquist EJ, D'Annunzio R, Gerrand A, MacDicken K, Achard F, Beuchle R, Brink A, Eva HD, Mayaux P, San-Miguel-Ayanz J, Stibig H-J (eds) FAO forestry paper no. 169. Food and Agriculture Organization of the United Nations and European Commission Joint Research Centre, FAO, Rome

FSC (2003) Forest plantations. Fact sheet November 2003. http://plantations.fsc.org/docs/Briefings/Forest%20Plantations%20Fact%20Sheet%20November%202003%20(EN).pdf

Gerber J-F (2011) Conflicts over industrial tree plantations in the South: who, how and why? Glob Environ Chang 21(1):165–176

Helms JA, Prüller R, Haddon BD, Lund HG (2003) The need to say what you mean and mean what you say. IUFRO conference on information interoperability and organization for national and global forest information services, September 17–19, 2003. XII World Forestry Congress, Quebec City, Quebec, Canada. http://www.iufro.org/download/file/318/184/archive-interoperability-full_doc

ITTO (2012) Annual review and assessment of the world timber situation 2012. International Tropical Timber Organization, Yokohama, 196 p

Jackson R et al (2005) Trading water for carbon with biological carbon sequestration. Science 310:1944–1947

Lambin EF, Meyfroidt P (2011) Global land use change, economic globalization, and the looming land scarcity. Proc Natl Acad Sci U S A 108(9):3465–3472

Overbeek W, Kröger M, Gerber J-F (2012) An overview of industrial tree plantation conflicts in the global South. Conflicts, trends, and resistance struggles. EJOLT Report No. 3, 100 p

Paquette A, Messier C (2010) The role of plantations in managing the world's forests in the Anthropocene. Front Ecol Environ 8(1):27–34

Sawyer J (1993) Plantations in the tropics: environmental concerns. IUCN Publications, Gland, Switzerland and Cambridge, UK in collaboration with UNEP and WWF, 83 p

Sohngen B, Tennity C, Hnytka M, Meeusen K (2008) Global forestry data for the economic modeling of land use. GTAP working paper, no. 41

van Dijk A, Keenan R (2007) Planted forests and water in perspective. For Ecol Manage 251(1–2):1–9

Varmola M, Del Lungo A (2003) Working paper FP/25 Planted Forests Database (PFDB): structure and contents

Whiteman A (2014) Global trends and outlook for forest resources. In: Fenning T (ed) Challenges and opportunities for the world's forests in the 21st century. Springer, Dordrecht, pp 163–211

Chapter 2
An Overview of Forest Management Planning and Information Management

Marc E. McDill

2.1 Introduction

Managing any complex operation requires planning. While this seems obvious enough, it is valuable to consider specifically why planning is important, especially in the context of the management of industrial forest plantations. Most importantly, plans specify what will be done, when, by whom, and for what purpose. The plan ensures that everyone within the organization knows what needs to be done, and it provides a basis for holding both the organization and the individuals within the organization accountable for what is accomplished. Plans should also be the basis for the allocation of scarce resources within an organization, so nearly everyone in the organization has a stake in the planning process. Furthermore, a plan communicates to external stakeholders, such as stockholders and the public, what the organization is doing, what it expects to do in the future, and why. Having a good plan and demonstrating that the plan can and will be implemented gives an organization credibility. Conversely, failing to plan, or having a plan but not following it, increases the likelihood of inefficiency, frustration, and a lack of credibility.

Ideally, a management plan also explains the rationale for why the organization has chosen to do what it says it will do. Thus, a good plan always begins by outlining the goals, objectives and constraints that directed the plan. Key management issues and challenges are identified. Some discussion of alternative management strategies that were considered might also be included, along with an evaluation of the pros and cons of each strategy. Such a discussion shows that the organization has considered its options and helps make the case that the strategy chosen really is the

M.E. McDill (✉)
Department of Ecosystem Science and Management, Pennsylvania State University, University Park, PA, USA
e-mail: mmcdill@psu.edu

J.G. Borges et al. (eds.), *The Management of Industrial Forest Plantations: Theoretical Foundations and Applications*, Managing Forest Ecosystems 33,
DOI 10.1007/978-94-017-8899-1_2, © Springer Science+Business Media Dordrecht 2014

one that makes the best use of the organization's resources and does the best job of achieving the organization's goals. When there are competing objectives, it is also important to show that the planners recognize the tradeoffs among these competing objectives and to provide some understanding of how conflicts among objectives were resolved.

Planning is especially important in forestry operations because many management goals take a long time to achieve and require the coordination of actions over a long period of time. Forests cannot be created in a day, a week, or even a year. Mature plantations that are ready to harvest are the result of a series of management decisions over many years and often decades. Even having seedlings that are ready to be planted is typically the result of years of research on seedling improvement and at a minimum requires the existence of a nursery. Furthermore, many forestry activities have long-term consequences. Regeneration decisions influence stand conditions for the entire rotation; competition control and thinning treatments change the structure of a stand until the end of the rotation; and the decision to harvest a stand now means that the stand will not be available for harvest again for many years.

Strategic plans are often viewed as a waste of time because they promise to achieve lofty goals that no one really believes can be accomplished. Such plans are almost inevitably ignored and left to collect dust on shelves. In order to avoid this fate, plans should include specific, quantified targets that can be achieved. Rather than sitting on shelves, plans need to play a central role in the daily management of operations, and managers need to be held accountable to meeting the goals specified in the plan. This is only possible if the planners recognize resource constraints, that they accept that there are trade-offs among competing objectives, and that they make some difficult decisions. When done properly, this can produce substantial benefits to an organization, as a thoughtful, carefully-considered plan can be a powerful management tool. When employee incentives are tied to achieving or exceeding the targets set in the plan, the goals and objectives of the plan tend to get accomplished. If the goals are not feasible, however, this just creates frustration and sets the organization up for failure.

While plans should have specific, quantified targets, there also needs to be flexibility in how they are implemented. It is impossible to anticipate everything that might happen, and when unexpected things happen organizations have to have the flexibility to adapt to changing circumstances. Ideally, planning is an ongoing process. However, if a plan is too flexible there is no accountability, and excessive flexibility can lead to a sense that the organization is not really committed to achieving its strategic objectives. It is therefore useful to have a clear distinction between minor plan revisions – which are done to address unanticipated events, changes in operating constraints or short-term market fluctuations – and writing a new plan. It should be relatively easy to revise a plan to deal with normal changes in operating conditions as long as the revisions enhance the organization's ability to meet its strategic goals and as long as the revisions are not merely providing cover for a failure to meet strategic goals.

Over time, even the goals and objectives of an organization evolve, and as unanticipated events accumulate and conditions change, eventually it becomes necessary to realign the plan with the current management situation, and plans should be rewritten. Some organizations re-plan on a fixed cycle, but ideally re-planning should occur whenever sufficient change has occurred that the existing plan is no longer appropriate. Distinguishing between strategic, tactical and operational planning – discussed in more detail later in this chapter – is useful in this context. Strategic plans should be revised relatively infrequently, and only when there is a need to reassess the strategic position of the organization. On the other hand, tactical and operational plans, whose purpose is primarily to find more efficient ways to achieve the organization's strategic goals, should be very flexible and able to respond to changing information and conditions.

A variety of analytical tools have been developed over time to assist forest managers in making better planning decisions in virtually all aspects of planning. This book is intended to introduce plantation managers to a wide variety of these tools. This chapter provides a brief introduction to some of the more important planning tools used in plantation management. All of these tools are discussed in more detail in later chapters.

What You Will Learn in This Chapter

- The planning process
- Different levels of forest planning
- Basic data requirements for forest planning
- How to formulate a non-spatial forest planning problem as a linear program
- How to formulate a spatially-explicit forest planning problem as a mixed-integer linear program

2.2 The Planning Process

The process used to develop a plan is as important as having one. Developing a good plan requires communication among all of the groups who will be involved in implementing the plan and who have an interest in the resulting plan. The planning process is a good time for an organization's administrators to communicate their goals and objectives to lower-level managers and for those managers to communicate to administrators the constraints they face and the resources they need to accomplish various tasks. Different people within an organization also have different ideas of how the organization can meet its objectives, and this is a good time for these competing ideas to be aired, discussed and assessed.

Planning involves six key steps: (1) goal setting, (2) identifying alternatives, (3) evaluating alternatives, (4) plan selection, (5) plan implementation, (6) monitoring, and (7) re-planning (Fig. 2.1). The most important step is identifying the goals of the organization; as Lewis Carroll famously said: "If you don't know where

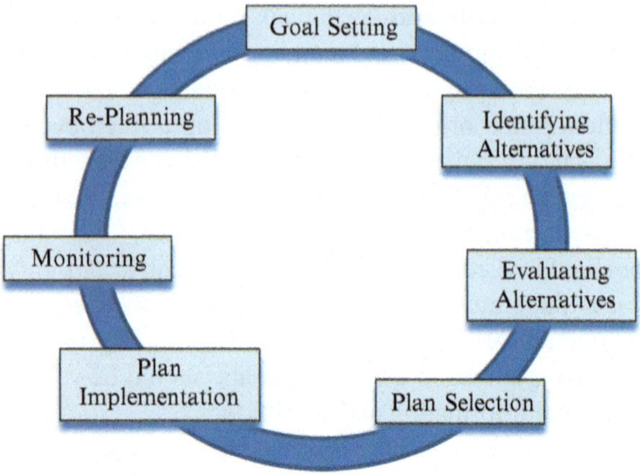

Fig. 2.1 The planning process

you are going, any road will get you there." For industrial plantations the primary goal is typically to maximize profits. Even with this goal, however, there is often a trade-off between short-term and long-term profits, and a business needs to decide how it will prioritize short vs. long-term profits. Furthermore, plantation managers also have other kinds of goals. For example, forest plantations are often managed in order to provide a steady supply of wood fiber to a mill. This goal can, and often does, conflict with maximizing profits from plantation management alone. In addition, forestry organizations also often have a goal of operating in a way that is environmentally sensitive and sustainable. Again, this goal can conflict with the goal of maximizing profits – especially short-term profits. Thus, the process of setting goals involves not just identifying goals, but establishing a framework for evaluating the trade-offs among competing goals and for choosing the right balance among goals.

People are creatures of habit and tend to do things the way they have always done them until they are forced to change. The phrase "thinking outside the box" refers to the ability to identify and be open to alternative ways of doing things. The planning process is an ideal place for this kind of thinking. An organization is much more likely to find the best way to achieve its goals if it considers a variety of alternative pathways for getting there. Being aware of alternative approaches also gives the organization more flexibility for reaching its goals in a world where conditions are always changing. This phase of the planning process is therefore very important. It is useful during this stage to be as open-minded as possible and to consider a wide range of alternatives. In the context of plantation management, management alternatives include such things as species or genotype choices, planting intensities, the timing and type of fertilization applications, thinning regimes, competition control, rotations, as well as land acquisition, choosing areas to set aside to address environmental constraints and objectives, investments in nurseries, and research activities. Different options regarding the timing of these activities also represent different management alternatives. The timing of activities is very important. Because forestry activities involve costs and benefits that occur over relatively long time frames, it is necessary to consider the time value of money when assessing the timing of alternative management activities. Management alternatives that are otherwise very similar and that differ only in the timing of costs and benefits are not equivalent. Indeed, in forestry determining the appropriate timing of activities is often a fundamental aspect of management planning.

After a variety of alternative management actions have been identified the next step is to evaluate alternative activities and to select the alternatives that are most effective in meeting the organization's goals. Ideally, the cost and contribution of each alternative to each management objective is quantified along with the timing of activities. It is also necessary to quantify the resources needed to carry out each management activity. Once the costs and benefits of each alternative have been evaluated relative to each of the goals, the next step is to select feasible sets of management alternatives that produce the greatest net return for the organization. Of course, when the organization has multiple objectives, it is also necessary to evaluate the trade-offs between objectives to identify the right balance among them. Techniques such as linear programming and multi-objective programming, discussed later in this chapter and in this book, are extremely useful for identifying the set of management alternatives that achieves the highest level of attainment for various goals while staying within constraints related to budgets, personnel, land, and other resources. In the case of a single objective, and when a model can safely be assumed to be an accurate representation of the real-world management problem, the model can identify the best set of management activities. However, when there are multiple objectives, or when there are multiple ways to model the real-world situation, or when there is uncertainty about the inputs or results of various management actions – which is almost always true – then management planning models should be viewed as tools for exploring the implications of various management decisions. Models are decision-making tools, not decision makers.

Once the various alternatives have been considered and thoroughly analyzed, the decision-makers must choose a set of management activities to implement that are then documented in the organization's plan. If it has been carefully constructed, the plan is feasible, the resources needed to accomplish the plan are known, and the people responsible for carrying out the different activities outlined in the plan have been identified and empowered with the resources they need to get the job done. The next step is to implement the plan.

But planning does not stop there. As any good plan includes specific activities with quantified inputs and outputs, it should be possible to continuously monitor the degree to which the goals identified in the plan are being achieved. On a periodic basis, there should be a deliberate comparison of the levels and outcomes of management activities with the targets specified in the plan. Inevitably, at least some goals will not be met. This can occur for a variety of reasons: (1) unexpected events occur that are outside the control of the organization, (2) planned activities do not occur at the planned level, or (3) inputs or outputs associated with some activities do not occur as planned.

Examples of unexpected events in plantation management include fires, insect and disease outbreaks, bad weather, and equipment failures. To the extent that these events are unavoidable, the organization needs to be flexible and adaptive and be able to adjust the plan to allow objectives to be met as closely as possible under the circumstances. For example, in the case of fires, some areas that were not scheduled to be harvested may have to be harvested to make up for areas that burned before they were harvested. However, sophisticated planning can also be done to reduce the likelihood of these undesirable unexpected events and to increase the organization's flexibility for coping with unexpected events. For example, forest management models have been developed that account for expected losses due to fire and even to design forested landscapes to reduce the likelihood of fire (e.g., Acuna et al. 2010; Wei 2012).

When planned activities do not occur or do not occur at the planned level, this is likely an indication of some type of management failure. One of the advantages of having a good plan is that it can be a powerful tool for enforcing accountability within an organization. Clear, quantified objectives provide a clear basis for employee incentives and consequences. On the other hand, objectives need to be realistic and attainable and individuals charged with carrying out the plan need to be given the resources necessary for achieving the components of the plan for which they are responsible. If planned activities did not occur because higher-level managers did not fill critical positions needed to carry out those activities, the lower-level managers tasked with accomplishing those activities can hardly be held accountable.

Frequently management activities are implemented according to plan, but inputs or outputs do not match those anticipated in the plan. For example, plantation management models typically utilize sub-models such as growth and yield models that are very useful – in fact, essential – for planning. Nevertheless, these models can never be perfectly accurate, especially at the individual stand level. If they are well constructed, they are at least unbiased, but it is impossible to accurately predict the development of every stand. The quality of these models is highly dependent on having copious quantities of site-specific data, something that is seldom available. Ideally, data collected from monitoring plan implementation can also be used over time to recalibrate the models used in planning so that future planning efforts can be based on more accurate models. Thus, plan monitoring should provide information for the continuous development of knowledge and for the improvement of the tools used for management planning. Ideally, data needed for plan development and monitoring are collected on an on-going basis as management occurs and are integrated into the organization's information management system.

As discussed in the previous section, organizations have to be flexible and able to deviate from plans to adapt to unexpected events. Over time, plans are modified, adjusted and amended. Furthermore, over time goals and objectives must be adjusted as market conditions and regulatory environments change and new management challenges emerge. Eventually, the cumulative effect of these changes is that the plan becomes increasingly less relevant. At this point, it becomes necessary to start the planning cycle again. Ideally, much has also been learned in the process of implementing the plan and monitoring plan implementation so that planners and managers have a better idea of what is feasible with a given set of resources. New management activities have been tried and existing activities have been improved so that managers have a more diverse and effective toolbox. In addition, planning tools, such as growth and yield models, and techniques have been improved so that each successive plan is better than the previous one.

2.3 Planning Levels

Forest management planning happens at many different scales. In our own lives some decisions, such as choosing a career, are strategic, while others, such as choosing what tasks to do on any given day, are operational. Forest planning is no different. Decisions such as whether to acquire a large tract of land are strategic, while decisions such as deciding what mix of equipment to use to harvest a particular stand are operational. Just as a person does not need to know exactly how each day will be spent in order to decide what career to pursue, forest planners do not need to know exactly how each stand will be managed in order to determine whether it is worth it to buy a large tract of forestland. Nevertheless, whether the decision turns out to be a good one depends ultimately on how the stands within the tract are managed, so strategic and operational decisions must ultimately be linked.

Forest planning levels differ primarily in terms of their spatial and temporal scales. In the spatial dimension, foresters often differentiate between the stand level and the enterprise or compartment level. The temporal scale of forest planning can range from days or weeks to decades or even centuries. Forest planning problems occur on a continuum from the highly strategic level to a very detailed operational level. This continuum is often conveniently, if somewhat arbitrarily, stratified into three general categories: operational, tactical and strategic. This section discusses the fundamental distinction between stand-level and enterprise-level planning. It also provides a brief discussion of hierarchical planning as it applies to forest planning.

2.3.1 Stand-Level Versus Enterprise-Level Planning

A "stand" is typically the unit of forest land at which silvicultural regimes are applied. Stands are generally homogeneous in terms of their age structure and species composition. In forest plantations, stands are contiguous even-aged areas that were established at the same time and where a single silvicultural management regime is applied. As such, the stand is in most cases the same thing as a "treatment unit." Terms such as "compartment" and "management unit" often refer to groups of stands of varying ages and species composition that are managed collectively.

However, these terms are somewhat ambiguous and mean different things in different organizations. The term "landscape" is an ecological term that is also somewhat ambiguous but that, in the case of forested landscapes, typically refers to an area that encompasses multiple stands. The term "forest" is also similarly ambiguous, but in the context of forest planning it generally refers to an area consisting of multiple stands. Similarly, in this context, the term "enterprise-level" planning refers to planning for a large area of land that is owned by a single organization and managed for a common strategic purpose such as providing wood fiber for a mill.

Stand-level management decisions are those that relate to the design and selection of a silvicultural regime for a particular stand. Stand-level decisions include deciding on the type of site preparation to do, selecting a species or seedling type to plant, selecting the planting intensity, decisions on competition control, choosing the timing, type and intensity of thinnings, and determining when to harvest and regenerate the stand. Considerable research has been done on selecting optimal management regimes for individual stands (see Chap. 5). This research is very important and provides the foundation for understanding forest management decisions. However, whenever a stand is part of a larger management unit the optimal management regime for that stand in the context of the larger unit may not be the same as the optimal management regime if the stand were managed independently. This is because larger-scale management objectives likely will not be adequately addressed if each stand is managed as an independent unit. The classic example of this is when a forest has an unbalanced age-class distribution – e.g., when it consists mostly of older stands with very few younger stands. In this case, if each stand is harvested when it is financially mature the forest will initially produce an abundant supply of timber, but eventually, when all the mature stands have been cut, the harvest level will drop and stay low until the stands harvested earlier are regenerated and have time to again reach maturity. In this case, the enterprise-level goal of producing a steady supply of timber will not be met. The only way to meet this goal is to hold some of the mature stands long past their financial maturity so they can be harvested gradually over time until the stands that are harvested now can be regenerated and regrow.

When enterprise-level concerns supersede stand-level concerns, it is generally best to develop a wide range of silvicultural alternatives for each stand and to select the silvicultural regime for each stand in the context of an enterprise-level planning model. In other words, in enterprise-level planning, silvicultural considerations determine the management alternatives that are feasible; they do not drive management decisions.

2.3.2 Enterprise-Level Planning

Enterprise-level planning involves decision-making at many different levels. Because of the long time periods required for trees to mature and because potentially

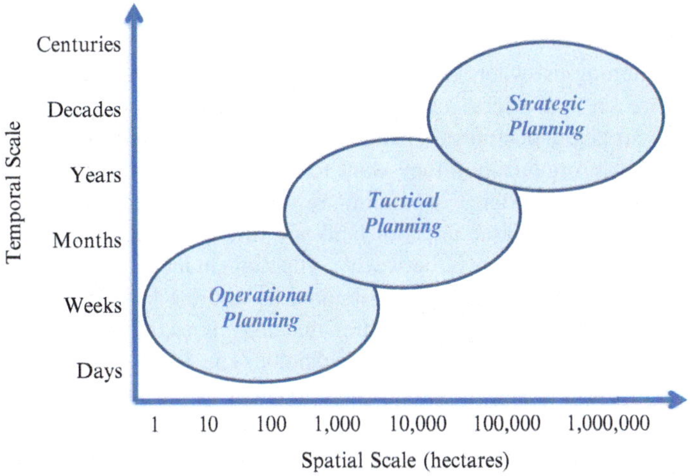

Fig. 2.2 Planning level versus temporal and spatial scale. Problems with smaller scope often involve greater detail and may be just as large and complex as problems with larger scope

large areas of land are involved in plantation management, some forest planning objectives can only be achieved over large spatial and temporal scales. Other planning problems do not involve such large areas or long planning horizons, yet may be just as complex. While nearly all decisions made by an organization are in some way interrelated, planning typically occurs at different levels with different degrees of detail at each level. Strategic decisions that involve long time frames and large spatial scales typically require less detailed information. For example, strategic forest management decisions can, in most cases, be made without addressing precisely how each stand will be managed, what kind of logging system will be used, and how products will be transported from each stand to the mill. Shorter-term decisions such as these can be made at a different level. Because the scope of tactical and operational planning problems is smaller, both spatially and temporally, it is possible to incorporate much more detailed information when making these decisions. As such, the size and complexity of the planning problem – as measured, for example, by the number of variables or constraints – may be just as challenging at the tactical and operational levels as at the strategic level (Fig. 2.2).

While planning at different scales can and should be done somewhat independently, there needs to be some degree of interaction between the different planning levels. Obviously, goals from the strategic plan need to drive tactical and operational planning, so information must flow downward in the form of goals, objectives and targets. However, in order for strategic plans to be feasible, they need to be informed by tactical and operational requirements and constraints, which means information must also flow upward from the tactical and operational levels to the strategic level. Determining how to effectively link different planning levels is an important planning problem in itself.

2.3.2.1 Strategic Planning

Strategic planning is where the broad, long-term goals and objectives of an enterprise are set and where general strategies for achieving them are developed. Essentially, strategic planning is where an enterprise determines who they are and what kind of organization they want to be. Major investment and divestment decisions, such as building, expanding or closing manufacturing facilities or acquisitions or sales of large areas of land, are fundamentally strategic decisions. Decisions involving trade-offs between competing management values, such as short-term profits versus long-term sustainability and the level of commitment of the organization to environmental values that may not contribute directly to the company's bottom line, are strategic decisions. The level of investment that a company will make in research and development is an example of a management decision that reduces short-term profitability with the expectation of increasing the long-term profitability and sustainability of the enterprise. Decisions about how much area and which areas to set aside for environmental protection or decisions not to apply certain types of practices in some areas because of potentially negative environmental impacts are indicative of the organizations commitment (or lack thereof) to environmental values, particularly when they involve significant opportunity costs to the organization.

Strategic decisions that are unique to forest management include setting long-term harvest targets, potentially by species and product type, and identifying the desired future forest structure, including the target species composition and age-class distribution of the forest. Because forests take years, if not decades, to develop, these types of decisions have to be made in the context of a plan that considers the long-term impact of management activities. As such, strategic forest management planning problems typically have planning horizons that are at least one rotation long and possibly more. Strategic planning problems also typically involve planning for the entire land holdings of the enterprise, which may include thousands, if not tens of thousands of hectares.

Because of the large spatial and temporal scope of strategic planning problems, it may not be possible or even useful to include detailed spatial information in the model. For example, there may be very little value in planning exactly where certain management activities will occur 10, 20 or 50 years from now. As such, similar stands that may not be spatially contiguous are often aggregated into larger analysis areas for the purposes of strategic planning. Thus, strategic plans often do not have a high degree of spatial specificity. To the degree that this spatial information is not necessary for addressing long-term strategic objectives, this may be perfectly appropriate.

2.3.2.2 Tactical Planning

Tactical planning addresses medium-term, medium-scale goals. The division of planning problems into three categories – strategic, tactical and operational – is somewhat arbitrary, so what constitutes a medium-scale problem is also somewhat arbitrary. Tactical planning basically involves any type of planning that is broader than operational planning and not quite broad enough to be strategic. Typically, tactical planning is done to determine in more detail but at smaller spatial and temporal scales how the broad goals from a strategic plan can be accomplished as efficiently as possible. Thus, operational plans are often made for subsets of the enterprise's land base and cover shorter time frames. Tactical plans are usually spatially explicit, determining exactly where and when certain activities are to take place.

An example of a tactical plantation management planning problem might be a spatially-explicit harvest scheduling model that schedules specific silvicultural practices for a forest compartment in order to achieve product flow goals for the next 2–20 years. The product flow goals for the tactical model may come from disaggregating goals obtained from a non-spatial strategic model with a longer planning horizon and/or a larger spatial scale. Because the tactical model is spatially explicit, it may include constraints such as green-up restrictions that were not explicitly addressed in the strategic model. Tactical planning problems may also include decisions on road construction, maintenance and closing.

2.3.2.3 Operational Planning

Operational planning addresses very short-term, detailed issues such as which stands will be harvested in the next weeks or months, by whom, with what equipment, what products will be produced, and to which mill will they go, etc. Personnel and equipment constraints are seldom included in tactical or strategic models, but they are common in operational planning. The temporal scale of operational planning is usually very short, days to months. On the other hand, the spatial scale of operational planning problems may be as small as a single stand or as large as the entire enterprise. An example of single-stand operational planning is determining the layout of the landings and skid trails for a single harvesting operation. A larger-scale operational planning problem might include scheduling stands for harvest over a 2-month period that would include scheduling the equipment used in each operation, including the cost and time required to move equipment between activities. The objective of the model might be to minimize the cost of meeting specific material requirements for multiple processing facilities owned by the company.

2.3.3 Linkages Between Different Planning Levels

While it is useful to separate strategic, tactical and operational planning problems, it is also important to recognize the relationships between the different levels of planning. The sum of all operational decisions should add up so that tactical goals are met, and, similarly, tactical decisions should be made in such a way that strategic goals are met. Because lower-level decisions often do not consider longer planning horizons, it is especially important to ensure that short-term decisions do not result in a failure or in inability to meet long-term goals. Because short-term planning models typically try to find the cheapest, easiest – i.e., most "efficient" – way to meet short-term objectives and do not explicitly recognize long-term goals, using them may lead to "cherry-picking" problems where the easiest, most lucrative activities are done first, leaving the harder, less profitable activities for later periods (e.g., McQuillan 1986; Pickens et al. 1990). Avoiding problems like this requires having carefully-constructed linkages between higher-level, longer-term planning tools and lower-level, shorter-term planning tools. Exactly how this should be done will vary from one planning context to another, but the first step is to be aware of the potential problem.

Lower-level, shorter-term planning tools also need to inform higher-level, longer-term planning tools. Because strategic planning models often leave out considerable detail they may ignore many tactical and operational constraints. When these constraints are ignored, solutions that appear feasible at the strategic level may not be feasible on the ground. A classic example of this would be a non-spatial strategic harvest scheduling model that does not include green-up constraints. Because some constraints have been ignored, it is entirely possible – even likely – that the strategic solution will not be feasible on the ground. Furthermore, the infeasibility of the solution may not be noticed when tactical models are run because of the shorter time horizons of those models. The infeasibilities may only become apparent after several years or even decades of implementing the models.

Hierarchical planning involves many pitfalls. These pitfalls have led many to conclude that the best thing to do is to include everything in one model. This is not the solution. This approach results in either (1) planning things in too much detail for too long a planning horizon, or (2) using too short a planning horizon. As stated earlier, some decisions are fundamentally strategic while others are operational. It is not necessary to know all the details to make good strategic decisions, nor is it necessary or even desirable, to make detailed operational plans for time periods that are far in the future. The challenge of hierarchical planning is determining how to make good linkages between different planning levels so that operational decisions are not short-sighted and so that strategic plans are operationally feasible.

2.4 Data Requirements, Acquisition, and Management

Forest management planning requires a lot of information. First, the forest resource itself has to be described, including the delineation and description of each management unit, or treatment unit. Next, the management options for each treatment unit – i.e., silvicultural prescriptions – have to be described. For each prescription, information is needed regarding how the state of the stand will evolve over time, what management actions will need to occur, costs that will be incurred, and outputs that will be produced. Growth and yield models are crucial for producing this type of information. If the planning problem involves providing raw material to one or more mills, then the transportation system connecting each management unit with each mill has to be described. Even if supplying a mill is not an explicit management objective, if access is limited for some treatment units, then this information, including the expected road segments that would need to be constructed to provide access, will be needed. Finally, all management and resource constraints have to be described. Assembling and managing of all of the data that can be required for planning may be as large a task as developing the planning model itself. Ideally, all of this information is stored and managed within the organization's enterprise information system and is continually updated. Having such a system is essential if planning is to be integrated into the decision-making processes of the organization.

2.4.1 Land Classification

The first step in forest management planning is to obtain information on the forest resource that is being managed. The first step in this process, if this has not been done already, is to delineate the organization's forest resources into treatment units. These units are typically called stands by foresters and generally range in size from a few hectares to as much as a hundred hectares. Treatment units should be relatively uniform in terms of species composition, age, stocking, and site quality, and it should be possible to apply a single silvicultural prescription to the entire area. Treatment units are typically contiguous, but they do not have to be. Treatment units may consist of multiple discontinuous units as long as a single management prescription can be applied to the entire area. Delineation of management units is typically done within a geographical information system (GIS). While delineation of management units can be a difficult and subjective activity in natural forests, it is usually straightforward in plantation forests.

Once the area has been delineated into management units, it is also necessary to describe the forest within each management unit. Ideally, information on timber attributes can be derived from a stand-level inventory where plots have been measured in each management unit. Such data often are not available, and in that case stand attributes have to be inferred from plots in similar stands or from models. Furthermore, information should be compiled on past management activities for each management unit, including the planting date, the species planted, and the planting density, and date and type of treatments such as thinnings, release treatments, or prescribed burning. If information is available on yields from past harvests, this can be useful in measuring the site's potential productivity. In addition to timber-based stand attributes, some relevant information may be obtained from existing GIS data sets. Such data may include roads and other access information, soils, slope and aspect. The presence of or proximity to environmental features that may result in management constraints could also be relevant information. For example, streams may require buffer zones with limited management options. In countries with such restrictions, the presence of or proximity to known habitat for threatened or endangered species could also be important information for planning.

2.4.2 Management Options

Forest management planning essentially boils down to the problem of selecting a prescription for each treatment unit so that the management goals and objectives of the forest enterprise are achieved as optimally as possible within the resource constraints of the organization. These treatments need to be diverse in order to give the planning process the flexibility to achieve various forest-wide management goals and objectives. This means that a set of alternative management options must be developed and considered for each treatment unit. Options can vary in terms of the final harvest period (rotation age), thinning regimes, release and fertilization treatments, etc. Foresters have a tendency to think at the stand level and may only want to consider the one prescription that they think is most appropriate for that management unit. However, as discussed in Sect. 2.3.1, forest-wide objectives often trump stand-level considerations, so individual stands may need to be managed quite differently from the way they would be managed if they were an independent ownership. Thus, it is best to consider a wide range of possible management options for each stand, with different final rotation ages, different thinning sequences, etc.

For each management unit/treatment combination, it is necessary to quantify the resources that will be needed, the outputs that will be produced, and the timing of all inputs and outputs. Resources include costs, personnel time and machine-hours. Outputs include various species and sizes of roundwood or chips, as well as environmental outputs such as habitat conditions, carbon storage and water flows. Projections of timber outputs are usually obtained from growth and yield models (discussed in Chap. 3). With a large number of management units and a large number of potential prescriptions for each management unit, it is generally essential to have a computer program that can simulate different prescriptions for each unit based on the unit's attributes and to populate a database with parameters quantifying the inputs and outputs for each prescription for each management unit. These programs are generally referred to as matrix generators, as their function is to develop the matrix of parameters used to develop the actual planning model. Simulating stand development and management outputs usually requires local growth and yield models, local cost parameters, and other local information, so in many cases a locally-appropriate matrix generator will not be available off the shelf and will need to be developed. This is often one of the biggest challenges in developing forest management models and may require some significant investment.

2.4.3 The Transportation Network

The cost of harvesting and transporting roundwood or chips from the forest to the mill is often a significant component off the delivered cost of the wood. Forest stands are often geographically widely distributed, and there may be multiple mills with different wood requirements to which products must be delivered. As such, part of the planning process, especially at the tactical and operational levels, may involve determining where the wood harvested from each management unit will be delivered and how it will be transported to that location. Alternate modes of transportation, such as large vs. small trucks, rail or barges, may have very different costs. In some cases, some management units may be inaccessible, and it may be necessary to build or improve roads in order to harvest or otherwise manage those units. In addition, moving harvesting equipment from one site to another can be expensive, and integrated firms that do their own harvesting may want to include these activities in their tactical and operational planning models. In these cases, it will be necessary to have data on the existing and potential transportation network that can be used to transport material from the forest to manufacturing facilities.

In many countries GIS databases on the existing road network already exist and can be freely downloaded or purchased. These databases may not include the smaller roads that can be used to access forest stands, however. For those components of the road network for which data are not available, a GIS layer can be developed either by digitizing roads from aerial photographs or by using global positioning systems (GPS) to capture data on those roads. Where roads do not exist to provide access to all management units, potential road links that could be built must be developed and included in the network.

2.4.4 Planning Constraints

Planning constraints include everything from the amount and type of products that the enterprise would like to produce to the amount of land to that should be set aside for conservation purposes. Planning constraints are often determined internally, although some constraints may result from the need to comply with regulations and legal restrictions. Ideally, planning constraints are identified up front so they can be included in planning models. Nevertheless, part of the planning process is to explore and evaluate the tradeoffs between different planning constraints. In fact, constraints often end up being objectives in a multi-objective planning problem.

A common type of planning constraint is the need to meet certain production targets. For example, if a forest land base is owned by an integrated forest products company, the company may expect to obtain a certain percentage of the fiber used in its mills from company-owned land. In other cases, the organization may only be concerned that the level of timber production or flow of revenue from the forest is stable and sustainable over time. Organizations also have resource constraints, such as personnel, equipment and budget constraints.

Other constraints reflect environmental objectives or regulations. For example, it is common to restrict harvesting activities within stream buffers and along roads or trails. Similarly, operational constraints may limit harvesting activities on steep slopes or on wet soils. In some cases, maximum clearcut size and green-up restrictions must be observed. Forest enterprises may choose or be required by law to provide certain amounts of forest meeting various habitat requirements.

All of these constraints need to be identified so they can be included in the planning models. With the exception of budget and personnel constraints, failure to recognize and include constraints tends to bias forest planning models to overestimate the harvest level that can be sustained. In this case, unsustainable harvesting is likely to be the result.

2.4.5 *Monitoring*[1]

Elzinga et al. (1998, p. 1) define forest monitoring as "the collection and analysis of repeated observations or measurements to evaluate changes in condition and progress toward meeting a management objective." A well-designed monitoring system is a key component in an adaptive management approach where managers learn from management activities and adjust management strategies over time to utilize new information that is gained from observing and measuring the outcomes of management activities (Fig. 2.1, Elzinga et al. (1998, pp. 1–2)). Monitoring systems give managers crucial feedback that either confirms that everything is going according to plan or indicates that some kind of corrective action is needed. Monitoring systems also provide key information for transparency and accountability within the organization and to outside stakeholders.

Forest inventories were originally developed with the goal of giving managers a point-in-time measure of their forest resources – especially in stands that were considered mature and available for harvest. This was accomplished largely through measuring temporary plots. In addition to measuring mature stands, temporary plots are often used to assess regeneration success in young stands. Eventually, systems of permanent plots were introduced by large forest enterprises, largely in response to the need to be able to measure growth and drain in order to project future forest conditions (Ware and Cunia 1962; Päivinen and Yli-Kojola 1989). Permanent plots are especially useful for the development of growth and yield models (Chap. 3). Today, permanent plots are a key tool for forest monitoring, as they are most useful for measuring not only current conditions, but changes and trends (Elzinga et al. 1998).

Modern forest monitoring systems address a wide range of concerns, including not only inventory and growth and yield, but insects and disease, invasive species,

[1]The author wishes to thank Dr. Laura Lietes, Penn State University, for her assistance in the research for this section.

carbon stocks, fuel loading for fire management, coarse woody debris and other wildlife habitat components, biodiversity, water quality and quantity, and even social indicators. The challenge with monitoring to address such a wide range of concerns is that the ideal sampling design for monitoring one resource (e.g., timber) may not be ideal for monitoring another (e.g., biodiversity). Thus, a comprehensive monitoring program will generally consist of several different inventory systems. Careful thought must therefore be given to how the data from the different inventory systems can be combined and analyzed within the context of an enterprise information management system. As long as data are geographically referenced, GIS can be a powerful tool for integrated analysis of such data.

Because monitoring systems typically rely on systems of permanent plots, it is important to give considerable thought up front to the design of monitoring systems, as data are much less useful and more difficult to work with when the sampling design changes over time. The value and usefulness of permanent plots increase substantially over time with each subsequent re-measurement. However, technology changes can result in new, less expensive methods of collecting data, making it worthwhile to change the way some data are collected. Furthermore, the questions that managers will want to use monitoring data to address often change over time, necessitating the need for protocol changes. Ideally, monitoring systems are designed to address a broad range of issues, recognizing that in the future the data will be used for uses that cannot be anticipated today. Because the data that are collected increase in value over time, monitoring systems require a long-term commitment within the enterprise to continue to provide the resources needed to continue to fund data collection efforts, to maintain the data that are collected, and to process the data so that managers receive critical feedback on management activities.

Several questions must be answered in order to design an effective monitoring system (Elzinga et al. 1998). The most important decision is choosing the resources that will be monitored. Long-term monitoring budgets are never sufficient to measure everything that may be important, so, as with any other management decision, tradeoffs have to be assessed. It is better to focus on a small number of resources and to measure them well enough to achieve statistically robust results than to measure too many things with too few observations to say anything with a sufficient level of statistical rigor. It is useful to think in terms of specific indicators that can be used to efficiently assess particular resource conditions. Another important question is how often permanent plots will be revisited. Many resources are unlikely to change significantly in 1 year, so visiting each plot annually may not be the best use of limited monitoring resources. On the other hand, if plots are revisited too infrequently, it may take too long to detect important trends. Finally, a sampling design must be selected for each resource. The sampling design, intensity and frequency may vary considerably for different resources. The U.S. Forest Service's Forest Inventory and Analysis (FIA, http://www.fia.fs.fed.us) system is an excellent example of a national forest monitoring system that can be adapted for use by industrial forest enterprises.

2.5 Planning Models

Virtually all industries today use mathematical models for planning. These models not only enable planners to find solutions to complex problems, but they also allow them to compare the implications of competing theories about how the real world functions and to evaluate trade-offs among competing solutions. These models enable decision-makers to make full use of a potentially vast array of information and data, and they often reveal important missing information that is needed to improve the decision-making process. Perhaps most important, mathematical models require planners, managers and decision-makers to be explicit about their assumptions and their objectives and they impose a level of rigor in the planning process that is unlikely to be achieved without them.

Mathematical models can take many forms and can be classified in many different ways. This book discusses a variety of models, including analytical models, statistical models, simulation models and optimization models. This section introduces a few basic mathematical optimization models that have been used in forest planning. The traditional harvest regulation models discussed first are analytical models that have been used in forestry for literally centuries. Within the past half century, however, more sophisticated optimization models have been used for forest planning. Computers have played a key role in the development and application of these models, as they often use large datasets and involve a vast number of calculations. A fundamental problem that many of these models address is the question of what treatments to apply to which stands at what time in order to achieve a profitable, consistent and sustainable harvest of timber while meeting organizational and environmental constraints. As the models used in forest planning have evolved, a key motivator of this evolution has been the desire to handle the more detailed information that is available today and to address the diversity of issues that arise in modern forest management applications. The basic models presented here are the foundation on which many forest management planning models in use today are built.

2.5.1 Traditional Harvest Regulation Methods

As populations, agriculture and economies expanded in Western Europe over the past millennium, the continent was gradually deforested (Kaplan et al. 2009). Yet wood was an essential resource for nearly every aspect of life – it was used for fuel, for housing, for making tools, wagons and ships, and for many other uses. By as early as the sixteenth and seventeenth centuries, these societies developed a growing awareness of the importance of ensuring the long-term sustainability of their wood supplies. John Evlyn's *Silva*, first published in 1664 (Evelyn 1664) and widely recognized as the first book on silviculture, was written at a time when England was experiencing one of these early wood supply crises (Nix 2014). It was in this context that the forestry profession first developed in Western European countries. At the time, the focus of the profession was to maintain sustainable supplies of wood, and the primary management objective was maximum sustained yield.

Sustained yield forestry, as it has been practiced by foresters for the past several centuries, has meant regulating the harvest of timber to ensure that a more-or-less constant amount of timber could be harvested each year. The most basic approach to sustained yield forestry is to harvest A/R hectares of forest every year, where A is the total area of forestland and R is the rotation period, or the time required for a stand of trees to reach maturity (however that is defined; c.f., Faustmann 1849; Gaffney 1957; Samuelson 1976; Newman 1988, Chap. 5 of this book). This approach is called Area Control because it is implemented by regulating the area of forest harvested each year. Over time, in the absence of significant unplanned stand-replacing disturbances, the application of Area Control techniques will generally produce a balanced age-class distribution with an equal area of forest land in each age-class up to the age where forests are considered mature and ready for harvest (i.e., age R). As summarized by Walker (1990):

> This concept, simply stated, is that the objective of forest management is to achieve and maintain a uniform distribution of age classes over a given forest area such that every year the oldest age class can be harvested and that the volume of this harvest will be equal to the annual growth of the entire forest. With the same number of acres in each age class and the oldest age class harvested each year, the forest will produce an even flow of harvested timber from year to year in perpetuity (p. 21).

For a variety of reasons, harvesting the same area each year may or may not result in a constant volume of harvest each year, so more sophisticated approaches were developed that regulated the volume of wood, rather than the area harvested. These approaches are referred to as Volume Control techniques (Chappelle 1966, Chap. 10 in Davis et al. 2001). In spite of their increasing sophistication, however, these models fail to address the many and diverse management objectives and constraints of modern forest management. Since the 1960s, forest managers have been developing more flexible and sophisticated forest management planning models based on mathematical optimization techniques, including linear programming (LP), mixed integer programming (MIP), simulation, and heuristics. Many of these techniques are discussed in this book. The following sections provide an introduction to linear programming (LP) and mixed integer programming (MIP) forest planning models. More detailed LP and MIP models, with examples, are presented in Chap. 6.

2.5.2 Non-spatial Methods – Linear Programming (LP)

Linear programming (LP) is a mathematical optimization method where the objective function and the constraints can be expressed as linear functions of the variables. The technique is very flexible and general and has been applied broadly to a wide variety of problems. This section assumes that the reader has already been exposed to basic LP concepts, including what an LP is, linear objective functions, linear constraints, feasibility, the feasible region, optimality, and duality. If not, the reader may want to consult one of the many introductory texts on LP, for example, Ignizio and Cavalier (1994), or one of the many (free) online introductions to the topic.

Some of the earliest applications of linear programming in forest planning include Curtis (1962), Kidd et al. (1966), Nautiyal and Pearce (1967), Ware and Clutter (1971), and Johnson and Scheurman (1977). Since then, LP has been widely applied by the forest products industry for forest planning, and it has been used extensively by the U.S. Forest Service for forest planning since the early 1980s (Iverson and Alston 1986). Weintraub et al. (2007) provides a good overview of the variety of applications of LP and other mathematical optimization techniques in agriculture, fisheries, forestry and mining.

This section describes a basic harvest scheduling model. It is important to note that there are many ways to formulate a forest planning LP model (c.f., Johnson and Scheurman 1977). The model formulation presented here is commonly used, but by no means the only one.

It is common, but not necessary, to combine similar management units into aggregates called *analysis areas*. For example, management units that have the same species composition, and that fall into a given age class and site class may be combined and treated as a uniform category of forestland. This simplifies the process of simulating prescriptions (see Sect. 2.4.2) and reduces the size of the LP model, as there are typically substantially fewer analysis areas than management units. The description of the model below assumes that management units have been aggregated into analysis areas. If this has not been done, then the management units are the analysis areas.

A fundamental feature of forest planning, and hence forest planning models, is that one must not only determine which treatment or treatments to apply to each management unit, but also when those treatments should be applied. Forest planning models typically define planning periods, and then schedule the period in which each activity should occur. Periods may be as short as a day or a week or as long as a decade, depending on the degree of temporal resolution that is desired. Generally, strategic planning models, which usually have long planning horizons, also have longer planning periods, and tactical and operational planning models with shorter planning horizons tend to use shorter planning periods.

The number of periods in a planning model is equal to the length of the planning horizon divided by the length of the planning periods. Combining long planning horizons with short planning periods can result in a very large numbers of periods. With more periods, more prescriptions must be defined, as there are more possibilities for the timing of activities. As a result, the size and complexity of the planning model tends to grow with the length of the planning horizon and increase as the length of the planning periods is shortened. Larger models are harder to formulate, to solve, and to interpret. Therefore, the selection of the length of the planning horizon and the planning periods are both very important decisions. A key criterion for whether the planning horizon is long enough is whether adding one more planning period to the planning horizon substantially changes the solution for the first one or two periods of the model. If so, the planning horizon is not long enough, as an arbitrary modeling decision such as the length of the planning horizon should not substantially influence management decisions. It is only necessary to consider decisions in the first one or two planning periods, as a new round of planning is likely to occur before more than one or two planning periods have passed. It is nevertheless important to plan for longer horizons because it is necessary to consider the long-term consequences of near-term actions.

When planning periods are relatively long, e.g., several years, it is important to be clear about at what time during the planning period activities are assumed to happen. For example, activities can be assumed to happen at the beginning, the midpoint or the end of the planning period. Which assumption is made can have a substantial impact on the values associated with different activities. For example, projected harvest volumes may be much higher at the end of the planning periods than at the beginning. Since in reality activities that are scheduled for a given period will happen at various times within the planning period, the least biased approach is to assume in the model that activities will happen at the midpoint of the planning period. This means, however, that even for harvests that occur in the first period, yields must be projected forward by one half period from the current state.

The first step in LP model formulation is to define the variables to be used. The key set of variables in the forest planning model are defined as follows:

x_{kj} = the area of analysis area k to be managed according to prescription j.

These variables assume that prescription j describes how the area will be managed for the entire planning horizon. Thus, using the terminology proposed by Johnson and Scheurman (1977), this is a Model I formulation. In LP models, variables can take on any continuous, non-negative value that is allowed by the constraints. Thus, the optimal value of variable x_{kj} may be fractional and not necessarily equal to the size of any combination of management units. Furthermore, when areas have been aggregated into analysis areas, the solution will not indicate which of the management units that make up analysis area k should be treated, only that a certain area within that analysis area should be treated with prescription j. This ambiguity can be advantageous in some cases as it gives the manager responsible for carrying out the plan some flexibility in how the solution should be applied on the ground. In the following section on spatially explicit models the variables will be defined differently to eliminate this ambiguity.

While other objectives could also be considered, the model presented here maximizes the net present value of the forest. The objective function is formulated as follows:

$$\max Z = \sum_{k \in K} \sum_{j \in J_k} a_{kj} x_{kj} \tag{2.1}$$

Where Z is the value of the objective function, K is the set of analysis areas, J_k is the set of prescriptions for analysis area k, and a_{kj} is the present value of the net revenue from managing one hectare of analysis area k according to prescription j. The values of the coefficients a_{kj} are calculated by simulating the various management options, as described above in Sect. 2.4.2. These present values should ideally include any costs and revenues associated with any treatments that are applied during the planning horizon as part of prescription j. In addition, this value should also include the present value of the ending value (i.e., the projected value at the end of the planning horizon) of hectares from analysis area k that are treated according to prescription j.

The first set of constraints are called *area constraints*, as they specify the amount of area available within each analysis area. The constraints are:

$$\sum_{j \in J_k} x_{kj} = A_k \quad \forall k \in K \tag{2.2}$$

Where A_k is the area in analysis area k and all other notation is as previously defined. This equation simply states that the area from analysis area k assigned to the various prescriptions available for this analysis area ($j \in J_k$) must equal the available area in this analysis area. There will be one of these constraints for each analysis area $k \in K$.

The optimal solution to a model with only objective function (2.1) and constraints (2.2) would assign all of the area in each analysis area to the prescription with the highest present value ($\max\{a_{kj} \in J_k\}$). This result can easily be proven by simply looking at the dual of the LP (Hoganson and Rose 1984). Intuitively, this result occurs because at this point the model has no forest-wide constraints.

Most forest-wide constraints can be classified as either (1) *quantity constraints*, or (2) *flow constraints* (Kent et al. 1991). Quantity constraints restrict the amount of some input, output or forest condition to be equal to, greater than, or less than some value in a given period or over a set of periods. For example, ending inventory volume constraints require the inventory volume at the end of the planning horizon to be greater than or equal to a specific amount. Flow constraints restrict the amount of some input, output or forest condition in one period to be similar to the amount of that input, output or forest condition in another period. For example, non-declining even flow constraints require the volume harvested in any given period to be greater than or equal to the volume harvested in the previous period.

A third category of forest-wide constraints, *accounting constraints*, are often used to facilitate the writing of quantity and flow constraints and to facilitate the interpretation of the solution (Kent et al. 1991). Accounting constraints do not actually restrict the feasible region of the model; they merely sum the quantity of some input, output or forest condition as a function of the variables x_{kj} and set the value of an *accounting variable* equal to that sum. Accounting constraints are formulated as follows:

$$\sum_{k \in K} \sum_{j \in J_k} v_{kjlt} x_{kj} = V_{lt} \tag{2.3}$$

Where V_{lt} is an accounting variable that represents the amount of input, output or condition l that is used, produced, or projected to exist in period t, and v_{kjlt} is the amount of input, output or condition l that is used, produced, or projected to exist in period t for every hectare of analysis area k that is managed according to prescription j. The values of the coefficients v_{kjlt} are calculated by simulating the various management options, as described above in Sect. 2.4.2.

The accounting variables V_{lt} make it easy to formulate quantity constraints:

$$V_{lt} \leq \geq or = \overline{V}_{lt} \tag{2.4}$$

Where $\overline{V}_{lt}$ is the maximum, minimum or target value of input, output or condition l in period t. The values of the parameters $\overline{V}_{lt}$ are determined by the organization's policy decisions, as discussed in Sect. 2.4.4.

Similarly, the accounting variables V_{lt} make it easy to formulate flow constraints:

$$V_{lt} - \delta_{lt} V_{l,t-1} \leq \geq or = 0 \tag{2.5}$$

Where δ_{lt} is one minus the allowable increase or one plus the allowable increase in the amount of input, output or condition l from period $t-1$ to period t. In the case of an allowable decrease, $\delta_{lt} \leq 1$ and the inequality would indicate a greater-than-or-equal-to constraint, in the case of an allowable increase, $\delta_{lt} \geq 1$ and the inequality would indicate a less-than-or-equal-to constraint. These constraints are usually formulated in pairs, with one constraint indicating the allowable increase and another indicating the allowable decrease. In some cases, the value of the input, output or condition is specified to be equal to the value in the previous period, in which case, $\delta_{lt} = 1$ and the constraint would be formulated as an equality.

One final set of constraints is needed that require that the values of all variables are non-negative:

$$x_{kj} \geq 0 \quad \forall k \in K \; and \; j \in J_k; V_{lt} \geq 0 \quad \forall l \in L \; and \; t \in \{1 \ldots T\} \tag{2.6}$$

Where L is the set of inputs, outputs and forest conditions for which accounting constraints are specified, and T is the number of planning periods. These constraints make explicit the requirement that no variables may take on negative values. Special techniques are required for those rare instances where variables should be allowed to take negative values (Ignizio and Cavalier 1994).

A wide variety of LP forest management planning formulations can be specified using Eqs. (2.1, 2.2, 2.3, 2.4, 2.5 and 2.6). It is worth noting that one common type of constraint, the minimum average ending age constraint, can be formulated as a special case of Eqs. (2.3) and (2.4) where the v_{kjlT} values are given by the age at the end of the planning horizon of hectares from analysis area k assigned to prescription j minus the target average ending age for the forest (or for a particular class of land, such as a site class), where the value of $\overline{V}_{lT}$ is 0, and Eq. (2.4) is specified as a greater-than-or-equal constraint.

It is also worth noting that quantity constraints can also be converted quite easily to objectives to produce a multi-criteria decision-making problem. For example, let

V_{ct} represent the total net carbon sequestration for the forest in time period t, and let V_c equal the sum of these variables, i.e., the total net carbon sequestration of the forest over the planning horizon. We can now add the maximization of V_c as a second objective function and through various techniques (c.f., Tóth and McDill 2009) construct a tradeoff curve between the objective of maximizing the discounted present value of the forest enterprise and the objective of maximizing the total net carbon sequestration of the forest over the planning horizon.

Once a forest management planning problem has been formulated, it can be solved using any of the many readily available LP solvers. Because formulations can be quite large, with potentially tens of thousands of variables, it may be quite helpful, if not essential, to have a program to interpret the results that will generate tables and charts summarizing the results and what they mean in terms of on-the-ground management operations. These report-writing programs are typically location-specific, like the matrix generators discussed in Sect. 2.4.2.

The solution to a linear programming model gives much more than the optimal variable values. Also included in the solution to a LP model are *reduced cost* values and *shadow prices*. Reduced cost values correspond to variables, and they are only non-zero when the variable value is zero. The reduced cost value indicates how much the objective function coefficient on the corresponding variable must be "improved" before the value of the variable will be positive in the optimal solution. In the case of a minimization problem, "improved" means "reduced;" in the case of a maximization problem, "improved" means "increased."

Shadow prices are associated with constraints, and shadow prices are only non-zero when constraints are binding. Shadow prices are also sometimes referred to as *dual prices*. The shadow price gives the improvement in the objective function if the corresponding constraint is relaxed by one unit. In the case of a less-than-or-equal constraint, such as a resource constraint, the dual price gives the value – in terms of the objective function – of having one more unit of the resource represented by that constraint. In the case of a greater-than-or-equal constraint, such as a minimum production level constraint, the dual price gives the cost – in terms of the objective function – of meeting the last unit of the minimum production target.

Reduced cost and shadow price information is sometimes referred to as sensitivity analysis information. These values indicate how sensitive the solution value is to the parameters in these constraints. Sensitivity information can be very useful for understanding the management situation being modeled. For example, the shadow prices corresponding to the area constraints indicate the value of having one more acre in each analysis area. These values can be extremely useful, for example, if the organization is considering acquiring or selling properties. One must be careful, however, in interpreting dual prices and reduced cost values; sometimes simply changing the way a constraint is written – without changing the actual physical interpretation of the constraint or the optimal solution – can change the shadow price associated with the constraint.

2.5.3 *Spatial Methods – Mixed-Integer Linear Programming (MILP)*

The LP models described in the previous section are generally not well suited for problems where the specific location of each management unit is important. The need to recognize the location and spatial arrangement of management units arises in a variety of forest management situations. One of the most widely studied problems is when there is a limit on the maximum size of a harvest opening (e.g., Thompson et al. 1973; Barahona et al. 1990; Jones et al. 1991; Murray and Church 1996; Murray 1999; McDill and Braze 2000; McDill et al. 2002; Gunn and Richards 2005; Goycoolea et al. 2005., Constantino et al. 2008). This problem is often referred to in the forest management literature as the adjacency problem, since the harvest of one or more adjacent management units may limit the feasibility of harvesting other adjacent management units, but a more appropriate name would be the maximum harvest opening size problem.

Other common forest management planning applications where it is important to know the location of the management units are where roads must be built in order to access all management units (e.g., Weintraub and Navon 1976; Kirby et al. 1986; Weintraub et al. 1995; Murray et al. 1998; Guignard et al. 1998; Richards and Gunn 2000; Andalaft et al. 2003) or where the transportation of products from the forest to one or more mills is modelled (e.g., Weintraub et al. 1996; Forsberg et al. 2005). In addition, knowing the spatial location and juxtaposition of different habitat types can be crucial for incorporating a variety of forest conservation and wildlife concerns into forest planning models (e.g., Bettinger et al. 1997; Hof and Bevers 2002; Rebain and McDill 2003a, b; Tóth et al. 2006; Tóth and McDill 2008, 2009).

The key difference between the spatially-explicit models described in this section and the non-spatial models in the previous section is that in the spatially-explicit models the primary variables are restricted to taking only two possible values: zero and one. Such variables are called *binary variables*. Problems with binary variables, that are otherwise similar to LP problems in that the constraints and objective function are linear functions of the variables, are referred to as mixed-integer linear programs or MIPs, for mixed-integer programs. The fact that these variables cannot take on a continuous set of possible values fundamentally changes the solution algorithms that can be used to solve these problems, and in fact, makes them immensely more difficult to solve. McDill and Braze (2001) provide a description of the branch and bound algorithm, which can often be used to solve MIPs to an exact optimal solution. More modern variations of the algorithm are also commonly referred to as branch and cut algorithms. In some cases, MIP problems cannot be solved to optimality with the branch and cut algorithm, and one must resort to heuristic techniques that in many cases can find near-optimal solutions very efficiently. In addition, a wide variety of heuristic algorithms have been used to solve forest management planning problems, including simulated annealing (Lockwood and Moore 1992), tabu search (Bettinger et al. 2002; Caro et al. 2003; Richards and Gunn 2003), and others (Boston and Bettinger 1999, 2002; Falcão and Borges 2001, 2002; Bettinger et al. 2002; Borges et al. 2002). Several off-the-shelf software packages, ranging from free to quite expensive, are available for solving MIPs. Most modern solvers use a combination of the branch and cut algorithm and heuristic methods. Nemhauser and Wolsey (1988) and Wolsey (1998) provide more in-depth treatments of integer programming concepts and techniques.

The primary variables in the spatially-explicit forest planning model presented in this section are defined as follows:

x_{kj} = a binary variable that takes a value of 1 when management unit k is scheduled to be managed according to prescription j; when the value is zero, management unit k should not be managed according to prescription j.

Note that one could relax the binary restrictions on the x_{kj}'s and simply require them to take values in the interval [0, 1]. In this case, the variable value would indicate the proportion of management unit k scheduled to be managed according to prescription j and the model would be an LP model, rather than a MIP model, and much easier to solve. However, certain constraints, e.g., maximum harvest area constraints, will only work if the binary restriction is enforced.

The objective function for the spatially-explicit, MIP model appears to be the same as for the LP model:

$$\max Z = \sum_{k \in K} \sum_{j \in J_k} a_{kj} x_{kj} \tag{2.7}$$

The differences are that K is now the set of management units, J_k is the set of prescriptions for management unit k, and a_{kj} is the present value of the net revenue from managing unit k according to prescription j. Note that the coefficients a_{kj} are no longer expressed on a per-hectare basis, rather they are on a per-management-unit basis.

The next set of constraints are analogous to *area constraints*. They specify that one and only one prescription can be applied to a management unit. The constraints are:

$$\sum_{j \in J_k} x_{kj} = 1 \quad \forall k \in K \tag{2.8}$$

These constraints are sometimes called *logical constraints*. There will be one of these constraints for each management unit $k \in K$.

As with the objective function, the *accounting constraints* in the spatially-explicit, MIP model appear to be the same as for the LP model:

$$\sum_{k \in K} \sum_{j \in J_k} v_{kjlt} x_{kj} = V_{lt} \tag{2.9}$$

The only difference is that the v_{kjlt} coefficients indicate the amount of input, output or condition l that is used, produced, or projected to exist in period t as a result of management unit k being managed according to prescription j. The purpose of the accounting constraints is the same in the MIP model as in the LP model: to sum the quantity of some input, output or forest condition as a function of the variables x_{kj} and set the value of the corresponding accounting variable (V_{lt}) equal to that sum. With the accounting variables defined and populated by the accounting constraints (2.9), *quantity constraints* and *flow constraints* can be formulated exactly the same as in the non-spatial LP model (see constraint sets (2.4) and (2.5)).

Finally, as with the LP model, nonnegativity constraints are needed for the continuous variables:

$$V_{lt} \geq 0 \quad \forall l \in L \; and \; t \in \{1 \ldots T\} \tag{2.10a}$$

And binary constraints are needed for the binary variables:

$$x_{kj} \in \{0, 1\} \quad \forall k \in K \; and \; j \in J_k \tag{2.10b}$$

The fact that the spatially-explicit model allows one to control not only how much of a given type of forest is treated in a certain way in a given period, but exactly where treatments will occur is very powerful. It makes it possible to plan treatments in a way that will create desired spatial configurations of forest conditions on the landscape. It makes it possible to design the transportation infrastructure that is needed to access each management unit in ways that are more economically efficient and more ecologically sensitive. Furthermore, it is likely that researchers and forest managers have only begun to explore the potential applications of these models. It is beyond the scope of this chapter to cover all these possibilities. The remainder of this section presents three different ways to formulate maximum harvest opening constraints in these models.

Early maximum harvest opening constraint formulations (e.g., Jones et al. 1991; McDill and Braze 2000) assumed that no two adjacent management units could be harvested simultaneously. These formulations are commonly referred to as unit restriction models, or URMs, after Murray (1999). The most obvious way to formulate this restriction is to write a constraint for each pair of adjacent management units in each time period, as follows:

$$\sum\nolimits_{j \in \Omega_{kt}} x_{kj} + \sum\nolimits_{j \in \Omega_{mt}} x_{mj} \leq 1 \quad \forall \{k, m\} \in \Phi, t \in \{1, \ldots, T\} \tag{2.11}$$

Where Ω_{kt} and Ω_{mt} are the sets of prescriptions for management units k and m, respectively, that involve clearcut harvests in period t, and Φ is the set of all pairs of adjacent management units in the forest. The constraints require that at most only one management unit in each pair of adjacent management units can be assigned to a prescription that involves a clearcut harvest in any period t. The type of constraint shown in Eq. (2.10) is commonly known as a *pairwise constraint* because the constraints are written for one pair of management units at a time.

It is often the case that there is more than one way to formulate any given restriction. In the case of MIP problems, how a restriction is formulated can have a substantial impact on how long it takes to solve a MIP problem, even to the extent that it determines whether the problem can be solved at all. A better way to formulate the URM restrictions is as follows:

$$\sum\nolimits_{k \in C} \sum\nolimits_{j \in \Omega_{kt}} x_{kj} \leq 1 \quad \forall C \in \Gamma, t \in \{1, \ldots, T\} \tag{2.12}$$

Where C is a set of mutually-adjacent management units (i.e., each management unit in C is adjacent to all of the other management units in C), and Γ is the set of all maximal sets of such mutually adjacent management units in the forest. A set C is maximal if it is not a subset of any other set of mutually-adjacent management units. The sets C are also known as cliques (Murray and Church 1996), and constraints of the form (2.11) are known as clique constraints. Solution times for problems formulated with adjacency constraints (2.11) will, on average, be shorter than for problems formulated with adjacency constraints (2.10). McDill and Braze (2000) showed this empirically, but the result can be expected for theoretical reasons as well, as the constraint sets (2.11) can be shown to give a tighter approximation of the convex hull of the set of feasible solutions to the problem (Wolsey 1998).

When the combined area of adjacent management units is less than the maximum harvest opening size, it may be allowable to harvest those units at the same time. In such cases, the URM may be too restrictive. The area restriction model, or ARM (Murray 1999), allows contiguous groups of management units to be harvested concurrently as long as their combined area is less than some maximum area. McDill et al. (2002) proposed what they called the Path formulation of the ARM:

$$\sum\nolimits_{k \in P} \sum\nolimits_{j \in \Omega_{kt}} x_{kj} \leq |P| - 1 \quad \forall P \in \Lambda, t \in \{1, \ldots, T\} \tag{2.13}$$

Where P represents a contiguous (connected) group of management units whose combined areas just exceed the maximum clearcut size, or a "path" in McDill et al.'s (2002) terminology. A more appropriate name for this type of constraints is cover constraints, and a better term for paths would be covers (Tóth et al. 2012). Λ represents the set of all minimal path sets in the forest. A path is minimal if the total area of the group drops below the maximum clearcut size if any management unit is removed. These constraints state that within each path group all but one of the management units may be harvested concurrently. Other formulations of the ARM have been proposed by McDill et al. (2002), Gunn and Richards (2005), and Constantino et al. (2008). None of these formulations has yet been shown to be clearly superior to another.

The presentation of the models in this chapter has been deliberately abstract. The purpose here has been to introduce the reader to some basic planning concepts. Chapter 6 presents more concrete example models that will help readers see how the tools described in this chapter can be applied to specific management problems. Readers are also encouraged to consult the many research papers cited in this and other chapters. These papers provide a wealth of example applications.

2.6 Summary

Good planning is essential to the credibility of an organization. Plans communicate both internally and externally the goals, objectives and constraints of the organization. Plans also provide a basis for accountability within an organization. They should have specific, quantified targets so that employee incentives can be tied to achieving or exceeding those targets. However, flexibility is also important in implementing plans in order to deal with unanticipated events. Planning is especially important in forestry operations because many management goals take a long time to achieve and require the coordination of actions over a long period of time.

The process used to develop a plan is as important as having one. The planning process should promote communication among all levels of the organization, with information about organizational goals flowing from upper management down and information about resource capabilities and constraints flowing up. Planning involves six key steps: (1) goal setting, (2) identifying alternatives, (3) evaluating alternatives, (4) plan selection, (5) plan implementation, (6) monitoring, and (7) re-planning. Ideally, planning should be approached as a continuous, ongoing process, rather than as a one-time event or a sequence of discrete planning efforts.

Planning occurs at different spatial and temporal scales. Stand-level management decisions are those that relate to the design and selection of a silvicultural regime for a particular stand. However, whenever a stand is part of a larger enterprise the optimal management regime for that stand in the context of the larger enterprise may not be the same as the optimal management regime if the stand were managed independently. Thus, enterprise-level management decisions often supersede stand-level management decisions. While nearly all decisions made by an organization are in some way interrelated, planning typically occurs at different levels with different degrees of detail at each level. Strategic planning is where the broad, long-term goals and objectives of an enterprise are set and where general strategies for achieving them are developed. Strategic planning problems often are not spatially explicit. Tactical planning addresses medium-term, medium-scale goals. Tactical planning problems are usually spatially-explicit and have shorter planning horizons. Operational planning addresses very short-term, detailed issues such as which stands will be harvested in the next weeks or months, by whom, with what equipment, what products will be produced, and to which mill will they go.

Forest management planning requires a substantial amount of information, including inventory information and information from growth and yield models. Simulations of a variety of management options for each management unit or analysis area provide critical information for planning. It may also be necessary to compile information about the transportation network that will be used to access each management unit and to transport products from the forest to the mill. Finally, various management constraints need to be identified, including production targets, personnel, equipment and budget constraints, and constraints that reflect environmental objectives or regulations.

Monitoring is a crucial component of a forest management planning system. Monitoring systems ensure that plans are being implemented as expected and that management activities are producing the expected outcomes. A good monitoring system enables managers to recognize when management actions are not achieving the desired result so that corrective action can be taken. Forest monitoring systems are generally based on a system of permanent plots, as permanent plots are best for identifying trends and change.

A variety of forest planning models have been developed that enable planners to find solutions to complex problems and allow them to compare the implications of competing theories about how the real world functions and to evaluate trade-offs among competing solutions. Two types of modern forest management planning models are presented in this chapter: non-spatial linear programming models and spatially-explicit mixed-integer programming models. Non-spatial linear programming models are commonly used for strategic planning where the goal is to ensure the long-term sustainability of the forest enterprise. Spatially-explicit mixed-integer programming models are commonly used for tactical and operational planning. They are potentially very powerful planning tools that make it possible to plan treatments in a way that will create desired spatial configurations of forest conditions on the landscape. They can also be used to model the transportation infrastructure that is needed to access each management unit in ways that are more economically efficient and more ecologically sensitive.

References

Acuna MA, Palma CD, Cui W, Martell DL, Weintraub A (2010) Integrated spatial fire and forest management planning. Can J For Res 40:2370–2383

Andalaft N, Andalaft P, Guignard M, Magendzo A, Wainer A, Weintraub A (2003) A problem of forest harvesting and road building solved through model strengthening and Lagrangean relaxation. Oper Res 51(4):613–628

Barahona F, Weintraub A, Epstein R (1990) Habitat dispersion in forest planning and the stable set problem. Oper Res 40(Suppl 1):S14–S22

Bettinger P, Sessions J, Boston K (1997) Using Tabu search to schedule timber harvests subject to spatial wildlife goals for big game. Ecol Model 42:111–123

Bettinger P, Graetz D, Boston K, Sessions J, Chung W (2002) Eight heuristic planning techniques applied to three increasingly difficult wildlife planning problems. Silva Fennica 36(2):561–584

Borges JG, Hoganson HM, Falcão AO (2002) Heuristics in multi-objective forest planning. In: Pukkala T (ed) Multi-objective forest planning. Kluwer Academic Publishers, Dordrecht, pp 119–151

Boston K, Bettinger P (1999) An analysis of Monte Carlo integer programming, simulated annealing, and Tabu search heuristics for solving spatial harvest scheduling problems. For Sci 45(2):292–301

Boston K, Bettinger P (2002) Combining Tabu search and genetic algorithm heuristic techniques to solve spatial harvest scheduling problems. For Sci 48(1):35–46

Caro F, Constantino M, Martins I, Weintraub A (2003) A 2-opt Tabu search procedure for the multiperiod forest harvesting problem with adjacency, greenup, old growth, and even flow constraints. For Sci 49(5):738–751

Chappelle DE (1966) A computer program for scheduling allowable cut using either area or volume regulation during sequential planning periods. U.S. Forest Service Research Paper PNW-33

Constantino M, Martins I, Borges J (2008) A new mixed-integer programming model for harvest scheduling subject to maximum area restrictions. Oper Res 56(3):542–551

Curtis FH (1962) Linear programming the management of forest property. J Forest 60:611–616

Davis LS, Johnson KN, Bettinger PS, Howard TE (2001) Forest management to sustain ecological, economic, and social values, 4th edn. McGraw-Hill, New York, 804 p

Elzinga CL, Salzer DW, Willoughby JW (1998) Measuring and monitoring plant populations. U.S. Bureau of Land Management Papers. Paper 17. http://digitalcommons.unl.edu/usblmpub/17. Accessed 9 Feb 2014

Evelyn J (1664) Sylva, or a discourse of forest trees and the propagation of timber. Available online in Nisbet, John. 1908. John Evelyn, Sylva, Or A Discourse of Forest Trees ... with an Essay on the Life and Works of the Author by John Nesbit Fourth Edition (1706), reprinted London: Doubleday & Co., 1908. Online edition, March 2007, last accessed 31 Jan 2014

Falcão AO, Borges JG (2001) Designing an evolution program for solving integer forest management scheduling models: an application in Portugal. For Sci 47:158–168

Falcão AO, Borges JG (2002) Combining random and systematic search heuristicprocedures for solving spatially constrained forest management scheduling problems. For Sci 48:608–621

Faustmann M (1849) On the determination of the value which forest land and immature stands pose for forestry. In: Gane M (ed) Martin Faustmann and the evolution of discounted cash flow. Paper 42. Oxford Institute, Oxford, England; 1968, 54 pp

Forsberg M, Frisk M, Rönnqvist M (2005) FlowOpt – a decision support tool for strategic and tactical transportation planning in forestry. Int J For Eng 16(2):101–114

Gaffney M (1957) Concepts of financial maturity of timber and other assets. A.E. Information series no. 62. Department of Agricultural Economics, NC State University, Raleigh, NC

Goycoolea M, Murray AT, Barahona F, Epstein R, Weintraub A (2005) Harvest scheduling subject to maximum area restrictions: exploring exact approaches. Oper Res 53(3):490–500

Guignard M, Ryu C, Spielberg K (1998) Modeling tightening for integrating timber harvest and transportation planning. Eur J Oper Res 111:448–460

Gunn EA, Richards EW (2005) Solving the adjacency problem with stand-centered constraints. Can J For Res 35:832–842

Hof J, Bevers M (2002) Spatial optimization for managed ecosystems. Columbia University Press, New York, 258 p

Hoganson HM, Rose DW (1984) A simulation approach for optimal timber management scheduling. For Sci 30(1):220–238

Ignizio JP, Cavalier TM (1994) Linear programming. Englewood Cliffs, Prentice Hall, 666 p

Iverson DC, Alston RM (1986) The genesis of FORPLAN: a historical and analytical review of forest service planning models. GTR INT-214. USDA Forest Service Intermountain Research Station, Ogden, UT, 31 p

Johnson KN, Scheurman EL (1977) Techniques for prescribing optimal timber harvest and investment under different objectives – discussion and synthesis. For Sci Mono 18:31

Jones JG, Meneghin BJ, Kirby MW (1991) Formulating adjacency constraints in linear optimization models for scheduling projects in tactical planning. For Sci 37(5):1283–1297

Kaplan JO, Krumhardt KM, Zimmermann M (2009) The prehistoric and preindustrial deforestation of Europe. Quaternary Sci Rev 28(27–28):3016–3034

Kent B, Bare BB, Field RC, Bradley GA (1991) Natural resource land management planning using large-scale linear programs: the USDA forest service experience with FORPLAN. Oper Res 39(1):13–27

Kidd WE, Thompson EF, Hoepner PH (1966) Forest regulation by linear programming – a case study. J Forest 64:611–613

Kirby MW, Hager WA, Wong P (1986) Simultaneous planning of wildland transportation alternatives. TIMS Stud Manag Sci 21:371–387

Lockwood C, Moore T (1992) Harvest scheduling with spatial constraints: a simulated annealing approach. Can J For Res 23:468–478

McDill ME, Braze J (2000) Comparing adjacency constraint formulations for randomly generated forest planning problems with four age-class distributions. For Sci 46(3):423–436
McDill M, Braze J (2001) Using the branch and bound algorithm to solve forest planning problems with adjacency constraints. For Sci 47(3):403–418
McDill ME, Rebain S, Braze J (2002) Harvest scheduling with area-based adjacency constraints. For Sci 48(4):631–642
McQuillan AG (1986) The declining even flow effect – non sequitur of national forest planning. For Sci 32(4):960–972
Murray AT (1999) Spatial restrictions in harvest scheduling. For Sci 45(1):45–52
Murray AT, Church RL (1996) Analyzing cliques for imposing adjacency restrictions in forest models. For Sci 42(2):166–175
Murray AT, Church RL, Weintraub A (1998) Operational forest planning and Steiner tree considerations. Geogr Syst 5:221–237
Nautiyal JC, Pearce PH (1967) Optimizing the conversion to sustained yield – a programming solution. J Forest 13(2):131–139
Nemhauser GL, Wolsey LA (1988) Integer and combinatorial optimization. Wiley, New York, xiv, 763 p
Newman DH (1988) The optimal forest rotation: a discussion and annotated bibliography. Gen Tech Rep SE-48. USDA Forest Service, Southeastern Forest Experiment Station, Asheville, 47 p
Nix S (2014) The history of forestry in Europe. http://forestry.about.com/od/foresthistory/p/The-Histroy-Of-Forestry-In-Europe.htm. Accessed 31 Jan 2014
Päivinen R, Yli-Kojola H (1989) Permanent sample plots in large-area forest inventory. Silva Fennica 23(3):243–252
Pickens JB, Kent BM, Ashton PG (1990) The declining even flow effect and the process of national forest planning. For Sci 36(3):665–679
Rebain SA, McDill ME (2003a) A mixed-integer formulation of the minimum patch size problem. For Sci 49(4):608–618
Rebain SA, McDill ME (2003b) Can mature patch constraints mitigate the fragmenting effect of harvest opening size restrictions? Int Trans Opera Res 10(5):499–513
Richards EW, Gunn EA (2000) A model and Tabu search method to optimize stand harvest and road construction schedules. For Sci 46(2):188–203
Richards EW, Gunn EA (2003) Tabu search design for difficult forest management optimization problems. Can J For Res 33:1126–1133
Samuelson P (1976) Economics of forestry in an evolving society. Econ Inq 14(Dec.):466–492
Thompson EF, Halterman BG, Lyon TJ, Miller RL (1973) Integrating timber and wildlife management planning. For Chron 49(6):247–250
Tóth SF, McDill ME (2008) Promoting large, compact mature forest patches in harvest scheduling models. Environ Model Assess 13(1):1–15
Tóth SF, McDill ME (2009) Finding efficient harvest schedules under three conflicting objectives. For Sci 55(2):117–131
Tóth SF, McDill ME, Rebain SA (2006) Finding the efficient frontier of a bi-criteria, spatially-explicit, harvest scheduling problem. For Sci 52(1):93–107
Tóth SF, McDill ME, Könnyu N, George S (2012) A strengthening procedure for the path formulation of the area-based adjacency problem in harvest scheduling models. Math Comput For Nat Res 4(1):27–49
Walker JL (1990) Traditional sustained yield management: problems and alternatives. Forest Chron 66(1):20–24
Ware GO, Clutter JL (1971) A mathematical programming system for management of industrial forests. For Sci 17:428–445
Ware KD, Cunia T (1962) Continuous forest inventory with partial replacement of samples. For Sci Mono 3, 40 p
Wei Y (2012) Optimize landscape fuel treatment locations to create control opportunities for future fires. Can J For Res 42:1002–1014

Weintraub A, Navon D (1976) A forest management planning model integrating silvicultural and transportation activities. Manag Sci 22(12):1299–1309
Weintraub A, Jones JG, Meacham ML, Magendzo A, Malchuck D (1995) Heuristic procedures for solving mixed-integer harvest scheduling-transportation planning models. Can J For Res 25:1618–1626
Weintraub A, Epstein R, Morales R, Serón J, Traverso P (1996) A truck scheduling system improves efficiency in the forest industries. Interfaces 26(4):1–12
Weintraub A, Miranda JP, Romero C, Bjørndal T, Epstein R (2007) Handbook of operations research in natural resources. Springer Science & Business Media, New York
Wolsey LA (1998) Integer programming. Wiley, New York, xviii, 264 p

Chapter 3
Forest Growth and Yield Models for Intensively Managed Plantations

Aaron R. Weiskittel

3.1 Introduction

Intensively managed plantations represent a significant financial investment that require detailed planning and forecasting to ensure an adequate return. Important issues in proper plantation management are initial planting density and vegetation control, the timing and intensity of mid-rotation management activities like thinning and fertilization, and the optimal rotation length. Across a forest ownership, these issues have both a temporal and spatial aspect, which can make them difficult to evaluate and more importantly, quantify. This is because tree growth and mortality are dynamic biological processes and sensitive to array of factors like genetics, climate, and disturbances (e.g. fire, insect or pathogen outbreaks). Consequently, forest growth and yield models are widely used tools in plantation management as they allow the projection and evaluation of future stand development, the influence of alternative management regimes, and potential impacts of changes in climate. They are used in both strategic and tactical forest management planning.

Forest growth and yield models have a long history of development and use, particularly in the last two decades due to the greater availability of personal computers to do both data analysis and complex simulations. Growth and yield models exist in a variety of forms as both conceptual and quantitative models are used for plantation management. Conceptual models are built from the extensive scientific literature that describes plantation response to management (e.g. Moores et al. 2007). Often conceptual models are difficult to apply because each plantation is unique due to its location and past management history. In addition, one major objective of sustainable plantation management is the ability to compare multiple alternative management strategies. Thus, quantitative models

A.R. Weiskittel (✉)
School of Forest Resources, University of Maine, Orono, ME, USA
e-mail: aaron.weiskittel@maine.edu

J.G. Borges et al. (eds.), *The Management of Industrial Forest Plantations: Theoretical Foundations and Applications*, Managing Forest Ecosystems 33,
DOI 10.1007/978-94-017-8899-1_3,

are widely used because they can be used to update and project forest inventories, compare alternative plantation management regimes, estimate sustainable harvest rates, and test important hypotheses regarding plantation growth and development. Quantitative models attempt to represent forests with mathematical equations that describe their behavior over time. However, understanding the various components and assumptions of quantitative models can be difficult, which is why they are often treated as black boxes.

In forest plantation management, various quantitative models are used. These models differ in terms of their temporal resolution (daily vs. annual vs. decadal), spatial scale (stand vs. individual tree), reliance on data (statistical vs. mechanistic), representation of competitive processes (distance-independent vs. distance-dependent), and degree of stochasticity. These differences have important implications for how useful they are for the plantation management planning process and understanding these tradeoffs is necessary for interpreting growth model limitations and output.

Plantation management activities vary from planting different genetic material, the selective removal of certain individuals (thinning), and altering soil nutrient availability (fertilization). Ascertaining the long-term effects of these management activities is difficult because of the dynamic nature of trees and high variability in the response of forests to management. In addition, new questions on the effective management of plantations are emerging like climate change, broader ecosystem management objectives, and increased demands for forest resource products. This suggests that models will continue to be an important component of the forest planning process.

The goal of this chapter is to explore various modeling approaches used for plantation management, provide a brief description of some example models, explore the ways that they have been used to aid the decision-making process, and make suggestions for future improvements.

What You Will Learn in This Chapter

- Key assumptions, components, and limitations of the different growth and yield models available for projecting intensively managed plantations
- Examples of widely used growth and yield models
- Approaches for representing various silvicultural activities widely used in intensively managed plantations in growth and yield models
- Necessary consideration in the development and use growth and yield models
- Strategies for improving future growth and yield models

3.2 Growth Modeling Approaches

As previously described, the modelling approaches used in forest plantation management differ widely in their general frameworks. One of the most significant distinctions is the way that the models treat forest processes. Based on this,

Weiskittel et al. (2011) divided forest growth and yield models into four broad categories: (1) statistical models; (2) mechanistic; (3) hybrid; and (4) gap. Statistical models describe the development of a plantation based on empirical data and mathematical equations parameterized from these data, while mechanistic models represent the key physiological processes such as light interception, photosynthesis, and respiration to predict growth. Hybrid models combine features of both statistical and mechanistic models to take advantage of strengths offered by each approach. Gap models are designed to explore long-term ecological processes generally for understanding interactions that control forest species succession (Bugmann 2001), but have seen limited use for plantation management and are not discussed here. Models that integrate the development of multiple forest stands, such as the landscape model LANDIS (Scheller et al. 2007), will not be covered in this chapter as they have seen limited application for plantation management.

Within any given model category, models differ in their resolution (both spatial and temporal), spatial-dependence, and degree of determinism. Spatial resolution refers to the basic unit for predictions, with the simplest being a whole-stand approach and the individual tree approach the most detailed. Temporal resolution is the basic time step for model predictions. Several mechanistic models have daily or even hourly time steps, while statistical models generally have 5- to 10-year temporal resolutions. Models also vary in their use of spatial information as distance-dependent or spatially explicit models require spatial location information; often individual tree x-y coordinates are needed. Distance-independent or spatially implicit models do not require this information. Finally, models differ in their use of stochasticity as some growth models incorporate some purely random elements and will give a different return value for any set of input values. This can be an important element of forest modeling as some relevant factors like climate and natural disturbances ultimately govern the growth and yield of a particular plantation can be random or unpredictable. However, a model with too many stochastic elements can make interpretation a challenge.

Each of the aforementioned growth and yield modeling approaches is described below.

3.2.1 Statistical

Statistical models depict trends observed in measurement plots that are established in forests. Consequently, statistical models are usually only as good as the data used to develop them. To be effective for modelling purposes, the data must cover the extremes of the population they are intended to represent, be extensive, and include measurements that likely describe the inherent variability of the observations. Due to regional differences, resolution of datasets, and the statistical approaches used, a vast number of statistical models currently exist. Most statistical models operate on 5- to 10-year time steps, but annualized models exist too (Weiskittel et al. 2007). Given the short rotation lengths and intensive management regimes, annualized models are likely the best option for plantations because they offer the greatest flexibility. In addition, most statistical models generally rely on site index, the average dominant height at a certain base age (generally 50 years), as a measure of potential site productivity (Skovsgaard and Vanclay 2008). Therefore, the largest differences in statistical models are their spatial resolution and treatment of competition. Based on their spatial resolution, three primary classes of statistical models exist: (1) whole-stand; (2) size-class; and (3) individual-tree. Given their wide use for forecasting plantations, these different classes are discussed separately below.

3.2.1.1 Whole Stand

Whole-stand models describe the stand in terms of a few values like total volume, basal area, or the number of individuals per unit of area and predict the change in these attributes over time. Whole-stand models are the simplest type of statistical model and have the longest history of development. One of the earliest examples of a whole-stand model in North America is the yield tables of Meyer (1929), which described growth in terms of stand age and site index. These yield tables were generalized into compatible growth yield equations that predicted changes in stand volume as a function of initial stand conditions and age (Buckman 1962; Clutter 1963; Moser 1972). Even today, yield tables and equations are still being widely developed for a variety of plantation species (e.g. Bermejo et al. 2004; Matsumura 2011) as they require minimal data, are easy to interpret, and of general interest to most managers.

Whole stand models are most appropriate for plantations when compared to other forest types. Although techniques have been developed to represent management activities with whole-stand models (Bailey and Ware 1983; Pienaar et al. 1985), they are not the most efficient approach, particularly when multiple thinnings are intended to be represented. However, whole-stand models continue to be developed using modern statistical techniques (Barrio-Anta et al. 2006; Castedo-Dorado et al. 2007a; García et al. 2011) as they are easy to use, relatively robust, and can be more accurate in long-term predictions (Cao 2006). For example, Vanclay (2010) presented a robust approach to predict dominant height, mean diameter, and mortality for plantations, even when data are sparse and remeasurements are not available. Some widely used whole-stand growth models for plantations still in use are DFSIM (Curtis et al. 1981), TADAM (García 2005b), Scube (García 2011; García et al. 2011), and GNY (MacPhee and McGrath 2006).

3.2.1.2 Size Class

Because a plantation can be comprised of trees of varying sizes, a size-class model divides each stand into multiple groups of similar-sized individuals, which are projected through time. Some of the most common size-class modeling approaches are stand-table projections (e.g. Trincado et al. 2003; Allen et al. 2011), matrix-based (Liang and Picard 2013), and diameter-distribution models (e.g. Qin et al. 2006). Stand-table projections and matrix-based approaches are similar in that the frequencies of trees in each cohort are projected through time by estimating the probability of moving from one group to another. A diameter-distribution approach uses statistical probability distributions to describe the frequency of trees in different size classes and their changes through time. The Weibull probability distribution has been commonly been used because it is flexible, relatively easy to integrate, and the parameters can be determined in multiple ways (Cao 2004). Often, stand-level models are linked with diameter-distribution models as it simplifies the calculations and still provides tree-level information (Weiskittel et al. 2011). This approach has been widely used to project plantations, particularly loblolly pine (e.g. Bailey 1980; Borders and Patterson 1990; Qin et al. 2006). Overall, the size class approach is a compromise between the stand- and tree-level approaches, which makes it probably the most useful approach for plantations with simplified and consistent management regimes.

3.2.1.3 Individual Tree

An individual-tree growth-and-yield model depicts the changes in each tree in a tree list representing a particular plantation. These models provide the highest resolution of predictions, but require the most data for both development and application. Since the individual tree is the focal point, these models are effective for representing both single and are mixed-species plantations (Porté and Bartelink 2002). In addition, these models are also effective for representing the effects of management, particularly complex thinning regimes. They have multiple components including individual tree diameter increment, height increment, and mortality equations, which are described in detail by Weiskittel et al. (2011).

One key distinction of statistical individual-tree models is whether they are distance-dependent or distance-independent. Distance-dependent models require the location of each tree included in the simulation to be known, while distance-independent models assume the trees are randomly distributed in the forest. Using tree location, distance-dependent models estimate competition indices such as size-distance (Opie 1968), area potentially available (Nance et al. 1988), and exposed crown surface area (Hatch et al. 1975). Distance-independent models represent competition using variables such as basal area in larger trees (Wykoff 1990), crown closure in higher trees (Hann et al. 2003), and relative tree diameter (Glover and Hool 1979). Most comparisons between the effectiveness of distance-dependent and distance-independent measures of competition for predicting growth have found distance-independent to be just as effective (Biging and Dobbertin 1995; Wimberly and Bare 1996; Corral Rivas et al. 2005). This suggests that knowledge of tree location is not worth the effort or expense of collecting that information, but emerging remote-sensing technologies like LiDAR may make it much easier to acquire this spatial information in the future. In addition, the use of distance-dependent growth models for plantations might be warranted if investigating plantations that involve various mixtures of genetic material and/or species as well as complex thinning regimes. This was recently demonstrated by Amateis and Burkhart (2012) who simulated the development of loblolly pine plantations with mixed genetic material across a range of initial densities and spacings.

Some key examples of distance-dependent, individual-tree models are PTAEDA (Burkhart et al. 1987, revised 2001), SILVA (Pretzsch et al. 2002), and TASS (Mitchell 1975), while ORGANON (Hann 2011), GLOBTREE (Soares and Tomé 2003), and FVS (Crookston and Dixon 2005) are some widely used distance-independent, individual-tree models for plantations. Individual-tree models have been widely modified to account for the effects of forest management activities like fertilization and thinning (Hann et al. 2003). Since the individual tree is the focus, the implementation of complex thinning regimes is relatively straightforward (Söderbergh and Ledermann 2003). However, significant errors can occur with individual tree growth models due to the potential for compounding error and often, unconstrained predictions (Cao 2006). These limitations can often be achieved by linking stand- and tree-level models (Yue et al. 2008; Zhang et al. 2010).

3.2.2 Mechanistic

Statistical models generally cannot be extrapolated to new situations that were not covered in the data used to develop them. Statistical models also commonly rely on site index, which is the dominant height at a specified reference age, to represent the potential productivity of a site. However, site index has several known problems (Skovsgaard and Vanclay 2008). Finally, most statistical models view climate as static and do not include it as part of the projections, which can be an important limitation for short-rotation plantations. In contrast, mechanistic models represent tree physiologically processes to avoid these limitations.

Although mechanistic models have a long history of development, they have been used primarily for research rather than forest management purposes (Mäkelä et al. 2000). This is because they often require extensive parameterization, rely on information not commonly available in forest inventories, and the output is often expressed in terms of little interest to forest managers, such as gross or net primary production (NPP). Regardless, several mechanistic models such as BIOME-BGC (Merganičová et al. 2005; Petritsch et al. 2007), CenW (Kirshbaum 2000), and MAESPA (Duursma and Medlyn 2012) have been developed to better understand the effects of forest management.

Most mechanistic models represent physiological processes at the whole-stand level because it simplifies the calculations and processes are better understood at this scale (Landsberg 2003b). Thus, differences between mechanistic models are in their temporal resolution, level of detail in physiological processes, and the representation of belowground processes. A monthly temporal resolution is commonly used because this type of climate information is widely available and some physiological processes scale better at this resolution. The limitation is that daily variation is not represented despite the fact that it can drive many physiological relationships.

Previous reviews have explored differences in various modeling approaches in representing specific physiological processes such as light interception (Wang 2003), photosynthesis (Medlyn et al. 2003), respiration (Gifford 2003), and carbon allocation (Lacointe 2000), while Weiskittel et al. (2011) as well as Landsberg and Sands (2011) provide a more detailed overview of mechanistic growth models. The representation of these physiological processes has varied from highly simplistic to very complex. A general standard in most mechanistic models used for forest management is to use the Beer-Lambert law to estimate light interception, the Farquhar et al. (1980) equation for photosynthesis, and to assume functional balance and allometric relations for carbon allocation (Le Roux et al. 2001). However, recent research indicates that using optimization approaches is more effective for predicting carbon allocation across a range of stand conditions (Mäkelä et al. 2008; McMurtrie and Dewar 2013). For belowground processes, some models treat the soil as a single layer and ignore most nutrient cycles (e.g. Running and Gower 1991), while others rely on very detailed models of soil processes (e.g. Kirschbaum and Paul 2002). Regardless of their temporal resolution or level of detail, most mechanistic models are highly sensitive to leaf area index (LAI) because it drives the within- and below-canopy microclimate, determines and controls canopy water interception, radiation extinction, transpiration, and carbon gas exchange (Bréda 2003).

Even today, basic physiological parameters are unavailable for several tree species, which can make using a mechanistic model challenging. An interesting alternative to parameterizing each individual equation used in a mechanistic model from the literature or with new data is the use of a Bayesian optimization technique. This technique has been demonstrated several times and often with promising results (Van Oijen et al. 2005; Svensson et al. 2008; Deckmyn et al. 2009). In this approach, Markov Chain Monte Carlo simulation is used to vary the model parameters and calibrate model predictions to observed data. The further application of this technique and increased availability of climate data should help increase the use of mechanistic models for representing forest management, particularly under climate change (Schwalm and Ek 2001). When properly parameterized, mechanistic models can be just as effective or even better than statistical models in short-term simulations (Miehle et al. 2009). However, mechanistic models can struggle with long-term projections because of the difficulty in representing mortality accurately (Hawkes 2000).

3.2.3 Hybrid

Hybrid models combine features of both statistical and mechanistic approaches. This approach relies on the robustness of statistical models, while increasing their ability to extrapolate and avoid limitations with site index. Hybrid models have been suggested as the most effective means for representing the effects of forest management because they provide output of interest to forest managers and avoid the heavy data requirements of most mechanistic models (Landsberg 2003a). Several hybrid models have been developed for plantations and single-species, even-aged stands including: CroBAS (Mäkelä 1997), DF.HGS (Weiskittel et al. 2010), CABALA (Battaglia et al. 2004), and SECRETS (Sampson et al. 2006). One widely used hybrid model is 3-PG (Landsberg and Waring 1997), which has been parameterized for a variety of forest types (Landsberg et al. 2003), particularly *Eucalyptus* plantations (Sands and Landsberg 2002; Almeida et al. 2004; Stape et al. 2004).

Three primary classes of hybrid model frameworks currently exist, namely: (1) statistical growth equations with a physiologically derived covariate; (2) statistical equations with a physiologically derived external modifier; and (3) allometric models. The degree of hybridization within each of these classes varies greatly, so an exact classification of a hybrid model is difficult. For example, Milner et al. (2003) linked the statistical Forest Vegetation Simulator (FVS) model and the mechanistic STAND-BGC model such that both models ran simultaneously in parallel and a user selected the degree of coupling.

An example of a statistical growth equation with a physiologically derived covariate is given in Baldwin et al. (2001), who related site index to NPP from a mechanistic model and allowed it to vary during a simulation. Henning and Burk (2004) provide an example of a statistical equation with a physiologically derived external modifier and found it improved projections. Allometric hybrid models rely on simplified representations of physiological processes and empirical equations that relate tree size to biomass. CroBAS (Mäkelä 1997) and 3-PG (Landsberg and Waring 1997) are two examples of allometric hybrid models. Both models use the light-use-efficiency concept to relate light interception to gross primary production (GPP), which avoids the complications of a detailed canopy photosynthesis equation. In addition, 3-PG avoids estimating respiration by assuming NPP is one-half of GPP, which has been supported by some empirical studies (Waring et al. 1998; Litton et al. 2007). Allometric equations are then used to convert typical forest inventory data into biomass and to estimate carbon allocation. However, using a mean tree approach like 3-PG to accomplish this can result in a significant bias as the diameter distribution becomes more varied (Duursma and Robinson 2003). To avoid this limitation, Mason et al. (2007, 2011) used the radiation sum obtained from 3-PG as a covariate in a statistical equation to model plantation growth across a range of management intensities. Several other researchers have modified 3-PG by linking it to a detailed soil model (Xenakis et al. 2008; Wang et al. 2011), refining estimation of the site fertility rating (Vega-Nieva et al. 2013), linking the model to data obtained from remote sensing (Waring et al. 2010), or altering the way wood density is calculated (Bryars et al. 2013).

Relative to purely statistical models, the degree of improvement achieved with a hybrid model has varied. At the stand level, hybrid models have been quite effective at improving predictions (Battaglia et al. 1999; Snowdon 2001; Dzierzon and Mason 2006), while less modest gains have been achieved at the individual tree level (Henning and Burk 2004; Weiskittel et al. 2010). The range of the reported improvements can vary widely at both the stand and tree levels because of the breadth of conditions covered by evaluation data, the length of the simulations, and differences in the adequacy of the statistical model. Interestingly, Henning and Burk (2004) found climate-dependent growth indices almost as effective as the mechanistically based ones, while Snowdon et al. (1998) found just the opposite. Regardless, the use of hybrid models will likely continue to increase in the future as the understanding of physiological processes improves and the complexity of questions facing plantation managers broaden.

3.3 Representing Plantation Management

One of the key uses of forest growth and yield models is projecting the long-term consequences of various plantation management decisions. This is an important use of models because installing and maintaining long-term field trials is expensive and time-consuming. In addition, not all treatment combinations can be replicated across the landscape. Finding an optimal management regime, therefore, is often dependent on projections obtained from a growth model. Thus, understanding how growth models represent different plantation management activities is crucial for deciding the credibility of the simulations.

Models represent plantation management activities to varying degrees as vastly different approaches have been employed in both tree- and stand-level models. One of the most significant decisions in representing the long-term response to forest management activities is deciding the type of response. Two of the most basic responses to management are generally considered Type 1 and 2 (Snowdon 2002). Type 1 increases growth, but does not change inherent site productivity, while Type 2 responses accomplish both. Although these two generalizations cover the majority of responses to management, a vast number of other responses have been observed.

Not all plantation management activities are explicitly represented in each model or the equations from the unmanaged plantations are simply assumed to be adequate for representing the treatment. For example, less than half of the commonly used models in the US Pacific Northwest explicitly represent fertilization, despite the wide use of this management activity in the region (Robinson and Monserud 2003). The implications of this are well illustrated by Johnson (2005) who compared the long-term projections of various management regimes from six commonly used growth models in the US Pacific Northwest. The conclusions of this analysis were that the models showed a wide range of responses to the treatments and no one model adhered to all of the general research findings on the treatments, which suggests that users need to be careful when making management decisions based on model simulations (Johnson 2005). Similar results to Johnson (2005) were also recently found for loblolly pine growth models in the southern US (Henderson et al. 2013).

This section focuses on the most common management activities in plantations including: (1) genetic improvements; (2) vegetation management; (3) thinning; and (4) fertilization. Approaches taken at both the stand and individual tree levels will be described.

3.3.1 Genetic Improvements

Geneticists have made significant strides in breeding commercial important tree species for certain attributes, particularly growth. Gains in growth of 5–25 % above unimproved stock have been reported for many species and even higher gains have been achieved (White et al. 2007; pp. 298–299). However, these gains are often obtained from individual trees on research trials with limited regional replication, which makes it difficult to understand the long-term implications of these gains at the stand level. Consequently, a variety of approaches have been used to represent genetic improvements in growth models.

3.3.1.1 Stand-Level

At the stand level, two primary approaches for representing stand response to genetic improvements have been used, namely (1) modification of the site index and (2) growth equations specific to a certain family or provenance. Buford and Burkhart (1987) simply modified site index values for different genetic material and indicated that no other major modifications were necessary to make models sensitive to genetic improvements. However, later studies suggested that simply adjusting the level of site index was not sufficient for representing genetic improvement in growth models (Xie and Yanchuk 2003; Adams et al. 2006). For example, Adams et al. (2006) was also one of the first to develop family-specific survival equations, while several other studies have presented growth equations with parameters specific to individual families (Knowe and Foster 1989; Sprinz et al. 1989). The limitation of this approach is that it isn't generalized and requires extensive datasets to obtain reliable parameter estimates. In contrast, the approach of genetic-gain multipliers requires relatively little modification to existing growth models. Carson et al. (1999) illustrated this approach for radiata pine in New Zealand as they developed modifiers for mean top height, basal area, and stocking.

3.3.1.2 Individual Tree

Individual tree growth models have been used to project stand-level gains using various hypotheses generated from different field trials (Rehfeldt et al. 1991; Hamilton and Rehfeldt 1994). However, the development of individual tree growth equations that incorporate the effects of genetic improvements are limited due to the lack of data and knowledge how to best model the changes (Talbert and Hyink 1988). Gould et al. (2008) used the growth modifier approach to estimate changes in Douglas-fir individual tree growth due to genetic improvements, which were found to be positively related to the seedlot's genetic worth for both diameter and height. Gould and Marshall (2010) demonstrated how these multipliers can be used in existing regional growth models and found that projected volume gains were nearly doubled when the genetic worth was increased from 5 % to 10 %.

Projecting the long-term influences of genetic improvements is still an open question due to the difficulty in assessing the influence of genetics on long-term mortality patterns. Continued improvement will likely occur due to the increased availability of data, but this often offset by the rapidly evolving nature of available genetic material. This will be important issue to resolve due to the continued investment in tree breeding programs and the greater application of clonal forestry. However, the limitation will continue to be trying to scale the individual tree plots used by most geneticists to the stand level.

3.3.2 Vegetation Management

The control of competing vegetation early in stand establishment is an effective method for significantly increasing long-term yields in a variety of forest types (Wagner et al. 2006). Representing the long-term influence of vegetation control in growth models is difficult because most models do not simulate non-tree growth and the treatment primarily influences early stand dynamics, which may not be recognized in most simulators that simulate diameter growth only at breast height. Richardson et al. (2006) as well as Mason and Dzierzon (2006) have reviewed the ways that the influence of vegetation management on growth have been modeled and found that it has been represented to varying degrees.

3.3.2.1 Stand-Level

Pienaar and Rheney (1995) modeled stand-level growth to vegetation control with dominant height and basal area equations that were dependent on time since treatment and other stand variables. Quicke et al. (1999) found that these equations were effective in modeling both the short- and long-term influence of vegetation control, but needed to be modified to account for initial planting-density effects. Mason and Milne (1999) used a difference form of the Pienaar and Rheney (1995) basal area equation, which was found to be more effective for capturing the effects of weed control, fertilization, and site preparation in radiate pine.

However, these previous studies only examined growth and did not model mortality. Mason et al. (1997) developed a stand-level model that predicted growth, survival, and size-class distributions for the first 5 years after planting and linked it with an existing growth model. Similar initial stand growth models exist for other regions (Belli and Ek 1988; Payandeh and Haig 1991), but significant biases can occur if they are improperly linked to existing growth and yield models (Mason and Dzierzon 2006). In addition, these previous studies generally accounted for vegetation control with a simple indicator variable rather than a continuous measure like percent control, which limits their ability to represent treatments not covered in their original datasets.

Knowe (1994b) developed a stand table projection system that projected the effects of vegetation management on early stand development, and the change in quadratic mean diameter was sensitive to the intensity of the treatment. Knowe and Stein (1995) expanded this model by developing equations to predict the influence of competing vegetation on survival, dominant height growth, and development of woody vegetation, which were updated by Knowe et al. (2005). These models illustrated the effectiveness of varying levels of vegetation control and provided tree lists that could be passed to another model at age 20. After 10 years of simulation, Knowe and Stein (1995) indicated that using the simpler stand-level approach was just as effective as a stand table projection, which require an initial tree list, in representing the influence of vegetation management on growth and survival.

3.3.2.2 Tree-Level

Individual tree growth models sensitive to vegetation control are relatively rare. Knowe (1994a) quantified the effects of vegetation control treatments on individual tree height-age and height-diameter relationships in young Douglas-fir plantations, which could be used in the stand table projection system of Knowe (1994b). Ritchie and Hamann (2006, 2008) presented individual tree diameter, height, and crown width increment equations for competing vegetation and young Douglas-fir that were sensitive to stand density and could be used to simulate varying levels of vegetation control. Vaughn (2007) modeled young tree growth and indicated that young Douglas-fir height, but not diameter growth, was sensitive to the amount of competing vegetation when expressed as a percent cover. Kimberley and Richardson (2004) presented a growth model for radiata pine that incorporated the influence of weed competition and seasonal growth patterns, but the parameters were only determined for one location in New Zealand. Westfall et al. (2004) developed a young stand growth model for loblolly pine that indicated that both tree diameter and height were highly sensitive to herbaceous weed control, but didn't evaluate the influence of varying levels of control.

Empirically modeling the influence of vegetation control is challenging because of the inherent high variability of early stand growth and the difficulty of quantifying the influence of competing vegetation on available site resources like moisture. In addition, most herbicide trials are commonly established on small plots or without tagged trees, which makes developing individual tree predictive equations difficult. Both Richardson et al. (2006) and Mason and Dzierzon (2006) suggest mechanistic and hybrid growth models might be the best approach for predicting the influence of vegetation management. Approaches for accomplishing this are illustrated by Watt et al. (2003) and Mason et al. (2007), which deserve additional attention.

3.3.3 *Thinning*

Thinning is one of the most commonly employed forest management activities and has been extensively studied. Thinnings come in a variety of forms and generally differ in their timing (precommercial vs. commercial), type (high vs. low vs. crop tree), and intensity, which can make it challenging to model. Most existing growth models address the influence of thinning to some degree, but the manner in how thinning effects are addressed differ appreciably.

3.3.3.1 Stand-Level

The representation of thinning at the stand-level is challenging because thinning treatments are designed to selectively remove individual trees, which greatly alters the residual stand structure. Initially, it was common to assume that the growth and yield of a thinned stand was similar to an unthinned stand of the same age, site index, and basal area (Clutter and Jones 1980; Cao et al. 1982; Matney and Sullivan 1982; Burkhart and Sprinz 1984). Bailey and Ware (1983) developed a thinning index based on the ratio of quadratic mean diameter of thinned trees to the quadratic mean diameter of all trees prior to thinning and used it as a modifier in a basal area growth equation. Instead of mean diameter, Pienaar et al. (1985) used the ratio of trees removed in the thinning to the ratio of residual trees in their basal area growth equations, but found predictions could be further improved by utilizing a suppression index.

When information on the thinning didn't exist, McTague and Bailey (1987) suggested that a basal area growth model based on diameter percentiles, which themselves were functions of age, stem density, site index, and indicators for thinning, was effective for projecting both unthinned and thinned stands. However, one factor that these previous studies didn't account for was the time since thinning. Hasenauer et al. (1997) developed a generalized basal area growth function and indicated that the growth of thinned stands slowly converged towards unthinned stands over time.

Repeated thinnings also have an influence on stand growth response. Knoebel et al. (1986) developed three different equations to account for growth before thinning, after the first thinning, and after the second and subsequent thinnings. The limitation of this approach is that the model parameters were estimated using three different datasets. Amateis (2000) also developed a basal area growth equation that was sensitive to the number of thinning entries, but estimated the influence by using an indicator variable rather than fitting separate equations. When compared to other approaches, Chikumbo et al. (1999) found a dynamic model dependent on stem density, basal area, and the number of thinned trees was much more effective at predicting basal growth response in thinned stands. In contrast to these previous studies, both Barrio-Anta et al. (2006) and Castedo-Dorado et al. (2007b) indicated that it was not necessary to directly account for thinning in stand basal area growth equations when the parameters were estimated using the generalized algebraic difference approach.

Although most stand-level studies have focused on basal area growth response to thinning, several other studies have looked at other aspects of stand response to thinning. Sharma et al. (2006) modeled the influence of thinning on dominant height growth in loblolly pine and found that growth was initially reduced in thinned stands, but eventually accelerated and exceeded growth in the unthinned stand. Amateis (2000) constructed a stand mortality equation that was sensitive to thinning and found that it was lower in thinned stands. Short and Burkhart (1992) developed a stand-level crown recession model dependent on dominant height, average crown ratio, age, and thinning variables that accounted for both intensity and time since treatment. Finally, Nigh (2013) avoided the need to model both basal area and height by simply assessing the influence of thinning on total volume growth.

3.3.3.2 Tree-Level

Modeling tree-level response to thinning has consisted of (1) fitting separate equations for thinned and unthinned trees and (2) developing modifiers. Amateis et al. (1989) illustrated the first approach as separate diameter and survival equations for thinned and unthinned trees were developed. Westfall and Burkhart (2001) found that separate equations for individual tree height increment were unnecessary if a distance-dependent measure of competition was included in the original equation.

Multipliers have commonly been used to model the influence of thinning on growth. An age-dependent thinning multiplier for diameter increment was found necessary in loblolly pine, even after the inclusion of a distance-dependent measure of competition (Westfall and Burkhart 2001). Hynynen (1995) found a thinning modifier was needed for a distance-independent diameter increment equation for Scots pine, but not height increment. In Douglas-fir and western hemlock, Hann et al. (2003) developed thinning modifiers for both diameter and height increment equations. Depending on the intensity of the thinning, the diameter growth modifier indicated that the thinning significantly increased growth for 15–20 years following thinning, while the height growth modifier suggested that height growth was reduced (Hann et al. 2003). In the case of multiple thinnings, Hann et al. (2003) discounted the basal area removed from previous thinnings and then added it to the basal area removed in the current thinning.

Predicting the long-term consequences of various thinning regimes in individual-tree growth models is difficult, but trying to implement the thinning properly in a simulation can also pose a significant challenge. Söderbergh and Ledermann (2003) reviewed the different thinning algorithms for five commonly used individual-tree growth models in Europe and found they could be grouped into six different categories based on how they were developed, the spatial resolution, and actual implementation in the model. Most thinning regimes can be applied using both an analytical algorithm and a distance-independent growth model, but representing geometric thinnings like strip thinnings can be difficult with this approach (Söderbergh and Ledermann 2003).

Representing the influence of thinning in growth models has ranged from assuming growth is the same as an unthinned stand of the same density to individual tree growth modifiers dependent on the intensity, time since treatment, and number of thinnings. Although diameter distribution equations have been modified to account for thinning (e.g. Murray and Von Gadow 1991; Álvarez González et al. 2002), the approach can predict trees to be present when they never existed or were removed in the thinning. Thus, an individual tree approach is recommended, particularly in mixed species forest where thinning regimes can favor the removal of a particular species. However, most of work on modeling the influence of thinning has occurred in single-species plantations and much different techniques than those presented here might be required in mixed species stands. Although the focus in this section has been on statistical models, both mechanistic and hybrid models have also been modified to account for thinning (Landsberg et al. 2005; Petritsch et al. 2007; Wang et al. 2011).

3.3.4 *Fertilization*

Fertilization is an important plantation management tool in several regions, particularly in the western (Talbert and Marshall 2005) and southern United States (Jokela et al. 2004). However, like vegetation management, it can be hard to predict empirically because of high variability and the limiting factors to growth not being reflected by site index. Response to fertilization is not only dependent on the amount and type of fertilizer, but the weather conditions at time of application, understory vegetation characteristics, and soil attributes, which are often not characterized in most statistical growth models.

3.3.4.1 Stand-Level

Ballard (1984) developed a simple model for predicting cumulative stand volume growth response to fertilizer in Douglas-fir. The model was dependent on time since treatment, the amount of fertilizer applied, stand composition, stocking, and site quality (Ballard 1984). Bailey et al. (1989) accounted for the effects of both nitrogen and phosphorus fertilization on loblolly pine stand basal area growth, density, and dominant height growth using differential yield equations. Likewise Amateis et al. (2000) developed equations that predicted stand response to fertilization as a function of several site and stand factors as well as the amount and type of fertilizer applied. Simply adjusting the site index has also been used to predict response to fertilizer, but this assumes that fertilization leads to a long-term increase in stand productivity (Carlson et al. 2008). This illustrates that the model form used to predict long-term response to fertilizer is important. Pienaar and Rheney (1995) indicated that an additive sub-model was sufficient for modeling the response, while Snowdon (2002) found that a quadratic model form was more effective.

An alternative to developing equations that include the effects of fertilization or simply adjusting the site index is to the use the age-shift method (e.g. Kimberley et al. 2004; Carlson et al. 2008). This method has also been used for modeling the response to vegetation management (Lauer et al. 1993; South and Miller 2007), and it simply involves reducing the amount of time that it takes a stand to reach a certain size when compared to an untreated one. This approach assumes that the shape of the growth curve is not changed by the treatment and that the stand development trajectory is only accelerated. Carlson et al. (2008) predicted the age-shift for dominant height, basal area, and volume of fertilized loblolly pine as a function stand attributes as well as the fertilizer rate and type. The estimated age-shifts were then used to adjust stand age in an existing growth equation and found to give superior predictions than growth equations that had directly included the effects of fertilizer (Carlson et al. 2008). The limitation of this approach is that it becomes difficult to estimate the age-shift for a stand which is simultaneously fertilized and thinned.

3.3.4.2 Tree-Level

Like at the stand-level, a modified site index has been used to predict individual tree response to fertilization (e.g. Daniels and Burkhart 1975). Shaffi et al. (1990) included simple indicator variables for thinning, fertilization, and their interaction in diameter growth models for grand fir and Douglas-fir. These equations were used to estimate a relative response to fertilization, which indicated that the response was greater in dense stands and for dominant trees (Stage et al. 1990). Moore et al. (1994) found that neither thinning nor fertilization changed the relative distribution of growth within a stand, so growth could be allocated to individual trees with the same equation used in untreated stands if an estimate of stand-level growth was available.

Modifiers have also been widely used to model response to fertilization. Hynynen et al. (1998) found both loblolly pine diameter and height increment to be sensitive to the amount, type, and time since fertilization. Hann et al. (2003) developed fertilization modifiers for Douglas-fir height to crown base, mortality, diameter increment, and height increment. Depending on the amount of fertilizer, diameter increment and height increment were predicted to increase (Hann et al. 2003). The influence of the modifier was nearly absent for diameter and height increment after 10 and 5 years, respectively (Hann et al. 2003). Hynynen et al. (2002) used a modifier that was dependent on tree species, site fertility class, soil type, and site temperature sum. In addition to growth, Hann et al. (2003) found that modifiers were also necessary for predicting the influence of fertilization on height to crown base and tree mortality. The height to crown base was predicted to remain lower for a fertilized tree until 4 years after the treatment, while mortality was expected to increase (Hann et al. 2003).

Although the direct influences of fertilization are brief (e.g., 2–10 years), it often has a significant long-term impact on growth. In the comparison of several growth models in the US Pacific Northwest, Johnson (2005) found that the model predictions differed the most when fertilizer was applied, particularly if combined with a thinning treatment. The models also differed whether the relative response to fertilization was greater in stands with a lower site index (Johnson 2005). These results highlight the difficulty in representing fertilization accurately and suggest that a more mechanistic approach might be more justified.

Plantation management significantly alters stand dynamics, and growth models need to be able to account for these changes. However, methods for representing forest management in growth models are not straightforward and are often limited by the type of data available. Approaches for representing plantation management have included modifying site index or stand age (e.g., age-shift), fitting separate equations by management regime, and using modifiers. At both the stand- and individual-tree levels, modifiers are likely the most logical way to represent management activities as they don't require modification of the existing equation, can be combined to represent multiple management activities, and are relatively easy to estimate. The ability to represent multiple management activities is particularly important as the response may be multiplicative. For example, Hann et al. (2003) found that the combination of fertilization and thinning significantly increased growth more than either treatment did alone. Consequently, improving the representation of plantation management in forest growth and yield models is critically important as rotations are likely to shorten due to more intensive and complex management regimes. The difficulty is in balancing the need for new data and the ability for the model to extrapolate.

3.4 Development and Use of Forest Growth Models

Regardless of whether the intent is to build a new growth model or evaluate an existing one, a strong empirical database of on-the-ground measurements is required. Data can come in a variety of forms, but it is generally from either permanent plots, temporary plots, or stem analysis. As discussed in Weiskittel et al. (2011), these different types of data have a range of advantages and disadvantages for growth modeling. In short, permanent plots are the most useful type of data as they avoid the need for growth reconstruction, but are expensive to maintain and generally require multiple remeasurements before becoming useful. Curtis and Marshall (2005) provide some general guidelines for establishing permanent plots. Some of the most important considerations with permanent plots include plot location, plot shape and size, measurement cycle, and the specific tree attributes.

For plantation growth modeling purposes, permanent plots that cover the full range of conditions, are large square or rectangular in shape (0.1–0.2 ha), measured at an interval consistent with the desired or current model projection interval, and have individually tagged trees with a consistent measurement protocol are the most desirable.

Temporary plots and stem analysis data offer the advantage of being less expensive and quicker to collect. As previously noted, there are some robust methods for using temporary plot data, particularly when the focus is at the stand level (Vanclay 2010). This is because of the strong relationship between plantation age and its various structural attributes like dominant height, basal area, and stocking. However, this relationship can be complicated by intensive management regimes. The use of stem analysis is often limited to individual trees and often in the case of tree ring analysis, the use of a single increment core obtained at breast height. As discussed in Weiskittel et al. (2011), the use of a single increment core for growth modeling should be avoided as there is significant within-tree variation in radial increment, which can produce significant negative prediction bias. The use of stem analysis for the development of height growth and site index curves can be problematic because a tree may not be dominant for the its entire lifespan, which can lead to very different developmental trajectories when compared to those of permanent plots (Raulier et al. 2003). García (2005a) presented a methodology for addressing this issue, but it requires having permanent sample plot data. Overall, temporary plot and stem analysis data are a good starting point for developing or evaluating a growth model, but their limitations have to be acknowledged and addressed if more detailed information on growth or the influence of management is desired.

Development of a new growth model is often a complex process that involves multiple steps and decisions. When existing models are available, users must evaluate whether a new model is truly necessary or if an existing model can be simply calibrated to a new situation. This can be done by comparing predictions to observed data, while simply benchmarking the model to other existing growth models (Henderson et al. 2013), against general growth and yield principles (Leary 1997), or assessing predicted tree size relations (Vospernik et al. 2010) when data aren't available. As discussed, Bayesian calibration methods (Van Oijen et al. 2005) are a useful way to calibrate an existing model without having to develop a new model. If an existing model is deemed unfit, additional decisions on model development are necessary, particularly the desired model type (statistical, mechanistic, hybrid) and scale (stand, size-class, individual tree). If a statistical model is desired, additional decisions on model form and model parameterization techniques are required. A variety of theoretical and empirical model forms exist (Kiviste et al. 2002), which strongly influence the model's ability to represent the data and more importantly, properly extrapolate beyond the data. Assessing model form fit is often tricky as there are a variety of metrics to examine and evaluate. As with model form, there are an array of model parameterization techniques available. Most modelers use parametric techniques like nonlinear regression to fit models, but nonparametric techniques like artificial neural networks are increasingly being used (Huang et al. 2012; Castro et al. 2013). Parametric options can be used to ensure proper behavior outside of the fitting dataset and generally require less data when compared to nonparametric approaches. Regardless of the methods used, extensive model testing and modification is required before being released to potential users.

Model end users must make several important considerations before using a growth model to make plantation management decisions. First, users must evaluate whether the desired growth models have been extensively tested and calibrated for the intended use. This requires a basic understanding of how the model works, the data used to construct it, and implicit assumptions. Too often, models developed for one region or species are simply used for a different region or species without consideration of the consequences of this decision. Even when used for the same species, models require significant calibration and modification when they are used outside of the developed region (e.g. Flewelling and Marshall 2008).

Other important considerations are collecting or generating the necessary input data, selecting the appropriate temporal and spatial scale, processing output, and dealing with uncertainty. Most growth models require the user to input basic stand information like basal area and stem density in the case of a stand-level model or a detailed tree list with species, diameter at breast height, and total height in the case of a tree-level model. If the necessary information is unavailable, techniques must be used to generate it. This can range from using imputation methods based on ground or remotely sensed data (Temesgen et al. 2003) to simply constructing hypothetical data. The consequences of using actual or generated data are rarely evaluated, but research has indicated that significant differences in projection accuracy can arise when data different than the model development data are used (Hann and Zumrawi 1991) and inaccuracy in the input data can have a larger influence on model predictions than changes in the underlying growth model equations (Mowrer and Frayer 1986).

Both the selected spatial and temporal scale can also influence model accuracy. Most models used in plantation management decision making are intended for stand-level projections, but are often used to make forest- or ownership-wide decisions. Generally, these decisions require the use optimization techniques to estimate allowable cut or schedule harvests, which complicates model use as they often need to be simplified for this use. This is commonly done by generating yield curves from various model simulations or developing a meta-model that predicts model output from simplified inputs. Both of these techniques often compress the observed variability, which may lead to non-optimal solutions in practice. Model accuracy also generally declines with increased projection lengths as prediction errors tend to accumulate (Kangas 1997). This is particularly true for individual tree growth models when compared to stand-level growth models (Mäkinen et al. 2008). Consequently, most growth projections should be limited to 10–30 years as longer projections lengths are generally extrapolation.

Growth models often provide a variety of output for users. This can range from simple tree lists to stand-level summaries with additional derived variables like total and merchantable volume, biomass/carbon, and quantity of specific forest products. Understanding how these various metrics are computed is important because significant and unexpected differences can arise due to the methodology used. Generally, it is best to use the simplest output provided by the model such as tree- or stand-level attributes like total basal area or stem density and then do conversion to desired metrics like biomass or merchantable volume rather than rely on internal model functions to do the calculations. Growth models also differ in how they summarize plot-level information to the stand level. Some models compute stand-level attributes based on the simulated averages of each individual plot in the stand, whereas other models project the average plot and just provide those attributes. Although the latter approach reduces the computational effort, the former approach ensures that within-stand variability is maintained during the projections and, more importantly, confidence intervals can be computed. This is particularly important as most models do not provide estimates of uncertainty with their projections due to their deterministic formulation. This is important because model uncertainty can often be quite significant and increase linearly with projection length (Mowrer 1991) due to the number of equations used during a typical simulation. Consequently, growth model output should not be interpreted to the nearest tenth decimal place even if that data resolution is provided. At best, growth models are tools designed to represent regional average trends and should always be treated with a degree of caution during the forest plantation planning process.

3.5 Summary

When compared to naturally-regenerated forests, plantations are relatively simple biological systems with fewer species, simpler stand structures, and more uniform management regimes. However, the development of accurate growth models for plantations is still a significant challenge due to the array of biotic and abiotic factors that influence their development as well as the continual evolving nature of their management. The various types of growth models discussed in this chapter (statistical, mechanistic, and hybrid) have all been used at both the stand and tree levels to project plantation development. As discussed, each of these various approaches has key advantages and disadvantages with the stand-level statistical approach seeing the widest application for plantations. This is because these type of models can achieve a high degree of accuracy, provide information relevant to plantations managers, and can be easily linked with other modeling approaches for improved resolution. Stand-level mechanistic and hybrid approaches are also seeing increased development and use due to the need to be account for climate variability in short-term projections and their theoretical ability to represent for more complex management regimes. This trend will likely continue with increased availability of improved physiological, climate, and soils data that these type of models require.

Models are tools designed to help inform plantation management decisions, but often require extensive testing and modification to represent the full array of management techniques employed in today's plantations. This is because the stand-level data on the influence of a particular management technique on plantation growth often lags behind its use, which is especially true for tree genetics. Therefore, models must make basic assumptions on how to represent these various management regimes, which suggests the need for more mechanistic approaches in growth modeling. Regardless of the management regime being simulated, growth model users must inform themselves on the basic model structure, the data used to parameterize it, and its suitability for the intended application. In addition, growth model users need to ensure that their input data is appropriately collected or generated, the output is being correctly computed and interpreted, and that there can be a tremendous amount of often unreported uncertainty associated with any given projection, particularly as the projection length increases.

Growth models will likely see continued and increased future use due to the need to better quantify return on investment in plantation management, the greater array of issues facing plantations (e.g., climate change, carbon neutrality, wood quality), and the wider geographic distribution of plantations globally. Future improvements in growth modelling will likely result from improved modeling techniques, input data, and computer hardware/software. Newer advanced statistical techniques like Bayesian spatial-temporal hierarchical models or nonparametric machine learning algorithms allow for better detection of trends in the data and more robust as well as local predictions. This, combined with increased data from national forest inventories and long-term experiments, should help improve growth model predictions. As discussed, the quantity and quality of input going into a growth model has important implications for the usefulness of model output. No growth model can overcome poorly collected and incomplete input data. The greater availability and quality of remote sensing technologies like LiDAR for plantation inventories should help improve growth model inputs. This also true for mechanistic models, which rely primarily on physiological parameters, climate data, and hard to quantify inputs like leaf area index. Near real-time observations of CO_2, water vapor, and weather from eddy flux towers and leaf area index estimates from satellites creates an array of opportunities for calibration of mechanistic growth models. Finally, much of the recent advances in growth models in the last two decades can largely be attributed to the power of modern day computers to store, process, and analyze data. This technology allows for more detailed, complex, and longer projections. Developing the software to handle these projections and provide realistic visualizations of the output allow users an improved opportunity to have a positive interaction with growth models. Unfortunately, maintaining the growth model software is often the most difficult aspect as technology and model user needs change rapidly. Advances in growth modeling software will hopefully allow for a greater number of simulations with detailed output and more sophisticated visualization, while improving the ability of model users to control and modify projections.

3.6 Problems

1. Provide example uses of forest growth and yield models for strategic and tactical forest plantation planning purposes.
2. Discuss a potential use and required output of a growth and yield model to be of relevance to a field forester, wildlife biologist, forest planner, financial inventor, and policy-maker.
3. In comparison to a naturally-regenerated stand, what attributes of a plantation would be necessary to quantify and incorporate into a growth model projection?
4. List the advantages and disadvantages of the three types of forest growth and yield models discussed in this chapter (statistical, mechanistic, and hybrid) as they relate to plantation management.
5. Describe situations where knowledge of individual tree attributes might be more effective than just stand-level attributes. In what type of situations would knowledge of individual tree spatial location in a plantation be important?
6. Draw stand-level basal area over age for a plantation with the same initial stem density and is: (a) unmanaged; (b) planted with genetically improved stock; (c) planted with normal stock and receives a mid-rotation thinning; (d) planted with normal stock and receives a mid-rotation fertilization; and (e) planted with genetically improved stocks and receives both a mid-rotation thinning and fertilization. What management regime would result in the greatest growth, highest total yield, and widest diameter distribution? What information would be required for a growth and yield model to make accurate projections in each of these scenarios?
7. What are the attributes of good forest growth and yield model? How would one quantify these attributes?
8. How should forest managers deal with the uncertainty implicit in any growth model projection?
9. Discuss potential strategies for improving predictions and the necessary data for a statistical, mechanistic, and hybrid growth model.
10. Define a list of criteria for evaluating the realism of growth model projections and outline strategies for addressing situations when growth models fail to meet expectations.

References

Adams JP, Matney TG, Land SB, Belli K, Duzan HW (2006) Incorporating genetic parameters into a loblolly pine growth-and-yield model. Can J For Res 36:1959–1967

Allen MG II, Coble DW, Cao QV, Yeiser J, Hung I (2011) A modified stand table projection growth model for unmanaged loblolly and slash pine plantations in east Texas. South J Appl For 35:115–122

Almeida AC, Landsberg JJ, Sands PJ (2004) Parameterisation of 3-PG model for fast-growing *Eucalyptus grandis* plantations. For Ecol Manag 193:179–195

Álvarez González JG, Schröder J, Rodríguez Soalleiro R, Ruíz González AD (2002) Modelling the effect of thinnings on the diameter distribution of even-aged Maritime pine stands. For Ecol Manag 165:57–65
Amateis RL (2000) Modeling response to thinning in loblolly pine plantations. South J Appl For 24:17–22
Amateis RL, Burkhart HE (2012) Extending the capabilities of an individual tree growth simulator to model non-traditional loblolly pine plantation systems for multiple products. In: Butnor JR (ed) Proceedings of the 16th biennial southern silvicultural research conference, e-General technical report SRS-156. U.S. Department of Agriculture Forest Service, Southern Research Station, Asheville, pp 7–9
Amateis RL, Burkhart HE, Walsh TA (1989) Diameter increment and survival equations for loblolly-pine trees growing in thinned and unthinned plantations on cutover site-prepared lands. South J Appl For 13:170–174
Amateis RL, Liu J, Ducey MJ, Allen HL (2000) Modeling response to midrotation nitrogen and phosphorus fertilization in loblolly pine plantations. South J Appl For 24:207–212
Bailey RL (1980) Individual tree growth derived from diameter distribution models. For Sci 26:626–632
Bailey RL, Ware KD (1983) compatible basal-area growth and yield model for thinned and unthinned stands. Can J For Res 13:563–571
Bailey RL, Burgan TM, Jokela EJ (1989) Fertilized midrotation-aged slash pine plantations-stand structure and yield prediction models. South J Appl For 13:76–80
Baldwin VC, Burkhart HE, Westfall JA, Peterson KD (2001) Linking growth and yield and process models to estimate impact of environmental changes on growth of loblolly pine. For Sci 47:77–82
Ballard TM (1984) A simple model for predicting stand volume growth response to fertilizer application. Can J For Res 14:661–665
Barrio-Anta M, Castedo-Dorado F, Diéguez-Aranda U, Álvarez González JG, Parresol BR, Rodríguez Soalleiro R (2006) Development of a basal area growth system for maritime pine in northwestern Spain using the generalized algebraic difference approach. Can J For Res 36:1461–1474
Battaglia M, Sands PJ, Candy SG (1999) Hybrid growth model to predict height and volume growth in young *Eucalyptus globulus* plantations. For Ecol Manag 120:193–201
Battaglia M, Sands P, White D, Mummery D (2004) CABALA: a linked carbon, water and nitrogen model of forest growth for silvicultural decision support. For Ecol Manag 193:251–282
Belli KL, Ek AR (1988) Growth and survival modeling for planted conifers in the Great Lakes region. For Sci 34:458–473
Bermejo I, Cañellas I, Miguel AS (2004) Growth and yield models for teak plantations in Costa Rica. For Ecol Manag 189:97–110
Biging GS, Dobbertin M (1995) Evaluation of competition indices in individual tree growth models. For Sci 41:360–377
Borders BE, Patterson WD (1990) Projecting stand tables: a comparison of the Weibull diameter distribution method, a percentile-based projection method, and a basal area growth projection method. For Sci 36:413–424
Bréda N (2003) Ground-based measurements of leaf area index: a review of methods, instruments and current controversies. J Exp Bot 54:2403–2417
Bryars C, Maier C, Zhao D, Kane M, Borders B, Will R, Teskey R (2013) Fixed physiological parameters in the 3-PG model produced accurate estimates of loblolly pine growth on sites in different geographic regions. For Ecol Manag 289:501–514
Buckman RE (1962) Growth and yield of red pine in Minnesota, Technical bulletin 1272. USDA, Washington, DC, p 50
Buford MA, Burkhart HE (1987) Genetic improvement effects on growth and yield of loblolly pine plantations. For Sci 33:707–724
Bugmann H (2001) A review of forest gap models. Clim Chang 51:259–305

Burkhart HE, Sprinz PT (1984) Compatible cubic volume and basal area equations for thinned old-field loblolly pine plantations. For Sci 30:86–93

Burkhart HE, Farrar KD, Amateis RL, Daniels RF (1987) Revised 2001. Simulation of individual tree growth and stand development in loblolly pine plantations in loblolly pine plantations plantations on cutover, site-prepared areas. Publication number FWS-1-87. Virginia Tech Department of Forestry, Blacksburg, p 28

Cao QV (2004) Predicting parameters of a Weibull function for modeling diameter distribution. For Sci 50:682–685

Cao QV (2006) Predictions of individual-tree and whole-stand attributes for loblolly pine plantations. For Ecol Manag 236:342–347

Cao QV, Burkhart HE, Lemin RC (1982) Diameter distributions and yield of thinned loblolly pine plantations, Publication no. FSW-1-82. Virginia Polytechnic Institute, School of Forestry and Wildlife Resources, Blacksburg, p 62

Carlson CA, Fox TR, Allen HL, Albaugh TJ, Cooperative FN (2008) Modelling mid-rotation fertilizer responses using the age shift approach. For Ecol Manag 256:256–262

Carson SD, García O, Hayes JD (1999) Realized gain and prediction of yield with genetically improved *Pinus radiata* in New Zealand. For Sci 45:186–200

Castedo-Dorado F, Diéguez-Aranda U, Álvarez González JG (2007a) A growth model for *Pinus radiata* D. Don stands in north-western Spain. Ann For Sci 64:453–465

Castedo-Dorado F, Diéguez-Aranda U, Barrio-Anta M, Álvarez González JG (2007b) Modelling stand basal area growth for radiata pine plantations in Northwestern Spain using the GADA. Ann For Sci 64:609–619

Castro RVO, Soares CPB, Leite HG, Souza AL, Nogueira GS, Martins FB (2013) Individual growth model for Eucalyptus stands in Brazil using artificial neural network. ISRN For 2013. Article ID 196832. 12 p. doi:http://dx.doi.org/10.1155/2013/196832

Chikumbo O, Mareels IMY, Turner BJ (1999) Predicting stand basal area in thinned stands using a dynamical model. For Ecol Manag 116:175–187

Clutter JL (1963) Compatible growth and yield models for loblolly pine. For Sci 9:354–371

Clutter JL, Jones EP (1980) Prediction of growth after thinning in old field slash pine plantations. Research paper SE-217. USDA Forest Service, p 19

Corral Rivas JJ, Álvarez González JG, Aguirre O, Hernández FJ (2005) The effect of competition on individual tree basal area growth in mature stands of Pinus cooperi Blanco in Durango (Mexico). Eur J For Res 124:133–142

Crookston NL, Dixon GE (2005) The forest vegetation simulator: a review of its applications, structure, and content. Comput Electron Agric 49:60–80

Curtis RO, Marshall DD (2005) Permanent-plot procedures for silvicultural and yield research, General technical report PNW-GTR-634. USDA Forest Service, Pacific Northwest Research Station, Portland, p 86

Curtis RO, Clendenen GW, DeMars DJ (1981) A new stand simulator for coast Douglas-fir: DFSIM user's guide, General technical report PNW-128. USDA Forest Service Pacific Northwest Research Station, Portland, p 79

Daniels RF, Burkhart HE (1975) Simulation of individual tree growth and stand development in managed loblolly pine plantations, Publication FWS-5-75. Virginia Polytechnic Institute and State University, Blacksburg, p 69

Deckmyn G, Mali B, Kraigher H, Torelli N, Op de Beeck M, Ceulemans R (2009) Using the process-based stand model ANAFORE including Bayesian optimisation to predict wood quality and quantity and their uncertainty in Slovenian beech. Silva Fennica 43:523–534

Duursma RA, Medlyn BE (2012) MAESPA: a model to study interactions between water limitation, environmental drivers and vegetation function at tree and stand levels, with an example application to [CO_2] × drought interactions. Geosci Model Dev 5:919–940

Duursma RA, Robinson AP (2003) Bias in the mean tree model as a consequence of Jensens' inequality. For Ecol Manag 186:373–380

Dzierzon H, Mason EG (2006) Towards a nationwide growth and yield model for radiata pine plantations in New Zealand. Can J For Res 36:2533–2543

Farquhar GD, von Caemmerer S, Berry JA (1980) A biochemical model of photosynthetic CO_2 assimilation in leaves of C_3 leaves. Planta 149:78–90

Flewelling JW, Marshall DD (2008) Calibration and modification for the Pacific Northwest of the New Zealand Douglas-fir silvicultural growth model, General technical report PNW-GTR-754. U.S. Department of Agriculture, Forest Service, Pacific Northwest Research Station, Portland, p 42

García O (2005a) Comparing and combining stem analysis and permanent sample plot data in site index models. For Sci 51:277–283

García O (2005b) TADAM: a dynamic whole-stand approximation for the TASS growth model. For Chron 81:575–581

García O (2011) A parsimonious dynamic stand model for interior spruce in British Columbia. For Sci 57:265–280

García O, Burkhart HE, Amateis RL (2011) A biologically-consistent stand growth model for loblolly pine in the Piedmont physiographic region, USA. For Ecol Manag 262:2035–2041

Gifford RM (2003) Plant respiration in productivity models: conceptualisation, representation and issues for global terrestrial carbon-cycle research. Funct Plant Biol 30:171–186

Glover GR, Hool JN (1979) A basal area ratio predictor of loblolly pine plantation mortality. For Sci 25:275–282

Gould P, Marshall DD (2010) Incorporation of genetic gain into growth projections of Douglas-fir using ORGANON and FVS. West J Appl For 25:55–61

Gould P, Johnson R, Marshall D, Johnson G (2008) Estimation of genetic-gain multipliers for Douglas-fir height and diameter growth. For Sci 54:588–596

Hamilton DAJ, Rehfeldt GE (1994) Using individual tree growth projection models to estimate stand-level gains attributable to genetically improved stock. For Ecol Manag 68:189–207

Hann DW (2011) ORGANON user's manual edition 9.0. Oregon State University, Department of Forest Engineering, Resources, and Management, Corvallis, OR, p 134

Hann DW, Zumrawi A-A (1991) Growth model predictions as affected by alternative sampling-unit designs. For Sci 37:1641–1655

Hann DW, Marshall DD, Hanus ML (2003) Equations for predicting height-to-crown base, 5-year diameter growth rate, 5-year height growth rate, 5-year mortality rate, and maximum size-density trajectory for Douglas-fir and western hemlock in the coastal region of the Pacific Northwest, Research contribution 40. Oregon State University, College of Forestry Research Laboratory, Corvallis, p 85

Hasenauer H, Burkhart HE, Amateis RL (1997) Basal area development in thinned and unthinned loblolly pine plantations. Can J For Res 27:265–271

Hatch CR, Gerrad DJ, Tappeiner JC (1975) Exposed crown surface area: a mathematical index of individual tree growth potential. Can J For Res 5:224–228

Hawkes C (2000) Woody plant mortality algorithms: description, problems, and progress. Ecol Model 126:225–248

Henderson JE, Grebner SDRDL, Munn IA (2013) A graphical comparison of loblolly pine growth-and-yield models. South J Appl For 37:169–176

Henning JG, Burk TE (2004) Improving growth and yield estimates with a process model derived growth index. Can J For Res 34:1274–1282

Huang J, Gao G, Guo F (2012) Forest growth simulation based on artificial neural network. Recent Adv Comput Sci Inf Eng Lect Notes Elect Eng 124:657–663

Hynynen J (1995) Predicting the growth response to thinning for Scots pine stands using individual-tree growth models. Silva Fennica 29:225–246

Hynynen J, Burkhart HE, Allen HL (1998) Modeling tree growth in fertilized midrotation loblolly pine plantations. For Ecol Manag 107:213–229

Hynynen J, Ojansuu R, Hökkä H, Siiplehto J, Salminen H, Haapala P (2002) Models for predicting stand development in MELA system, Research papers 835. Finnish Forest Research Institute, Helsinki, p 116

Johnson GP (2005) Growth model runoff II. Growth Model Users Group, Portland, p 61

Jokela EJ, Doughterty PM, Martin TA (2004) Production dynamics of intensively managed loblolly pine stands in the southern United States: a synthesis of seven long-term experiments. For Ecol Manag 192:117–130

Kangas A (1997) On the prediction bias and variance in long-term growth projections. For Ecol Manag 96:207–216

Kimberley MO, Richardson B (2004) Importance of seasonal growth patterns in modelling interactions between radiata pine and some common weed species. Can J For Res 34:184–194

Kimberley MO, Wang H, Wilks PJ, Fisher CR, Magesan GN (2004) Economic analysis of growth response from a pine plantation forest applied with biosolids. For Ecol Manag 189:345–351

Kirschbaum MUF (2000) CenW: a generic forest growth model. N Z J For 45:15–19

Kirschbaum MUF, Paul KI (2002) Modelling C and N dynamics in forest soils with a modified version of the CENTURY model. Soil Biol Biochem 34:341–354

Kiviste A, Álvarez G, Rojo A, Ruiz G (2002) Funciones de crecimiento de aplicación en el ámbito forestal (Growth equations to model forest development). Instituto Nacional de Investigación y Tecnología Agraria y Alimentaría, Madrid, p 190

Knoebel BR, Burkhart HE, Beck DE (1986) A growth model and yield model for thinned stands of yellow-poplar. For Sci Monogr 27:62

Knowe SA (1994a) Effect of competition control treatments on height-age and height-diameter relationships in young Douglas-fir plantations. For Ecol Manag 67:101–111

Knowe SA (1994b) Incorporating the effects of interspecific competition and vegetation management treatments in stand table projection models for Douglas-fir saplings. For Ecol Manag 67:87–99

Knowe SA, Foster GS (1989) Application of growth models for simulating genetic gain of loblolly pine. For Sci 35:211–228

Knowe SA, Stein WI (1995) Predicting the effects of site preparation and protection on development of young Douglas-fir plantations. Can J For Res 25:1538–1547

Knowe SA, Radosevich SR, Shula RG (2005) Basal area and diameter distribution prediction equations for young Douglas-fir plantations with hardwood competition: Coast Ranges. West J Appl For 20:77–93

Lacointe A (2000) Carbon allocation among tree organs: a review of basis processes and representation in functional-structural tree models. Ann For Sci 57:521–533

Landsberg J (2003a) Modelling forest ecosystems: state of the art, challenges, and future directions. Can J For Res 33:385–397

Landsberg JJ (2003b) Physiology in forest models: history and the future. For Biom Model Inf Sci 1:49–63

Landsberg JJ, Sands P (2011) Physiological ecology of forest production: Principles, processes and models, 4th edn. Academic, Maryland Heights, p 331

Landsberg JJ, Waring RH (1997) A generalized model of forest productivity using simplified concepts of radiation use efficiency, carbon balance and partitioning. For Ecol Manag 95:209–228

Landsberg JJ, Waring RH, Coops NC (2003) Performance of the forest productivity model 3-PG applied to a wide range of forest types. For Ecol Manag 172:199–214

Landsberg JJ, Mäkelä A, Sievänen R, Kukkola M (2005) Analysis of biomass accumulation and stem size distributions over long periods in managed stands of *Pinus sylvestris* in Finland using the 3-PG model. Tree Physiol 25:781–792

Lauer DK, Glover GR, Gherstad DH (1993) Comparison of duration and method of herbaceous weed control on loblolly pine response through midrotation. Can J For Res 23:2116–2125

Le Roux X, Lacointe A, Escobar-Gutiérrez A, Dizès S (2001) Carbon-based models of individual tree growth: a critical appraisal. Ann For Sci 58:469–506

Leary RA (1997) Testing models of unthinned red pine plantation dynamics using a modified Bakuzis matrix of stand properties. Ecol Model 98:35–46

Liang J, Picard N (2013) Matrix models of forest dynamics: an overview and outlook. For Sci 59:359–378

Litton CM, Raich JW, Ryan MG (2007) Carbon allocation in forest ecosystems. Glob Chang Biol 13:2089–2109

MacPhee B, McGrath TP (2006) Nova Scotia growth and yield model version 2: user's manual. Report FOR 2006–3, No. 79. Nova Scotia Department of Natural Resources, Halifax, Nova Scotia, p 15

Mäkelä A (1997) A carbon balance model of growth and self-pruning in trees based on structural relationships. For Sci 43:7–24

Mäkelä A, Landsberg J, Ek AR, Burk TE, Ter-Mikaelian M, Agren GI, Oliver CD, Puttonen P (2000) Process-based models for forest ecosystem management: current state of the art and challenges for practical implementation. Tree Physiol 20:289–298

Mäkelä A, Valentine HT, Helmisaari H-S (2008) Optimal co-allocation of carbon and nitrogen in a forest stand at steady state. New Phytol 180:114–123

Mäkinen A, Kangas A, Kalliovirta J, Rasinmäki J, Välimäki E (2008) Comparison of treewise and standwise forest simulators by means of quantile regression. For Ecol Manag 255:2709–2717

Mason EG, Dzierzon H (2006) Applications of modeling to vegetation management. Can J For Res 36:2505–2514

Mason EG, Milne PG (1999) Effects of weed control, fertilization, and soil cultivation on the growth of *Pinus radiata* at midrotation in Canterbury, New Zealand. Can J For Res 29:985–992

Mason EG, Whyte AGD, Wollons RC, Richardson B (1997) A model of the growth of juvenile radiata pine in the Central North Island of New Zealand: links with older models and rotation-length analyses of the effects of site preparation. For Ecol Manag 97:187–195

Mason EG, Rose RW, Rosner LS (2007) Time vs. light: a potentially useable light sum hybrid model to represent the juvenile growth of Douglas-fir subject to varying levels of competition. Can J For Res 37:795–805

Mason EG, Methol R, Cochrane H (2011) Hybrid mensurational and physiological modelling of growth and yield of *Pinus radiata* D.Don. using potentially useable radiation sums. Forestry 84:99–108

Matney TG, Sullivan AD (1982) Compatible stand and stock tables for thinned and unthinned loblolly pine stands. For Sci 28:161–171

Matsumura N (2011) Yield predictions for *Acacia mangium* plantations in Southeast Asia. FORMATH 10:295–308

McMurtrie RE, Dewar RC (2013) New insights into carbon allocation by trees from the hypothesis that annual wood production is maximized. New Phytol 199:981–990

McTague JP, Bailey RL (1987) Compatible basal area and diameter distribution models for thinned loblolly pine plantations in Santa Catarina, Brazil. For Sci 33:43–51

Medlyn B, Barrett D, Landsberg J, Sands P, Clement R (2003) Conversion of canopy intercepted radiation to photosynthate: a review of modelling approaches for regional scales. Funct Plant Biol 30:153–169

Merganičová K, Pietsch SA, Hasenauer H (2005) Testing mechanistic modeling to assess impacts of biomass removal. For Ecol Manag 207:37–57

Meyer WH (1929) Yields of second growth and fir in the Northeast, Technical bulletin 142. US Department of Agriculture, Washington, DC, p 52

Miehle P, Battaglia M, Sands PJ, Forrester DI, Feikema PM, Livesley SJ, Morris JD, Arndt SK (2009) A comparison of four process-based models and a statistical regression model to predict growth of Eucalyptus globulus plantations. Ecol Model 220:734–746

Milner KS, Coble DW, McMahan AJ, Smith EL (2003) FVSBGC: a hybrid of the physiological model STAND-BGC and the Forest Vegetation Simulator. Can J For Res 33:466–479

Mitchell KJ (1975) Dynamics and simulated yield of Douglas-fir. For Sci Monogr 17:1–39

Moore JA, Zhang L, Newberry JD (1994) Effects of intermediate silvicultural treatments on the distribution of within-stand growth. Can J For Res 24:398–404

Moores A, Puettmann K, Maguire DA (2007) The focus of intensive silvicultural research on coastal Douglas-fir over the last 20 years. West J Appl For 22:21–28

Moser JW (1972) Dynamics of an uneven-aged forest stand. For Sci 18:184–191

Mowrer HT (1991) Estimating components of propagated variance in growth simulation-model projections. Can J For Res 21:379–386

Mowrer HT, Frayer WE (1986) Variance propagation in growth and yield projections. Can J For Res 16:1196–1200

Murray DM, Von Gadow K (1991) Relationship between the diameter distributions before and after thinning. For Sci 37:552–559

Nance WL, Grissom JR, Smith WR (1988) A new competition index based on weighted and constrained area potentially available. In: Ek AR, Shifley SR, Burk TE (eds) Forest growth modelling and prediction, vol I, General technical report NC-120. USDA Forest Service, North Central Forest Experiment Station, St. Paul, pp 134–142

Nigh G (2013) Evaluating Douglas-fir and western hemlock volume growth in response to thinning and fertilization. N Z J For 43:9. doi:10.1186/1179-5395-43-9

Opie JE (1968) Predictability of individual tree growth using various definitions of competing basal area. For Sci 14:314–323

Payandeh B, Haig RA (1991) A management-oriented model for assessing early stand establishment. For Can 5:4

Petritsch R, Hasenauer H, Pietsch SA (2007) Incorporating forest growth response to thinning within Biome-BGC. For Ecol Manag 242:324–336

Pienaar LV, Rheney JW (1995) Modeling stand level growth and yield response to silvicultural treatments. For Sci 41:629–638

Pienaar LV, Shiver BD, Grider GE (1985) Predicting basal area growth in thinned slash pine plantations. For Sci 31:731–741

Porté A, Bartelink HH (2002) Modelling mixed forest growth: a review of models for forest management. Ecol Model 150:141–188

Pretzsch H, Biber P, Ďurský J (2002) The single tree-based stand simulator SILVA: construction, application and evaluation. For Ecol Manag 162:3–21

Qin J, Cao QV, Blouin DC (2006) Projection of a diameter distribution through time. Can J For Res 37:188–194

Quicke H, Glover G, Meldahl R (1999) Loblolly pine growth response to herbaceous vegetation control at different planting densities. Can J For Res 29:960–967

Raulier F, Lambert MC, Pothier D, Ung CH (2003) Impact of dominant tree dynamics on site index curves. For Ecol Manag 184:65–78

Rehfeldt GE, Wykoff WR, Hoff RJ, Steinhoff RJ (1991) Genetic gains in growth and simulated yield of Pinus monticola. For Sci 37:326–342

Richardson B, Watt MS, Mason EG, Kriticos DJ (2006) Advances in modelling and decision support systems for vegetation management in young forest plantations. Forestry 79:29–42

Ritchie MW, Hamann JD (2006) Modeling dynamics of competing vegetation in young conifer plantations of northern California and southern Oregon, USA. Can J For Res 36:2523–2532

Ritchie MW, Hamann JD (2008) Individual-tree height-, diameter- and crown-width increment equations for young Douglas-fir plantations. New For 35:173–186

Robinson AP, Monserud RA (2003) Criteria for comparing the adaptability of forest growth models. For Ecol Manag 172:53–67

Running SW, Gower ST (1991) FOREST-BGC, a general model of forest ecosystem processes for regional applications. II. Dynamic carbon allocation and nitrogen budgets. Tree Physiol 9:147–160

Sampson DA, Waring RH, Maier CA, Gough CM, Ducey MJ, Johnsen KH (2006) Fertilization effects on forest carbon storage and exchange, and net primary production: a new hybrid process model for stand management. For Ecol Manag 221:91–109

Sands PJ, Landsberg JJ (2002) Parameterisation of 3-PG for plantation grown *Eucalyptus globulus*. For Ecol Manag 163:273–292

Scheller RM, Domingo JB, Sturtevant BR, Williams JS, Rudy A, Gustafson EJ, Mladenoff DJ (2007) Design, development, and application of LANDIS-II, a spatial landscape simulation model with flexible temporal and spatial resolution. Ecol Model 201:409–419

Schwalm CR, Ek AR (2001) Climate change and site: relevant mechanisms and modeling techniques. For Ecol Manag 150:241–257

Shaffi B, Moore JA, Newberry JD (1990) Individual tree diameter growth models for quantifying within-stand response to nitrogen fertilization. Can J For Res 20:1149–1155

Sharma M, Smith M, Burkhart HE, Amateis RL (2006) Modeling the impact of thinning on height development of dominant and codominant loblolly pine trees. Ann For Sci 63:349–354

Short EA, Burkhart HE (1992) Predicting crown-height increment for thinned and unthinned loblolly pine plantations. For Sci 38:594–610

Skovsgaard JP, Vanclay JK (2008) Forest site productivity: a review of the evolution of dendrometric concepts for even-aged stands. Forestry 81:13–31

Snowdon P (2001) Short-term prediction of growth of *Pinus radiata* with models incorporating indexes of annual climatic variation. For Ecol Manag 152:1–11

Snowdon P (2002) Modeling Type 1 and Type 2 growth responses in plantations after application of fertilizer or other silvicultural treatment. For Ecol Manag 163:229–244

Snowdon P, Benson ML, Woollons RC (1998) Incorporation of annual climatic indices into models of *Pinus radiata* in a spacing experiment. New For 16:101–123

Soares P, Tomé M (2003) GLOBTREE: an individual tree growth model for *Eucalyptus globulus* in Portugal. In: Amaro A, Reed D, Soares P (eds) Modelling forest systems. CABI International, Wallingford, pp 97–110

Söderbergh I, Ledermann T (2003) Algorithms for simulating thinning and harvesting in five European individual-tree growth simulators: a review. Comput Electron Agric 39:115–140

South DB, Miller JH (2007) Growth response analysis after early control of woody competition for 14 loblolly pine plantations in the southern U.S. For Ecol Manag 242:569–577

Sprinz PT, Talbert CB, Strub MR (1989) Height-age trends from an Arkansas seed source. For Sci 35:677–691

Stage AR, Crookston NL, Shaffi B, Moore JA, Olson J (1990) Representing growth response to fertilization in the Prognosis model for stand development, Research note INT-392. USDA Forest Service Intermountain Research Station, Ogden, p 6

Stape JL, Ryan MG, Binkley D (2004) Testing the utility of the 3-PG model for growth of *Eucalyptus grandis* x *urophylla* with natural and manipulated supplies of water and nutrients. For Ecol Manag 193:219–234

Svensson M, Jansson P-E, Gustafson D, Berggren Kleja D, Langvall O, Lindroth A (2008) Bayesian calibration of a model describing carbon, water and heat fluxes for a Swedish boreal forest stand. Ecol Model 213:331–344

Talbert CB, Hyink D (1988) Implications of genetic improvement for the growth and yield of trees and stands. In: Ek AR, Shifley SR, Burk TE (eds) Forest growth modeling and prediction. USDA Forest Service, St. Paul, pp 562–571

Talbert C, Marshall D (2005) Plantation productivity in the Douglas-fir region under intensive silviculture practices: results from research and operations. J For 103:65–70

Temesgen H, LeMay VM, Froese KL, Marshall PL (2003) Imputing tree-lists from aerial attributes for complex stands of south-eastern British Columbia. For Ecol Manag 177:277–285

Trincado GV, Quezada PR, von Gadow K (2003) A comparison of two stand table projection methods for young *Eucalyptus nitens* (Maiden) plantations in Chile. For Ecol Manag 180:443–451

Van Oijen M, Rougier J, Smith R (2005) Bayesian calibration of process-based forest models: bridging the gap between models and data. Tree Physiol 25:915–927

Vanclay JK (2010) Robust relationships for simple plantation growth models based on sparse data. For Ecol Manag 259:1050–1054

Vaughn N (2007) An individual-tree model to predict annual growth of young stands of Douglas-fir (*Pseudotsuga menziesii* (Mirbel)) in the Pacific Northwest. M.S. thesis, University of Washington, Seattle, WA, p 106

Vega-Nieva DJ, Tomé M, Tomé J, Fontes L, Soares P, Ortiz L, Basurco F, Rodrígez-Soalleirod R (2013) Developing a general method for the estimation of the fertility rating parameter of the 3-PG model: application in Eucalyptus globulus plantations in northwestern Spain. Can J For Res 43:627–636

Vospernik S, Monserud RA, Sterba H (2010) Do individual-tree growth models correctly represent height: diameter ratios of Norway spruce and Scots pine? For Ecol Manag 260:1735–1753

Wagner RG, Little KM, Richardson B, McNabb K (2006) The role of vegetation management for enhancing productivity of the world's forest. Forestry 79:57–79

Wang YP (2003) A comparison of three different canopy radiation models commonly used in plant modelling. Funct Plant Biol 30:143–152

Wang W, Peng C, Zhang SY, Zhou X, Larocque GR, Kneeshaw DD, Lei X (2011) Development of TRIPLEX-Management model for simulating the response of forest growth to pre-commercial thinning. Ecol Model 222:2249–2261

Waring RH, Landsberg J, Williams M (1998) Net primary production of forests: a constant fraction of gross primary production? Tree Physiol 18:129–134

Waring RH, Coops NC, Landsberg JJ (2010) Improving predictions of forest growth using the 3-PGS model with observations made by remote sensing. For Ecol Manag 259:1722–1729

Watt MS, Whitehead D, Richardson B, Mason EG, Leckie AC (2003) Modelling the influence of weed competition on the growth of young *Pinus radiata* at a dryland site. For Ecol Manag 178:271–286

Weiskittel AR, Garber SM, Johnson GP, Maguire DA, Monserud RA (2007) Annualized diameter and height growth equations for Pacific Northwest plantation-grown Douglas-fir, western hemlock, and red alder. For Ecol Manag 250:266–278

Weiskittel AR, Maguire DA, Monserud RA, Johnson GP (2010) A hybrid model for intensively managed Douglas-fir plantations in the Pacific Northwest, USA. Eur J For Res 129:325–338

Weiskittel AR, Hann DW, Kershaw JA, Vanclay JK (2011) Forest growth and yield modeling. Wiley, West Sussex, p 411

Westfall JA, Burkhart HE (2001) Incorporating thinning response into a loblolly pine stand simulator. South J Appl For 25:159–164

Westfall JA, Burkhart HE, Allen HL (2004) Young stand growth modeling for intensively-managed loblolly pine plantations in southeastern US. For Sci 50:823–835

White TL, Adams WT, Neale DB (2007) Forest genetics. CABI Publishing, Cambridge, MA, p 500

Wimberly MC, Bare BB (1996) Distance-dependent and distance-independent models of Douglas-fir and western hemlock basal area growth following silvicultural treatment. For Ecol Manag 89:1–11

Wykoff WR (1990) A basal area increment model for individual conifers in the northern Rocky Mountains. For Sci 36:1077–1104

Xenakis G, Ray D, Mencuccini M (2008) Sensitivity and uncertainty analysis from a coupled 3-PG and soil decomposition organic matter decomposition model. Ecol Model 219:1–16

Xie C, Yanchuk A (2003) Breeding values of parental trees, genetic worth of seed orchard seedlots, and yields of improved stocks in British Columbia. West J Appl For 18:88–100

Yue C, Kohnle U, Hein S (2008) Combining tree- and stand-level models: a novel approach to growth prediction. For Sci 54:553–566

Zhang X, Lei Y, Cao QV (2010) Compatibility of stand basal area predictions based on forecast combination. For Sci 56:552–557

Chapter 4
Management of Industrial Forest Plantations

José Leonardo de Moraes Gonçalves, Luciana Duque Silva, Maurel Behling, and Clayton Alcarde Alvares

4.1 Planted Forest in the World

The total area of planted forest (composed of trees established through planting and/or through deliberate seeding of native or introduced species) is estimated to be 264 million hectares, corresponding to 6.6 % of the forest area. Since 1990 the planted forest area has steadily increased in all regions and subregions defined by FAO (2006). It increased by more than 3.6 M ha per year from 1990 to 2000, by 5.6 M ha per year from 2000 to 2005, and by 4.2 M ha per year from 2005 to 2010. Around 76 % of planted forests has production as their primary function.

Most of the planted forest was established through afforestation particularly in China. Three-quarters of all planted forests consist of native species while 25 % comprises introduced species. East Asia, Europe and North America reported the greatest area of planted forests, together accounting for about 75 % of global planted forest area. Among the ten countries with the highest annual increase in planted forest areas, China takes the first rank in the past 20-year period, followed by the United States of America, Canada and India. These four countries together account for an average annual increase in planted forests of 3.3 M ha over this period. Given this trend, a further rise in the planted forest area up to 300 M ha by 2020 can be anticipated. Planted forest expansion helped reduce the net loss of forest area to 5.2 M ha per year in the last decade, compared with the gross rate of loss through

J.L. de Moraes Gonçalves (✉) • L.D. Silva
ESALQ, Departamento de Ciências Florestais, USP, Universidade de São Paulo, Piracicaba, CEP 13418-900, SP, Brazil
e-mail: jlmgonca@usp.br

M. Behling
Embrapa Agrossilvipastoril – CPAMT, Sinop, MT, Brazil

C.A. Alvares
Instituto de Pesquisas e Estudos Florestais (IPEF), Forest Productivity Cooperative (FPC), Piracicaba, SP, Brazil

J.G. Borges et al. (eds.), *The Management of Industrial Forest Plantations: Theoretical Foundations and Applications*, Managing Forest Ecosystems 33,
DOI 10.1007/978-94-017-8899-1_4,

deforestation and natural causes, estimated at 13 M ha per year in the last decade (largely converted to agriculture or lost through natural causes) (FAO 2011).

In spite of data limitations it is evident that wood supply (particularly industrial roundwood) is shifting from natural forests to planted forests. It is thus foreseen that planted forests will increasingly contribute to the supply of the world's wood, fibre, fuel and non-wood forest products supply (as well as protecting soil and water resources and fulfilling other purposes) and that this shift may reduce the pressure on natural forests. The impact of this development on timber markets and environmental conservation should be considered by policy-makers, planners and forest managers.

The great majority of forest plantations are based on exotic species chosen for their capacity to grow rapidly and produce wood of appropriate quality. In some cases sites can support wood production with proper management, but elsewhere there are serious problems. The extent to which intensively managed plantations with relatively narrow genetic bases are susceptible to disease, insects, and natural disasters is not well evaluated. However, where there have been appropriate species-site matching and when management prescriptions are effective plantations usually remain healthy and productive (Nambiar 1996; Gonçalves et al. 2013).

The challenge is to develop forest plantations that are financially viable as well as ecologically sustainable. Management requirements to achieve sustainability differ, however, depending on site specific climatic and edaphic characteristics and the needs of the crop.

What You Will Learn in This Chapter

- Understand the principles, aims, strategies and practices relevant to sustainable forest management
- Have knowledge about silvicultural and management practices in fast-growing forest
- Have knowledge of Brazilian forestry as a case aimed at sustainable management of fast-growing eucalypt plantations
- Understand on the principles and uses of process-based model in the commercial forests

The remaining chapter has the following sections: Sect. 4.2 describes the silvicultural management practices; Sect. 4.2.1 focus on selection appropriate genotypes to sites; Sects. 4.2.2, 4.2.3 and 4.2.4 focus on soil preparation, residue management and application of fertilizers, respectively; Sect. 4.2.5 describes stocking; Sect. 4.2.6 focus on weed control; Sects. 4.2.7 and 4.2.8 focus on the impact of harvesting systems on soil; Sect. 4.2.8 describes the precision forestry; Sect. 4.3 describes the use of physiological growth models as a silvicultural tool. The chapter ends with open problems in Sect. 4.4 and References.

4.2 Silvicultural Management Practices

Forest plantations can be established to enhance the productivity of lands that have been degraded by deforestation and intensive agricultural disturbances. Selected species can be planted and harvested to meet age, size and quality specifications to match the needs of the final product. The development of improved forest planning and operations can minimize site impacts, increase utilization and minimize or avoid adverse environmental effects.

Some principles, aims, strategies and silvicultural practices must be followed for developing sustainable forest plantations based on holistic approach of the productive process (Fig. 4.1).

4.2.1 Selection Appropriate Genotypes to Sites

Most plantations consist of a limited number of fast growing and exotic genera including *Pinus*, *Eucalyptus*, *Cunninghamia*, *Populus*, *Acacia*, *Larix*, *Picea* and

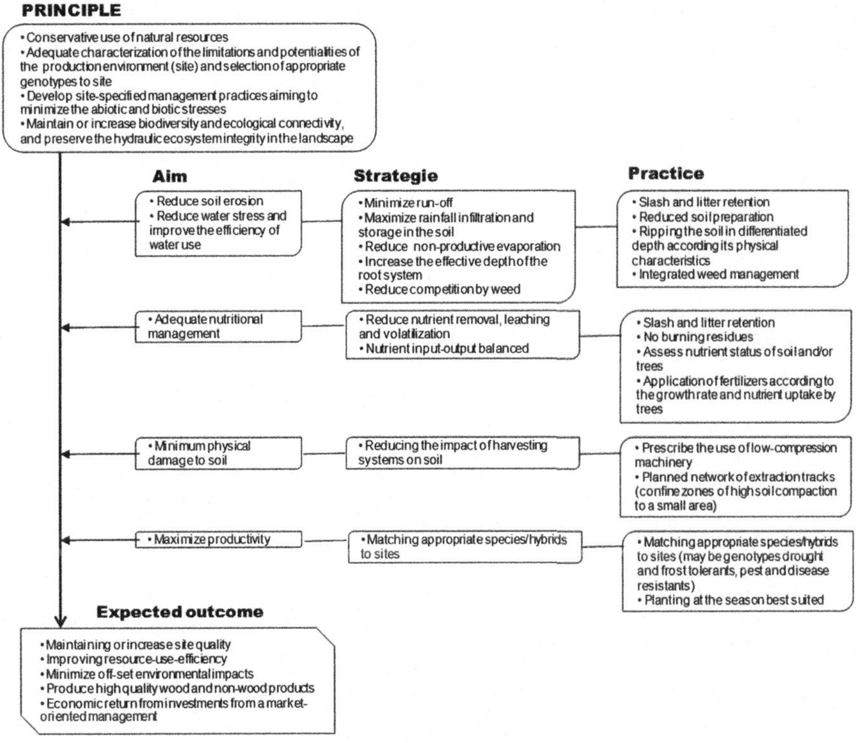

Fig. 4.1 Some silvicultural principles, aims, strategies and practices towards to sustainable management of forest plantations

Tectona, although there are efforts to identify appropriate local species (Brown et al. 1997; Evans and Turnbull 2010). With selection, breeding, and vegetative propagation, genetic gain has been rapid and further improvement is expected as provenances and even individual genotypes are matched to the site (Boyle et al. 1997).

The wide range of tree species and hybrids with different climatic and edaphic suitability associated with the easy propagation by seeds and cloning allow the adaptation of plantations to various environmental conditions. The possibility of using the wood of most planted species in a range of purposes has led large and small enterprises to establish plantations for multiple uses. The desirable characteristics in association with the accumulated knowledge on silviculture encourage the use of those species in most plantations (Evans and Turnbull 2010; Gonçalves et al. 2013).

The most important factors in the selective process for a genotype are wood characteristics, productivity level, susceptibility to pests and diseases, drought tolerance, especially in tropical regions, and cold tolerance in temperate regions (mostly without water deficit). In regions with pronounced seasonality and moderate to long drought periods, the planting of hybrid genotypes can be used, propagated by cutting. Under temperate conditions, the planting of single species predominates, propagated by seedling. Clonal plantations with interspecific hybrids have been fundamental for eucalypt adaptation in regions under water and nutritional stresses in Brazil and Southern China. Given the rapid advances in eucalypt breeding, regarding adaptation to water stress and resistance to diseases and pests, and the adoption of clonal propagation techniques, genotypes are rapidly becoming obsolete and are replaced by more productive ones after harvesting (Gonçalves et al. 2013).

Matching site and species can increase the amount of nitrogen fixed by some species. While there are prospects for judicious mixtures of N-fixing and non-N-fixing species in forest plantations, such a strategy also needs well developed site and stand management practices (Binkley and Giardina 1997; Bouillet et al. 2013). Because changes in species or gains from genetic improvement occur in parallel with operational plantations, they may make productivity comparisons between crop rotations difficult. However, genetic gain through tree breeding is not a panacea for sustained productivity. Genetic gains cannot be realized operationally unless plantation management is carried out under sound site-soil and stand management, and over-zealous claims about the benefit of genetic improvement need to be treated with caution (Nambiar 1996; Gonçalves et al. 2013). Improved genotypes have potential advantages but their additional demands on applications of fertilizers, pesticides or herbicides need to be taken into account – as does costs and can require special management skills.

Achievements in crop breeding have been made and efforts are still ongoing to achieve the following: tolerance to drought, severe cold, salinity, low soil fertility, water logging, resistance/tolerance to disease and pest, higher yielding genotypes, better water use efficiency (mainly for water scarce areas), better wood quality (Boyle et al. 1997; Evans and Turnbull 2010; Gonçalves et al. 2013).

4.2.2 Soil Preparation

Until 1980s, silvicultural practices regarding residue management and soil preparation in many forest plantations followed a typically agronomical pattern, as windrowing and/or residues burning and intense stirring of the topsoil. This was the sole recommendation for large forest plantations, regardless of climate, soil type and genotype. The perception in force was that the forest species needed intense soil preparation, with planting survival rates and productivity gains that would justify the operational costs (Gonçalves et al. 2008). Two of the great advances of the last thirty years in many countries were the understanding and abolition of burning as a form of land cleaning and the adoption of conservationist techniques for soil management, which culminated in the implementation of tillage systems with minimum soil disturbance.

The concern with the preservation of natural resources and the use of herbicides were the factors predisposing and permitting the adoption of minimum tillage. Weed control with herbicides was a crucial factor since this system has no ploughed land inversion (unlike the conventional system), so the weed seed bank remains on the topsoil rather than buried. This leads to an infestation of these plants, making manual control operationally and economically harder or unfeasible (Gonçalves et al. 2008).

One of the major factors limiting forest production in many regions is the susceptibility to drought. Despite the high and monsoonal rainfall received in parts of the tropics, soil water deficit can be a recurring constraint on productivity on many sites (Gonçalves et al. 1997; Landsberg and Waring 1997). In the semiarid and arid tropics, water stress may limit rates of growth below commercially viable levels. Soil management for conserving available water is a critical consideration. Soil preparation can help overcome the limitations of water resources for forest plantations in two ways. The first is due to the increase of rainfall infiltration, reducing runoff, thus increasing the water reserve in the soil. The second refers to the increase of soil effective depth when there are layers of physical impediment (Gonçalves et al. 2002; Stape et al. 2002). The increased infiltration is favored when compacted or hardened layers are disrupted, especially when crop residues are kept on the soil surface.

The reduction of bulk density through soil preparation in the planting row or hole leads to fast root growth and, consequently, increases fertilizer use efficiency through greater use of water and nutrients by adjacent seedlings. Even in friable soils, soil preparation ensures faster growth of juvenile roots, which experience lower physical soil resistance, resulting in energy saving and increased radial and longitudinal growth (Gonçalves et al. 1997). Rapid seedling establishment increases its competitive capacity to survive under water, cold and nutrient stress, besides providing greater protection to the soil. In flat and slightly undulating terrain, soil preparation may consist of ripping up to 30–40 cm depth. Regarding soil hardening and compaction, the ripping depth is usually 30–35 cm and the planting hole opening, either manual or mechanical, is limited to 25–30 cm. These depths may sometimes increase, depending on soil bulk density (Gonçalves et al. 2008).

Typically, cohesive soils show serious physical and chemical restrictions associated with high rainfall variability (monthly and interannually). Therefore, silviculture can become unviable if the soil preparation is not properly planned and performed (Souza 2002). Thus, soil preparation constitutes an essential practice for the plant to adapt to this kind of soil characteristics. In eucalypt plantations of Brazil on cohesive soils, for a similar amount of rainfall, soils of higher effective depth or clones with higher root penetrating capacity tend to be more productive. The nutritional restrictions are overcome by fertilizer application. High productivities have been achieved with a statistical model using two independent variables for decision-making on soil preparation: the rainfall rate, and root depth to the layer of physical impediment (Stape et al. 2002). In areas where rainfall is higher and evenly distributed, the ripping depth is lower. If the soil has a fragipan or hardpan between depths of 50 and 80 cm, the ripping depth will be in the range 60–90 cm; should the layer with physical impediment be above 80 cm, the ripping depth may reach 110 cm. These soils should be prepared during the humid season to facilitate implement penetration and provide seedlings with a long period for root growth (Sasaki et al. 2007).

4.2.3 Residue Management

At plantations under tropical and subtropical conditions, maintenance of forest residues on soil is undoubtedly important for sustaining long-term site productivity. Ideally, should be left at least 30 % of the soil surface covered with crop residues after harvest. It requires residue from the previous crop as the main resource (thus burning is discouraged). Retention of crop residues helps reducing erosion, improving water infiltration and therefore moisture conservation, maintain soil organic matter levels (Gonçalves et al. 2013). Carbon dioxide emissions are thus reduced.

Depending on whether the harvesting system is either semi-mechanized (chainsaw) or mechanised (i.e. harvester/forwarder and feller-buncher/skidder), the distribution of residues and the impact on soil are quite distinct. Harvesting debris can be distributed throughout the field randomly or orderly, in this case, in order to favour subsequent operations of soil preparation, sprouting control and planting. In mechanized harvesting in which residues are randomly scattered over the terrain, residues are to be chopped or removed from the soil preparation and planting rows to prevent bushing and quality and yield reduction of the equipment for soil preparation and other cropping practices. Usually, during soil preparation, the principle of minimum interference in the residues arrangement or characteristics is adopted in order to reduce cost and avoid soil disturbance. For this purpose there are some resources that can be used: (i) introduction of mechanical supplies to soil preparation implements (i.e. cutting disk); (ii) combining clean-rail with ripper, to move the residues from the ripping row; (iii) elevation of the wheeled tractor chassis and/or use of special tyres (i.e. bigger, sometimes using a belt conveyor); (iv) change of planting spacing to allow operations in strips with fewer obstacles (Gonçalves et al. 2002). The use of these procedures must be well planned because, depending on residue and soil conditions, the quality and yield of soil preparation and other operations may decay. Due to interdependence among silvicultural practices, their planning and implementation must be combined and synchronized to create a chain of beneficial effects of some procedures in subsequent operations. Thus, the harvesting system and residues disposition on the ground have a high influence on the operation efficiency (yield, cost and environment damages) of stand replanting and quality in the long term.

Intensity of logging and site preparation methods affect the types and amount of slash remaining on the site. While the bole wood is usually of greatest economic interest, other biomass may be removed as well. Tops may be utilized for firewood, piled within the plantation or burned in order to improve regeneration, facilitate weed control or lower fire hazard. Nutrient distribution in trees is affected by age, species and site, but generally the foliage and branches contain a major portion of the nutrients, especially compared to their total biomass (Fölster and Khanna 1997; Gonçalves et al. 1997). The litter layer is an important source of nutrients and organic matter and their fate during forest management has major impacts on nutrient cycling, especially in short rotation forestry (O'Connell and Sankaran 1997).

Experiments established as part of a CIFOR research network (Bouillet et al. 2000; Gonçalves et al. 2000a; O'Connell et al. 2000; Sankaran et al. 2000; du Toit et al. 2000; Xu et al. 2000; Laclau et al. 2010; Versini et al. 2012) indicate that the presence of different amounts of litter and logging residues on the soil surface increase eucalypt productivity at different levels related to some extent to water and nutrient availability. Residue retention in some sites increased water and nutrient status. This was associated with reduced losses of nutrient and organic matter and the maintenance of crucial soil physical properties such as porosity and thereby permeability, root growth, infiltration, and aeration (Gonçalves et al. 2002; Stape et al. 2002; Xu and Dell 2002).

Gonçalves et al. (2004) conducted an experiment to assess the effect of site management practices of minimal and intensive soil cultivation on plant growth, soil fertility, nutrient cycling and nutrition of a *Eucalyptus grandis* stand. The climate of the area is of the Cwa type (Köppen classification), i.e. mesothermic with dry winters. The soil type of the area is classified as a typical Oxisol, loamy (200 g kg^{-1} clay) and dystrophic. Different soil preparation and residue management resulted in pronounced growth effects. The treatment in which all residues were retained on soil surface showed the highest growth at 6.4 years of age. The lowest growth occurred when all residues were removed. Removal of bark and slash caused a reduction in productivity of 40 m^3 ha^{-1} (14.5 %) of stem volume. When compared to the treatment in which all residues were retained and that in which all residues were removed, the reduction reached 101 m^3 ha^{-1} (36.5 %). Furthermore, in the same treatments, Silva et al. (1997) found that weed occurrence was much lower when residues were maintained on soil surface, obstructing, therefore, the germination of weed seed bank in the soil.

Effects of slash burning on forest production have been a subject of many investigations and varying conclusions and controversies. There is clear evidence now that slash burning will lead to loss of organic matter and nutrients, especially N (Gonçalves et al. 2008). Nitrogen loss may be less critical for plantations of N-fixing species although the soil N level in *Acacia* plantations in general is not higher than those at eucalypt or pine sites (Tiarks and Ranger 2008). In Brazil, although slash burning reduced mineral N by more than 50 % compared to slash retention, growth under the burning treatment remained relatively high throughout the rotation (Gonçalves et al. 2008).

4.2.4 Application of Fertilizers

The loss of nutrients and organic matter is a very significant factor in soil chemical degradation (Fölster and Khanna 1997). Loss of nutrients varies with management practices and can have a dramatic effect on the growth of the next stand. Nutrients lost by removal or leaching can be replaced by fertilization. Since investments in fertilizers are relatively high for most forest producers, fertilization should be combined with other silvicultural practices (i.e. soil preparation, residue management and weed control) to reduce fertilizer demands in the short or long-term (Nambiar and Kallio 2008). This is especially true on sites at tropical and subtropical regions where the major limitations to using soils for short rotation tree crops are low nutrient reserves, poor nutrient retention ability and susceptibility to drought (Gonçalves et al. 2013). Using nutrient amendments to balance harvest removals is thus consistent with the goals of sustainable forestry. Nutrient budgets can be used to gain understanding of the mechanisms underlying system function and may allow declining sustainability trends to be detected early (Paustian et al. 1990; Gonçalves et al. 2008). Adequate nutrient supplies and balance resulting from fertilization can also improve crop vigor, and reduce incidence of disease and the need for fungicides (Almeida et al. 2010b).

Application of fertilizers is widely used in forest plantations across the world. It now forms a crucial component of most plantation management systems and plays a key role in allowing plantation forestry to be a profitable enterprise. There are significant yield gains in response to fertilization in most forest plantations (Gonçalves et al. 2013). Regardless of weather conditions, the magnitude of the response depends on the nutritional demand of the genotype and on the availability of soil nutrients. Especially in soils with low fertility, continuous nutrient removal by crops consecutively increases the potential response to fertilizer application (Gonçalves et al. 1997, 2013; Laclau et al. 2005, 2010). Gains in productivity attributed to mineral fertilizers are quite variable and high, but in general, they represent at least 30–50 % on average (Gonçalves 2011).

Although continuous fertilization throughout crop rotation can be used to maximize tree growth, recent results showed that forests adequately fertilized at early stages, which produced an adequate canopy structure, are highly efficient in using nutrients through the biogeochemical cycles, becoming little responsive to further fertilization (Stape et al. 2004; Laclau et al. 2010). Nutrient cycling reduces tree dependence on net nutrient supply from soil reserves. Mobile nutrients (namely N, P and K) are redistributed in the plant increasing efficiency for biomass production. Biogeochemical cycles lead to a smaller dependence on nutrient reserves in the mineral soil at the end of the rotation, through intense recycling processes of retranslocation, foliar leaching and litter decomposition (Laclau et al. 2003, 2010; Barros et al. 2004; Gonçalves et al. 2008). Before leaf area index (LAI) peak is reached, responses to fertilizer are very common (Cromer et al. 1995; Herbert and Schonau 1989; Gonçalves et al. 2004; Laclau et al. 2010). When a response to fertilizer occurs, increased rates of nutrient uptake lead to an LAI increase, which can prolong leaf retention and increase photosynthetic efficiency (Cromer et al. 1995; Binkley et al. 2004; Laclau et al. 2009). Increased nutrient uptake also affects partitioning in forest plantations by increasing allocation to foliage in detriment of roots, reducing the root: shoot ratio (Bouillet et al. 2002; Mello et al. 2007; Laclau et al. 2010). After canopy closure, internal and external nutrient cycles become more important, as light and water become more limiting due to intraspecific competition (Grove et al. 1996; Gonçalves et al. 2000b, 2008; Laclau et al. 2010).

Nutritional stages of a forest stand can be divided into before, during, and after canopy closure. Understanding these stages and nutrient cycling is essential for the adequate planning of fertilizer application (rate, method, and time). Fertilizer recommendation should preferably be adjusted at local level to the most representative species and soil types, based on field experimentation, and should allow optimization of financial returns. Fertilization should be performed during the initial stage of tree establishment, from the planting to canopy closure. The most frequent and most significant responses to fertilizers are to N, P, K and B. Normally, for sandy and water-deficient soils, responses to fertilizers are more common (Gonçalves et al. 2008; Gonçalves 2011).

Especially for fast-growing species, phosphorous doses can be thoroughly applied at planting, since this nutrient is little movable and relatively little soluble. N and K doses should be divided into one or two surface applications (Gonçalves et al. 2008). Doses of up to 50 kg ha^{-1} of N and K_2O may be applied thoroughly in one single surface application, once risks of leaching are low (Maquère et al. 2005; Laclau et al. 2010).

4.2.5 Stocking

The appropriate stocking rate (planting density) promotes optimum growth rates and efficient plantation management. Defining the initial spacing for plantations is essential because it determines the amount of natural resources available for each tree growth. Spacing greatly affects production, moreover, it has several implications regarding silvicultural, technological and economic aspects, as it influences growth and survival rates of trees, crown and branch amount, wood quality, bark amount, age of harvesting as well as harvesting processes and forest management, and therefore, production costs (Gonçalves et al. 2013).

The recommended tree planting design is 3–4 m between rows and 1–3 m between trees, giving an initial stocking rate, usually, of 1,000–1,800 trees per hectare. These densities encourage the canopy to close rapidly, reducing weed problems and improving tree form and branching. If the goal is to produce solid timber for sawmill this planting density is high enough to allow trees with exceptional form and vigor to be selected in elapsing of thinning plan of the stand. Closer spacing promotes faster development of the LAI, which increases light interception and photosynthesis. The dynamics of LAI can be used to characterize pre- and post-canopy closure of developmental stages of forest plantations (Landsberg and Waring 1997), which are affected in different ways by silvicultural practices. During the pre-closure phase, trees tend to be more responsive to cultivation, fertilizers and weed control. After canopy closure, intra-specific competition for resources becomes strong. Species with denser crowns, commonly, are more responsive to spacing variations than less dense crowns. This behavior is directly associated with the intra-specific capacity of species to compete for light, water and nutrients (Gonçalves et al. 2013). Wide spacing may also be used in water catchments to increase water yield in the site (Lima et al. 2012). Higher stocking will increase mean annual increment and shorten the time to the maximum.

4.2.6 Weed Control

A major limiting factor to plant productivity is weeds. Good SLM practices can reduce the weed infestation considerably by providing cover by crops, residues and mulch, and by minimum soil disturbance. On grazing land the control of undesirable species should be a key focus. In forests the problem of invasives is a great concern.

Forest plantations are very sensitive to weed competition at earlier stages of growth. A reduction in plant survival and growth may result from competition for light, water and nutrients, because weeds use larger volumes of soil than young tree seedlings (Gonçalves and Barros 1999). Wood production and economic benefits of managing weeds during establishment have been widely demonstrated for eucalypt plantations (Zen 1987; Pitelli and Marchi 1991; Toledo et al. 2000, 2003; Tarouco et al. 2009). Competition for water can be intense as indicated by stomatal conductance measured in contrasting weed control treatments (Silva et al. 2000; Lima et al. 2003). However, once established, trees may be able to uptake water in deeper soil layers than most annual herbaceous species.

Weed control prior to planting, both in areas of afforestation and replanting in the system of minimum cultivation of the soil, is essential for the removal of weeds that vegetatively propagate and for the reduction of their seed bank (Gonçalves et al. 2004). The procedures are usually carried out combining mechanical and chemical methods, using total-action herbicide (nonselective), such as glyphosate. During the planting of forest stands, most production systems apply pre-emergence herbicides at a one-meter strip on the crop row. After planting, the post-emergence control of weeds is performed by spraying herbicides. In the application, special care must be taken to avoid the drift to leaves and stems of the cultivated plants because it can cause phytotoxic effects of reduced growth (Salgado et al. 2011).

4.2.7 Reducing the Impact of Harvesting Systems on Soil

The effect of soil compaction and other soil disturbances can be severe if operations are not properly managed. Inappropriate harvesting systems have the potential to severely and adversely affect soil and water values. Physical degradation may be soil compaction caused by the use of heavy equipment and/or loss of soil structure when litter layers are disturbed. In Cameroon, hand clearing increased soil bulk density from 1.16 to 1.30 g cm^{-3} in the surface 10 cm (Ngeh et al. 1995). Use of heavy tractors and removal of all logging residue not only increased the bulk density of the surface to 1.53 g cm^{-3}, but significantly increased bulk density to the 55 cm depth. However, not all logging and site preparation treatments impact physical properties negatively. Nine years after planting, the physical properties of soil in plantations of *Eucalyptus spp.* and *Paraserianthes falcataria* were similar to soils compared to soils in nearby undisturbed and selectively logged dipterocarp forests in East Kalimantan (Wenzel et al. 1995). Bulk densities were slightly higher under the eucalypts and saturated hydraulic conductivities were higher at 30 cm depths under both plantation species. Apparently, these changes in physical properties were too small to affect the growth or distribution of tree roots (Murach et al. 1998). These findings encourage or prescribe the use of low-compression machinery and a planned network of extraction tracks, because this confines zones of high soil compaction to a small portion of the plantation area (e.g. Beets et al. 1994). The effects of planting, tending and harvesting on the physical properties of the soils will be highly dependent on the soils and the type of equipment used.

Delimbing and debarking the stem at the stump and avoidance of fire significantly reduces nutrient exports in above-ground biomass (Gonçalves et al. 2000b). Rotation length is another variable that affects soil, because longer rotations reduce the frequency of major disturbance during harvesting, wood production per unit of nutrient exported will decrease with an increase in rotation length, and the average concentration of most nutrients decreases with tree age. Hence, overall nutrient-use efficiency measured as wood production per unit of nutrient accumulated can be increased by prolonging forest rotations and using practices that lead to nutrient and organic matter retention (Gonçalves et al. 2004).

Gonçalves et al. (2008) presented an estimated nutrient budget for sites of short-rotation eucalypt plantations under a set of management scenario. Management practices that include burning of the forest floor and slash, and removal of wood and bark have the highest impact on nutrient depletion. The impact varies between nutrients. For instance, N availability might fall to a critical level beyond the third rotation, that of P beyond the first rotation, and that of K beyond the third rotation.

The scenario which does not include residue burning or removal (litter, slash and bark) and include nutrient addition through fertilization provides the long-term option. In this case, there may be enough supplies of N, P, K, Ca and Mg to support more than seven rotations (more than 50 years), indicating a large effect of these practices on nutrient stocks. Fertilizer application is a practical and economically proven strategy for sustaining production in the long term.

4.2.8 Precision Forestry

Precision Forestry (PF) can be defined as the planning and conduction of the silviculture and management activities located on forest and of operations to improve the quality and forest products utilization, reduce losses, increasing profit and keep quality of the environment, which are based on prior knowledge that covering the spatial and temporal variability of the production factors and own productivity (Vettorazzi and Ferraz 2000; Taylor et al. 2002; Gonçalves and Alvares 2005). It is a great tool that aims to provide a silviculture customized for each unit area. It is a new model of forest plantation management, where the stands are treated geographically point to point, ie, the total area is divided into fractional units differentiated by a quality index site (Brandelero et al. 2007). The adoption of PF presupposes the use of information technologies such as the Global Navigation Satellite Systems (GNSS), geoprocessing techniques, database on Geographic Information System (GIS), statistics and geostatistics, remote sensing, in addition to use of machinery and implements for data collection or perform services for localized and varied-rate applications of inputs.

Such tools are becoming more available and applicable to different stages of fast growth forest plantation, and with proven gains in both the increment of wood, and in reducing the use of inputs. Although these initiatives are reduced coverage, these technologies are being increasingly sought after by large forestry companies.

PF may have applications in so much landscape level analyzes including several farms, such as applications within a single stand. Thus, GIS techniques with the use of GNSS, geostatistical and statistics have provided subsidies for the identification and correlation of variables that affect the forest productivity (Luu et al. 2013) by storing, processing, crossing and overlapping of GIS data. These data is viewed and managed together with digital maps of vegetation, topography, soil, forest sites productive capacity so on, which allows the forest engineer recognize and visualize in an integrated the distinct characteristics of production and preservation forest areas (Pallett 2005; Dalaqua 2010; Alvares and Gonçalves 2011).

The PF as well as the Precision Agriculture can have basically three forms of data collection to processing the information: visual or manual collection of information; embedded systems and remote sensing. The manual measurement or visual taking information occurs, for example, in operations of surveying mortality, notes to pests and diseases, weed infestation and nutrition. This way of PF although not use high technology is still valid, with high benefit/cost ratio, since the PF does not mean automation or informatization of the activities. The embedded systems comprise electronic devices installed in farm and forestry implements, which are different from remote sensing systems because they presented physical contact with the process (Molin 2001; Castro and Vieira 2008). Remote sensing has the advantage of covering large areas with relatively low cost (Aronoff 2005), through of digital image sensors on airborne, such as orthorectified aerial photography, videography (Alvares et al. 2010), Light Detection and Ranging (LiDAR) (Andersen et al. 2006; Moskal et al. 2009; Packalén et al. 2011) and satellite images with high temporal resolution, i.e. MODIS, and high spatial resolution as Ikonos, Geoeye and QuickBird (Wulder et al. 2005).

Major forestry companies have massive amounts of cadastral data base like wide cartographic base, aerial photos, satellite imagery, digital and printed maps, GPS tracking and points, field worksheets, database, and to control all these assets and supporting to decision-making is necessary have structured a geographic information system, which contributes to the creation of maps and reports to better understand the relationship of spatial variables through the processing of spatial information management and effective data storage. In the forestry sector, these systems are generally divided into three databases interrelated on a single geodatabase: (i) geographic (land use, altimetry, hydrograph, permanent preservation areas and legal reserves and the boundaries of properties), (ii) forestry cadastre and operations (climate, soils, inventory, nursery harvest, trials) and (iii) strategic and financial (budget, production orders, costs, materials, contracts, pricing, and customers) (Dalaqua 2010, 2011; Schoeninger 2011).

Thus, the PF has enabled the adoption of some management practices and streamlined other difficult to operationalization through compilation of organized information. There has achieved higher accuracy and technical and financial detail of recommendations silvicultural increasingly specific which has resulted in better adaptation to the plant environment, and reducing the occurrence of negative environmental impacts. At the same time, there is the optimization of operating earnings, the rational use of inputs and reduction of production costs. Higher the environmental contrasts, as in most regions of rugged relief, the greater the potential gains of technological capability and productivity (Gonçalves and Alvares 2005).

The biggest difficulty to work with this system is related to the large amount of information, usually derived from heterogeneous sources with large spatial and temporal variability, which creates the need for complex information systems to process the data. Moreover, successful cases are going on in many private companies worldwide. In these companies routinely use PF for planning and mechanized and automated activities.

PF techniques have been used for: (i) planning integrated management of areas for forest production through geospatialization of forest compartments and areas of environmental conservation and preservation, as well as for creating spatial models of forest productivity; (ii) precise application of inputs and practices for forestry quality control, namely the use of inputs at fixed or variable ratios (e.g. fertilizers, herbicides), and assessments of soil preparations practices (e.g. regularity of subsoiling depth), of weeds reinfestation after control practices, and of operational yields; (iii) geospatialization and monitoring of areas of adverse weather (susceptibility to droughts, frosts and fires) (Taylor et al. 2002; Gonçalves and Alvares 2005; Pallett 2005; Alvares and Gonçalves 2011). In general, the greater the physical contrasts, the higher the potential gains of technological and environmental capacity with the application of PF. Nevertheless, the use of PF techniques is still limited in implementation and management of forest plantations. One hindrance to use of these technologies is the lack of skilled professionals.

4.3 The Use of Physiological Growth Models as a Silvicultural Tool

Over the years, efforts have been made to understand, define and quantify the determinant processes of tree growth to simulate, predict and improve productivity of forest plantations. Accurate predictions of forest growth are necessary to estimate the productive potential of a particular forest, determining the effects of climate changes on productivity and enabling exploration of new areas for the expansion of forest planting (Mäkela et al. 2000; Sands et al. 2000; Almeida et al. 2010a; Grace and Basso 2012). There are different classification proposals to group the models according to similar characteristics. Since the purpose of this topic is to show practical aspects of modeling as a tool for silvicultural management, a simple and generalist classification is showed, based on two groups: empirical models and mechanistic models (Fig. 4.2) (Mäkela et al. 2000).

Empirical models (EM) based on inventory data and tree mensuration are used in most forestry companies and continue to be valuable tools for forest

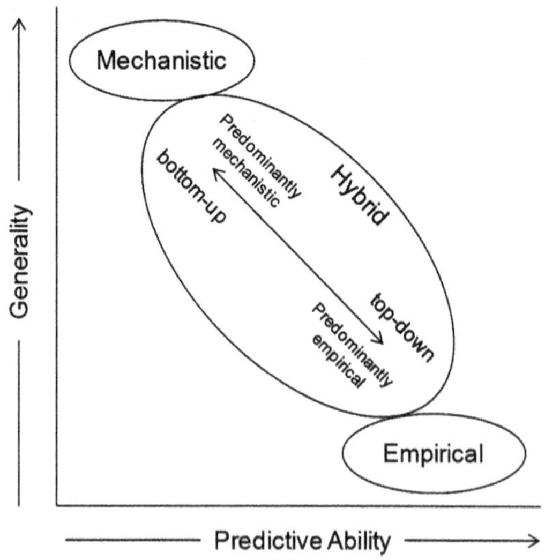

Fig. 4.2 The relations between model generality and predictive ability versus approach to model formulations (Adapted from Reed 1999)

management. However, they do not provide answers to some critical questions that arise from the planning process, especially regarding factors affected by climate and silviculture. These models are adjusted to local conditions, are site-specific, making extrapolations or simulations impossible for conditions other than those used in the productivity model because they are guided by a reductionist view. Therefore, issues related to forest sustainability, considering the temporal expansion or the regional impacts of management practices, may not be approached (Stape 2002; Alvares 2011).

The mechanistic models, known as process-based model (PBM), sometimes called ecophysiological models, simulate the tree growth patterns in terms of biophysical mechanisms, defining tree growth in response to environmental conditions and management practices. They have great capacity of extrapolation, although many times with less accuracy (Landsberg and Gower 1997; Johnsen et al. 2001). They have predictive, spatial and temporal capacity and constitute an important tool in analyzing the effects on different scenarios for a particular set of climate, soil, genetic material and management applications. They allow the optimization of some production factors, both from a standpoint of physical production and from an economic one, besides allowing expectations of environmental impacts in production process. However, PBM are more complex and tend to show more difficulties for adjustment when compared to empirical models. These models are mostly applied in research. Still, they need to be simplified to ensure practical use. The large number, the type of input variables as well as the necessity of careful parameterization are the main difficulties for their execution (Landsberg and Waring 1997; Battaglia and Sands 1998).

Hybrid models (HM) seek to ease restrictions imposed by the two previous categories combining the empirical approach with the based on processes (Kimmins et al. 1990; Landsberg and Sands 2011). Actually, there are no models totally empirical. The knowledge about physiological processes must always be taken into account. Moreover, there are no pure mechanistic models, since we do not have full knowledge about the different processes that determine forest productivity (Almeida et al. 2010a). In bottom-up models, closer to mechanistic models, growth is synthesized by calculating the actions and interactions of physiological processes that contribute to them. Top-down models, more empirical, are based on simplified formulations of main physiological growth processes driven by a control variable (e.g. radiant energy) and external influences (Landsberg 1986). Models that present behavior according to biological hypotheses of universal application have more generality (extrapolation capacity) and inspire greater confidence to the user (Fig. 4.2).

The current understanding of many processes is still too limited to allow the construction of models totally based on processes. Many of the ecophysiological models use a series of empirical or semi-empirical equations in their structures. This method is valid, but most of these models are hybrid models. They contain casual and empirical elements at the same hierarchical level for improving the predictions of models based on empirical processes. The estimated empirical relations from inventory data are used in hybrid models to compensate incomplete knowledge

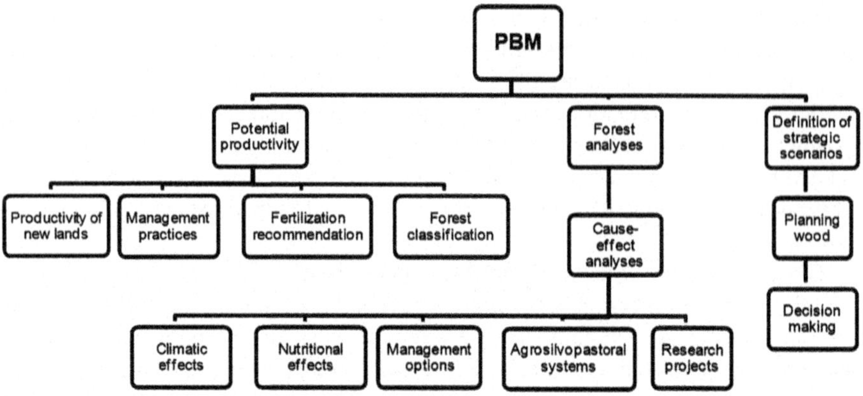

Fig. 4.3 Diagram showing the general areas of applications of process-based model (PBM) to fast-growing plantations for commercial purposes. There are three areas of applications, each of which breaks into specific applications (Adapted from Almeida et al. 2004)

about some mechanisms (e.g. carbon allocation, relation between growth and longevity) and the partial ability of predictive capacity (Mäkela et al. 2000). Hybrid models require fewer input variables and show lower restrictions to extrapolation of their results. In other words, due to the complexity of some processes involved in tree growth, these models bring advantages regarding the procedural approach and the necessary simplification (no less scientific) to processes not yet fully understood.

To sum up, a good model must combine mechanicalness (complexity) and empiricism (simplicity), require small quantity of information, which must be easy to obtain and have enough flexible structure to allow the introduction of new information or ideas as the scientific knowledge improves (Mäkela et al. 2000). Obviously, models may not be used to simulate the growth response of forest plantations to climate changes, if variables such as temperature, precipitation and other environmental variables are not used as input variables and/or directional in their structure (Levy et al. 2004; Grace and Basso 2012).

The use of ecophysiological models integrates multidisciplinary knowledge to describe forest growth based on the growth processes of plants (photosynthesis, respiration, transpiration, allocation, and nutrition and falling of leaves and branches) and the existing natural resources (availability light, water and nutrients). Forestry companies have great interest in their use to provide information about climate effects and silvicultural practices for wood production (Landsberg and Waring 1997, 2004; Stape 2002; Almeida, et al. 2004; Alvares 2011; Landsberg and Sands 2011). The ecophysiological models have spatial and temporal predictive capacity, when properly calibrated and validated, may provide reliable estimates that allow to refine and interpret complex research activities, discussing policy issues and developing forest management tools (Fig. 4.3). In the forestry sector, these models have been widely used : (i) to analyze forest production and study cause/effect relations of growth, allowing to answer how the climate change, especially rising temperatures and atmospheric CO_2 concentration, can affect productivity, (ii) to identify environmental factors that limit forest growth, (iii) to assess the impacts of intensive silvicultural practices on water dynamic at drainage basin of planted forest areas, (iv) to evaluate the long-term forest productivity and improve forest management to obtain greater productivity, (v) to develop strategic scenarios to plan in the long-term the products supply, (vi) to estimate the potential productivity to verify economic viability of investments in areas already planted, helping the decision making process to purchase new areas, (vii) to assess the risks of changes and variation of climate and external factors (e.g. droughts and diseases) on forest production and (viii) to determine how forest components interact and affect crop productivity and how livestock components integrate crop-livestock-forest (Agrosilvopastoral).

Choosing the model that will be used in the forest growth and productivity estimation depends on the available information for its calibration, validation and the level of details and accuracy required by the wielder (Taylor et al. 2009). In the last two decades, there has been considerable progress in the use of forest models based on processes underpinned by computer technology advances (Landsberg and Gower 1997; Almeida et al. 2004). Examples of ecophysiological models that may be used to predict forest productivity are LINKAGES, CENTURY, MAESTRO, BEX, BIOMASS, FOREST-BGC, TREGROW, PnET, G'DAY, 3-PG GOTILWA+, ProMOd, CABALA etc., which vary in complexity (number of parameters) and simulation steps (from time to annual). Another feature used is the coupling between models to achieve greater robustness and predictive capacity of hybrid models (Fontes et al. 2010), e.g. 3-PG + CENTURY (Zhao et al. 2009), 3-PG + FULLCAM (Richards and Evans 2004), PTAEDA2 + MAESTRO (Baldwin et al. 2001).

There are numerous reasons, undoubtedly important, to incorporate ecophysiological models to the largest number of forest management activities, particularly growth predictions, which are important parts in silvicultural planning. The full use of ecophysiological models as forest management tools still needs further development. Many interactions among species, soil and atmospheric processes are still poorly understood. Those factors include the role of below-ground biomass (the "hidden environment" of forests), the effect of nutrient availability, factors that affect decomposition of soil organic matter and specific answers to climate changes, factors namely the rise in atmospheric CO_2 concentration, drought periods and adaptation to these changes (Landsberg and Sands 2011). The fast development of research is constantly applied to improve the understanding and the capacity of producing robust models and accurate results (Fontes et al. 2010). Current efforts focus on the use of field data at large density, through deployment of flux tower networks, with high precision (e.g. using covariance eddy techniques) to validate the models to a vast range of local conditions and different structures of ecosystem. This high quality and quantity of available data will allow to highlight important processes that are not fully described today (Landsberg and Sands 2011). It is also essential to perform continuous validation of these models, the most effective way to increase credibility to the use of ecophysiological models.

Case Study – Sustainable Eucalypt Plantations in Brazil

Eucalypt plantations fulfill multiple functions in landscapes in different Brazilian ecosystems. The Brazilian experience has shown that despite repeated short-rotation cropping, continuous gains in productivity of eucalypts are possible. The increase rate has been steady for over 40 years, indicating the large-scale productivity gain through improved genotype and silviculture. Even so, there are a number of risks associated with intensive, short-rotation, high yielding eucalypt plantations. Those risks must be carefully assessed and managed. The search, test and selection of appropriate genotypes and site management practices are imperative to sustain productivity and maintain environmental services of these forests for generations to come.

Background

Organised forestry in Brazil began in the late 1960s, stimulated by a government policy that granted tax incentives which subsidised afforestation programs from 1967 to 1989 to develop an internationally-competitive wood-based industry, managed by the private sector. In the beginning, productivity was relatively low ranging from 10 to 30 m^3 ha^{-1} $year^{-1}$ (Ferreira 1992; Campinhos 1999). High rate of segregation caused by uncontrolled hybridizations, major changes in vigor and wood quality showed that the seed sources were not adequate. Consequently, the phenotypic variability of the most planted species, namely *E. grandis* (over 70 % of the planted area), *E. urophylla* and *E. saligna* were very large (Ferreira 1992). Seeking productivity increase and wood quality improvement, new species and new provenances of *Eucalyptus*, such as *E. grandis* (Coff's Harbour, Australia), and *E. urophylla* (Timor and Flores, Indonesia) were introduced. At that time, the ecological zoning of exotic species (Golfari 1974), the selections of superior site-specific provenances and progenies in addition to the establishment of areas of improved seed production played a critical role. Concurrently, in 1979, the first clonal plantations were established in the State of Espirito Santo (Ferreira 1992). All these advances led the area of eucalypt plantations to surpass that of pine plantations by the end of the 1970s.

The period between the 1980s and the 2000s was marked by the consolidation of the Brazilian forest sector, involving mainly breeding programs, productivity increases, expansion of cropped areas, diversification of the use of products, increase of competitiveness and concerns with social and environmental issues (ABRAF 2012). New eucalypt populations emerged from interspecific crosses, especially between *E. grandis* and *E. urophylla*

(continued)

(continued)

seeking segregation of individuals with high productivity and desirable wood quality (Lemos 2012). Superior individuals planted on commercial scale in various regions were selected by cloning techniques. In addition to forest improvement, there was significant advance of silvicultural practices, especially regarding production of high-quality seedlings, use of minimum tillage, integrated control of weeds, pests and diseases, judicious fertilization recommendation, and effective control of forest fires (Gonçalves et al. 2008). Thus, at the beginning of the 21st century, mean annual increments exceeded 30 m^3 ha^{-1} $year^{-1}$ roundwood.

Currently, planted forests in Brazil total about 6.9 million ha, from which 4.9 million ha is planted with eucalypt (around 25 % of world plantation), 1.6 million ha with pine, and 0.42 M ha with other species. Roundwood consumption of forest plantations totaled 170.1 million m^3 in 2011, eucalypt plantation accounted for 80.6 % of this total. Most eucalypt plantations are managed in short rotations (6–8 years) and are established in regions with water, nutritional and frost stresses of low to high degrees. The mean annual increment is 40 m^3 ha^{-1} $year^{-1}$ roundwood, ranging from 25 to 60 m^3 ha^{-1} $year^{-1}$ depending on the level of environmental stress (ABRAF 2012).

Silvicultural Practices

The wide range of eucalypt species and hybrids with different climatic and edaphic aptitudes associated with the easy propagation by seeds and cloning allow the adaptation of plantations to various tropical and subtropical regions in Brazil. The desirable characteristics in association with the accumulated knowledge on eucalypt silviculture encourage the use of this genus in most plantations even more. Under tropical conditions, in regions with pronounced seasonality and moderate to long drought periods, the planting of hybrid genotypes predominates, propagated by cloning (Table 4.1). Under subtropical conditions, the planting of single genotypes predominates, propagated by seed. Clonal plantations with interspecific hybrids have been fundamental to adapt eucalypt in regions with water and nutritional stresses. Clones from hybrids are more advantageous in relation to seed-originated seedlings of pure species, because they can combine silvicultural traits such as wood quality and site adaptability (Fonseca et al. 2010).

Genotypes propagated by cloning, due to the small genetic diversity, are little plastic relatively genotypes propagated through seeds. Hence, they have a higher risk of genotype-environment incompatibility (Gonçalves et al. 2013). Thus, the specific allocation of a genotype to a site should be well tested, based on field and laboratory tests. Given the rapid advances in

(continued)

(continued)

Table 4.1 Climate type, mean annual rainfall, temperature and actual evapotranspiration, dry season, the main species and hybrids recommended for planting and expected average productivity (roundwood with bark) of the planted areas with eucalypts

Type climate	Mean annual rainfall (mm year^{-1})	Mean annual temperature (°C)	Mean actual evapotranspiration (mm year^{-1})	Dry season: Number of months	Dry season: Water deficit[a] (mm year^{-1})	Species/hybrid[b]	Mean annual increment (m^3 ha^{-1} year^{-1})
Cfa, Cfb	1,500–2,500	13–20	500–1,000	0–2	0–50	EUG, Egr, Eur, Esa, Cci, Edu, Ebe, EUG	35–60
Cwb, Am	1,000–1,800	18–20	800–1,100	2–3	50–100	EUG, Egr, Eur	35–45
Cwa, Aw	1,000–1,800	20–24	1,000–1,200	3–4	100–200	EUG, Eur, EGC, EUC, Eca, Ete	35–45
Aw	1,100–2,000	24–26	1,100–1,500	4–6	200–400	EGC, EUC, ETB, Eca, Ete, Ebr, EUT	25–35
As, BSh	700–1,500	23–27	600–1,000	>6	>400	Planting is not feasible	

Source: Gonçalves et al. (2013)

[a]According to the soil water balance proposed by Thornthwaite and Matter (1955)

[b]*Egr E. grandis, Esa E. saligna, Eur E. urophylla, Cci Corymbia citriodora, Ete E. tereticornis, Ebr E. brassiana; Ebe E. benthamii, Edu E. dunnii, EGU Eucalyptus urophylla x grandis* (urograndis), *EGC E. grandis* x *camaldulensis* (grancam), *EUC E. urophylla* x *camaldulensis* nurocam), *ETB E. tereticornis* x *brassiana, EUG E. urophylla* x *globulus, EUT E. urophylla* x *tereticornis*

(continued)

(continued)

eucalypt breeding, regarding adaptation to water stress and resistance to pests and diseases, and the adoption of clonal propagation techniques, genotypes rapidly become obsolete and are replaced by more productive ones after harvesting.

Until the late 1970s, silvicultural practices regarding residue management and soil preparation followed a typically agronomical pattern, as windrowing and/or residues burning and intense turning of the topsoil. The need to reduce soil erosion, nutrient output and costs has led to a progressive increase in the use of minimum cultivation practices, including slash and litter retention, during the last decades in Brazil (Gonçalves et al. 2008). The minimum cultivation system prescribes the maintenance of plant residues (litter and harvest residues) on the soil, followed by soil preparation in planting rows or pits. Currently, it is estimated at over 85 % of the plantations are established in this system. The most commonly used implements in managed areas in minimal cultivation systems are the ripper (work depth >30 cm) and the pit digger. The latter is used on very steep slopes (heavily undulated and mountains) (Silva et al. 2002) or where there are many physical obstacles to using the ripper, such as in areas under intercropping with many thick stumps. On steep slopes higher than 30–35 % of slope (depending on the irregularity of the terrain), mechanical soil preparation is not feasible, thus it is restricted to the manual opening of planting holes (20 cm × 20 cm × 20 cm). The operational yields obtained with the ripper are higher than those using a pit digger (Gonçalves et al. 2008).

The distance between trees (spacing), or the number of trees planted per hectare (stocking), is one of the most important silvicultural decisions for the establishment of an eucalypt plantation. The choice of planting density depends on edaphoclimatic conditions of the site, requirements of the timber market and the many purposes of the plantation. Consumption and water use efficiency are greatly affected by tree spacing. In an experiment conducted in the municipality of Santa Bárbara, southern region in Brazil, Leite et al. (1999) assessed the internal rainfall, evapotranspiration and soil water status (Typical Clayey Oxisol) cropped with *Eucalyptus grandis* (32–38 months old), with stand densities ranging from 500 to 5,000 ha^{-1} plants. The canopy rainfall interception (18–21 %) increased linearly with higher stocking (due to higher LAI), while evapotranspiration rates were not affected. Soil moisture was higher with lower stocking, highlighting the important role spacing plays in the efficient use and management of soil water status. The results showed that consumption and water use efficiency are higher in wider spacing, mainly because of lower rainfall interception by the canopy in closer spacing, where water is lostby evaporation.

(continued)

(continued)

For many years, seedling plantations were established using stocking ranging from 2,200 to 1,600 trees ha^{-1} ($3.0 \times 1.5 - 2.0$ m spacing), which anticipated loss rates up to 10 % due to mortality and poor growth of some trees (Stape et al. 2001). In drier regions of Brazil, water availability can impose serious limitations to forest growth and reduce stocking. For instance, in some plantations in the Brazilian northeast, the initial densities of 1,600–2,500 trees ha^{-1} can decrease to 900–1,000 of living trees ha^{-1} at 7 years of age (Gonçalves et al. 1997). Currently, concerning the production of roundwood over 7 years of growth, it was concluded that there is not a point for ideal spacing, but a range. Within this range, there is compensation for the growth rate, without major changes in productivity (Gonçalves et al. 2013).

Recent studies have shown that fertilizer prescription in eucalypt plantations in Brazil are well calibrated and that the greatest limitation to additional productivity gains is related to water deficit. For example, in an experimental research network, Stape et al. (2010) examined the potential growth of clonal *Eucalyptus* plantations at eight locations in southeast and northeast Brazil by manipulating the supply of nutrients and water. With no fertilization or irrigation, mean annual increments of roundwood were about 28 % lower (33 m^3 ha^{-1} year^{-1}) than yields achieved with current operational rates of fertilization (46 m^3 ha^{-1} year^{-1}). Fertilization beyond current operational rates used by the companies did not increase growth, whereas irrigation raised growth by about 30 % (to about 62 m^3 ha^{-1} year^{-1}). Gonçalves et al. (2008) proposed classes of expected responses and recommendations for N-P-K fertilization of eucalypts based on soil organic matter and clay contents, resin-extractable P, and exchangeable K. The recommendations for N are 60, 40 and 20 kg ha^{-1} depending on whether soil organic matter concentrations are 0–20, 21–50 and >50 g kg^{-1}, respectively. The authors considered that organic matter and clay contents, besides relating to the availability of N, P and K, directly affect potential productivity through water status. The extreme rates of N, P and K are of 60, 70 and 120 kg ha^{-1}, respectively. Normally, in average rotations of 7 years, up to 2 Mg ha^{-1} of lime, 60–80 kg ha^{-1} of N, 60–80 kg ha^{-1} of P_2O_5, 140–160 kg ha^{-1} of K_2O, 1–5 kg ha^{-1} of B are applied, depending on local water deficit, and 1 kg ha^{-1} of Cu and Zn. Fertilizers are applied in synchrony with plant growth and, therefore, the nutrients are rapidly uptake.

4.4 Problems

The role of forest plantations in relation to the sustainable management of natural resources remains a contentious issue in many parts of the world. That is because regardless of a significant advance in scientific understanding of management and resource conservation interactions based on centuries of research, the expansion of the forest plantations will occur to a large extent in degraded areas and in areas susceptible to high degrees of water, thermal and salt stresses, thus more susceptible to biotical stresses. Uncertainty, and in some cases confusion, persists because of difficulties sometimes in translating research findings between countries and regions, between different scales of planting, between different species/provenances, and between different forest management regimes. There has also been a failure to effectively communicate results to policy makers and planners and to challenge entrenched views.

Questions that rise up:

- Can forest plantations be grown indefinitely for rotation after rotation on the same site keeping the same or increasing yields without serious risk to their well being and health?
- Is there a technology that can work in the long term or are there inherent flaws biologically which will eventually lead to insuperable problems for such silviculture?
- What is the optimal level of productivity for a given site: one in which the economic return is optimized or one in which ecosystem services are optimized? Or are they equivalent?

References

ABRAF – Brazilian Association of Forest Plantation Producers (2012) ABRAF statistical year book – base year 2011. Brasília, p 150

Almeida AC, Landsberg JJ, Sands PJ, Ambrogi MS, Fonseca S, Barddal SM, Bertolucci FL (2004) Needs and opportunities for using a process-based productivity model as a practical tool in Eucalyptus plantations. For Ecol Manag 193:167–177

Almeida AC, Siggins A, Batista TR, Beadle C, Fonseca S, Loos R (2010a) Mapping the effect of spatial and temporal variation in climate and soils on Eucalyptus plantation production with 3-PG, a process-based growth model. For Ecol Manag 259:1730–1740

Almeida JCR, Laclau JP, Gonçalves JLM, Ranger J, Saint-André L (2010b) A positive growth response to NaCl applications in *Eucalyptus* plantations established on K-deficient soils. For Ecol Manag 259:786–1795

Alvares CA (2011) Mapping and edaphoclimatic modeling of productivity of eucalyptus plantations in the south of São Paulo State. PhD thesis, University of Sao Paulo, Piracicaba, Brazil, 279 pp

Alvares CA, Gonçalves JLM (2011) Geotecnologia aplicada à silvicultura de precisão e aos modelos ecofisiológicos. In: Encontro Brasileiro de Silvicultura, 2. Campinas, São Paulo State, Brazil. Proceedings… Piracicaba: PTSM/IPEF/ESALQ/FUPEF, 2011, pp 3–8

Alvares CA, Campoe OC, Stape JL, Polizel JL, Nouvellon Y, Laclau JP, Silva Filho DF, Foltran E (2010) Estimating biometric variables at the end of the rotation of *Eucalyptus* stands using high resolution multispectral aerial images. In: Seminário de Atualização em Sensoriamento Remoto e Sistemas de Informações Geográficas Aplicados à Engenharia Florestal, 9. Curitiba, Paraná State, Brazil. Proceedings… Curitiba: UFPR, 2010, CD-ROM

Andersen H, Reutebuch SE, McGaughey RJ (2006) A rigorous assessment of tree height measurements obtained using airborne lidar and conventional field methods. Can J Remote Sens 32:355–366

Aronoff S (2005) Remote sensing in agriculture. In: Aronoff S (ed) Remote sensing for GIS managers. ESRI Press, Redlands, pp 338–350

Baldwin VC, Burkhart HE, Westfall JA, Peterson KD (2001) Linking growth and yield and process models to estimate impact of environmental changes on growth of loblolly pine. For Sci 47:77–82

Barros NF, Neves JCL, Novais OF (2004) Mineral fertilizer recommendations for eucalypt plantations. In: Gonçalves JLM, Benedetti V (eds) Forest nutrition and fertilization. IPEF, Piracicaba, pp 269–305

Battaglia M, Sands PJ (1998) Process-based forest productiv-ity models and their application in forest management. For Ecol Manag 102:13–32

Beets PN, Terry T, Manz J (1994) Management systems for sustainable productivity. In: Dyck WJ, Cole DW, Comerford NB (eds) Impacts of forest harvesting on long-term site productivity. Chapman & Hall, London, pp 219–246

Binkley D, Giardina C (1997) Nitrogen fixation in tropical forest plantations. In: Nambiar EKS, Brown AG (eds) Management of soil, water and nutrients in tropical plantation forests, Australian Centre for International Agricultural Research (ACIAR), monograph 43. Australian Centre for International Agricultural Research (ACIAR), Canberra, pp 297–337

Binkley D, Stape JL, Ryan MG (2004) Thinking about efficiency of resource use in forests. For Ecol Manag 193:5–16

Bouillet JP, Nzila JP, Laclau JP, Ranger J (2000) Effects of site management on *Eucalyptus* plantations on the equatorial zone, on the coastal plains of the Congo. In: Nambiar EKS, Cossalter C, Tiarks A, Ranger J (eds) Site management and productivity in tropical plantation forests: workshop proceedings, 7–11 December 1999, Kerala, India. CIFOR, Bogor

Bouillet J-P, Laclau J-P, Arnaud M, Thongo A (2002) Changes with age in the spatial distribution of roots of a *Eucalyptus* clone in the Congo. Impact on water and nutrient uptake ability. For Ecol Manag 171:43–57

Bouillet J-P, Laclau J-P, Gonçalves JLM, Voigtlaender M, Gava JL, Leite FP, Hakamada R, Mareschal L, Mabiala A, Tardy F, Levillain J, Deleporte P, Epron D, Nouvellon Y (2013) *Eucalyptus* and *Acacia* tree growth over entire rotation in single- and mixed-species plantations across five sites in Brazil and Congo. For Ecol Manag 301:89–101

Boyle TJB, Cossalter C, Griffin AR (1997) Genetic resources for plantation forestry. In: Nambiar EKS, Brown AG (eds) Management of soil, water and nutrients in tropical plantation forests, Australian Centre for International Agricultural Research (ACIAR), monograph 43. Australian Centre for International Agricultural Research (ACIAR), Canberra, pp 25–63

Brandelero C, Antunes MUF, Giotto E (2007) Silvicultura de precisão: nova tecnologia para o desenvolvimento florestal. Ambiência 3:269–281

Brown AG, Nambiar EKS, Cossalter C (1997) Plantations for the tropics – their role, extent and nature. In: Nambiar EKS, Brown AG (eds) Management of soil, water and nutrients in tropical plantation forests, Australian Centre for International Agricultural Research (ACIAR), monograph 43. Australian Centre for International Agricultural Research (ACIAR), Canberra, pp 1–23

Campinhos E (1999) Sustainable plantations of high-yield Eucalyptus trees for production of fiber: the Aracruz case. New For 17:129–143

Castro BFD, Vieira GR (2008) Panorama da silvicultura de precisão no Brasil. In: Encontro Brasileiro de Silvicultura, 1. Curitiba, Parana State, Brazil. Proceedings... Piracicaba: PTSM//IPEF//ESALQ//FUPEF, 2008, pp 125–134

Cromer RN, Smethurst P, Turnbull C, Misra R, LaSala A, Herbert A, Dimsey L (1995) Early growth of *Eucalyptus* in Tasmania in relation to nutrition. In: Potts BM, Borralho NMG, Reid JB, Cromer RN, Tibbits WN, Raymond CA (eds) Eucalypt plantations: improving fibre yield and quality. CRC for Temperate Hardwood Forestry, Hobart, pp 331–335

Dalaqua RM (2010) Improving sustainability planning in Brazilian eucalyptus forests. In: Eredics P (ed) Mapping forestry. ESRI Press, Redlands, pp 73–76

Dalaqua RM (2011) GIS na Suzano Celulose e Papel. Encontro de Usuários Esri Brasil, 2. São Paulo, São Paulo State, Brazil. Proceedings... ESRI and Imagem: www.euesri.com.br/2011

du Toit B, Esprey LJ, Job RA, Fuller GF, Dovey SD (2000) Effects of site management in *Eucalyptus grandis* plantations in South Africa. In: Nambiar EKS, Cossalter C, Tiarks A, Ranger J (eds) Site management and productivity in tropical plantation forests: workshop proceedings 21–30 December 1999, Kerala, India. CIFOR, Bogor, pp 11–20

Evans J, Turnbull J (2010) Plantation forestry in the tropics, 3rd edn. Oxford University Press, New York, 467 p

FAO (2006) Global planted forests thematic study: results and analysis, by A. Del Lungo, J. Ball and J. Carle. Planted forests and trees working paper 38. Rome, Italy, 159p. Also available at www.fao.org/forestry/site/10368/en

FAO (2011) State of the world's forests 2011. Food and Agriculture Organization of the United Nations, Rome, 165p

Ferreira M (1992) Melhoramento e a silvicultura intensiva clonal. IPEF 45:22–30

Fölster H, Khanna PK (1997) Dynamics of nutrient supply in plantation soils. In: Nambiar EKS, Brown AG (eds) Management of soil, water and nutrients in tropical plantation forests, Australian Centre for International Agricultural Research (ACIAR), monograph 43. Australian Centre for International Agricultural Research (ACIAR), Canberra, pp 339–378

Fonseca SM, Resende MDV, Alfenas AC, Guimarães LMS, Assis TF, Grattaplaglia D (2010) Manual Prático de Melhoramento Genético do Eucalipto. Editora UFV, Viçosa

Fontes F, Bontemps J-D, Bugmann H, Van Oijen M, Gracia C, Kramer K, Lindner M, Rötzer T, Skovsgaardl JP (2010) Models for supporting forest management in a changing environment. For Syst 19:8–29

Gonçalves JLM (2011) Fertilização de plantação de eucalipto. In: Encontro Brasileiro de Silvicultura, 2. Campinas, São Paulo State, Brazil. Proceedings... Piracicaba: PTSM/IPEF/ESALQ/FUPEF, 2011, pp 85–113

Gonçalves JLM, Alvares CA (2005) A Silvicultura de Precisão e as exigências ambientais. Visão Agríc 4:80–82

Gonçalves JLM, Barros NF (1999) Improvement of site productivity for short-rotation plantations in Brazil. Bosque 20:89–106

Gonçalves JLM, Serrano MIPS, Mendes KCFS, Gava JL (2000a) Effects of site management in a Eucalyptus grandis plantation in the humid tropics: São Paulo, Brazil. In: Nambiar EKS, Cossalter C, Tiarks A, Ranger J. (eds) Site management and productivity in tropical plantation forests: workshop proceedings, 7–11 December 1999, Kerala, India. CIFOR, Bogor, pp 3–9

Gonçalves JLM, Barros NF, Nambiar EKS, Novais RF (1997) Soil and stand management for short-rotation plantations. In: Nambiar EKS, Brown AG (eds) Management of soil, nutrients and water in tropical plantations forests, ACIAR monograph 43. ACIAR, Canberra, pp 379–417

Gonçalves JLM, Stape JL, Benedetti V, Fessel VAG, Gava JL (2000b) Reflexos do cultivo mínimo e intensivo do solo em sua fertilidade e nutrição das árvores. In: Gonçalves JLM, Benedetti V (eds) Nutrição e fertilização florestal. IPEF, Piracicaba, pp 1–57

Gonçalves JLM, Stape JL, Wichert MCP, Gava JL (2002) Manejo de resíduos vegetais e preparo de solo. In: Gonçalves JLM, Stape JL (eds) Conservação e cultivo de solos para plantações florestais. IPEF, Piracicaba, pp 131–204

Gonçalves JLM, Stape JL, Laclau JP, Smethurst P, Gava JL (2004) Silvicultural effects on the productivity and wood quality of eucalypt plantations. For Ecol Manag 193:45–61

Gonçalves JLM, Stape JL, Laclau JL, Bouillet JP, Ranger J (2008) Assessing the effects of early silvicultural management on long-term site productivity of fast-growing eucalypt plantations: the Brazilian experience. South For 70:105–118

Gonçalves JLM, Alvares CA, Higa AR, SILVA LD, Alfenas AC, Stahl J, Ferraz SFB, Lima WP, Brancalion PHS, Hubner A, Bouillet JPD, Laclau JP, Nouvellon Y, Epron D (2013) Integrating genetic and silvicultural strategies to minimize abiotic and biotic constraints in Brazilian eucalypt plantations. For Ecol Manag 301:6–27

Grace PR, Basso B (2012) Offsetting greenhouse gas emissions through biological carbon sequestration in North Eastern Australia. Agric Syst 105:1–6

Grove TS, Thomson BD, Malajczuk N (1996) Nutrition physiology of eucalypts: uptake, distribution and utilization. In: Attiwill PM, Adams MA (eds) Nutrition of eucalypts. CSIRO, Australia, pp 77–108

Herbert MA, Schonau APG (1989) Fertilising commercial forest species in Southern Africa: research progress and problems (part 1). S Afr For J 151:58–70

Johnsen K, Samuelson L, Teskey R, Mcnulty S, Fox T (2001) Process models as tools in forestry research and management. For Sci 47:2–8

Kimmins JP, Comeau PG, Kurz W (1990) Modelling the interactions between moisture and nutrients in the control of forest growth. For Ecol Manag 30:361–379

Laclau J-P, Deleporte P, Ranger J, Bouillet J-P, Kazotti G (2003) Nutrient dynamics throughout the rotation of *Eucalyptus* clonal stands in Congo. Ann Bot 91:879–892

Laclau J-P, Ranger J, Deleporte P, Nouvellon Y, Saint-Andre' L, Marlet S, Bouillet J-P (2005) Nutrient cycling in a clonal stand of *Eucalyptus* and an adjacent savanna ecosystem in Congo 3. Input–output budgets and consequences for the sustainability of the plantations. For Ecol Manag 210:375–391

Laclau J-P, Almeida JCR, Gonçalves JLM, Saint-Andre L, Ventura M, Ranger J, Moreira RM, Nouvellon Y (2009) Influence of nitrogen and potassium fertilization on leaf life span and allocation of above-ground growth in *Eucalyptus* plantations. Tree Physiol 29:111–124

Laclau J-P, Ranger J, Gonçalves JLM, Maquere V, Krusche AV, M'Bou AT, Nouvellon Y, Saint-Andre' L, Bouillet J-P, Piccolo CM, Deleporte P (2010) Biogeochemical cycles of nutrients in tropical *Eucalyptus* plantations: main features shown by intensive monitoring in Congo and Brazil. For Ecol Manag 259:1771–1785

Landsberg JJ (1986) Physiological ecology of forest production. Academic, Sydney, p 198

Landsberg JJ, Gower ST (1997) Aplications of physiological ecology to forest management. Academic, San Diego, 354p

Landsberg J, Sands P (2011) Physiological ecology of forest production: principles, processes and models. Elsevier, Amsterdam

Landsberg JJ, Waring RH (1997) A generalized model of forest productivity using simplified conceps of radiation-use efficiency, carbon balance and partitioning. For Ecol Manag 95:209–228

Landsberg JJ, Waring RH (2004) Top-down models and flux measurements are complementary methods of estimating carbon sequestration by forest: illustrations using the 3-PG. In: Mencuccini M, Moncrieff J, Mcnaugton K, Grace J (eds) Forest at land atmosphere interface. CABI Publishing, Cambridge, MA, pp 37–50

Leite FP, Barros NF, Novais RF, Sans LMA, Fabres AS (1999) Relações hídricas em povoamento de eucalipto com diferentes densidades populacionais. Rev Brasil Cienc Solo 23:9–16

Lemos CCZ (2012) Aprimoramentos, teste e uso do modelo 3-PG em plantios clonais de Eucalyptus no nordeste do estado de São Paulo. Ph.D. thesis, University of Sao Paulo, Piracicaba, Brazil, 171 pp

Levy PE, Cannell MGR, Friend AD (2004) Modelling the impact of future changes in climate, CO_2 concentration and land use on natural ecosystems and the terrestrial carbon sink. Global Environ Chang 14:21–30

Lima WP, Jarvis P, Rhizopoulou S (2003) Stomatal responses of *Eucalyptus* species to elevated CO_2 concentration and drought stress. Sci Agric 60:231–238

Lima WP, Laprovitera R, Ferraz SFB, Rodrigues CB, Silva MM (2012). Forest plantations and water consumption: a strategy for hydrosolidarity. Int J For Res 2012, Article ID 908465: 8

Luu TC, Binkley D, Stape JL (2013) Neighborhood uniformity increases growth of individual *Eucalyptus* trees. For Ecol Manag 289:90–97

Mäkela A, Landsberg J, Ek AR, Burk TE, Ter-Mikaelian M, Agren GI, Oliver CD, Puttonen P (2000) Process-based models for forest ecosystem management: current state of the art and challenges for practical implementation. Tree Physiol 20:289–298

Maquère V, Laclau J-P, Gonçalves JLM, Krushe AV, Piccolo MC, Gine MF, Ranger J (2005) Influence of fertilizer inputs on soil solution chemistry in eucalypt plantations established on Brazilian sandy soils. In: Proceedings of the workshop on management of tropical sandy soils for sustainable agriculture: a holistic approach for sustainable development of problem soils in the tropics, Khon Kaen, Thailand. International Union of Soil Science, Khon Kaen, pp 466–471

Mello SLM, Gonçalves JLM, Gava JL (2007) Pre- and post-harvest fine root growth in *Eucalyptus grandis* stands installed in sandy and loamy soils. For Ecol Manag 246:186–195

Molin JP (2001) Precision agriculture – the management of variability. ESALQ/USP, Piracicaba, 83p

Moskal LM, Erdody T, Kato A, Richardson J, Zheng G, Briggs D (2009) LiDAR application in precision forestry. In: Popescu S, Nelson R, Zhao K, Neuenschwander A (eds) Silvilaser, 9. College Station, Texas, USA. Proceedings... (CD-ROM) College Station, October 2009

Murach D, Ruhiyat D, Iskandar E, Schulte A (1998) Fine root inventories in dipterocarp forest and plantations in East Kalimantan, Indonesia. In: Schulte A, Ruhiyat D (eds) Soils of tropical forest ecosystems – characteristics, ecology and management. Springer, Berlin, pp 186–191

Nambiar EKS (1996) Sustained productivity of plantation forests is a continuing challenge to tree improvement. In: Dieters MJ, Matheson DG, Harwood CE, Walker SM (eds) Tree improvement for sustainable tropical forestry. Proceedings QFRI-IUFRO conference, Caloundra, QLD, Australia, 27 October–1 November 1996. Queensland Forestry Research Institute, Gympie, pp 6–18.

Nambiar EKS, Kallio MH (2008) Increasing and sustaining productivity in tropical forest plantations: making a difference through cooperative research partnership. In: Nambiar EKS (ed) Site management and productivity in tropical plantation forests: workshop proceedings, 22–26 November 2004, Piracicaba, Brazil, and 6–9 November, Bogor, Indonesia. Center for International Forestry Research, Bogor, Indonesia, pp 205–227

Ngeh P, Furtey PA, Grace J (1995) Effect of land clearing methods on soil physical and chemical properties in a lowland tropical forest at Mbalmayo, Cameroon. In: Proceedings international congress on soils of tropical forest ecosystems. 29 October–3 November 1995, Balikpapan, Indonesia, vol 2. Mulawarman University Press. Samarinda, pp 103–122

O'Connell AM, Sankaran KV (1997) Organic matter accretion, decomposition and mineralisation. In: Nambiar EKS, Brown AG (eds) Management of soil, water and nutrients in tropical plantation forests, Australian Centre for International Agricultural Research (ACIAR), monograph 43. Australian Centre for International Agricultural Research (ACIAR), Canberra, pp 443–480

O'Connell AM, Grove TS, Mendham D, Rance SJ (2000) Effects of site management in eucalypt plantations in Southwestern Australia. In: Nambiar EKS, Cossalter C, Tiarks A, Ranger J (eds) Site management and productivity in tropical plantation forests: workshop proceedings 61–71 December 1999, Kerala, India. CIFOR, Bogor, pp 11–20

Packalén P, Mehtätalo L, Maltamo M (2011) ALS-based estimation of plot volume and site index in a eucalyptus plantation with a nonlinear mixed-effect model that accounts for the clone effect. Ann For Sci 68:1085–1092

Pallett RN (2005) Precision forestry for pulpwood re-establishment silviculture. S Afr For J 203:33–40

Paustian K, Andrén O, Clarholm M, Hansson AC, Johansson G, LagerloÈf JK, Lindberg T, Pettersson R, Sohlenius B (1990) Carbon and nitrogen budgets of four agro-ecosystems with annual and perennial crops, with and without N fertilization. J Appl Ecol 27:60–84

Pitelli RA, Marchi SR (1991) Interferência das plantas invasoras nas áreas de reflorestamento. In: Seminário técnico sobre plantas daninhas e o uso de herbicidas em reflorestamento, 3, 1991, Belo Horizonte, Minas Gerais State, Brazil. Proceedings... Belo Horizonte: 1991, pp 1–11

Reed DD (1999) Ecophysiological models of forest growth: uses and limitations. In: Amaro A, Tomé M (eds) Empirical and process-based models for forest tree and stand growth simulation. Edições Salamandra Novas Tecnologias, Lisboa, pp 305–311

Richards GP, Evans DMW (2004) Development of a carbon accounting model (FullCAM Vers. 1.0) for the Australian continent. Aust For 67:277–283

Salgado TP, Alves PLCA, Kuva MA, Takahashi EN, Dias TCS, Lemes LN (2011) Sintomas de intoxicação Inicial de *Eucalyptus* proporcionados por subdoses de glyphosate aplicadas no caule ou nas folhas. Planta Daninha 29:913–922

Sands PJ, Battaglia M, Mummery D (2000) Application of process-based models to forest management: experience with PROMOD, a simple plantation productivity model. Tree Physiol 20:383–392

Sankaran KV, Chacko KC, Pandalai RC, Kallarackal J, Somen CK, Sharma JK, Balagopalan M, Balasundaruan M, Kumaraswamy S, Sankar S, Gilbes RJ, Grove TS, Mendham D, O'Connel AM (2000) Effects of site management on eucalypts plantations in the monsoonal tropics – Kerala, India. In: Nambiar EKS, Cossalter C, Tiarks A, Ranger J (eds) Site management and productivity in tropical plantation forests: workshop proceedings 51–60 December 1999, Kerala, India. CIFOR, Bogor, pp 11–20

Sasaki CM, Gonçalves JLM, Silva AP (2007) Ideal subsoiling moisture content of Latosols used in forest plantations. For Ecol Manag 243:75–82

Schoeninger ER (2011) Geoprocessamento na Klabin. Encontro de Usuários Esri Brasil, 2. São Paulo, São Paulo State, Brazil. Proceedings... ESRI and Imagem: www.euesri.com.br/2011

Silva CR, Gonçalves JLM, Folegatti BS, Stape JL, Gava JL (1997) Weed proliferation in eucalypt stands established under minimum and intensive site preparation. In: IUFRO conference on silviculture and improvement of eucalyptus. Salvador, Bahia State, Brazil. Proceedings... Colombo, EMBRAPA-CNPF, pp 234–241

Silva CR, Camargo FRA, Jacob WS, Iijima J (2002) Preparo de solo em áreas acidentadas do Vale do Paraíba do Sul – SP. In: Gonçalves JLM, Stape JL (eds) Conservação e cultivo de solos para plantações florestais. IPEF, Piracicaba, pp 245–258

Silva W, Silva AA, Sediyama T, Freitas RS (2000) Absorção de nutrientes por mudas de duas espécies de eucalipto em resposta a diferentes teores de água no solo e competição com plantas de *Brachiaria brizantha*. Cienc Agrotec 24:147–159

Souza AJ (2002) Preparo de solos coesos para cultura do eucalipto no extremo sul da Bahia. In: Gonçalves JLM, STAPE JL (eds) Conservação e cultivo de solos para plantações florestais. IPEF, Piracicaba, pp 297–311

Stape JL (2002) Production ecology of clonal Eucalyptus plantations in northeastern Brazil. PhD thesis, Colorado State University, Ft. Collins, USA, 225 pp

Stape JL, Gonçalves JLM, Gonçalves AN (2001) Relationships between nursery practices and field performance for Eucalyptus plantations in Brazil. New For 22:19–41

Stape JL, Andrade S, Gomes AN, Krejci LC, Ribeiro JA (2002) Definição de métodos de preparo de solo para silvicultura em solos coesos do litoral norte da Bahia. In: Gonçalves JLM, Stape JL (eds) Conservação e cultivo de solos para plantações florestais. IPEF, Piracicaba, pp 259–296

Stape JL, Binkley D, Ryan MG (2004) *Eucalyptus* production and the supply, use and efficiency of use of water, light and nitrogen across a geographic gradient in Brazil. For Ecol Manag 193:17–31

Stape JL, Binkley D, Ryan MG, Fonseca SRL, Takahashi EN, Silva CR, Hakamada SR, Ferreira JM, Lima AM, Gava JL, Leite FP, Silva G, Andrade H, Alves JM (2010) The Brazil Eucalyptus potential productivity project: influence of water, nutrients and stand uniformity on wood production. For Ecol Manage 259:1684–1694

Tarouco CP, Agostinetto D, Panozzo LE, Santos LS, Vignolo KG, Ramos LOO (2009) Períodos de interferência de plantas daninhas na fase inicial de crescimento do eucalipto. Pesq Agropec Bras 44:1131–1137

Taylor SE, Veal MW, Grift TE, Mcdonald TP, Corley FW (2002) Precision forestry: operational tactics for today and tomorrow. In: International meeting of the Council on Forest Engineering, 23, Corvallis. Oregon State University Press, Corvallis

Taylor AR, Chen HYH, Vandamme L (2009) A review of forest succession model and their suitability for forest management planning. For Sci 55:23–36

Thornthwaite CW, Mather JR (1955) The water balance. Publications in climatology. In: Centerton NJ (ed). Drexel Institute of Technology, Laboratory of Climatology, 86

Tiarks A, Ranger J (2008). Soil properties in tropical plantation forests: evaluation and effects of site management: a summary. In: Nambiar EKS (ed) Site management and productivity in tropical plantation forests: workshop proceedings, 22–26 November 2004, Piracicaba, Brazil, and 6–9 November, Bogor, Indonesia. Center for International Forestry Research, Bogor, Indonesia, pp 191–204

Toledo REB, Victoria Filho R, Pitelli RA, Alves PLCA, Lopes MAF (2000) Efeito de períodos de controle de plantas daninhas sobre o desenvolvimento inicial de plantas de eucalipto. Planta Daninha 18:395–404

Toledo REB, Victoria Filho R, Alves PLCA, Pitelli RA, Lopes MAF (2003) Faixas de controle de plantas daninhas e seus reflexos no crescimento de plantas de eucalipto. Sci For 64:78–92

Versini A, Nouvellon Y, Laclau J-P, Kinana A, Ranger J, Zeller B, Mareschal L, Epron D (2012) The manipulation of organic residues affects tree growth and heterotrophic CO_2 efflux in a tropical *Eucalyptus* plantation. For Ecol Manag 301:79–88

Vettorazzi CA, Ferraz SFB (2000) Silvicultura de precisão: uma nova perspectiva para o gerenciamento de atividades florestais. In: Borém A, Giudice MP, Mantovani EC, Queiroz DM, Ferreira LR, Valle FXR, Gomide RL (eds) Agricultura de Precisão. UFV, Viçosa, pp 65–75

Wenzel WW, Unterfrauner H, Schulte A, Ruhiyat D, Simorangkir D, Kuraz V, Blum WEH (1995) Physical properties of soils under different forest land use in East Kalimantan, Indonesia. In: Proceedings international congress on soils of tropical forest ecosystems, 29 October–3 November 1995, Balikpapan, Indonesia, vol 1. Mulawarman University Press, Samarinda, pp 135–153

Wulder MA, Hall RJ, Franklin SE (2005) Remote sensing and GIS in forestry. In: Aronoff S (ed) Remote sensing for GIS managers. ESRI Press, Redlands, pp 351–362

Xu DP, Dell B (2002) Nutrient management of eucalypt plantations in South China. In: International symposium on eucalypt plantations. Presentation papers & Abstracts. 139–159 September 2002, Guangdong, P.R. China. RITF, Guangdong, pp 1–6

Xu DP, Yang ZJ, Dell B, Gong M (2000) Effects of site management in Eucalyptus urophylla plantations in Guangdong province, China. In: Nambiar EKS, Cossalter C, Tiarks A, Ranger J (eds) Site management and productivity in tropical plantation forests: workshop proceedings 31–40 December 1999, Kerala, India. CIFOR, Bogor, pp 11–20

Zen S (1987) Influência da matocompetição em plantas de *Eucalyptus grandis*. IPEF Série Téc 12:25–35

Zhao M, Xiang W, Peng C, Tian D (2009) Simulating age-related changes storage and allocation in Chinese fir plantation growing in southern China using 3-PG model. For Ecol Manag 257:1520–1531

Chapter 5
Economics and Management of Industrial Forest Plantations

Luis Diaz-Balteiro, Carlos Romero, Luiz C.E. Rodriguez, Silvana Ribeiro Nobre, and José G. Borges

5.1 Introduction

When planning a book devoted to the management of industrial forest plantations, it is indispensable to incorporate into it the main economic analysis tools which should be applied in any management plan for these forest systems. This chapter focuses on these issues, proposing them from the perspective of a decision-maker which aims to obtain information either on the profitability of its investments, on the optimal rotation which should be associated to these plantations, or on how to apply some tools based on operational research which could help it in its decision making. That is why Chap. 5 presents some obvious links both to Chap. 2, and, essentially, to Chap. 6, since some of the results obtained with the methodologies which will be developed will be constituted as inputs of the models set up in subsequent chapters.

Moreover, this economic view should be incorporated into all the stages associated with the life of a forest plantation, even before the latter has been initiated. Indeed, and not wanting to be exhaustive, the decision-maker has to assess if or not it wishes to undertake this type of investment. It should also ask itself where

L. Diaz-Balteiro (✉)
Departamento de Economía y Gestión Forestal, ETS Ingenieros de Montes, Technical University of Madrid, Madrid, Spain
e-mail: luis.diaz.balteiro@upm.es

C. Romero
ETS Ingenieros de Montes, Technical University of Madrid, Madrid, Spain

L.C.E. Rodriguez
Department of Forest Sciences, University of São Paulo, São Paulo, Brazil

S.R. Nobre
Department of Forest Management, Atrium Forest Consulting, São Paulo, Brazil

J.G. Borges
Forest Research Centre, School of Agriculture, University of Lisbon, Lisbon, Portugal

J.G. Borges et al. (eds.), *The Management of Industrial Forest Plantations: Theoretical Foundations and Applications*, Managing Forest Ecosystems 33,
DOI 10.1007/978-94-017-8899-1_5,

this should be done, given that it must seek the place most favourable to investors. This location need not be selected only for technical criteria (i.e., greater growth of forest stands). Following this reasoning, the decision-makers of this afforestation project should ask themselves if it is better to buy the land or, on the contrary, to rent it. If these initial steps have been fulfilled, then the species or clone used should be chosen not only as a function of its technical characteristics but also in accordance with a minimum of economic parameters. The next decision to be made is if to begin the plantation all over the land at the same time, or to programme it over the time. It will also be important to optimize the investments to be made throughout the life of the plantation, as well as to define, a priori, the most appropriate rotation age, into which economic criteria should also be incorporated. In this latter decision, the outputs which it is expected to obtain from this investment are also included. In short, this chapter aims to help readers to acquire a good level of knowledge on a set of basic economic tools, permitting them to answer the questions posed above.

Before beginning to set out the basic principles of the investment analysis applied to these forest systems, it is necessary to indicate the main hypotheses underlying throughout this chapter. Thus, firstly, it should be pointed out that the plantations dealt with in this book are, in most of the cases, of a private and not public nature. Namely, in accordance with what was proposed in Chap. 1, we shall focus on plantations which are the property of industries, of private non-industrial owners (NIPF), of communities of owners, of institutional investors, etc., but not on plantations which can be financed with public funds, unless the latter are exclusively with timber outputs. Therefore, plantations which, for example, have as an objective prevention against erosion, the fixation of riverbeds, the enrichment of forest systems for landscape reasons, for biodiversity, etc. will not be analysed in this book. However, for the management of these plantations a large part of the methodologies expounded in this book could be used, but the objectives pursued in their handling are notably different from those contemplated herein. Furthermore, it should be made clear that, on some occasions, a private owner can provide a public good without being able to obtain any financial reward. This is the case of the capture of carbon associated with plantations, which can be computed in the Kyoto agreements in Europe. The increase in carbon sequestered by these plantations does not revert to the benefits of their owners, although from the point of view of society it can be considered as being a cost avoided when fulfilling the objectives fixed by the E.U. at a country level. Finally, it is necessary to mention that this precision regarding the private/public origin of the plantations does not prevent the private plantations from receiving possible grants or subsidies proceeding from the public sector.

In addition, in this chapter, diverse methodologies will be presented starting from the hypothesis that we find ourselves before a determinist situation. With the aim of facilitating an understanding of the methods expounded, no aspects related to risks and uncertainty will be considered. This does not exclude the fact that when managing these forest systems some natural phenomena affecting the growth of these stands do exist (frost, drought, pests, etc.) as well as episodes associated with forest fires. In another direction, from an economic point of view, there is uncertainty related to how the evolution of the prices of the final products, as well as of the costs predicted throughout the life of the plantation can affect the analyses made. Without being too exhaustive, other aspects like those related to the evolution of technology, and, in general, of the innovation in these plantations, as well as certain regulatory aspects (Zinkhan et al. 1992, p. 28) can also condition their future development. Having arrived at this point, it could appear that not considering these conditioning circumstances in any decision making process would be a clear weakness in the exposition of this chapter. However, it has been preferred to group the methodologies in order to deal with risk in industrial plantations in Chap. 10.

Another hypothesis implicit in the development of this chapter is that, in principle, we are tackling problems of a mono-criterion nature, i.e. where it is only aimed to focus on a single objective. This initial simplification does not imply that a plantation does not produce more than one output, but given that the casuistry may be high, it is assumed that, habitually, it is aimed to optimize only one objective. This objective is usually linked, as we shall see in the next sections, to the economic nature of these investments. As will be seen throughout this book, this hypothesis is consistent with what will be explained in Chap. 6 referring to linear programming, although some applications developed with the aid of the multi-criteria theory will be dealt with in different chapters. These methodologies permit the decision-maker to integrate several objectives simultaneously. When illustrating these facts, it is easy to think of plantations in which there is clearly a well-defined output (timber, pulp, biomass, fungi, barks, juices, etc.), but on some occasions for these stands, the object of the analysis can also produce more than one output (timber and carbon, as will be seen in Chap. 14; timber and biomass, timber and cattle, etc.), although sometimes some of them are mutually exclusive outputs. On the other hand, it is not necessary for the objectives of the decision-maker to be associated with perfectly differentiated outputs, but they have to do with characteristics related to how those outputs have been produced. For instance, one of the objectives does not have to be the maximization of timber production but only to achieve a minimal production, which should be sufficient to supply the capacity of an industry, and, in turn, ensure a homogeneous exploitation throughout the time. Moreover, in this book it will be presumed that the decision-maker is assimilated in a single individual. Namely, it is assumed that it is not necessary to use procedures, like the group decision-making theory, which permit the aggregation of the preferences of possible decision-makers involved in the management of these forest systems.

Other aspects, which should be borne in mind when making any economic evaluation of a plantation, are those related both to public authorities, and to some conditioning circumstances or constraints existing in the market. That is to say, certain public policies which may affect the different outputs that it is expected to produce should be taken into account. Aspects like possible grants or subsidies should be considered at any stage of the project. In the same way, it is recommended from the beginning to integrate into the analysis the different payments associated with the diverse taxes that could be levied on the plantations being studied. The institutional context can also affect the plantation (positively or negatively) by means of various legislation measures, which should be the subject of a careful study. Finally, the consequences of including plantations under a certification system should be meticulously analysed, as will be shown in Chap. 15 of this book.

In spite of the complexity, which can be gleaned from the previous paragraphs in relation to the multiple factors and decisions to be made in the management of an industrial plantation, readers should be aware that when they finish this chapter they will have learnt key economic concepts to address industrial forests management planning.

What You Will Learn in This Chapter

- Basic principles of the investment analysis applied to forest stands.
- How to include a discount rate in this type of analysis.
- Criteria for assessing the profitability of the possible investment alternatives.
- Understanding different approaches for calculating the financial optimal forest rotation.
- Principles of forest valuation.
- Basic aspects of dynamic programming and its use to optimize stand management planning decisions.

5.2 Principles of Investment Appraisal

In this section the basic concepts of the investment analysis, which will serve as a basis for explaining the tools most usually employed for analysing the profitability of the investments, will be set out.

5.2.1 *Basic Concepts*

First, when investing in a forest plantation, it is necessary to characterize their common elements. Thus, any investment project is described by the following three basic parameters.

(a) Initial outlay or investment payment (K). This figure represents the monetary units that the investor or owner of the plantation must pay out in order to establish the investment. Thus, for a plantation, this outlay will cover expenses derived from ground preparation, planting, fencing, brushing, etc.
(b) The life of the investment (n); that is, the number of years during which the investment will be generating returns. For a plantation n represents the so called rotation age of the respective stand.
(c) The returns generated by the investment throughout its life (R_j). These returns can be measured according to an accounting optic as cash-flows (i.e., receipts

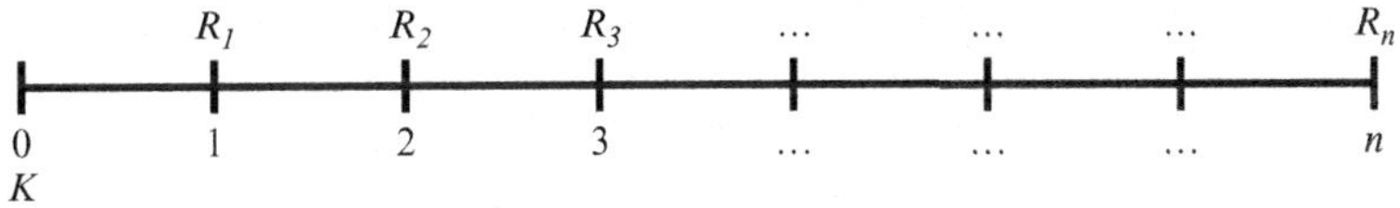

Fig. 5.1 Parameters characterizing an investment process

minus payments) or from an economic optic as profits (revenues minus costs). The accounting optic is more adequate for private investments and the economic one for public investments. Normally the ownership of plantations is private, so the recommended approach is the accounting one. Thus we will account for payments such as those derived from pruning and cleaning operations, road construction, maintenance, etc. The only receipts to be taken into account will be due to the sale of the timber obtained with the final cutting, plus in some few cases the sales revenue derived from thinning operations.

The above parameters characterizing an investment are usually represented in a temporal diagram like the one shown in Fig. 5.1. The returns (cash-flows) are represented in the upper part of the time axis, and the outlay or investment payment in the lower time of this axis. As can be verified in this Figure it has been assumed that the cash-flows are produced at the end of the year. This is the hypothesis habitually considered, although there would be no problem in presuming, for example, that the cash-flows occur at the beginning of each year.

5.2.2 *Discount Rate*

From an economic point of view, a forestry plantation underlies a temporal production function instead of a classic instant production function. In fact, a plantation requires a relatively long period of time since its birth until its final cutting; that is, the maturing of the underlying investment is rather long, although lower than other forest systems. Thus, the life of the investment or planning horizon is normally comprised between 6 and 20 years depending on the species and habitat. This temporary character of the plantations implies that the underlying cash-flows are generated at different moments of time.

The above commented temporary character of the receipts as well as in the payments generated by the plantations oblige to undertake a normalization or standardization process. In fact, it is obvious that it is not the same to receive or to pay R monetary units now that as in 3 or 5 years time. In short, a basic economic rationality tell us that any economic agent prefers to receive money now rather than in some years time. In order to understand the temporal preference for money it is enough to remark that there is a market for money. In this market, any economic agent can borrow or lend money according to the law of the compound interest at a certain interest rate of i per cent. Therefore, for any economic agent is the same thing to receive R monetary units at the current moment or to perceive $R(1+i)^n$ monetary units in n years time.

This simple argument leads us to the necessity of discounting the future cash-flows associated with the management of a plantation. In other words, the present value of a return R to be received in n years is less than the value R itself. The respective reduction will be determined by a discount factor, that, according to the compound interest law is equal to $R/(1+i)^n$. This process is called discounting. Inversely, if the decision maker seeks at the future value of the money, a compounding (or capitalization) process will be provided. Thus, the future value of R monetary units compounded with compound interest in n years will be equal to $R(1+i)^n$.

Our problem now is the right choice of the discount rate i. This important question has to different answers depending on whether the ownership of the plantation is private or public. In other words, the discount rate to be chosen will depend on the preferences for this type of consumption for a particular private owner in the former case or on the preferences of the society as a whole for current consumption in the latter case.

From a conceptual point of view, the determination of the discount rate within a private ownership scenario has an easier treatment. Thus, within a private ownership context, the discount rate is a subjective measure of the loss or opportunity cost incurred by an individual for receiving a certain amount of money some years later instead of receiving it at the current moment. Consequently, the private discount rate is a financial rate whose correct value is susceptible to being estimated by the interest rate for which this individual or firm can borrow or lend money in the capital markets. In the case of industrial plantations, the discount rate chose must reflect the cost of capital of this firm.

When the referential context is not a private owner of the plantation but society as a whole, the conceptualization as well as the measurement of the discount rate is more complex. In this case, the discount rate is called social discount rate and its correct estimation is a complex and controversial topic, and in view of the orientation of this book this issue will not be analysed. However, readers interested in this topic can consult, among others, Sen (1967) or Kula (1984, 1985). In the works by Kula appear estimations of the social time preferences rate for countries like Canada, United States and Great Britain. These figures oscillate between 5 % for Canada and United States and around 4 % for Great Britain. Finally, it is important to be aware that the discount is necessary but, on the other hand, it damages the welfare of future generations. Moreover, the larger the discount rate chosen, the more the current consumption is stimulated in detriment of any future one. Low discount rates are beneficial for the welfare of future generations, and obviously the opposite is true.

Throughout our presentation we have worked with compound interest according to a year's accumulation process, which leads to the idea of a discrete discount rate. However, in many cases, it is more operational to resort to a continuous

Table 5.1 Equivalences between discrete and continuous discount rates

i(%)	i′ = ln(1 + i) (%)
5	4.88
6	5.83
7	6.77
8	7.70
9	8.62
10	9.53
11	10.44

discount rate. Continuous and discrete interest rates are related in the following way. If the interests are accumulated yearly, an initial capital C will be transformed after n years in $C(1+i)^n$. However, if the accumulation of interests is made, for instance, monthly, we will apply a interest rate $i/12$ each month, and the initial capital C will be transformed after n years in $C(1+i/12)^{12n}$. In general, if we accumulate interests m times per year, the initial capital C will be transformed after n years in $C(1+i/m)^{mn}$.

It is said that the interest is continuous when $m \to \infty$; that is, the interests are accumulated instantly or in a continuous way. Hence, in that situation we need to calculate the following mathematical limit:

$$Z = \lim_{m \to \infty} \left(1 + \frac{i}{m}\right)^{mn} \tag{5.1}$$

The calculation of the limit given by (5.1) leads initially to an indeterminate of the type 1^{∞}. However, it is relatively easy to demonstrate that by taking Neperian logarithms in (5.1) and by applying the L'Hôpital rule, the mathematical limit given by (5.1) is equal to:

$$Z = e^{ni} \tag{5.2}$$

Hence, the discrete discount factor $(1+i)^{-n}$ is equivalent to the continuous discount factor e^{-ni}. Now a link between the discrete discount rate i and the continuous discount rate i' will be determined. For that, we start by establishing the following identity:

$$(1+i)^n = e^{ni'} \tag{5.3}$$

By taking Neperian logarithms in (5.3), we have:

$$\ln(1+i)^n = i' \tag{5.4}$$

Therefore, to use a discrete discount rate i' is equivalent to use a continuous discount rate $\ln(1+i)$. In Table 5.1 the respective equivalences for different value of the discount rate appear.

Data from Table 5.1 clearly show that results obtained with the continuous or with discrete discount rate are very close, especially when these rates are small. Therefore, the choice of the continuous instead of the discrete discount rate is not due to the financial logic, but to operational reasons related with the optimization calculus to be made in this chapter and in the following ones.

5.2.3 *Inflation*

The analysis made in the previous sections is based on an unrealistic hypothesis: that the cash-flows do not vary monetarily throughout the life of the project. That is to say, it has been presumed that these cash-flows are not affected by inflation. However, when assessing any type of investment, it is essential to take into account the effects which can be caused by inflation. In this sense, it should be remembered that both the cash-flows and the discount rate employed may be affected by it.

Two approaches are usually used when integrating inflation into an analysis of the profitability of an industrial plantation. The first one would be to work with cash-flows measured at current prices, i.e. prices which incorporate inflation and a nominal discount rate. A nominal discount rate would be equal to, for example, the capital cost necessary to undertake the investment. The second way to integrate inflation consists of working with cash-flows in which the effect of inflation has been discounted. Thus, this would mean working with constant prices. In the same way, the discount rate to be used would be exempt from inflation and in this case would be denominated real discount rate.

When deciding which of the two approaches is the most recommendable one, the implications associated with each of them should be borne in mind. Working with current prices and nominal discount rates implies having a precise knowledge of how inflation will evolve in the future. On the contrary, when working with real prices and real discount rates we are assuming the implicit hypothesis that inflation will affect receipts and future payments in the same way. Having arrived at this point it should be remembered that, on many occasions, the evolution of the price of products derived from the timber industry throughout the different time lapses has not increased in real terms (Buongiorno et al. 2003), which is not likely to have occurred with the costs associated with a plantation.

Although finance handbooks mostly recommend the first commented approach (e.g. Peak and Neale 2006), it is necessary to remember that the duration of a project associated with a plantation is not equivalent to most of the financial products, so that making predictions of its evolution for many years ahead could be very risky and would therefore justify working with constant prices and real discount rates. However, there are situations (e.g., biomass plantations), in which it could be recommendable to use current prices and nominal discount rates due to the scant duration of the project.

Also, the relationship between a nominal discount rate and a real discount rate has already been formulated by Fisher (1930). If we denominate as in the nominal discount rate, i_r the real discount rate and p the inflation rate, the following equation should be yielded:

$$(1 + i_n) = (1 + i_r) \cdot (1 + p) \tag{5.5}$$

Regrouping terms in (5.5) the following equation is obtained:

$$i_n = i_r + p + i_r \cdot p \tag{5.6}$$

Given that, generally, the product of i_r by p is a very small figure, it can be accepted that, in the absence of any risk, the nominal discount rate is equal to the real discount rate plus the expected inflation

$$i_n = i_r + p \tag{5.7}$$

5.2.4 Decision Making Issues

Once the main components of the investment analysis have been introduced, this section should be finalized by pointing out some steps to follow when handling this information, especially if the methodologies developed in the following section are going to be applied.

The first step would be to compile with the greatest precision possible the information necessary to estimate the different cash-flows. This involves, for example, in addition to calculating the initial payment of the plantation and how it is accrued, seeing if there is a possibility of the existence of grants or incentives of any kind. Furthermore, it is necessary to be consistent in the estimation of receipts and future payments. These should be measured by means of current or constant monetary units, but homogeneously. Similarly, a criterion should be set up in relation to taxes when computing the different cash-flows. These could be measured before or after tax, but always maintaining the same criterion. Also, the hypotheses associated with the final moment of the plantation should also be defined appropriately, in accordance with its characteristics.

As well as these considerations, it should be remembered that one of the main decisions to be made is that of correctly selecting the discount rate to be employed. Unless, due to legal requisites, we find that there is an obligatory discount rate, for all the other cases, it is advisable to start from two basic premises. The first, as affirmed by Wagner (2012) is that there is no single rate which can be applied for, focusing on the context of this book, all plantations. The second one is that the discount rate should be estimated independently for each case, taking into account the characteristics of each plantation, the objectives associated with it, as well as the investor's preferences. Further and as previously mentioned, the same hypotheses established in relation to the cash-flows should be formulated when considering nominal or real discount rates. On the other hand, and although this issue will be discussed in other sections of this book, up to now a determinist analysis has been presumed. If the decision-maker is capable of evaluating a certain risk when making predictions on the elements constituting the investment, an additional factor associated with the existence of these types of risk can be introduced into Eq. 5.7. Finally, and especially when dealing with investment projects in plantations with a long rotation age, one possible indicator for justifying the discount rate to be used could be the long-term public debt rate (Campbell and Brown 2003).

5.3 Decision Making Principles I: Financial Analysis

The starting point in the economic analysis of a plantation is to analyse the financial profitability of the underlying investment. For instance, the optimal rotation age of a plantation will be defined in the next sections as the life of the plantation for which the net present value of the underlying investment is maximized. Hence, and given the purpose of providing a book which is to some extent self-contained, in this section, some concepts and basic principles of the financial investment analysis will be presented.

Once the investment process underlying a forestry plantation has been defined, we will introduce the usual methods used to evaluate its profitability. The most intuitive method to evaluate the profitability of an investment is to calculate an algebraic sum of the returns or cash-flows (accounting approach) and the outlay of the investment, as appears in the following expression:

$$NPV = \frac{R_1}{(1+i)} + \frac{R_2}{(1+i)^2} + \ldots \frac{R_j}{(1+i)^j} + \ldots \frac{R_n}{(1+i)^n} - K \tag{5.8}$$

or what is tantamount to:

$$NPV = \sum_{j=1}^{n} \frac{R_j}{(1+i)^j} - K \tag{5.9}$$

The above expressions correspond to the concept or criterion of profitability known as **net present value** (*NPV*) of the investment. It should be noted that the *NPV* is a measure of the profits associated with the investment process taking into account the temporary dimension through the application of the discount rate i. In short, if the *NPV* is larger than zero the investment is financially viable and its figure indicates us the respective profits. However, if the *NPV* is negative the investment is not financially viable and its figure shows us the respective loss. Hence, $NPV \geq 0$ is a necessary although no sufficient condition for undertaking an investment process. According to the above definition the *NPV* is an indicator of the absolute profitability associated with an investment. Thus, if the *NPV* of a certain plantation is of 2,000$ per hectare, that figure means the total discounted profit associated with this investment.

Absolute profitability is an important indicator, but we also need another indicator of a relative profitability; that is, what are the profits per monetary unit invested or per monetary unit of investment payment? The most straightforward way to build this indicator consists of dividing the *NPV* among the investment payment K. In this way, the following ratio Profit/Investment is obtained:

$$Q = \frac{NPV}{K} \tag{5.10}$$

Thus, if for a certain plantation the ratio Q takes the value 0.75 dollars, this means that for each dollar invested the plantation has generated a discounted profit of 0.75 dollars. The relative profitability of an investment can be measured in an alternative way. In order to explain this, we shall interpret the investment metaphorically as a loan that the investor makes to an abstract entity (the plantation in our case). The investor lends the investment (the plantation) K monetary units in the current moment. The investment commits itself to pay back at the end of each year and during n years the annuities $R_1, R_2, \ldots R_j, \ldots R_n$. Once the investment process is interpreted in this way, it might be useful to find out the metaphorical or hypothetical interest rate paid by the investment to the investor. This interest rate will represent an indicator of the financial efficacy of the investment from the point of view of the investor. If we represent this interest rate by λ, the following equation must hold:

$$K = \sum_{j=1}^{n} \frac{R_j}{(1+\lambda)^j} \tag{5.11}$$

The value of λ holding Eq. 5.11 is known as the **internal rate of return** (*IRR*) of the investment. This rate is named as internal because for its correct determination we have used only variables that are internal to the investment process itself.

On the other hand, if we compare expressions (5.9) and (5.11), it is clearly deduced that the value of λ holding (5.11), which by definition is the internal rate of return of the respective investment, has the property of making the corresponding net present value zero. That is, if we use λ as discount rate the *NPV* is zero.

The concept of an internal rate of return allows us to provide a new definition of the financial viability of an investment. Thus, we can now say that an investment is viable if, an only if, the internal rate of return λ is higher than the interest rate i, for which the investor can obtain financial resources in capital markets. In fact, in this situation the investment can be undertaken with a loan for K monetary units to be paid back according to an interest rate i, this circumstance leaving a profit to the investor equal to $(\lambda - i)$ per monetary unit invested. In short, if:

$\lambda < i$: the investment is not viable, it is more profitable to lend the K monetary units at am interest rate i, but if:

$\lambda > i$: the investment is viable from a financial point of view.

On the other hand, expressions (5.9) and (5.11) which are valid for calculating the net present value and the internal rate of return for a discrete discount rate, are transformed into the following expressions for the continuous case:

$$NPV = \int_0^n R(j)e^{-ij}\,dj - K \tag{5.12}$$

$$K = \int_0^n R(j)e^{-\lambda j}\,dj \tag{5.13}$$

Although in the following section the differences existing with regard to the consideration of the VAN or of the TIR in order to obtain an economically optimal rotation will be described, this section should finish by making a brief comparison between both methods. Thus, some authors recommend the use of the NPV instead of IRR, due to, among other reasons, it being easy to use (Gunter and Hanley 1984), whereas others (Wagner 2012) detail the weaknesses associated with the election of the IRR as a criterion when deciding which investment is the best. In short, it can be said that both criteria can be valid when we are analysing the profitability of a plantation, but if the analysis is focussed on mutually exclusive investments (e.g. only one and no other forest plantation is undertaken) different results can be obtained employing one method or the other. On the other hand, given that the IRR tends towards infinity when the payment of the investment becomes zero, it is better not to use this criterion when, for example, there are substantial grants for setting up a plantation. However, and despite these problems, the IRR in many cases supplies the decision-maker with useful information, especially in quantifying the maximum cost of the capital so that the investment can be profitable (Götze et al. 2008).

5.3.1 *Series of Payments*

When discounting or capitalizing the different cash-flows associated with the life of a plantation, it should be taken into account that some financial mathematics tools exist which can simplify the analysis, especially when those cash-flows are repeated throughout the planning horizon. This situation is fairly common in forest plantations since there are receipts (derived from fellings, thinnings and grants) which are produced periodically or annually, or payments (taxes, cultural treatments, etc.) which occur either annually or periodically.

The basic idea is that, depending on the problem and as will be seen in later sections, it could be of interest to take some cash-flows to the future or discount them, and these cash-flows can refer to a single year, can be annual or can be periodic. In addition, the group of cash-flows can be finite or perpetual. Taking as a basis the work of Gunter and Hanley (1984) in Fig. 5.1, the formulae used have been compiled both considering a discrete discount rate and a continuous discount rate (shaded in Fig. 5.2). Next, the cited formulae are briefly explained.

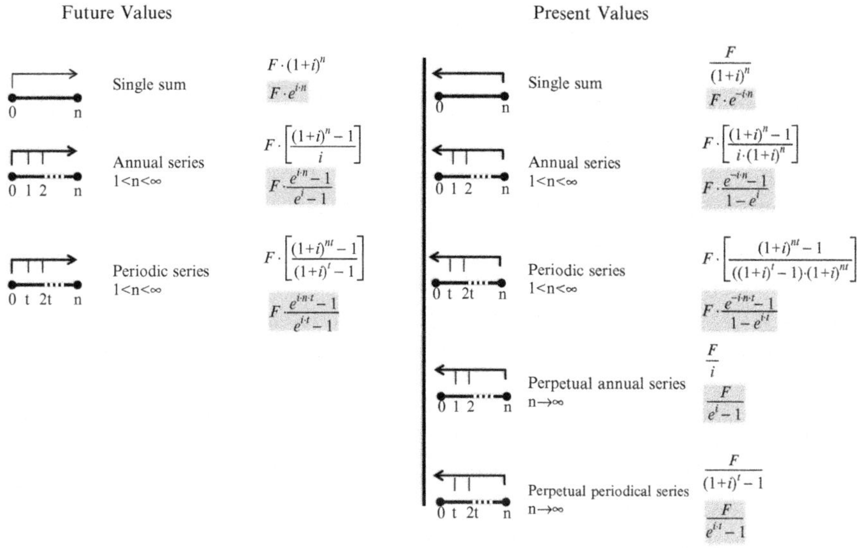

Fig. 5.2 Present values and future values expressions

Except for the formulae already introduced, which permit the calculation of a future or present value, the rest of the expressions contained in Fig. 5.1 are obtained from the expression which permits the calculation of the sum of a geometric progression:

$$S = a\frac{r^m - 1}{r - 1} \tag{5.14}$$

Where a is the first term of the expression, r the rate, and m the number of terms of which the cited progression consists. Applying this formula would give us the discount or update factor which then would have to be multiplied by the cash-flow which is repeated. In Fig. 5.1 the cited cash-flow has been denominated as F.

The expressions contained in Fig. 5.1 enable us to resolve the situations which present themselves most habitually when it is necessary to sum up the cash-flows adequately. However, it should be emphasized that these expressions are based on two basic hypotheses. The first one is that it has been presumed that the cash-flows are produced at the end of the year and the sums refer to the first day of each year. The second hypothesis, related to the previous one, assumes that the first cash-flow of the series is not produced at the initial moment but in the first year, or, if it is periodic, in a subsequent year. For that reason, if it is desired to apply the previous expressions, they would have to be corrected, taking into account this possibility. As an illustration, in Bettinger et al. (2009) some examples of the current value of annual or periodic series with a cash-flow at the initial moment and a discrete discount rate are shown. However, we encourage any interested reader to obtain the new expressions using the Eq. 5.14. Finally, it should be anticipated that some of the formulae included in Fig. 5.1 are very much employed in aspects like forest valuations. To be specific, those which show the current value of periodic perpetual series constitute the basis for calculating what in future sections will be denominated the expected value of the land.

5.3.2 Integrating Risk

The analysis made up to now has exclusively admitted a determinist starting point, in which no risks or uncertainties existed. As an easily understandable result, these assumptions are not realistic, and less so in investments associated with industrial plantations. Indeed, in addition to the parameters analysed in this chapter (discount rate, price of timber, life of the plantation, costs, etc.), there are other variables in the model whose variation can drastically affect the economic viability of the plantation. Precisely, we refer to risks associated with fires, pests, etc. Although in Chap. 10 these issues will be looked into further, it would seem to be necessary to end this section pointing out the convenience, when assessing economically these plantations, of using a simple tool, the sensitivity analysis, which permits one to ease that premise of certainty initially admitted.

The sensitivity analysis allows a verification of how the results associated with the economic indicators defining an investment (VAN, TIR, etc.) vary when some of the parameters enabling the calculation of the cited indicators are modified. Namely, this consists of re-calculating the profitability indicator selected by modifying the initial parameter and making all the other parameters keep constant (*ceteris paribus*). For instance, if we have any doubts about the value of the discount rate to be used because the investment may present a greater risk than that initially foreseen, it is easy to re-calculate the net current value with a higher discount rate which integrates a risk premium. If the investment continues to be profitable, this would mean that the decision-maker could justify its realization. This same reasoning can be employed for prices, for costs, for the life of the investment, etc., and provides a simple way of assessing the impact of modifying an initial hypothesis associated with the value of the cited parameters. Also, this technique enables the calculation of the variation interval of the starting out data with the aim of establishing when an investment is viable.

5.4 Decision Making Principles II: Optimal Rotation

The forest rotation age of a timber stand as a plantation means the "optimal" life of the stand. Obviously there are many "best" or optima rotation ages according to the criterion used to define optimality. The most traditional criterion used in the forestry literature, since the nineteenth century has been the net present value of the underlying investment. Hence, by definition the optimal rotation age is the life of

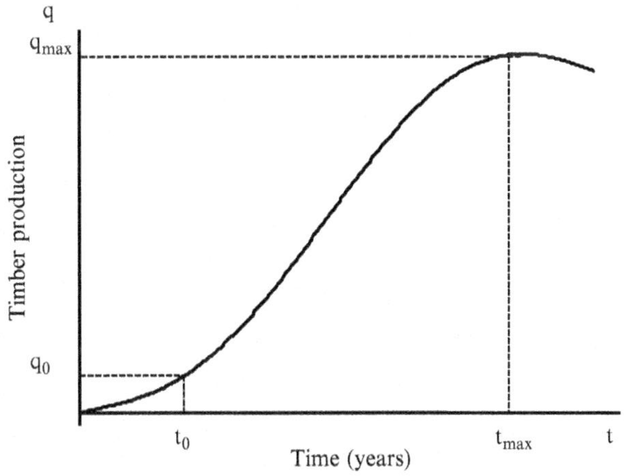

Fig. 5.3 Growth curve or temporal production function for a plantation

the timber or plantation for which the corresponding net present value is maximized. Before developing this idea that will lead us to the best solutions from an economic point of view, in this section and as a first step, the optimal rotation age will be analyzed from a purely technical point of view without introducing any economic parameters into the analysis.

The basic ingredient for our analysis is a growth curve or temporal production function that maps the timber stock q with the age of the stand t, such as:

$$q = f(t) \tag{5.15}$$

being $f'(t) \geq 0$ and $f''(t) \geq 0$

$f'(t) \geq 0$, implies positive marginal productivities; i.e., the stock of timber monotonically increases with the age of timber, until a certain critical age t_{max} is achieved. From this age onwards, the stock of timber decreases. The stock of timber corresponding to the age t_{max} is usually called **technical maximum.**

$f''(t) \geq 0$, implies geometrically the concavity of the growth curve and economically the monotonically decreasing character of the marginal productivities (i.e., the classic law of the diminishing returns).

Figure 5.3 shows a typical growth curve or temporal production function holding the above properties and valid for a single tree or for a plantation.

Now we are going to determine the value of t or the rotation age at which it is suitable to cut down the tree or the plantation. That is, to establish the rotation age that utilizes the underlying technology in a better way. To achieve this purpose it seems sensible to search for the age t, for which the average productivity achieves a maximum value. In this way, the timber sustainable yield will be maximized. In other words, following this strategy the amount of timber harvested per year will be optimized. The maximization of the average productivity $\overline{q}$, implies maximizing the following expression:

$$\bar{q} = \frac{f(t)}{t} \tag{5.16}$$

A first order condition to maximize (5.16) is given by, the following equation:

$$\frac{d\bar{q}}{dt} = \frac{d}{dt}\left[\frac{f(t)}{t}\right] = \frac{f'(t)t - f(t)}{t^2} = 0 \tag{5.17}$$

From (5.17) it is straightforward to obtain the following necessary condition to determine an optimal rotation age from a technical point of view:

$$f'(t) = \frac{f(t)}{t} = \bar{q} \tag{5.18}$$

That is, the optimal rotation age from a technical point of view implies harvest the plantation at an age at which its average productivity coincides with the marginal productivity. It is interesting to remark that this solution implies the maximization of the sustainable yield. For this reason, this solution also receives the name of **technical optimum**. From a management point of view this property facilitates the calculation of this optimal rotation. Indeed, with only a production table at one's disposal, the calculation of this rotation is immediate since it is produced where the mean growth and the current growth are equalized.

This technical optimum can be interpreted geometrically as the point of tangency between the growth curve $q = f(t)$ and a straight line crossing the coordinates origin (see Fig. 5.3). In fact, the tangents of a beam of straight lines passing through the origin of coordinates ($\mathrm{tg}\alpha_1$, $\mathrm{tg}\alpha_2$, ...) represent the respective average productivities. Hence, as we know larger angles imply larger tangents and consequently larger average productivities. In short, the commented point of tangency (see Fig. 5.4) corresponds to the best rotation age from a technical perspective.

5.4.1 Forest Rotation Age: The First Economic Solutions

The rotation age established in the previous section is of a very limited interest for the private or public owner of the plantation. Obviously, the owner of the plantation wants to maximize the profits derived from the management of the plantation instead of maximizing the timber income obtained. To achieve this, we will need to introduce into the analysis economic parameters like timber prices, discount rate, plantation costs, etc., that complement the technological information provided by

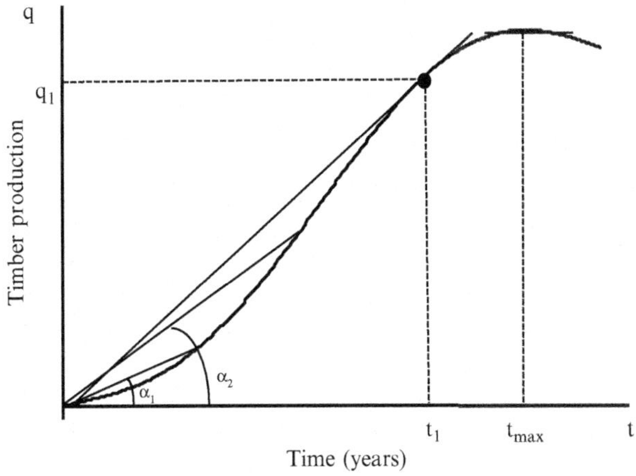

Fig. 5.4 Graphical representation of the optimum technical rotation age

the growth curve. The first approximations to the determination of the optimal forest rotation age from an economic discipline perspective were made by Fisher (1930) and Hotelling (1925). They propose defining the optimal forest rotation of a plantation as the age for which the underlying net present value is maximized. This basic idea leads to the following optimization problem:

$$Max\ NPV(t) = P(t)f(t)e^{-it} - K \tag{5.19}$$

where *P*(*t*) is the timber price that is expressed as a function of time.

By calculating the first derivative of (5.19) and by implementing simple algebraic operations, the following first order condition is obtained:

$$\frac{P'(t)}{P(t)} + \frac{f'(t)}{f(t)} = i \tag{5.20}$$

The value of *t* holding (5.20) represents the economic optimal rotation age of the plantation, and is usually known as the Fisher-Hotelling or just the Fisher rotation age. Condition (5.20) has a clearly economic interpretation. In fact, the owner of the plantation will postpone felling if and only if the sum of the relative growth of the timber price and of the timber stand itself surpasses the current interest rate. Some basic conclusion can be derived from this rather simple model. In fact, if the real price of the timber grows, the felling condition is more difficult to be uphold, which will lead to longer rotation ages. On the contrary, if the interest rate *i* grows it will be easier to uphold the felling condition, and this will lead to a shorter rotation age for the plantation.

5.4.2 Forest Rotation Age: The Faustmann Solution

The above solution to the determination of the rotation age of a plantation is an elegant one and is of some economic interest, although in its current formulation there is an important flaw that must be corrected. In fact, Fisher and Hotelling did not take into account that the existence of the stand implies an opportunity cost to the owner of the plantation, namely, increasing the rotation age for example by 1 year, what implies a cost (or land rent) equivalent to the income that the owner would have obtained by felling the stand and dedicating the soil to an alternative use. This type of drawback or insufficiency can be remedied following two seemingly different, although analytically equivalent, approaches.

The first approach was proposed by Samuelson (1976) and basically consists of the explicit introduction of the land rent or opportunity cost into the expression that calculates the net present value of the respective investment. Thus, we have the following expression:

$$Max\ NPV(t) = P(t)f(t)e^{-it} - K - R\int_0^t e^{-it}dt \tag{5.21}$$

The other approach was proposed by the German forester Faustmann in 1849, and basically consists of considering a chain of infinite plantation cycles or by using a financial language to consider a chain of infinite investments. In this way, there is no omission with the land rent or opportunity cost due to the existence of the current plantation. This approach leads to the following optimization problem:

$$Max\ NPV(t) = \left[P(t)f(t)e^{-it} - K\right]\left[1 + e^{-it} + e^{-2it} + e^{-3it} + \ldots\right] \tag{5.22}$$

By adding the geometric progression corresponding to the second term of (5.22) and by implementing simple arithmetic operations, we have:

$$Max\ NPV(t) = \frac{P(t)f(t)e^{-it} - K}{1 - e^{-it}} \tag{5.23}$$

Expression (5.23) is known in the forest economics literature as "Faustmann Formula", although this distinguished researcher did not propose any procedure for its maximization. According to Samuelson approach, by solving the integral in (5.21), and by calculating the first derivative, the following first order condition is obtained:

$$P'(t)f(t) + P(t)f'(t) - iP(t)f(t) - R = 0 \tag{5.24}$$

Now taking into account that the land value V can be estimated by capitalizing its income R (i.e., $V = R/i$), expression (5.24) turns into:

$$P'(t)f(t) + P(t)f'(t) = i\left[P(t)f(t) + V\right] \tag{5.25}$$

Before interpreting the new condition of equilibrium given by (5.25), the optimal forest rotation age from the seemingly alternative approach proposed by Faustmann will be investigated. To achieve this purpose the first derivative of the *NPV*(t) with respect to t in expression (5.23) is calculated. In this way, we have:

$$P'(t)f(t) + P(t)f'(t) - iP(t)f(t) - i\frac{P(t)f(t)e^{-it} - K}{1 - e^{-it}} \tag{5.26}$$

It should be noted that the last term of (5.26) without considering the discount rate i coincides with the net present value of the underlying investment given by (5.23); that is, the land value obtained by discounting an infinite chain of investments. Therefore, Samuelson's approach that makes the land income explicit and the Faustmann proposal based upon the consideration of infinite plantation cycles lead to the same solution or rotation age.

This solution represents the foundations of forest economics in general and of the economic management of plantations in particular, and it is known as the Faustmann-Pressler-Ohlin (FPO) solution or rotation age. The contribution of each author to the FPO solution is the following. As mentioned before Faustmann, in 1849, proposed the correct formulation based on infinite cycles of plantation. Pressler in 1860 provides the first solution to the underlying optimization problem. Finally, Ohlin independently from Pressler's work, in 1921, provides a better solution to the problem. Löfgren (1983), and Johansson and Löfgren (1985) are excellent presentations of the historical aspects related with the determination to the FPO solution.

Furthermore, it should be remembered that Faustmann's idea of defining the optimal rotation to the one which maximizes its profitability for an infinite number of rotations, as has been shown in expression (5.22), permits the introduction of a concept widely used in forest valuation, the *land expectation value* (*LEV*). The latter is the term on the left in the expression (5.23). As will be seen in a future section, the usefulness of *LEV* is that it enables one to assign a value to a land which is bare but with the possibility of accommodating certain types of plantations. The latter's value would precisely be the *LEV* associated with the payments and collections foreseen for the hypothetical plantation.

Finally, The FPO formula can be easily generalized for a situation where the annual expenditure for administration G, as well as plus the payments associated with the final felling Q are relevant. In this case, expression (5.16) turns into the following one:

$$Max\ NPV(t) = P(t)f(t)e^{-it} - K - (R + G)\int_0^t e^{-it}dt - Qe^{-it} \tag{5.27}$$

By solving the integral in (5.27), and by calculating the first derivative, the following new first order condition is obtained:

$$P'(t)f(t) + P(t)f'(t) - iP(t)f(t) - G - R + iQ = 0 \tag{5.28}$$

Equation (5.28) is just a generalization of Eq. (5.26). It is interesting to note, that the inclusion of management payments as well as payments for the final felling in the Faustmann formula (expression (5.24) leads to the equilibrium condition given by Eq. (5.28). In the next section we will see how, losing a little of mathematical elegance, the Faustmann approach can be considerably generalized. In short, we will interpret Faustmann just as a logic system and we can incorporate into this logic every significant characteristic of the plantation.

5.4.3 Extensions to FPO Solution

This section will deal with two topics that compliment the material presented up to now. These topics are.

(a) To implement static comparative analysis in order to predict the influence of changes in the main parameters characterizing the plantation in the rotation age.
(b) To generalize the Faustmann approach.

The static comparative analysis starts with a general function $Z\ (x_1, x_2, \ldots x_n)$ and then the influence on the value achieved by function *Z* of variations in one of the variables is investigated. For instance, what is the influence on *Z* of variations of the value of variable x_1 is, keeping constants the values achieved by other n-1 variables. In our case, we work with the function *Z*(*t*, *K*, *P*, *i*) in order to determine the relation between the rotation age t and the plantation cost *K*, timber price *P* and discount rate *i*. After some tedious mathematical calculations (see for details Romero 1997, pp. 157–159) it is possible to demonstrate the following relationships:

$$\frac{dt}{dk} \geq 0 \quad \frac{dt}{dp} \geq 0 \quad and \quad \frac{dt}{di} \leq 0$$

The interpretation of the above results is interesting and quite straightforward. Thus, any increase (decrease) in the plantation costs or in the timber price will increase (decrease) the respective rotation ages. On the contrary, an increase (decrease) in the value of the discount rate will decrease (increase) the respective rotation age. The results derived from a comparative static analysis are of a qualitative nature. If we want to quantify these effects some more sophisticated calculus has to be undertaken. In the next section with the help of a case study this type of quantification will be shown.

The Faustmann approach can be considered as a net present value maximization logic, and in this way it can be generalized to any specific silvicultural context. Thus, let us assume the following additional ingredients with respect to the ones introduced up to now:

C_h = cash-flows from thinnings during the rotation ($h = h_1, h_2 \ldots$). The sub-periods $h_1, h_2, \ldots$ represent the years for which according to the silvicultural plan thinnings will be made. For instance, for a plantation of *Pinus radiata* in Spain, a thinning is scheduled every 5 years starting the 15th year, then we have, $h_1 = 15$, $h_2 = 20$, $h_3 = 25$, etc.

Y_s = Operational forestry expenditures during the rotation ($s = s_1, s_2, \ldots$). The sub-periods $s_1, s_2, \ldots$ represent the years for which according to the silvicultural plan several operational forestry activities will be carried out. For instance, in the year 5th a cleaning is scheduled, in the 10th year a pruning, etc.

By implicitly assuming the land rent, that is, by considering infinite plantation cycles, the Faustmann logic leads to the maximization of the following expression:

$$\left[P(t)f(t)e^{-it} + \sum_{\forall h} C_h e^{-ih} - K - G\int_0^t e^{-it}dt - \sum_{\forall s} Y_s e^{-is} \right] \cdot \left(1 + e^{-it} + e^{-2it} + \ldots\right) \quad h = h_1, h_2, \ldots s = s_1, s_2, \ldots \tag{5.29}$$

By solving the integral in Eq. (5.29), summing the corresponding geometric progression, and doing simple arithmetic operations, we have:

$$\left[P(t)f(t)e^{-it} + \sum_{\forall h} C_h e^{-ih} - K - G\left(1 - e^{-it}\right) - \sum_{\forall s} Y_s e^{-is} \right] \cdot \left(1 - e^{-it}\right)^{-1} \quad h = h_1, h_2, \ldots s = s_1, s_2, \ldots \tag{5.30}$$

Expression (5.30) can be called "Generalized Faustmann Formula". It is obvious that it is extremely difficult to obtain the value of t that maximizes (5.33) by standard optimization techniques based on differential calculus. However, it is relatively easy to approximate this optimal value of t by resorting to numerical calculus techniques. In this way, we can obtain the Faustmann rotation age and consequently the maximum net present value for the owner of the plantation.

Another way to extend the previous formulae would be to include possible grants and other outputs in the analysis, assuming a joint production context. How to integrate these elements into the above model is shown in what follows (Diaz-Balteiro and Romero 2001). Thus, besides the above payments and receipts might be also necessary to introduce the financial subsidies provided by the current European forestry policy. In this sense there are three categories of aids. A maintenance premium P_m received during the first 5 years of the plantation cycle; a compensatory premium P_c received during the first 20 years of the plantation cycle and a subsidy K_1 to mitigate afforestation costs. Taking into account all these components, and being aware that the different subsidies are perceived only in first plantation cycle, the *NPV* attached to the investment will be equal to:

$$NPV = \frac{P(t)f(t)e^{-it} - K - G\alpha - \sum_{\forall s} Y_s e^{-s} + \sum_{\forall l} C_l e^{-il}}{1 - e^{-it}} + K_1 + P_m\beta + P_c\gamma$$

with:

$$\alpha = \frac{e^{(-i1)} \cdot \left(e^{(-it)} - 1\right)}{\left(e^{(-i1)} - 1\right)}$$

$$\beta = \frac{e^{(-i1)} \cdot \left(e^{(-i5)} - 1\right)}{\left(e^{(-i1)} - 1\right)}$$

$$\gamma = \frac{e^{(-i1)} \cdot \left(e^{(-i20)} - 1\right)}{\left(e^{(-i1)} - 1\right)} \quad (5.31)$$

The optimal forest rotation as well as the profitability of the different plantations is obtained by maximising expression (5.31). However, when we are considering a joint production process, like timber-carbon captured, the above procedure is not applicable. Neither is the extension of Faustmann formula introduced by Hartman (1976) applicable since this approach requires the estimation of a flow of services measured in monetary terms. To circumvent this problem we have followed in this chapter a methodology proposed by Romero et al. (1998). With this purpose in mind a subsidy *A* per ton of carbon captured will be introduced. In the same way, a tax of *A* will be levied for each ton of carbon emitted to the atmosphere. Taking into account this new context Eq. 5.31 turns into:

$$NPV = \frac{P(t)f(t)e^{-it} - K - G\alpha - \sum_{\forall s} Y_s e^{-is} + \sum_{\forall l} C_l e^{-il} + A\sum_{\forall r} C_a e^{-ir} - A\sum_{\forall v} C_e e^{-iv}}{1 - e^{-it}} + K_1 + P_m\beta + P_c\gamma$$

with:

$$\alpha = \frac{e^{(-i1)} \cdot \left(e^{(-it)} - 1\right)}{\left(e^{(-i1)} - 1\right)}$$

$$\beta = \frac{e^{(-i1)} \cdot \left(e^{(-i5)} - 1\right)}{\left(e^{(-i1)} - 1\right)}$$

$$\gamma = \frac{e^{(-i1)} \cdot \left(e^{(-i20)} - 1\right)}{\left(e^{(-i1)} - 1\right)} \quad (5.32)$$

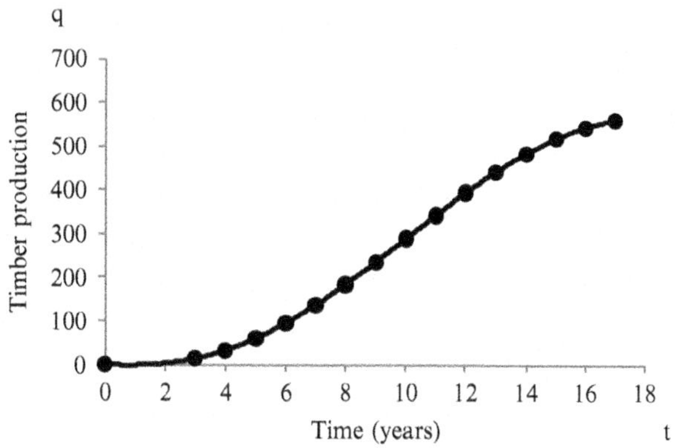

Fig. 5.5 Growth curve for a plantation of poplars ("Campeador clone") in the Province of León (Spain)

Where C_a represents the carbon captured when the age of the stand is *r*. C_e represents the carbon emitted in the year *v*. It should be noted that we have only considered the carbon stored in the marketable timber in the final felling as well as in the thinnings. However, we have not considered the carbon captured in other types of biomass or the variation of carbon in the soil. No taxes will be considered. In all the cases studied, the NPV, the optimal forest rotation as well as the amount of carbon captured will be calculated for the private optimum (corresponds to the Faustmann optimum) and the environmental optimum (corresponds to the maximum capture of carbon). In short, simply by proceeding to maximize the expressions (5.31) or (5.32) the maximum NPV would be reached and the age at which this circumstance occurs would be the economically optimal rotation. Finally, Chap. 14 will deal in detail with the integration of the carbon capture in the management of industrial plantations.

5.4.4 *Some Illustrations*

The ideas and methods presented in the preceding sections will be illustrated with the help of a case study. We will deal with a poplar plantation in Spain, province of León, for the "Campeador clone". We start from the following growth curve or temporal production function (Diaz-Balteiro and Romero 1994):

$$q = f(t) = -0.00221t^4 - 0.20458t^3 + 7.53805t^2 - 27.2853t + 34.384 \qquad R^2 = 0.99 \tag{5.33}$$

where, *q* measures cubic meter of timber per hectare and *t* represent the age of the plantation in years. Figure 5.5 shows graphically the growth curve given by (5.37). As a first step in our exercise, we have calculated the technical maximum. As it was explained before for this we need to equate to zero the marginal productivity (or current growth), that is:

$$f'(t) = -0.00884t^3 - 0.613743t^2 + 15.0761t - 27.2853 = 0 \tag{5.34}$$

By solving (5.38) we obtain $t = 17.6$ years. By substituting this value of t in the growth curve (5.37) a timber volume of 561.77 m^3/ha is obtained; that is, the maximum possible timber production is around 562 m^3/ha and it is obtained for a rotation length of approximately 18 years.

Now the technical or biological optimum will be calculated. As a first step, we calculate the average productivity (average growth), that is equal to:

$$q = \frac{f(t)}{t} = -0.00221t^3 - 0.2458t^2 + 7.53805t + 34.384t^{-1} - 27.2853 \tag{5.35}$$

By equating Eqs. 5.34 and 5.35, we obtain a value of t of 14.6 years. By substituting this value of t in the growth curve a timber volume q of 505.73 m^3/ha is obtained; that is, the technical or biological optimum implies a timber volume of around 506 m^3/ha and it is obtained for a rotation age of around 15 years. This solution implies a maximum sustainable yield of 506/15 $\cong$ 34 m^3/year.

Now we shall calculate the Fisher-Hotelling rotation age as a first economic approximation. Thus, we assume a discount rate i of the 3 %, and a timber price constant (i.e., $P(t) = P \Rightarrow P'(t) = 0$) which is a common assumption for short-term economic analysis. For these data, Eq. 5.20 turns into:

$$\left(-0.00884t^3 - 0.613743t^2 + 15.0761t - 27.2853\right) = 0.03 \cdot \left(-0.00221t^4 - 0.20458t^3 + 7.53805t^2 - 27.2853t + 34.3841\right) \tag{5.36}$$

By solving (5.36), a value of t equal to 16.35 years is given. By substituting this rotation age in the growth curve given by (5.33) a timber volume of 551.22 m^3/ha is obtained. If we assume now that the plantation costs K are equal to 1,500$/ha, and the timber price P equal to 70$/ha, then the corresponding maximum net present ***NPV**** value will be equal to:

$$70 \cdot 551.22e^{-0.03 \cdot 16.35} - 1500 = 20300.55\ \$/\ \text{ha} \tag{5.37}$$

The next step in our exercise consists in calculating the generalized FPO rotation age. For that, we need the following additional data:

Land rent (R) = 400$/ha
Annual expenditure for administration (G) = 150$/ha
Felling costs (Q) = 500$/ha

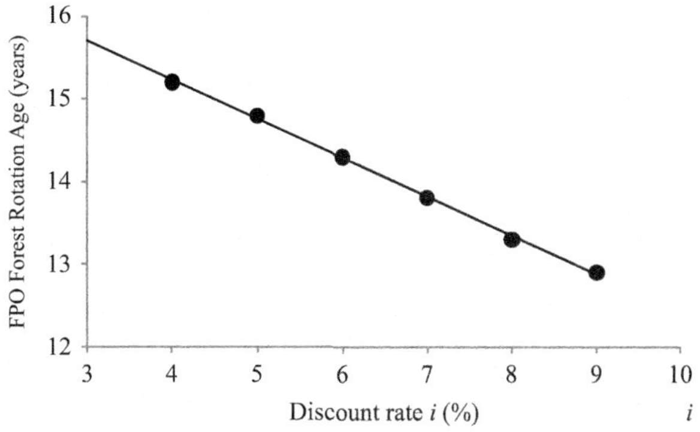

Fig. 5.6 Relationship between discount rate i and FPO rotation age

By substituting these data in the FPO equilibrium condition given by Eq. (5.28), we have:

$$70 \cdot \left(-0.00884t^3 - 0.613743t^2 + 15.0761t - 27.2853\right) - 0.03 \cdot 70 \cdot \left(-0.00221t^4 - 0.20458t^3 + 7.53805t^2 - 27.2853t + 34.3841\right) - 150 - 400 + 0.03 \cdot 500 = 0 \tag{5.38}$$

By solving (5.38), a value of t equal to 15.71 years is obtained. By substituting this rotation age in the growth curve given by (5.33) a timber volume of 538.39 m^3/ha is achieved. The maximum ***NPV**** associated with the generalized FPO rotation age will be equal to:

$$70, 538.39 \cdot e^{-0.03 \cdot 15.71} - 1,500 - 550 \frac{\left(1 - e^{-0.03 \cdot 15.71}\right)}{0.03} - 500 \cdot e^{-0.03 \cdot 15.71} = 13,095.55\$/\text{ ha} \tag{5.39}$$

When we compare the two optimum values for the respective net present values, we should take into account that in the FPO solution we have deduced the land rent, the annual expenditure costs and the felling costs. Thus, the second figure for the net present value is significantly lower than the first figure obtained.

As a last step in our practical exercise we now have to implement a sensitivity analysis to the FPO solution obtained. In other words, we shall test the sensitivity of the solution to changes in the values of some key parameters. First, we shall study the influence on the FPO forest rotation age of changes in the value of the discount rate i. For that purpose, we have increased the value of i from the initial figure of 0.03 up to a maximum value of 0.09. Figure 5.6 shows the results obtained. It can be verified that, according to the theoretical prediction, increases in the value of the discount rate i generates decreases in the FPO rotation age. Quantitatively this influence is fairly moderate for this particular case study. By using an economic term, we can say that the FPO rotation age is very inelastic with

Table 5.2 FPO rotation age for a discount rate of 0.03 and different values of the land rent R and the timber price P.

Land rent(R) / Timber price(P)	200	300	400	500	600
50	15.79	15.62	15.44	15.25	15.06
70	15.96	15.84	15.71	15.59	15.46
90	16.04	15.95	15.85	15.76	15.67

respect to variations in the value of the discount rate i. Thus, the FPO rotation age ranges from a maximum of 15.71 years for the lowest value of i (0.03) to a minimum of 12.86 years for the highest value of i (0.09). Which means that an increase in the discount rate i of 200 % has generated a decrease in the FPO rotation age of only 22.16 %.

Finally and mainly to illustrate the enormous possibilities of the sensitivity analysis, we have calculated the FPO rotation age for a discount rate i of 0.03, but for different values of the land rent R and timber price P. The results obtained are shown in Table 5.2.

From the results shown in Table 5.2 two main conclusions are obtained. First, the basic theoretical predictions are corroborated. In fact, when the value of the land rent increases the FPO rotation ages decreases. In the same way, when the timber prices increases the FPO rotation ages decreases. Second, the FPO rotation age obtained is very robust with respect to variations in the values of both the parameters chosen. It can be said that the difference in the values of the different rotation ages shown in Table 5.2 have an arithmetical meaning but are not significantly different from a forestry perspective. Obviously, there are many more type of robustness analysis that might be implemented. However, the simple exercise undertaken can provide to the readers with a clear idea about the enormous practical interest of this type of approach.

5.4.5 Some Critical Issues in Determining Economic Optimal Rotation

The optimal rotation analysis will be completed in this section by showing some considerations, which have not been included under previous sections. The first idea is based on the fact, for some authors, that the correct way to obtain an economically optimal rotation would be to maximize the IRR associated with the plantation being analyzed. Indeed, Boulding (1935) proposed the maximization of the IRR as the criterion for determining the optimal rotation length. The Boulding rotation age raised some interest some time ago but nowadays is of very little practical interest. In fact, the IRR measures the relative profitability of an investment, but it does not provide any information about the absolute profitability. It seems rather obvious, that it is not economically sound to make a decision by resorting to the maximization of the relative profitability instead of in the maximization of the absolute profits.

Moreover, in those cases where the plantation costs are very small or even zero (natural regeneration), then as K tends to zero, the IRR trends to infinite which has no economic meaning. Lastly, it should be noted that the Boulding rotation age is always shorter than the other economic solutions obtained so far.

On the other hand, in the optimal rotation analysis shown previously it can be verified how the economically optimal rotations are usually shorter than technically optimal ones. Equalling the expressions (5.22) and (5.27) it is possible to obtain conditions obliging both rotations to be equal, and, hence, to deduce under what range of values of the parameters intervening in this equality, the circumstance of the economic rotation being longer than the technically optimal one is not fulfilled. From an empirical point of view, this hypothesis is true for most plantations, save one exception that should be mentioned. In effect, this idea is not carried out in the case of the growth of the plantations being so high that this implies technically optimal plantations only a few years old. In Rodríguez et al. (1997) an example of this circumstance is shown.

The analysis carried out to calculate the economically optimal rotation started out implicitly from the hypothesis that the plantations all proceeded from seed (high forest). However, in some species such as *Eucalyptus sp.* it is usual in many countries to take advantage of shoot reappearance as an additional cycle, which can be repeated on several occasions. This leads us to a dynamic problem in which both the rotation in each cycle and the optimal number of cycles before taking up the plantation should be optimized. In the literature this possibility has been studied in works by Medema and Lyon (1985) and Tait (1986). In a more recent work, in which both pulp production and carbon capture were integrated, an optimal rotation in eucalyptus plantations in Spain and Brazil (Diaz-Balteiro and Rodriguez 2006) was calculated. The results show different responses in both countries and, unlike previous works, no shorter rotations were observed in the cycles associated with forests compared to rotations of plantations coming from seed.

5.5 Decision Making Principles III: Forest Valuation

Once the investment analysis and optimal forest rotation concepts have been reviewed, this section goes on to show the basic aspects of forest valuation. When making decisions in relation to a plantation, it is indispensable to find out its value, either for patrimonial or tax issues or for a possible sale or purchase. Likewise, if different investment alternatives are compared beyond the forestry ones themselves, it is necessary to find out with precision the value of a possible investment in lands of a forestry nature. The tools enabling us to find out this value are included in what is known as forest valuation. Although they are sometimes taken as synonyms, some countries distinguish the methods for making a valuation as against what is strictly speaking an appraisal (Davis et al. 2001, p. 396). In any case, forest valuation can be defined as that discipline which aims to estimate the value of a forest estate, in agreement with a set of hypotheses and a specific objective.

In this book we shall delimit this last definition since its reason for being is practically circumscribed to the outputs associated with the forest plantations presenting a market price. This signifies that, except for the case of carbon, no other positive externalities associated with this type of forest stands will be taken into account.

5.5.1 Forest Valuation Methods

Although this idea can be further enlarged, in principle it can be said that the valuation of a forest plantation can be approached from two completely different perspectives. On one hand, this would be through its market value, and, on the contrary, on other occasions, these valuations are made by calculating the value associated with the capitalization of any future income estimated as being produced in that plantation. The market value represents the value at which the stand or the forest (stand plus land) can be sold or purchased immediately. The capitalization value, as will be seen in the next section, is the value discounted from the flow of products which can be obtained from the forest throughout the time. In classic agricultural valuation texts, methods which provide a market value are usually called synthetic methods, whereas those which provide a value from the capitalization of revenue are denominated analytical methods.

In order to calculate a market value of a plantation it is necessary to possess some knowledge of a series of transactions which have been carried out recently in properties similar to the one it is aimed to value. If this information exists, it is possible to calculate this market value through its comparison with other recent sales. However, as affirmed by Healy and Bergquist (1994), the difficulty in finding this information means that this method is hardly ever used even when it produces what is really being sought: a market value associated with a property. This circumstance is habitual in markets where it is not possible to speak with any precision of a perfect competition market. There are several reasons invalidating this assumption: the properties are not homogeneous, normally few buyers or sellers concur, the sale and purchase prices are often not known, and transactions are usually infrequent.

Capitalization methods start out from the hypothesis that the plantation will provide, indefinitely, a flow of receipts and payments. Thus, in order to apply this method the structure of the receipts and payments associated with the production function being employed should be known. This method is applied both to the valuation of the stand and to the products which can be obtained from the forest land, and it is necessary, as well as estimating the prices and future costs, to fix the discount rate to be used.

There are other methods for making these types of valuations. For example, the residual value criterion is sometimes used, deducting from the sale price of the final product (timber, pasture, etc.) all the costs incurred since the plantation began.

Another method which could potentially be applied would be that of the replacement cost. Namely, this involves estimating what it would cost to replace the plantation to be valued with other similar ones. This type of valuation would only be of any sense in very young stands, and, frequently, after a fire has been produced. In another direction, the case of institutional investors in forest plantations (TIMOs) obliges the use of financial techniques like the capital asset pricing model in order to make the valuations of this type of assets (Zinkhan et al. 1992, chap. 5).

5.5.2 Land Expectation Value (LEV)

Given the difficulties in using market data for a valuation, analytical methods are those most frequently used in practice. These methods are based on the potential production capacity of a piece of land for accommodating a plantation, so that one should talk about an expected value of the land as a function of that production capacity for the species being analyzed. In addition to the quality of the stand, the factors influencing the value of forest land would be the rotation considered, the discount rate selected, the intensity and cost of the management actions (related to volume and quality of the products obtained) and the market price of the forest products.

The estimation of the value of a piece of land for a regular forest stands is known by the name of Land Expectation Value (*LEV*) and corresponds to the net present value of bare soil perpetually devoted to the production of timber of the same species, with the same type of management and subject to the same constraints (ecological, legal, administrative and economic). Bearing all these hypotheses in mind, the expected value of the soil will be defined by starting from (5.31). Thus, by taking the same notation as for receipts and payments, first, the net present value associated with a plantation (*npv*) is defined from the following expression:

$$npv = P\ (t)\ f(t)e^{-it} - K - G\alpha - \sum_{\forall s} Y_s e^{-i\,s} + \sum_{\forall l} C_l e^{-i\,l} \tag{5.40}$$

Given that it is intended to obtain the value of an infinite series of plantations with one same rotation fixed initially, the result would be a perpetual and periodic series. However, the corresponding expression is not found in Fig. 5.1 since we started off from a stream of future flows where the first of these is found in the year zero. To solve this problem, although there are several ways to tackle it, in this case formula (5.14) has been applied. In any case, the land expectation value would be:

$$LEV = \frac{npv}{e^{it} - 1} \tag{5.41}$$

Note that the above equation is a similar expression to that which was employed for defining the Faustmann optimal rotation (5.23). On the other hand, if a discrete

Table 5.3 Revenues and cost of a teak plantation (From Hallett et al. 2011)

Age	Activity	Revenue ($/ha)
0	Plant	−3,500
3	Pre-thinning	960
6	First thinning	1,500
10	Second thinning	4,050
15	Third thinning	10,150
25	Final harvest	85,500

discount rate is used, taking *npv′* associated with the same plantation but calculated taking a discrete discount rate, the expression which would have to be applied is the following:

$$LEV = \frac{npv'(1+i)^t}{(1+i)^t - 1} \tag{5.42}$$

The final *npv* can be a negative value. This leads one to think that, from a strictly financial point of view, and according to those values of the parameters included in the analysis (prices, costs, discount rate, etc.), the owner will not be incentivized to invest in tree-covered areas, unless there is a grant attached. Also, due to the habitual dilated action horizons typical of investments in forest plantations, the land expectation value is highly sensitive to the discount rate selected. Logically, the lower the type of interest is, the higher the land expectation value. On the other hand, the idea of the *LEV* permits a comparison between possible alternatives which do not present the same rotation age and are not circumscribed to the existence of a plantation whose principal output is timber. If the plantation is orientated towards, for example, some non-timber forest products, the concept remains totally valid.

5.5.3 Some Examples

In this section an example is presented in which the *LEV* concept is illustrated. This starts from the work of Hallett et al. (2011) in which a series of receipts and payments associated with teak plantations (*Tectona grandis* L.f.) in Mexico, is shown in Table 5.3. It is aimed to calculate the value of the land which could accommodate this plantation. For this purpose, it is assumed that the rotation age is of 25 years.

If it is now desired to calculate the *LEV*, it is essential to fix the discount rate to be used. Following recent works analysing the profitability of forest plantations in Mexico (Téllez et al. 2008; Zamudio et al. 2010) a real discount rate of 5 % has been taken. With these data, the calculation of the net current value associated with a rotation of the plantation (*npv*) is immediate:

$$npv = -3{,}500 + 960e^{-3\cdot 0{,}05} + 1{,}500e^{-6\cdot 0{,}05} + 4{,}050e^{-10\cdot 0{,}05} +$$
$$+ 10{,}150e^{-15\cdot 0{,}05} + 85{,}500e^{-25\cdot 0{,}05} =$$
$$= 30{,}185\$/ha \quad (5.43)$$

Next, the land expectation value is obtained by applying the expression (5.41):

$$LEV = \frac{npv}{1 - e^{-it}} = \frac{30{,}185}{1 - e^{-25\cdot 0{,}05}} = 42{,}305\$/ha \quad (5.44)$$

In short, the value per hectare of this land reaches 42,305$/ha according to the assumptions with regard to the receipts, payments, rotation and discount rate previously introduced. Note that the difference between *LEV* and *npv* (42,305−30,185 = 12,121$/ha) would be the value of the sum of the infinite rotations which would be produced in this property as from the first rotation. Finally, to illustrate the sensitivity of the LEV towards changes in the discount rate, it can be verified that when the discount rate is doubled (10 %), the LEV is reduced to 9,595$ha.

5.5.4 Plantation Valuation

When making a valuation of a forest plantation, its casuistry may be very high. The following are some of the most common cases starting from the hypothesis that we are, in principle, talking about regular stands. Besides, in Management Planning in Action 5.1 some appraisal approaches applied in different countries are shown.

Management Planning in Action 5.1: The Importance of Discounting Cash Flows to Forest Appraisers

Timberland Investment Management Organizations (TIMOs) and Real Estate Investment Trusts (REITs) manage forest properties. The main difference between these two institutions is the composition of its portfolio: the REITs have in their portfolio other assets, like real state, while TIMOs are exclusively focused on the forest assets.

TIMOs handle funds through the acquisition, management and sale of forest assets. High demand for forestland in the U.S. has made the best opportunities quickly scarce in the country and therefore some U.S. investors began to look into international opportunities to diversify their assets. Brazil, for instance, has quickly become one of the main destinations for these investors. In 2008, foreign and Brazilian TIMOs had around 180 thousand hectares in the country.

(continued)

(continued)

Annually TIMOs hire specialized professionals to determine the value of their properties and to inform their customers whether the property is depreciating or not. The need for annual updates of the value of the property requires standardized methods. These methods, when applied to the valuation of forest assets, are referred to as Forest Appraisal Approaches.

In USA, the Uniform Standards of Professional Appraisal Practice (USPAP)[1] promotes the standardization of the appraisal process and serves as the reference for professional forest appraisers. The European counterpart, with its operational headquarters in London, is the International Valuation Standards Council (IVCS),[2] which focuses on the production of international standards and promotes the strengthening of the valuation profession. In 2013, for instance, IVCS distributed a draft document with the organization's proposals for a future standard on valuation of forests (IVSC 2013), as a response to "the need for technical guidance to assist both professional appraisers and users in understanding the application of those principles to the valuation of interests in forests and forestry operations". In Brazil, a similar standard, issued in 2004 (ABNT NBR 14653-3) by the Brazilian Association of Technical Standards (ABNT), has been used as the reference for the valuation of Rural Assets.

Considering the current interest on investments in forest assets in Brazil, and the predominant use of the USPAP standards to evaluate these assets, a brief description of these methods is provided. Basically, and according to these standards, the appraiser evaluates the forest based on the results of three alternative methods: (i) the sales approach method, which compares and adjusts transactions made in the same region of the forest being evaluated; (ii) the cost approach method, which adds the costs of forming the current standing forest; and (iii) the income approach method, which considers all discounted future net benefits derived from the forest to determine a present value. The concepts presented in this chapter are essential to understand the income approach.

---

[1] http://www.appraisalfoundation.org/

[2] http://www.ivsc.org/

5.5.4.1 Valuation of a Mature Regular Forest

In this case the valuation of the stand and of the land can be made independently. Leaving aside the synthetic methods due to the lack of reliable data, the valuation is carried out by measuring the timber existing in the plantation and multiplying this volume by the market price corresponding to each case. Note that it can be a price in the forest, in the industry, etc. The land's valuation is made by calculating the LEV according to the indications in the expressions (5.41) or (5.42). Thus, the total value is the sum of the value of the stand and of the land.

5.5.4.2 Valuation of an Immature Regular Forest

An immature regular forest does not have any actual value as it does not have the minimum age necessary for producing saleable products yet. Therefore, should be a joint valuation of the land and the stand. Intuitively one could think that an immature stand would present a higher value than that of the deforested forest but how much higher? The initial value of the forest at the beginning of the rotation is the value of the soil (bare soil), to which the value of the stand is added as its growth is completed.

One procedure for calculating the valuation of a plantation in the first stages of its life consists of capitalizing the flow of revenue and payments from its current age up to its felling age (*RC*), adding it to the soil expectation value (*LEV*) and discounting from that flow a number of years equal to those remaining for the final felling, so that the net updated value of the soil and of the stand of an immature mass at the age of e years (NPV_e) is obtained. The algebraic expression of this sum would be:

$$NPV_e = \frac{RC + LEV}{e^{-i(t-e)}} \tag{5.45}$$

Where this sum of cash-flows up to rotation is defined as:

$$RC = \sum_{t-e}^{t} R_k e^{\,i(t-k)} - \sum_{t-e}^{t} C_k e^{\,i(t-k)} \tag{5.46}$$

Other authors (Gunter and Hanley 1984) propose carrying out this calculation by a slightly different procedure. In synthesis, this would be the capitalization of all the cash-flows from the current moment up to *t* years later (i.e. a complete rotation). Once calculated at this age the difference between the revenues and payments conveniently capitalized (*b*), it is considered as if it were a periodic and perpetual series, since it is assumed to be repeated indefinitely. In short, it would mean applying a formula equivalent to that which has been used to define the LEV:

$$NPV_e = \frac{b}{1 - e^{-it}} \tag{5.47}$$

Where that term *b* is defined as:

$$b = \sum_{k=e}^{t+e} R_k e^{\,i((t+e)-k)} - \sum_{k=e}^{t+e} C_k e^{\,i((t+e)-k)} \tag{5.48}$$

5.5.5 Impact of Forest Taxes on Stand Values and Rotations

Finally, it is necessary to add that the analysis carried out up to now on the optimal rotation as a forest valuation has excluded the taxes to which both the cash-flows associated with a plantation and its ownership might be subjected. When talking about taxes we find ourselves with a highly varied casuistry, which is modified from country to country, and even there are differences at a regional level. However, and due to its undoubted importance, the literature on this subject is extensive as is shown in some review works (Amacher 1997). In these works, the most direct relationship consists of how the economically optimal rotation can be modified when taxes are introduced into the analysis. Thus, among others, the works of Chang (1983), Gamponia and Mendelsohn (1987), or Amacher et al. (1991) deal with by how much the economically optimal rotation is lengthened or shortened in terms of endogenous factors of the stand (age, site index, etc.), or of the type of tax (whether it be the value of the property, of the land, the value of the trees standing, so much percent of the value of the annual growth of the stand, etc.) and of the moment of its application. Although it is not aimed to encompass all the possible cases which might exist, one example of how a type of tax could be introduced into the cited analysis is shown next:

$$LEV = \frac{P(t)f(t)e^{-it} - \omega e^{-it} - K - G\alpha - \sum_{\forall s} Y_s e^{-is} - K'}{1 - e^{-it}} + K' + K_1$$

with:

$$\alpha = \frac{e^{(-i1)} \cdot \left(e^{(-it)} - 1\right)}{\left(e^{(-i1)} - 1\right)} \tag{5.49}$$

Equation 5.49 shows a simpler case than that shown previously, like the one indicated in Eq. 5.31 but in this expression the term ω has been introduced which would represent the total amount which has to be paid for different taxes. This amount will depend on several factors; on one hand those relative to the type of tax and its modality, and, on the other, motivated both by gross returns and by the specific situations of each owner. With this modification, the optimal rotation could be calculated, or could modify more precisely this expression so that each of the possible situations is reflected.

5.6 Stand-Level Management Planning

Industrial forests management planning is typically developed at regional and forest-level spatial scales as it encompasses the coordination of stand management activities to address concerns with the sustainability of the ecosystem and its product flows (Chaps. 2 and 6). Stands i.e. land units that are homogeneous according to ecological, economic and development criteria (Chap. 2) are thus the building blocks of industrial forests. Management options are most often implemented at the stand-level spatial scale. Nevertheless, in countries where the property is fragmented, management planning itself is conducted at stand level. This was highlighted by Borges et al. (2014) in a recent review of forest management problems that are prevalent world-wide. Forest management experts representing 17 countries from Europe, South America, Africa and Asia involved in the FORSYS Cost Action (http://fp0804.emu.ee/) did include stand-level management planning in the list of the most important problems that foresters have to face in their countries (Borges et al. 2014). Stand-level decisions are receiving renewed attention because the call for an ecosystem based management approach has increased the complexity of planning (Rose et al. 1995).

In fact, most often the supply of raw material to the forest industry is partially or totally provided by ownerships other than the industry itself. Stand-level management planning is reported by Borges et al. (2014) as typically encompassing a single decision-maker and a single financial return objective that depends most often on the supply of wood products. Further, the supply of timber to the forest industry is frequently associated to even-aged management practices. Thus, the solution of stand-level management planning encompasses the simultaneous optimization of regeneration, thinning and rotation options. In the case of coppice systems it may include also the definition of the number of coppice cycles to include in a rotation.

Modelling approaches thus need to be tailored to address these specific features. They take as input the prescriptions developed using stand-level growth and yield and silviculture models (Chaps. 3 and 4). Prescriptions correspond to the decision variables. i.e., the set of strategies available for the manager to make a choice. The choice is made with the support of finance models (Sect. 6.4). Nevertheless, the number of strategies is typically very large and its financial assessment requires numerical approaches. Several numerical methods for optimization of even-aged stand management have been discussed in the literature (Chap. 2). In this chapter we will illustrate the application of one method – dynamic programming – with a very simple numerical example from Borges and Falcão (1999).

Dynamic Programming (DP) is a programming technique that is very useful for stand-level optimization as it helps to avoid the problem of needing to enumerate and evaluate all possible management options (Hoganson et al. 2008). For that purpose, it decomposes the problem into a series of smaller, inter-related problems. DP has been used extensively to address thinning options for individual stands (e.g. Amidon and Akin 1968; Brodie et al. 1978; Haight et al. 1985; Arthaud and Klemperer 1988; Guo and Peyron 1995; Pelkki 1997). Diaz-Balteiro and Rodriguez (2006) used DP to address eucalypt stands management planning in Spain and in Brazil. DP further makes possible to recognize, the stochastic nature of the problem (Norstrom 1975; Haight and Smith 1991; Gunn 2005; Ferreira et al. 2011, 2012). DP formulations can be solved both forward and backward through time with each solution approach offering different advantages (Hoganson et al. 2008).

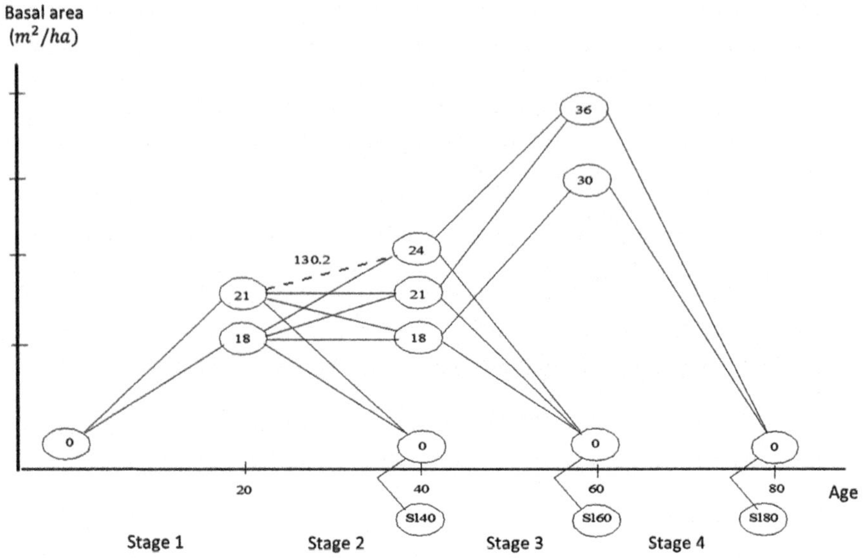

Fig. 5.7 Dynamic programming scheme

Our example stand is in Leiria National Forest (LNF), a pine forest in Portugal (Fig. 7.1). Volume growth was estimated with a simplified version of a model developed by Falcão (1998). Silvicultural options include:

- Natural regeneration.
- Thinning at 20 and at 40 years of age removing at least 1 m^2/ha of basal area (BA). Residual BA values after a thinning may be 18, 21 or 24 m^2/ha.
- Clearcut at 40, 60 or 80 years of age

The prescriptions defined according to these options may be represented in a DP network (Fig. 5.7). This is defined by nodes (states) that characterize the inventory of the stand at each age and by arcs that link nodes and correspond to decisions. The BA corresponds to the state descriptor as it characterizes the inventory in this example.

The decisions are displayed as arcs and occur at end of the stage. A dotted arc is displayed to illustrate the correspondence between arcs and silvicultural options in the text. Nodes above the xx axis identify the basal area of the stand at the beginning and end of the stages. Nodes below the xx axis are bare land nodes used to include rotations after the first (from Borges and Falcão 1999).

According to the silvicultural options and the growth and yield model, the stand BA may be 18 or 21 m^2/ha at 20 years. In the case of the latter, the stand may be thinned when it reaches 40 years to leave 24 m^2/ha of BA. This management option is represented by the dotted arc in Fig. 5.7 and corresponds to a harvest of 130.2 m^3/ha. The value of arcs in a DP network usually corresponds to the net economic returns associated with the management options.

In this example, stages or sub-problems are defined by stand ages (e.g. 0, 20, 40, 60 and 80 years), the ages when management options may be implemented. A 20-years interval is used to define each stage. In order to avoid rounding-off errors associated with arcs that correspond to harvests, it will be assumed that management options occur at the end of the stage. A path in the DP network, starting in the initial node at age 0 and ending in a node with BA = 0 at ages 40, 60 or 80 corresponds to a feasible prescription.

The optimization of the stand-level problem thus corresponds to the identification of the optimal path in the DP network. Both forward or backward through time approaches may be used. If the net present value is used to compare prescriptions and one single rotation is considered the forward approach may be described as:

$$\text{Determine } F^*_{Rv}(0) \ \ \forall Rv \tag{5.50}$$

with

$$F^*_n(BA_n) = \max_{BA_{n-1} \geq BA_n - G_{n-1}(BA_{n-1})} \Big(R_{n-1}(BA_{n-1} + G_{n-1}(BA_{n-1}), BA_n) + F^*_{n-1}(BA_{n-1}) \Big) \tag{5.51}$$

$$F_1(0) = 0 \tag{5.52}$$

$$R_{n-1}(BA_{n-1} + G_{n-1}(BA_{n-1}), BA_n) = P_A * Vol(BAR_{n-1}(BA_{n-1}, BA_n)) \tag{5.53}$$

$$BA_n = BA_{n-1} + G_{n-1}(BA_{n-1}) - BAR_{n-1}(BA_{n-1}, BA_n) \tag{5.54}$$

with

$n = 1, 2, 3, 4$ identifies the stage, which is é characterized by stand age.
$Rv = 2$, 3 and 4 identifies the stages at the end of which a clearcut may occur.
BA_n = the node (state) in the beginning of stage n. It corresponds to the basal area per hectare in the stand at the beginning of the stage n.
$G_{n-1}(BA_{n-1})$ = BA growth in stage n−1.
$BAR_{n-1}(BA_{n-1}, BA_n)$ = BA per hectare removed by a harvest at the end of stage n−1 to leave a residual BA_n
$Vol(BAR_{n-1}(BA_{n-1}, BA_n))$ = volume harvested at the end of stage n−1.
P_A = discounted price ($/m^3).

Equation 5.50 expresses the stand-level management objective of maximizing net present value. Equation 5.51 reflect the DP recursive relations and compute the value of the optimal path arriving to nodes BA_n, taking into account the sum of the discounted net return associated with the management policy implemented at the end of stage n with the value of optimal policies associated with the nodes BA_{n-1} to which BA_n is linked. Equation 5.52 states the initial condition, the value of the initial node is 0. Equation 5.53 are the return functions, they compute the discounted net return resulting from the harvest at the end of stage n−1, the value of the corresponding arcs. Finally, Eq. 5.54 corresponds to the DP transition functions that define the relations between states (nodes) in consecutive stages.

In order to apply the forward solution approach we started by computing the arc values, i.e. the values of the return function for all management options (Table 5.4). We considered a constant price 8×10^3\$/m^3 and a discount rate $i = 0.04$. No regeneration costs were considered.

The data needed to develop both DP solution approaches may be organized as follows (Table 5.4). Each line in the table represents one of the 18 arcs in the DP network. Column (2) presents BA values at the beginning of the 20-years period that defines the stage. It thus presents the values of the state descriptors in nodes at the beginning of the stage. Column (3) presents BA values just before a management option is implemented at the end of the 20-years period. These values correspond to the sum of the value in column (2) with the BA growth value computed by the growth and yield model. Column (4) presents the values of BA removed by the management option associated with the arc that connects the node described in Column (2) and the node described in Column (5). The latter presents BA values at the end of the 20-years period that defines the stage. It thus presents the values of the state descriptors in nodes at the end of the stage. These are computed by taking the difference between the values in Columns (3) and (4). Column (6) displays the volume harvested as computed by an equation where BA removed is the dependent variable. Column (7) presents the value of the discounted price at end of the stage. Finally the values in Column (8) result from the product of values in Columns (6) and (7). They correspond to the values of the DP return function or arc values (Eq. (5.53))

The forward solution approach generally only applies to situations involving a full rotation. It starts by labelling the node at the start of the planning horizon (time = 0) as having a zero value and then solves forward through time, stage-by-stage. Nodes values represent the maximum net present value from time zero to the time associated with the corresponding stage (Hoganson et al. 2008). The approach may be illustrated as follows (Table 5.5).

We start by identifying the departure and the arrival nodes of each arc in the DP network. They correspond to Columns (2) and (3) in Table 5.5. We also identify the management option associated to each arc (Column 5) as well as the resulting financial return (Column 6). The latter is taken from Column (8) in Table 5.4 Afterwards the solution approach proceeds as follows. The initial condition (Eq. 5.52) provides the value of the first node in the network. The sum of the value of the departure node (0 in Column (4)) and the values of the arcs that connect it with the nodes at the end of the first stage (92.5 and 46.3, respectively, in Column (6)) provide the values of these two arrival nodes (92.5 and 46.3, respectively, in Column (7)).

We have now the solution of the first stage (sub problem). In the second stage, ending nodes may be arrived from different departure nodes (Fig. 5.6). For example two arcs lead to node representing BA equal to 24 m^2/ha at 40 years of age. The first

Table 5.4 Financial returns associated to management options (DP arc values)[a] (Adapted from Borges and Falcão 1999)

Stage n (1)	Initial BA BA_n (2)	BA before the harvest $BA_n + G_n(BA_n)$ (3)	BA removed $BAR_n(BA_n, BA_{n+1})$ (4)	Residual BA BA_{n+1} (5)	Volume harvested $Vol(BAR_n (BA_n, BA_{n+1})$ (6)	Discounted price P_A (7)	Net present value NPV (8)
1	0	24	6	18	25.3	$8/(1+i)^{20}$	92.5
1	0	24	3	21	12.7	"	46.3
2	21	42.7	24.7	18	172.2	$8/(1+i)^{40}$	287
2	21	42.7	21.7	21	151.2	"	251.9
2	21	42.7	18.7	24	130.2	"	216.9
2	21	42.7	42.7	0	301.5	"	502.4
2	18	40.3	22.3	18	155.5	"	259.2
2	18	40.3	19.3	21	134.5	"	224.1
2	18	40.3	16.3	24	113.5	"	189.1
2	18	40.3	40.3	0	284.8	"	474.4
3	24	36	0	36	0	$8/(1+i)^{60}$	0
3	24	36	36	0	325.0	"	247.2
3	21	36	0	36	0	"	0
3	21	36	36	0	325.0	"	247.2
3	18	30	0	30	0	"	0
3	18	30	30	0	284.8	"	216.6
4	36	43.7	43.7	0	454.0	$8/(1+i)^{80}$	157.6
4	30	39.1	39.1	0	406.9	"	141.2

[a]BA, Volume and Net Present Value units: m^2/ha, m^3/ha and 10^3 \$/ha, respectively

Table 5.5 Solution of the stand management planning problem by the DP forward approach. Shaded cells indicate the optimal path in the case of a 80 years rotation (Adapted from Borges and Falcão 1999)

Stage n	BA_n	BA_{n+1}	$F_n^*(BA_n)$	Management option	$R_n(BA_n+G_n(BA_n),BA_{n+1})$	$F_{n+1}^*(BA_{n+1})$
(1)	(2)	(3)	(4)	(5)	(6)	(7)
1	0	18	0	Thinning	92.5	92.5
1	0	21	0	Thinning	46.3	46.3
2	18	24	92.5	Thinning	189.1	281.6
	21	24	46.3	Thinning	216.9	
2	18	21	92.5	Thinning	224.1	316.6
	21	21	46.3	Thinning	251.9	
2	18	18	92.5	Thinning	259.2	351.7
	21	18	46.3	Thinning	287	
2	18	0	92.5	Clearcut	474.4	566.9
	21	0	46.3	Clearcut	502.4	
3	21	36	316.6	-	0	316.6
	24	36	281.6	-	0	
3	18	30	351.7	-	0	351.7
3	18	0	351.7	Clearcut	216.6	568.3
	21	0	316.6	Clearcut	247.2	
	24	0	281.6	Clearcut	247.2	
4	30	0	351.7	Clearcut	141.2	492.9
	36	0	316.6	Clearcut	157.6	

(2) and (3): m^2/ha; (4), (6) and (7): 10^3\$/ha

starts at node 18 m^2/ha at 20 years of age. The second has as its departure node 21 m^2/ha at 20 years of age. In this case, the DP recursive function (Equation 51) determined the value of node 24 m^2/ha at 40 years of age as the value of the path ending in the node that leads to a higher net return. The value of the path ending in this node and passing through node 18 m^2/ha at 20 years of age is equal to the sum 92.5 (value of the departure node in Column (4)) and 189.1 (value of the arc in Column (6)). The value of the path ending in this node and passing through node 21 m^2/ha at 20 years of age is equal to the sum 46.3 (value of the departure node in Column (4)) and 216.9 (value of the arc in Column (6)). Thus the value of the ending node is 281.6 (Column (7)). The values of the other nodes at the end of the second stage are computed the same way. This completes the solution of the second stage (sub problem).

The forward solution approach ends when all ending nodes values have been computed (Column (7) in Table 5.5). These represent the net present value of the optimal paths reaching each node. For example, in the case of node 0 at 80 years of age the value of that path is 492.9×10^3\$/ha. It corresponds to the value of Eq. (5.50) for Rv equal to 4 and to the net present value associated with a 80 years rotation. That path is traced backwards through the DP network and corresponds to the optimal prescription for a 80 years rotation (Table 5.5):

- Thinning at 20 years leaving a residual BA of 18 m^2/ha. The volume harvested is 25.3 m^3/ha (Table 5.4) and the discounted return is 92.5×10^3\$/ha.
- Thinning at 40 years leaving a residual BA of 18 m^2/ha. The volume harvested is 155.3 m^3/ha (Table 5.4) and the discounted return is 259.2×10^3\$/ha.
- No harvest at 60 years so that the BA may reach 30 m^2/ha.
- Clearcut at 80 years. The volume harvested is 406.9 m^3/ha (Table 5.4) and the discounted return is 141.2×10^3\$/ha.

The identification of the optimal paths (prescriptions) associated to 40 and 60 years rotations is similar. The values of Eq. (5.50) for Rv equal to 2 (40 years rotation) and 3 (60 years rotation) are, respectively, $F_2^*(0) = 566.9 \times 10^3$\$/ha and $F_3^*(0) = 568.3 \times 10^3$\$/ha (Table 5.5). Nevertheless, the values of the recursive function associated with the three DP ending nodes associated with each of the three rotations are not sufficient for selecting the optimal stand management plan. Planning horizons need to normalized (Sect. 6.4). For that purpose perpetual series are considered (Equation 55):

$$F_{Rv}^*(0)(1+i)^{Rv*20} / \left((1+i)^{Rv*20} - 1\right) \tag{5.55}$$

with:

$(1+i)^{Rv*20}$ = factor to compute the value of the optimal path at the rotation age.
$(1+i)^{Rv*20} - 1$ = factor to compute the present value of a perpetual series of revenues occurring every $Rv * 20$ years

Equation 5.55 corresponds to the Faustmann formula. In our example, its values are 716.2, 628.0 and 514.9×10^3\$/ha in the case, respectively of rotations of 40, 60 and 80 years. The optimal rotation age is thus 40 years. Opportunity costs associated with the delay of future rotations suggest that a rotation of 40 years is selected rather than a 60 years rotation although the latter leads to higher returns during the rotation. The node values for all of the bare land nodes Sl40, Sl60 and Sl80 (Fig. 5.6) at the end of a full rotation correspond to these estimates by the Faustmann formula.

The optimal prescription by the backward through time approach is the same. Yet the information provided is different. The method may be described as follows:

$$\text{Determine } F_1^*(0) \ \forall SlRv \tag{5.56}$$

with:

$$F_n^*(BA_n) = \max_{BA_{n+1} \leq BA_n + G_n(BA_n)} \left(R_n\left(BA_n + G_n\left(BA_n\right), BA_{n+1}\right) + F_{n+1}^*\left(BA_{n+1}\right) \right) \tag{5.57}$$

$$F_{Rv}(SlRv) \text{ is known} \forall SlRv \tag{5.58}$$

$$R_n\left(BA_n + G_n\left(BA_n\right), BA_{n+1}\right) = P_A * Vol\left(BAR_n\left(BA_n, BA_{n+1}\right)\right) \tag{5.59}$$

$$BA_n = BA_{n+1} - G_n\left(BA_n\right) - BAR_n\left(BA_{n+1}, BA_n\right) \tag{5.60}$$

with:n, Rv, BA_n, $G_n(BA_{n+1})$, $BAR_n\left(BA_n, BA_{n+1}\right)$, $Vol\left(BAR_n\left(BA_n, BA_{n+1}\right)\right)$, P_A as described earlier and $SlRv = Sl40$, $Sl60$ and $Sl80$ identify the bare land nodes.

Equation 5.56 expresses the stand-level management objective of maximizing net present value of a perpetual series of rotations. The value of the initial node thus corresponds to the net present value of a perpetual series of optimal rotations. Equation 5.57 reflect the DP recursive relations and compute the value of the optimal path from nodes, taking into account the sum of the discounted net return associated with the management policy implemented at the end of stage $n+1$ with the value of optimal policies associated with the nodes to which is linked. Equation 5.58 states the initial condition, the value of bare land nodes *SlRv* must be known to start the solution process. Equation 5.59 are the return functions, they compute the discounted net return resulting from the harvest at the end of stage n, the value of the corresponding arcs. Finally, Eq. 5.60 corresponds to the DP transition functions that define the relations between states (nodes) in consecutive stages.

After the computation the DP arc values (Table 5.4), the backward through time approach uses an estimate of the bare land value at the end of the rotation. This estimate is used to value all the bare land nodes associated with each of the possible rotation ages (Hoganson et al. 2008). Each of these nodes in the DP network is valued as the estimated bare land value discounted the number of years corresponding with the point in time associated with the stage. This solution approach solves the network backward through time, with the last node labelled being the bare land node at the start of the planning horizon (Hoganson et al. 2008). Nodes values represent estimates of the maximum net present value that can be achieved from that node to infinity. The approach may be illustrated as follows (Table 5.6).

Again, we start by identifying the departure and the arrival nodes of each arc in the DP network. They correspond to Columns (2) and (3) in Table 5.6. We also identify the management option associated to each arc (Column 6) as well as the resulting financial return (Column 7). The latter is taken from Column (8) in Table 5.4. Afterwards the solution approach proceeds as follows. The initial condition (Eq. 5.58) provided the value of the bare land nodes at the end of the rotation. The estimate used was 716.2×10^3\$/ha. Thus the estimated values of S140, S160 and S180 were, respectively 149.3, 68.1 and 31.1×10^3\$/ha (Column (4).

Table 5.6 Solution of the stand management planning problem by the DP backward approach. Shaded cells indicate the optimal path for an estimate of bare land value at the end of the rotation equal to 716.2 × 103$/ha (Adapted from Borges and Falcão 1999)

Stage n	BA_{n+1}	BA_n	*SIRv*	$F^*_{n+1}(x)$	Management option	$R_n(BA_n + G_n(BA_n), BA_{n+1})$	$F^*_n(BA_n)$
(1)	(2)	(3)	(4)	(5)	(6)	(7)	(8)
1	18	0	-	623.7	Thinning	92.5	716.2
	21	0	-	651.7	Thinning	46.3	
2	18	18	-	284.7	Thinning	259.2	
	21	18	-	315.3	Thinning	224.1	
	24	18	-	315.3	Thinning	189.1	
	0	18	-	149.3	Clearcut	474.4	623.7
2	18	21	-	284.7	Thinning	287	
	21	21	-	315.3	Thinning	251.9	
	24	21	-	315.3	Thinning	216.9	
	0	21	-	149.3	Clearcut	502.4	651.7
2	0	-	*SI40*	149.3 (a)	2nd, 3rd,... rotations	-	-
3	30	18	-	172.3	-	0	
	0	18	-	68.1	Clearcut	216.6	284.7
3	36	21	-	188.7	-	0	
	0	21	-	68.1	Clearcut	247.2	315.3
3	36	24	-	188.7	-	0	
	0	24	-	68.1	Clearcut	247.2	315.3
3	0	-	*SI60*	68.1 (b)	2nd, 3rd,... rotations	-	-
4	0	30	-	31.1	Clearcut	141.2	172.3
4	0	36	-	31.1	Clearcut	157.6	188.7
4	0	-	*SI80*	31.1 (c)	2nd, 3rd,... rotations	-	-

(2) and (3):m^3/ha; (4), (6) and (7): 10^3\$/ha; (5) $x = BA_{n+1}$ or *SLRv* in the case of, respectively, nodes that characterize the state of the stand over the rotation and bare land nodes (Fig. 5.7); (a) $149.3 = 716.2*(1+i)^{40}$; (b) $68.1 = 716.2*(1+i)^{60}$; (c) $31.1 = 716.2*(1+i)^{80}$

The values of nodes at the end of each rotation are the same as the values of the bare land nodes to which they are connected, as the connecting arcs are not associated with a harvest. The values of departure nodes in the beginning of stage 4 (30 and 36 in Column (3)) are computed by adding the value of the arrival node (31.1 in Column (5)) and the values of the connecting arcs (141.2 and 157.6 in Column (7)) for, respectively nodes 30 and 36. The solution to the last stage (subproblem) is presented in Column (8): 172.3 and 188.7. In stage 3, the values of each departure node 18, 21 and 24 is computed by maximising the sum between the values of the corresponding arrival nodes (Column (5) and the values of the arcs that connect it to these nodes (Column (7)). For example, in the case of node 24, the value in Column (8), 315.3 corresponds to the highest value in the pair $(0 + 188.7)$ and $(247.2 + 68.1)$.

The backward solution approach ends when the value of the bare land node at the beginning of the rotation is computed. The values in Column (8) correspond to the optimal path out of each node, i.e. the optimal sequence of management options if the stand is in that state at that age. Accordingly, the value of the bare land node at the beginning of the rotation ((716.2 in Column (8)) corresponds to the value associated with the perpetual series of optimal rotations. This optimal solution may be identified by tracing the optimal path forward through time starting with the node at the start of the planning horizon:

- Thinning at 20 years leaving a residual BA of 18 m^2/ha. The volume harvested is 25.3 m^3/ha (Table 5.4) and the discounted return is 92.5×10^3\$/ha.
- Clearcut at 40 years. The volume harvested is 248.8 m^3/ha (Table 5.4) and the discounted return is 474.4×10^3\$/ha.
- A perpetual series of 40 years rotations with a value equal to 149.3×10^3\$/ha.

The solution by the backward approach is the same as the solution found by the forward approach as the estimate of the bare land node at the end of the rotation was exact. In general, the difference between the value of the bare land node at the start of the planning horizon and the value of the bare land node at the end of the rotation represents the estimated value of the first rotation. The estimated value of the first rotation can be used to re-estimate the land value (Hoganson et al. 2008). Ferreira et al. (2011) demonstrated that the iterative process to re-estimate the bare land value at the end of the rotation converges to the correct value.

Nevertheless the information provided by the both approaches is different, the forward approach indicates the optimal path to arrive to a node while the backward approach indicates the optimal path out of a node. For example in the case of node 21 at 40 years the value by the forward approach is 316.6×103\$/ha according to the thinning regime to follow to get 21 m^3/ha at 40 years. The value by the backward approach is 651.7×103\$/ha and it corresponds to the optimal policy of clearcutting at 60 years and having a series of 40 years rotations afterwards. Hoganson et al. (2008) discuss in detail de advantages and disadvantages of each approach in the case of stand-level management planning.

5.6.1 *Summary*

This chapter shows the economic basics associated with the management of forest plantations. First, under reasonable and commonly accepted hypotheses, we have formalized the basic principles of investment analysis applied to this type of forest systems. Thus, concepts such as the discount rate have been presented, indicating the differences between applying a discrete and a continuous discount rate. Besides, we discuss how to integrate inflation in our analysis. This financial analysis is completed by introducing the concepts of net present value and internal rate of return, which are the indicators used to address the absolute and relative profitability, respectively, of an investment in a plantation. Additionally, we illustrate different cases focusing on the way to compute the expected receipts and payments in the future. After analysing the financial profitability of the underlying investment in a plantation, the next step is to calculate the optimal forest rotation age of a plantation. Thus, the optimal rotation age is defined as the life of the timber or plantation for which the corresponding net present value is maximized. However, before developing this idea that will lead us to the best solutions from an economic point of view, the optimal rotation age is analysed from a purely technical point of view without introducing any economic parameters into the analysis. After this, several solutions has been showed in this chapter, focusing on the FPO solution. The approach basically consists, following Samuelson (1976) in the explicit introduction of the land rent or opportunity cost into the expression that calculates the net present value of the respective investment. The other approach was proposed by the German forester Faustmann in 1849, and basically consists of considering a chain of infinite plantation cycles or by using a financial language to consider a chain of infinite investments. In this chapter has been show than both approaches lead to the same solution or rotation age. Besides, some extensions and considerations to FPO solution has been approached in this chapter. The following section deals with the basic aspects of forest valuation. When making decisions in relation to a plantation, it is indispensable to find out its value. For example, if different investment alternatives are compared beyond the forestry ones themselves, it is necessary to find out with precision the value of a possible investment in lands of a forestry nature. The estimation of the value of a piece of land for a regular forest stands is known by the name of *Land Expectation Value* (LEV) and corresponds to the net present value of bare soil perpetually devoted to the production of timber of the same species, with the same type of management and subject to the same constraints. Bearing in mind that as we make a valuation of a forest plantation, its casuistry may be very high, some frequent cases has been addressed in this chapter. Finally, in the last section of this chapter, an operations research technique as dynamic programming is introduced. This method is very useful in order to optimize stand management planning decisions such as planting intensity and thinning timing and intensity.

5.6.2 Problems

1. We try to establish the economic viability of a plantation of *Quercus ilex* L. mycorrhizated with *Tuber melanosporum* Vitt. Such plantations can provide promising results according to the price is reached currently on the market by

some kind of truffles (600€/kg). The information available (Marco 2011) is summarized below:

Costs	
Plantation costs	14,000€/ha
Annual costs	200€/ha
Truffle Yields	
Year 1–4	0 kg
Year 5	1 kg
Year 6	2 kg
Year 7	4 kg
Year 8	6 kg
Year 9	8 kg
Year 10	10 kg
Year 11	14 kg
Year 12	18 kg
Year 13	22 kg
Year 14	25 kg
Year 15	28 kg
Years 16–35	30 kg
Year 36	28 kg
Year 37	25 kg
Year 38	23 kg
Year 39	21 kg
Year 40	19 kg
Year 41	17 kg
Year 42	15 kg
Year 43	13 kg
Year 44	11 kg
Year 45	9 kg

Some questions:

(a) Determine the net present value of the investment associated with the plantation, assuming a real discount rate of 4 %, and a price of truffles of 300€/kg (assuming that already includes the picking cost).
(b) From which minimum truffle price would be viable this plantation?

2. We know the growth function of a eucalyptus plantation in the state of São Paulo (Brazil). This function has the following mathematical expression: $V(t) = 751,336 * e^{(-6,0777/t)}$.

   If the price of pulpwood is 20€/t, the plantation cost is 1,500€/ha and with a real discount rate of 8 %, you must calculate:

   (a) The technical optimum rotation
   (b) The Fisher-Hotelling rotation age
   (c) The FPO rotation age
   (d) The Boulding rotation age

Besides, you must calculate the economic optimal rotation (FPO rotation) when the pulpwood price is 20€/t and, finally, repeat the calculation of economic optimal rotation if the discount rate rises to 12 %.

3. Calculate the economic optimal rotation for this poplar plantation (*Populus* sp.), taking into account these information:

Costs	
Plantation costs	1,412€/ha
Year 1	150€/ha
Year 2	150€/ha
Year 3	150€/ha
Year 4	225€/ha
Year 5	225€/ha
Year 6	120€/ha
Year 7	120€/ha
Year 8	120€/ha
Year 9	120€/ha
Annual costs	30€/ha

The real discount rate to use would be a 5 %. Besides, is known that the volume function is obtained following this equation:

$$V(t) = -0,1712^{*}t^{3} + 4,9112^{*}t^{2} - 13,613^{*}t + 3,1151$$

4. It is necessary to calculate the value of a regular *Pinus radiata* plantation, with 15 years old and a final cut at 30 years, using a discrete discount rate of 5 %. Operations to be performed (cost per hectare) are:

Year (t)	Payments (€/ha)	Receipts (€/ha)
0	1,500	
5	100	
10	150	
20		600 (first thinning)
25		1,200 (second thinning)
30		12,000 (final cut)
All years	30	

5. Calculate the value for the plantation shown in Table 5.3, if this plantation is 10 years old.
6. A *Gmelina arborea* plantation in Central America has the following estimates for the costs, revenues and yield (adapted from Hughell 1991; Rojas et al. 2004). The unitary price ($/m^3) in the final cutting is about 50$/m^3)

Costs	
Plantation costs	650$/ha
Annual costs	45$/ha
Pre-commercial thinning (year 4)	120$/ha
Pruning (year 2)	100$/ha
Pruning (year 4)	100$/ha
Pruning (year 6)	100$/ha
Revenues	
Pre-commercial thinning (year 8)	200$/ha
Pre-commercial thinning (year 12)	250$/ha
Yield table	
Year	volume (m^3/ha)
2	28.9
4	72.8
6	154.5
8	177.0
10	239.3
12	221.5
14	258.1

You must calculate the optimal rotation (FPO rotation) if discount rate raises up to 9 %.

7. In the previous exercise, you must calculate the profitability (NPV) of this plantation if a payment for carbon captured would be established. The price for the carbon is around 5$/t, and the carbon emitted by thinnings has not be considered in this analysis.
8. A forest company that sells Christmas trees wish to know the annual income obtained (it is supposed an annual income could be produced every year) over the years of their investment. The owner estimates that this investment will obtain an NPV of 10,000€ for a rotation of 7 years, by using a discount rate of 6 %.

References

Amacher G (1997) The design of forest taxation: a synthesis with new directions. Silva Fenn 31:101–119

Amacher G, Brazee R, Thompson T (1991) The effects of forest productivity taxes on rotation age and initial stand investment. For Sci 37:1099–1118

Amidon EL, Akin GS (1968) Dynamic programming to determine optimum levels of growing stock. For Sci 14:278–291

Arthaud GJ, Klemperer WD (1988) Optimizing high and low thinnings in loblolly pine with dynamic programming. Can J For Res 18:1118–1122

Bettinger P, Boston K, Siry JP, Grebner DI (2009) Forest management and planning. Academic Press, Burlington

Borges JG, Falcão A (1999) Programação dinâmica e gestão de povoamentos com estrutura regular e composição pura. Aplicação à Mata Nacional de Leiria. Revista Florestal XII(1/2):69–82

Borges JG, Nordström EM,Garcia-Gonzalo J, Hujala T, Trasobares A (eds) (2014) Computer-based tools for supporting forest management. The experience and the expertise world-wide. SLU University Press (in press)

Boulding KE (1935) The theory of a single investment. Q J Econ 49:475–494
Brodie JD, Adams DM, Kao C (1978) Analysis of economic impacts on thinning and rotation for Douglas-fir, using dynamic programming. For Sci 24:513–522
Buongiorno J, Zhu S, Zhang D, Turner J (2003) The global forest products model. Structure, estimation and applications. Academic Press, San Diego
Campbell H, Brown R (2003) Benefit-cost analysis. Cambridge University Press, New York
Chang S (1983) Rotation age, management intensity and the economic factors of timber production: do changes in stumpage price, interest rate, regeneration cost, and forest taxation matter? For Sci 29:257–277
Davis LS, Johnson KN, Bettinger PS, Howard TE (2001) Forest management, 4th edn. McGraw-Hill, New York
Diaz-Balteiro L, Rodriguez LCE (2006) Optimal rotations on Eucalyptus plantations including carbon sequestration – a comparison of results in Brazil and Spain. For Ecol Manag 229:247–258
Diaz-Balteiro L, Romero C (1994) Rentabilidad económica y turnos óptimos de choperas en España (*Economic profitability and financial rotations of poplar plantations in Spain*). Investig Agrar Sist Recur For 3:43–56
Diaz-Balteiro L, Romero C (2001) Forest management and carbon captured: analytical aspects and policy implications. Investig Agrar Sist Recur For Fuera de Serie 1:153–165
Falcão A (1998) DUNAS – a growth model for the national forest of Leiria. In: Proceedings of the IUFRO workshop empirical and process-based models for forest tree and stand growth simulation, Oeiras, 21–27 Setembro 1997
Faustmann M (1849) Gerechnung des werteswelchenWaldbodensowienochnichthaubareholzbestäncefüe die Waldwirts-chafbesitzen. Allg Forst Jagdztg 25:441–455. English version: J Forest Econ 1995, 1:7–44
Ferreira L, Constantino M, Borges JG (2011) A stochastic approach to optimize Maritime pine (*Pinus pinaster* Ait) stand management scheduling under fire risk. An application in Portugal. Ann Oper Res. doi:10.1007/s10479-011-0845-z
Ferreira L, Constantino M, Borges JG, Garcia-Gonzalo J (2012) A stochastic dynamic programming approach to optimize short-rotation coppice systems management scheduling. An application to eucalypt plantations under wildfire risk in Portugal. For Sci 58:353–365
Fisher I (1930) The theory of interest. Macmillan, London
Gamponia V, Mendelsohn R (1987) The economic efficiency of forest taxes. For Sci 33:367–378
Götze U, Northcott D, Schuster P (2008) Investment appraisal. Methods and models. Springer, Berlin
Gunn E (2005) A neuro-dynamic programming approach to the optimal stand management problem. In: Bevers,M, Barrett TM comps. Systems analysis in forest resources: proceedings of the 2003 symposium. General technical report PNW-GTR-656. U.S. Department of Agriculture, Forest Service, Pacific Northwest Research Station, Portland OR, pp 265–272
Gunter JE, Hanley HL (1984) Essentials of forestry investment analysis. Oregon State University Bookstores, Corvallis, OR 97339
Guo B, Peyron J-L (1995) Optimiser la sylviculture à lon terme des peuplements forestiers équiennes grâce au logiciel Sylopt. Rév For Fr n° sp. 120–130
Haight RG, Smith WD (1991) Harvesting loblolly pine plantations with hardwood competition and stochastic prices. For Sci 37:1266–1282
Haight RG, Brodie JD, Dahms WG (1985) A dynamic programming algorithm for optimization of Lodgepole pine management. For Sci 31:321–330
Hallett JT, Díaz-Calvo J, Villa-Castillo J, Wagner MR (2011) Teak plantations: economic bonanza or environmental disaster? J For 109:288–292
Hartman R (1976) The harvesting decision when a standing forest has value. Econ Inq 16:52–58
Healy MJ Jr, Bergquist K (1994) The sales comparison approach and timberland valuation. Apprais J 62:587–595
Hoganson HM, Borges JG, Wei Y (2008) Coordinating management decisions of neighboring stands with dynamic programming. In: von Gadow K, Pukkala T (eds) Designing green landscapes. Managing forest ecosystems, vol 15. Springer, Dordrecht, pp 187–214

Hotelling H (1925) A general mathematical theory of depreciation. J Am Stat Assoc 20:340–353
Hughell D (1991) Modelo preliminar para la predicción del rendimiento de *Gmelina arborea* Roxb. En América Central. Silvoenergía (C.R.) No. 44, pp 1–4
IVSC (2013) The valuation of forests – exposure draft. International Valuation Standards Council, London. 29 pp. http://www.ivsc.org/sites/default/files/Forestry%20TIP%20Exposure%20Draft_0.pdf
Johansson PO, Löfgren KG (1985) The economics of forestry and natural resources. Basil Blackwell, Oxford
Kula E (1984) Derivation of social time preferences rates for the United Sates and Canada. Q J Econ 99:873–883
Kula E (1985) Derivation of social time preferences rates for the United Kingdom. Environ Plann A 17:199–212
Löfgren KG (1983) The Faustmann-Ohlin theorem: a historical note. Hist Polit Econ 15:261–264
Marco R (2011) Propuesta de incorporación del aprovechamiento trufero en el paraje de La Umbría, Codes (Guadalajara). Trabajo Fin de Carrera. ETS Ingenieros de Montes, Madrid (not published)
Medema EL, Lyon GW (1985) The determination of financial rotation ages for coppicing tree species. For Sci 31:398–404
Norstrom CJ (1975) A stochastic model for the growth period decision in forestry. Swed J Econ 77:329–337
Ohlin B (1921) Till fragan om skogarnas omloppstid. EkonomiskTidskrift 22:89–113. English version, in J Forest Econ,1995 1:89–114
Peak R, Neale B (2006) Corporate finance and investment decisions & strategies, 5th edn. FT Prentice Hall, London
Pelkki M (1997) The effects of neighborhood storage size on dynamic programming solutions. For Sci 43:387–395
Pressler MR (1860) Aus der Holzzuwachlehre (zweiterArtikel). Allg Forst Jagdztg 36:173–191. English version, in J For Econ 1995, 1:45–87
Rodríguez LCE, Santos Bueno AR, Rodrigues F (1997) Rotações de eucaliptos mais longas: análise volumétrica e económica (*Longer eucalypt rotations: volumetric and economic analysis*). Sci For 51:15–28
Rojas F, Arias D, Moya R, Meza A, Murillo O, Arguedas M (2004) Manual para productores de melina (*Gmelina arbórea*) en Costa Rica. FONAFIFO, San José
Romero C (1997) Economía de los Recursos Ambientales y Naturales. Alianza Editorial, Madrid
Romero C, Ríos V, Díaz-Balteiro L (1998) Optimal forest rotation age when carbon captured is considered: theory and applications. J Oper Res Soc 49:121–131
Rose DW, Borges JG, Pelkki MH (1995) Forest management planning based on stand level decisions. North J Appl For 12(3):133–142
Samuelson PA (1976) Economics of forestry in a evolving society. Econ Inq 14:466–492
Sen AK (1967) The social time preference rate in relation to the market rate of interest. Q J Econ 81:112–124
Tait DE (1986) A dynamic programming solution of financial rotation ages for coppicing tree species. Can J For Res 16:799–801
Téllez E, González M, De los Santos HM, Fierros AM, Lilieholm RJ, Gómez A (2008) Rotación óptima en plantaciones de eucalipto al incluir ingresos por captura de carbono en Oaxaca, México (*Optimal timber rotation lengths in eucalyptus plantations including revenues from carbon capture in Oaxaca, México*). Rev Fitotec Mex 31:173–182
Wagner JE (2012) Forestry economics. A managerial approach. Routledge, New York
Zamudio FJ, Romo JL, Cervantes JOA (2010) Evaluaciónfinanciera y de riesgo de una plantación forestal comercial en Zihuateutla, Puebla. (*Financial and risk assessment of a commercial forest plantation in Zihuateutla, Puebla*). Rev Chapingo 16:69–78
Zinkhan FC, Sizemore WR, Mason GH, Ebner TJ (1992) Timberland investments. A portfolio perspective. Timber Press, Portland

Chapter 6
Strategic Management Scheduling

José G. Borges, Jordi Garcia-Gonzalo, Susete Marques, Victor A. Valdebenito, Marc E. McDill, and André O. Falcão

6.1 Introduction

Strategic management planning plays a key role in the development of forest schedules as the temporal dimension is a determinant characteristic of all forestry production systems. Strategic or long-term management planning typically encompasses temporal horizons extending over more than 10 years so that forestry economic and biological processes may be adequately acknowledged (Chap. 2). The importance of this planning level was highlighted in a recent review of forest management problems that are prevalent world-wide (Borges et al. 2014b). Forest management experts representing 26 countries from Europe, North and South America, Africa and Asia involved in the FORSYS Cost Action (http://www.cost.eu/domains_actions/fps/Actions/FP0804) did include strategic management planning in the list of the most important problems that foresters have to face in their countries (Borges et al. 2014b).

Long term stand-level problems tend to be more important in countries where the forest is mostly privately owned and highly fragmented (Borges et al. 2014b). This spatial scale structures a problem with specific features and that requires specialized solution techniques (Chap. 6). In this chapter we will focus on long-term management planning at the regional and forest-level spatial scales (Chap. 2). The

J.G. Borges (✉) • J. Garcia-Gonzalo • S. Marques
Forest Research Centre, School of Agriculture, University of Lisbon, Lisbon, Portugal
e-mail: joseborges@isa.ulisboa.pt

V.A. Valdebenito
School of Forest Resources, Pennsylvania State University, University Park, PA, USA

M.E. McDill
Department of Ecosystem Science and Management, Pennsylvania State University, University Park, PA, USA

A.O. Falcão
Faculdade de Ciências, University of Lisbon, Lisbon, Portugal

J.G. Borges et al. (eds.), *The Management of Industrial Forest Plantations: Theoretical Foundations and Applications*, Managing Forest Ecosystems 33,
DOI 10.1007/978-94-017-8899-1_6, © Springer Science+Business Media Dordrecht 2014

latter typically encompasses several contiguous stands and is more frequent in the case of industrial plantations. Nevertheless, in the case of the vertically integrated forest industry its land base is often dispersed over a region. Further, the industry is often concerned with the supply of raw material from other ownerships within a regional framework.

Socioeconomic development and demographic trends have led to an increasing awareness of the externalities that derive from the use of forests for commercial purposes over extended planning horizons. Strategic management scheduling of industrial forests is thus increasingly framed by demands other than by the forest industry (Chap. 2). As a consequence, the long term sustainability of an industrial forest is currently assessed by a set of indicators (Chap. 13) to characterize both the supply of industrial product flows and the ecological/economic features of its land base. Moreover, the forest industry competitiveness often depends on the success of certification processes where a wide range of ecological and socioeconomic indicators are considered (Chap. 15).

In this context, strategic or long term management scheduling of industrial forests has to address objectives that range from traditional products' even-flow goals to stock control and landscape structure targets. Borges et al. (2014b) reported that even when strategic management planning addresses only market wood products thus excluding all other goods and services, forest managers often target multiple objectives rather than a single one. For that purpose, forest managers sometimes sequence land allocation and harvest scheduling decisions. For example, in Brazil, land allocation decisions in areas managed by the industry are often made prior to harvest scheduling decisions. The legal framework prescribes the amount of land that a rural property has to preserve for environmental purposes or to maintain uncultivated, protecting soils and water streams (usually referred in Brazil as *APPs – Permanent Preservation Areas*), or to reserve as a precautionary measure to maintain a constant stock of wood and forest resources (also referred in Brazil as *RLs – Legal Reserves*) (Rodriguez and Nobre 2013). Nevertheless, even when uses are segregated in space the management of the industrial plantation may still be framed by concerns other than with the sustainability of the product flow to the factory.

The spatial context of harvest scheduling decisions is typically addressed at tactical and operational planning scales (Chap. 2). The importance of locational specificity in a management plan, i.e. the definition of the spatial location where management options are actually to be implemented, grows when tactical and operational concerns are to be addressed (Chap. 7). Nevertheless the possibility of balancing strategic and tactical goals may be facilitated if long-term management planning does address locational specificity. The management of the industrial plantation is often framed by spatial considerations that require the acknowledgement of neighboring relations between stands. If these are totally ignored by strategic management planning it may even more difficult to reconcile the long-term and medium and short-term perspectives and schedules. Accordingly, in countries where strategic forest-level management planning targeting only wood products is reported as a prevalent problem, locational specificity of the long-term solution is typically required (Borges et al. 2014b). In fewer cases it was reported that neighborhood relations should also be fully acknowledged.

The practice of strategic management scheduling of industrial plantation thus focus on the structuring, the representation and the solution of a range of long-term planning problems. These may be classified according to dimensions that include the spatial scale, the number of objectives as well as of products and the spatial context. In any case, thoughtful development of all stages of the decision process requires information and knowledge about the structure of the planning process, the land base that is the object of the planning exercise, the models available to make projections of forest products and the models available to generate prescriptions that may be implemented in each stand (Chap. 2). Last but not the least it requires the application of forest economic and finance models (Chap. 5).

Problem structuring issues may be better addressed in the framework of the development of information systems that may encapsulate the data and the model base needed to support the integration of temporal planning levels (Chap. 9). In this chapter we will address the representation of industrial forest strategic forest management problems as well as the interpretation of its solution. We will start with a simple harvest scheduling model aiming at the maximization of economic returns while addressing concerns with the sustainability of a product flow (Problem 1). This policy scenario will be expanded to reflect concerns with the volume in the ending inventory and with the average carbon stock (Problem 2). Further concerns with the environmental impacts of clearcuts will be addressed in Problem 3. The integration of road building and maintenance decisions within strategic forest management scheduling will be discussed as Problem 4.

Several modeling approaches will be considered to represent and solve each problem. In the case of Problems 1 and 2 we will start with linear programming formulations, the technique most widely used in long-term forest management scheduling (Chap. 2). Mixed integer programming formulations will be used to ensure the locational specificity of the solutions to Problems 1 to 3. We will present a goal programming representation of the multiple objective policy scenarios in Problems 1 to 3. A meta-heuristic – simulated annealing – will be considered as an alternative integer solution approach in the case of Problems 1 to 3. Further, a Pareto frontier approach will be used to illustrate how the setting of management planning targets may benefit from *a priori* display of trade-offs between objectives. The reader is referred to Chap. 2 for an introduction to each modeling approach.

6.1.1 Example Forest and Applications

In this chapter, an example forest will be used to illustrate all problems and how they may be addressed by each modeling approach. Several applications of the modeling approaches to strategic management planning of industrial plantations will be briefly described in six Management Planning in Action boxes.

Our example forest encompasses a set of 16 stands from Leiria National Forest (LNF), a pine forest in Portugal (Fig. 6.1). Land classification led to stand areas that range from 28.5 to 31 ha (Table 6.1). Long-term planning typically extends over the

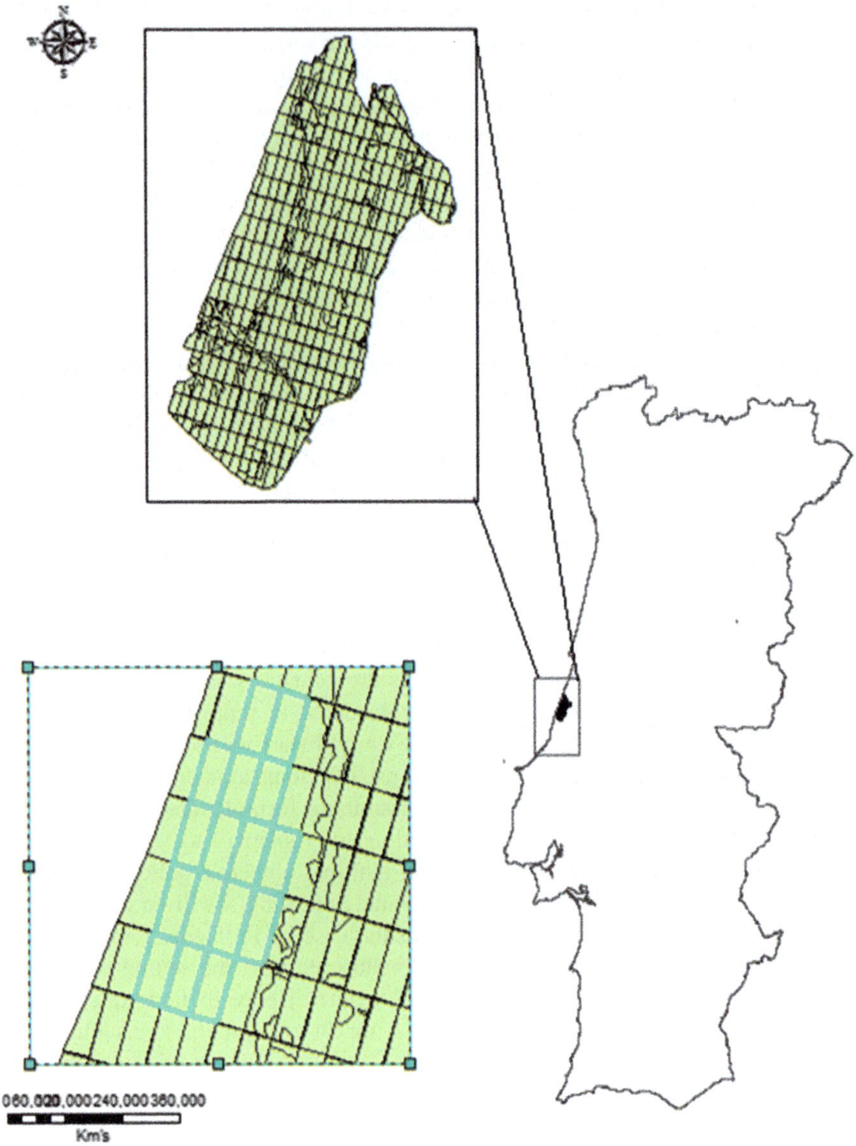

Fig. 6.1 Example forest with 16 pine stands from Leiria National Forest in Portugal

number of years corresponding to 1.5 rotations (Chap. 2). In the case of LNF the planning horizon extends over 100 years (Tomé et al. 2010; Garcia-Gonzalo et al. 2013). Nevertheless, for illustration purposes we will consider a four 10-year period planning horizon.

Table 6.1 Current inventory in the 16 stands example forest

Stand	Age (years)	Area (ha)	Vol (m^3)
1	43	31	268.9
2	43	29.8	268.9
3	43	30	268.9
4	43	28.9	268.9
5	53	29.2	331.3
6	53	29.8	331.3
7	53	29.8	331.3
8	53	30	331.3
9	33	29	193.6
10	33	29.5	193.6
11	33	29.6	193.6
12	33	30.5	193.6
13	23	28.5	107.5
14	23	28.9	107.5
15	23	29	107.5
16	23	29.7	107.5
Total		**473.2**	
Average	**38**	**29.6**	**225.3**

These stands were assigned LNF inventory plot information (Table 6.1). Stand age ranges from 23 to 53 years. The volume ranges from 107. 5 to 331.3 m^3/ha in the case of younger and older stands, respectively. A prescription writer was used to generate 3–5 prescriptions for each stand. For simplicity, the prescriptions involve only clearcuts and it is assumed that stands regenerate naturally. Prescriptions are available that involve clearcuts in alternative years of the same planning period.

An empirical growth and yield model was used to estimate the values of volume harvested in each planning period and at the ending inventory as well as of the carbon stock, resulting from the assignment of each prescription to each stand (Table 6.2). The corresponding revenues were computed considering a constant price equal to 15.5 €/m^3. Their present value in year 1 of the planning horizon was computed considering a discount rate equal to 3 % (Table 6.2).

What You Will Learn in This Chapter

- The dimensions of strategic forest management planning in the case of industrial plantations.
- How to build a model that may represent a strategic forest management planning problem.
- How to analyze the potential of operations research techniques to address several long-term management planning problems.
- How to interpret the long-term schedules provided by several modeling approaches.

Table 6.2 Prescription summaries for the 16 stands example forest

		FMA 1					FMA 2					FMA 3					FMA 4					FMA 5				
Stand	Period	Age	Vol	Carb	DR	HY	Age	Vol	Carb	DR	HY	Age	Vol	Carb	DR	HY	Age	Vol	Carb	DR	HY	Age	Vol	Carb	DR	HY
1	**1**	2	307.4	61.3	3,874.5	8	52		90.2			52		90.2			52		90.2			52		90.2		
	2	12		2.4			7	335.7	20.1	3,649	13	2	361.4	72.6	3,389.3	18	62		105.8			62		105.8		
	3	22		20.6			17		9.2			12		2.4			7	385.0	22.9	3,114.7	23	2	406.6	81.8	2,837.2	28
	4	32		47.2			27		33.9			22		20.6			17		9.2			12		2.4		
	Age			32					27					22					17					12		
	NPV			5,019.4					4,636.6					4,241.3					3,849.7					3,471.2		
	Vol EI			181.8					140.0					96.6					54.3					20.7		
	VEI			1,144.9					987.6					852.0					735.0					634.0		
2	**1**	2	320.0	63.8	4,032.6	8	52		93.8			52		93.8			52		93.8			52		93.8		
	2	12		2.5			7	349.4	20.9	3,797.9	13	2	376.2	75.6	3,527.7	18	62		110.1			62		110.1		
	3	22		21.4			17		9.6			12		2.5			7	400.8	23.8	3,241.9	23	2	423.2	85.2	2,953.0	28
	4	32		49.1			27		35.3			22		21.4			17		9.6			12		2.5		
	Age			32					27					22					17					12		
	NPV			5,224.3					4,825.9					4,414.4					4,006.8					3,612.9		
	Vol EI			189.2					145.8					100.6					56.5					21.5		
	VEI			1,191.7					1,028.0					886.7					764.9					659.9		
3	**1**	2	310.6	61.9	3,914.0	8	52		91.1			52		91.1			52		91.1			52		91.1		
	2	12		2.4			7	339.1	20.4	3,686.2	13	2	365.1	73.3	3,423.9	18	62		106.9			62		106.9		
	3	22		20.8			17		9.3			12		2.4			7	389.0	23.1	3,146.5	23	2	410.8	82.7	2,866.2	28
	4	32		47.7			27		34.3			22		20.8			17		9.3			12		2.4		
	Age			32					27					22					17					12		
	NPV			5,070.7					4,683.9					4,284.6					3,888.9					3,506.6		
	Vol EI			183.6					141.5					97.6					54.8					20.9		
	VEI			1,156.7					997.7					860.7					742.4					640.4		

4	**1**	2	320.0	63.8	4,032.6	8	52		93.8			52		93.8			52		93.8			52		93.8		
	2	12		2.5			7	349.4	21.0	3,797.9	13	2	376.2	75.6	3,527.9	18	62		110.1			62		110.1		
	3	22		21.4			17		9.6			12		2.5			7	400.8	23.8	3,241.9	23	2	423.2	85.2	2,953.0	28
	4	32		49.1			27		35.3			22		21.4			17		9.6			12		2.5		
	Age			32					27					22					17					12		
	NPV			5,224.3					4,825.9					4,414.4					4,006.8					3,612.9		
	Vol EI			189.2					145.8					100.6					56.5					21.5		
	VEI			1,191.7					1,028.0					886.5					764.9					659.9		
5	**1**	7	342.5	20.7	5,004.0	3	2	368.8	74.1	4,648.0	8	62		108.0												
	2	62		9.4			12		2.4			7	392.9	23.4	4,271.4	13										
	3	27		34.6			22		21.0			17		9.4												
	4	37		60.8			32		48.2			27		34.6												
	Age			37					32					27												
	NPV			6,229.8					5,702.4					5,183.5												
	Vol EI			225.1					185.5					142.9												
	VEI			1,225.8					1,054.4					912.1												
6	**1**	7	349.4	21.0	5,104.1	3	2	376.2	75.6	4,740.9	8	62		110.1												
	2	62		9.6			12		2.5			7	400.8	23.8	4,356.8	13										
	3	27		35.3			22		21.1			17		9.6												
	4	37		62.0			32		49.1			27		35.3												
	Age			37					32					27												
	NPV			6,354.4					5,819.5					5,287.2												
	Vol EI			229.6					189.2					145.8												
	VEI			1,250.3					1,078.6					930.4												

(continued)

Table 6.2 (continued)

Stand	Period	FMA 1					FMA 2					FMA 3					FMA 4					FMA 5				
		Age	Vol	Carb	DR	HY	Age	Vol	Carb	DR	HY	Age	Vol	Carb	DR	HY	Age	Vol	Carb	DR	HY	Age	Vol	Carb	DR	HY
7	**1**	7	334.2	20.7	4,882.0	3	2	359.8	74.5	4,534.6	8	62		108.6												
	2	62		9.13			12		2.4			7	395.3	22.8	4,167.2	13										
	3	27		34.8			22		21.1			17		9.5												
	4	37		61.1			32		48.4			27		34.8												
	Age			37					32					27												
	NPV			6,077.9					5,566.2					5,057.1												
	Vol EI			219.6					181.0					139.4												
	VEI			1,043.6					1,031.6					889.9												
8	**1**	7	342.5	20.6	5,004.0	3	2	368.8	74.1	4,648.0	8	62		108.0												
	2	62		9.4			12		2.4			7	392.9	23.4	4,271.4	13										
	3	27		34.6			22		21.0			17		9.4												
	4	37		60.8			32		48.2			27		34.6												
	Age			37					32					27												
	NPV			6,229.8					5,705.4					5,183.5												
	Vol EI			225.1					185.5					142.9												
	VEI			1,225.8					1,057.4					912.1												
9	**1**	42		73.4			42		73.4			42		73.4			42		73.4			42		73.4		
	2	2	307.8	63.5	2,985.9	18	52		93.4			52		93.4			52		93.4			52		93.4		
	3	12		2.4			7	347.6	20.9	2,812.2	23	2	374.3	75.2	2,612.1	28	62		109.6			62		109.6		
	4	22		21.3			17		9.6			12		2.4			7	398.8	23.7	2,400.4	33	2	421.1	84.7	2,186.6	38
	Age			22					17					12					7					2		
	NPV			3,868.3					3,573.3					3,268.7					2,966.8					2,675.1		
	Vol EI			100.1					56.2					21.4					2.3					0.0		
	VEI			882.4					761.1					656.6					566.4					488.5		

10	**1**	42		73.7			42		73.7			42		73.7			42		73.7			42		73.7		
	2	2	325.5	63.8	3,000.6	18	52		93.8			52		93.8			52		93.8			52		93.8		
	3	12		2.5			7	349.4	21.0	2,826.0	23	2	376.2	75.6	2,624.93	28	62		110.1			62		110.1		
	4	22		21.5			17		9.6			12		2.5			7	400.8	23.8	2,412.3	33	2	423.2	85.2	2,197.3	38
	Age			22					17					12					7					2		
	NPV			3,887.4					3,590.9					3,284.8					2,981.4					2,688.3		
	Vol EI			100.6					56.5					21.5					2.3					0.0		
	VEI			886.8					764.9					3,284.8					569.1					491.0		
11	**1**	42		72.6			42		72.6			42		72.6			42		72.6			42		72.6		
	2	2	316.3	62.9	2,956.5	18	52		92.5			52		92.5			52		92.5			52		92.5		
	3	12		2.4			7	344.2	20.7	2,784.5	23	2	370.6	75.5	2,586.3	28	62		108.5			62		108.5		
	4	22		21.1			17		9.5			12		2.4			7	394.9	23.5	2,376.8	33	2	417.0	83.9	2,165.0	38
	Age			22					17					12					7					2		
	NPV			3,830.2					3,538.1					3,236.4					2,937.6					2,937.6		
	Vol EI			99.1					55.7					21.2					2.3					0.0		
	VEI			873.7					753.7					650.1					560.8					772.6		
12	**1**	32		72.3			32		72.3			32		72.3			32		72.3			32		72.3		
	2	2	323.2	62.6	3,031.2	18	52		92.0			52		92.0			52		92.0			52		92.0		
	3	12		2.4			7	342.5	20.6	2,770.6	23	2	368.8	74.1	2,573.5	28	62		108.0			62		108.0		
	4	22		21.0			17		2.4			12		2.4			7	392.9	23.3	2,365.0	33	2	414.9	83.5	2,154.5	38
	Age			22					17					12					7					2		
	NPV			3,811.2					3,520.5					3,220.3					2,923.0					2,635.6		
	Vol EI			98.6					55.4					21.1					2.3					0.0		
	VEI			780.0					749.9					646.8					558.0					2,635.6		

(continued)

Table 6.2 (continued)

		FMA 1					FMA 2					FMA 3					FMA 4					FMA 5				
Stand	Period	Age	Vol	Carb	DR	HY	Age	Vol	Carb	DR	HY	Age	Vol	Carb	DR	HY	Age	Vol	Carb	DR	HY	Age	Vol	Carb	DR	HY
13	**1**	32		49.6			32		49.6			32		49.6			32		49.6							
	2	42		74.4			42		74.4			42		74.4			42		74.4							
	3	2	323.1	64.4	2,254.6	28	52		94.8			52		94.8			52		94.8							
	4	12		2.5			7	352.8	21.2	2,123.4	33	2	379.9	76.3	1,972.3	38	62		111.2							
	Age			12					7					2					62							
	NPV			2,920.9					2,698.2					2,468.1					2,386.9							
	Vol EI			21.7					2.4					0.0					390.6							
	VEI			666.3					574.7					495.8					2,386.9							
14	**1**	32		49.3			32		49.3			32		49.3			32		49.3							
	2	42		74.1			42		74.1			42		74.1			42		74.1							
	3	2	321.5	64.1	2,243.7	28	52		94.3			52		94.3			52		94.3							
	4	12		2.5			7	351.1	21.1	2,113.1	33	2	378.0	75.9	1,962.8	38	62		110.7							
	Age			12					7					2					62							
	NPV			2,906.8					2,685.1					2,456.1					2,375.3							
	Vol EI			21.6					2.4					0.0					388.2							
	VEI			663.1					572.0					493.3					2,375.3							
15	**1**	32		49.8			32		49.8			32		49.8			32		49.8							
	2	42		74.8			42		74.8			42		74.8			42		74.8							
	3	2	324.7	64.8	2,265.6	28	52		95.2			52		95.2			52		95.2							
	4	12		2.5			7	354.5	21.3	2,133.7	33	2	381.7	76.7	1,981.9	38	62		111.8							
	Age			12					7					2					62							
	NPV			2,935.1					2,711.3					2,480.1					2,398.5							
	Vol EI			21.6					2.4					0.0					392.0							
	VEI			669.5					577.6					498.2					2,398.5							

16	**1**	32		49.1			32		49.1			32		49.1			32	49.1
	2	42		73.7			42		73.7			42		73.7			42	73.7
	3	2	320.0	63.8	2,232.8	28	52		93.8			52		93.8			52	93.8
	4	12		2.5			7	349.4	21.0	2,102.8	33	2	376.2	75.6	1,953.2	38	62	110.1
	Age			12					7					2				62
	NPV			2,892.6					2,672.0					2,444.2				2,363.8
	Vol EI			21.5					2.3					0.0				386.3
	VEI			659.8					569.2					491.0				2,363.8

Units: Vol, m^3 ha^{-1}, Carb, Mg ha^{-1}, DR, € ha^{-1}, NPVt, € ha^{-1}, Vol EI, m^3 ha^{-1}, VEI, € ha^{-1}

6.2 Volume and Area Control in Long Term Industrial Forest Management Planning (Problem 1)

The demand of timber has played a major role in defining strategic forest management planning problems as well as in early modeling approaches to address it (Chap. 2). Ecological and socioeconomic conditions prevailing in nineteenth century Europe, e.g. extensive deforestation of areas close to urban centers and limited transportation technology, prompted the development of management models emphasizing the stability of timber supply (Borges and Hoganson 1999). Theoretical concepts such as "sustainable yield" and "normal forest" (Alves 1984, Chap. 2) emerged in order to address that society economic requirements. Biological or financial parameters contributed explicitly to time the harvest (Chap. 5). Socioeconomic factors were implicitly considered by area or volume harvested control methods (Chap. 2) that targeted the sustainability of timber flows over extended planning horizons.

The area control method aims to get to the maximum sustainable timber even flow by balancing the age-class distribution as quickly as possible. It achieves this goal by simply harvesting *A*/*r* hectares each period where *A* is the forest area and *r* is the number of periods in the rotation. The area to be harvested in each period may be adjusted to take into account the productivity of each hectare in the forest. In the worst-case scenario, this will result in a regulated forest in *r* periods. The problem with area control is that during the periods where the age-class distribution of the initial forest is being regulated – which can be a long time if *r* periods is very long – the volume harvested each period can vary dramatically. The more unbalanced the initial age-class distribution, the larger this problem will be.

Volume control methods were developed to try to produce a more even flow of volume while still eventually producing a regulated forest. In area control, the forester cuts a constant area of forest each period and hopes to produce a steady flow of timber, sometimes not too successfully. With volume control, the forester cuts a certain volume each period and hopes to eventually produce a regulated forest. With area control, the question of how much area to cut each period is straightforward. With volume control, the problem of determining how much to cut each period is less obvious.

An approach that would be analogous to the area control solution would be to simply cut the long term sustainable yield (LTSY) each period. Unfortunately, depending on the initial age-class distribution of the forest, this approach could result in severe overcutting, eventually reducing the inventory of the forest to zero, or cut so little that it would take a very, very long time to regulate the forest. For example, the Hundeshagen method of determining the volume to cut adjusts the LTSY by the ratio of the current total forest inventory over the total forest inventory that will be present when the forest is regulated. The logic of this is that if the forest currently has more inventory than it will have once it is regulated, then more volume should be cut to reduce the excess inventory. Conversely, if the forest currently has less inventory than it will have once it is regulated, then less volume should be cut to allow the inventory to build up.

Mathematically, Let I_0 be the current inventory (at time 0) and let I_{Reg} be the inventory that will be present in the forest when it is regulated. The volume to cut in the first period, H_1, according to the Hundeshagen method is:

$$H_1 = \frac{I_0}{I_{\mathrm{Reg}}} LTSY \tag{6.1}$$

In many cases, if this volume is harvested each period, one will achieve a fairly constant volume of timber over time and the forest will eventually be regulated. However, depending on the initial age-class distribution of the forest, this formula can lead to overcutting, followed by undercutting, followed by overcutting, until the age-class distribution approaches a somewhat balanced state, and the harvest level fluctuations settle into a more constant level. Other volume control formulas have been developed, some as general rules and others for very specific situations, but none of them ideal for all situations. The main problem with classical approaches, however, is that all of them, when confronted with the detailed inventory information and varied conditions and constraints of modern forestry, fail to handle the complexity of industrial forest management planning.

Furthermore, more sophisticated approaches are available that can handle much of this complexity. Management science and computational capacity developments enhance management planning processes by providing the ability to further analyze alternative strategies through the use of mathematical programming or simulation (Chap. 2). Automation provided the means to process huge amounts of data and enabled the use of these more sophisticated techniques. As a result, the potential for the definition of more sound strategic schedules for industrial forests has been increased. Therefore, we now turn our attention to the application of modern techniques introduced in Chap. 2 to address Problem 1.

6.2.1 Linear and Integer Programming

While many different objective functions can be defined for harvest scheduling models, and even combined in a single, multi-objective model, the two most common objectives are maximizing the discounted net revenues from the forest or minimizing the cost of managing the forest. We will consider the former for illustration purposes as most planning models developed for industrial forest plantations use the discounted net revenues objective function. Moreover, to enhance readability we will list a sub-set of the decision variables when describing each equation.

In the case of our example forest, a model to address sustainability concerns with the supply of volume through both volume and area control may be described by Eqs. 2.1, 2.2, 2.3, 2.4, 2.5, and 2.6 in Chap. 2. The LP formulation of Problem 1 (F1) will thus include (a) an objective function that expresses the management objective of maximizing the forest net present value subject to (b) a set of area constraints to ensure that the area managed in each stand does not exceed the area available, (c) a set of accounting equations to determine the volume harvested in each period, (d) a set of accounting equations to determine the area harvested in each period, (e) a set of volume control constraints, (f) a set of area control constraints and (g) a set of non-negativity constraints.

(a) Objective function Z. In this abbreviated form it displays the decision variables x_{kj}, i.e. the area of stand k assigned to prescription j, for stands 1, 2 and 16:

$$MAX\ Z = 5.0x_{11} + 4.6x_{12} + 4.2x_{13} + 3.9x_{14} + 3.5x_{15} + 5.2x_{21} + 4.8x_{22} + 4.4x_{23} + 4.0x_{24} + 3.6x_{25} + \cdots + 2.9x_{161} + 2.7x_{162} + 2.4x_{163} + 2.3x_{164} \quad (6.2)$$

(Eq. 2.1 in Chap. 2)

The coefficients of the decision variables c_{kj} correspond to the net present value per hectare associated with prescription j for stand k. They include the value of the ending inventory. For example the net present value resulting from the assignment of an hectare of stand 1 to its prescription 1 is equal to 5.0×10^3 €/ha (c_{11}).

This value was computed by adding the discounted return associated to the volume harvested in year 8 of planning period 1 to the value of ending inventory. The former was computed by multiplying the volume harvested – v_{111} (307.4 m^3/ha) in Table 6.2 – by the stumpage price – 15.5 €/m^3 – and by the discount factor – $((1/(1+0.03)^7)$ – and is thus equal to, approximately, 3.9×10^3 €/ha. The latter was computed assuming a perpetual series of optimal rotations. In the case of stand 1, the optimal rotation is 50 years. Thus the Soil Expectation Value (SEV) (Chap. 5) was computed by first multiplying v_{111} (30.7×10 m^3/ha) in Table 6.2) – by the stumpage price – 15.5 €/m^3 to get the revenue resulting from selling the stumpage from the second rotation in year 58. This was multiplied by the discount factor $((1/((1+0.03)^{50}-1))$ to get the present value of a perpetual series of revenues occurring every 50 years (1.4×10^3 €/ha). As the initial year of this series corresponds to the 8th year of the planning horizon that value must be discounted further 7 years. The present value of ending inventory is thus estimated as 1.4×10^3 €/ha times $1/(1+0.03)^7 = 1.1 \times 10^3$ €/ha. In summary, approximately, $c_{11} = 3.9 + 1.1 = 5.0 \times 10^3$ €/ha (5019.4 €/ha in Table 6.2).

The maximization is subject to

(b) The set of area constraints stating that the sum of the stand area assigned to each prescription cannot exceed the total stand area (Eq. 2.2 in Chap. 2) (6.3)

$$\begin{aligned} &x_{11} + x_{12} + x_{13} + x_{14} + x_{15} = 31 \\ &x_{21} + x_{22} + x_{23} + x_{24} + x_{25} = 29.8 \\ &\ldots \\ &x_{161} + x_{162} + x_{163} + x_{164} = 30 \end{aligned} \quad (6.3)$$

(c) The set of accounting equations to determine the volume harvested in each period H_1 to H_4 (Eq. 2.3 in Chap. 2). Each Eq. (6.4) includes all decision variables that involve a harvest in the corresponding period.

$$
\begin{aligned}
&30.74x_{11}+32.00x_{21}+30.16x_{31}+32.00x_{41}+\ldots+34.25x_{81}+36.88x_{82}-H1=0\\
&33.57x_{12}+36.14x_{13}+34.94x_{22}+37.62x_{23}+\ldots+31.63x_{111}+32.32x_{121}-H2=0\\
&38.50x_{14}+40.66x_{15}+40.08x_{24}+42.32x_{25}+\ldots+32.47x_{151}+32.00x_{161}-H3=0\\
&39.88x_{94}+42.111x_{95}+40.08x_{104}+42.32x_{105}+\ldots+34.94x_{162}+37.62x_{163}-H4=0
\end{aligned}
\tag{6.4}
$$

The coefficients of the decision variables v_{kjt} correspond to the volume harvested per hectare in period t when prescription j is assigned to stand k. For example, as we have just checked, the volume harvested in period 1 from stand 1 if it is assigned to its prescription 1 (v_{111}) is equal to 30.7 × 10 m^3/ha (Table 6.2).

(d) The set of accounting equations to determine the area harvested in each period AH_1 to AH_4. Each equation (6.5 – analogous to Eq. 2.3 in Chap. 2) includes all decision variables that involve a harvest in the corresponding period.

$$
\begin{aligned}
&x_{11}+x_{21}+x_{31}+x_{41}+\cdots+x_{81}+x_{82}-AH1=0\\
&x_{12}+x_{13}+x_{22}+x_{23}+\cdots+x_{111}+x_{121}-AH2=0\\
&x_{14}+x_{15}+x_{24}+x_{25}+\cdots+x_{151}+x_{161}-AH3=0\\
&x_{94}+x_{95}+x_{104}+x_{105}+\cdots+x_{162}+x_{163}-AH4=0
\end{aligned}
\tag{6.5}
$$

(e) The set of volume control constraints (6.6 – Eq. 2.4 in Chap. 2). It expresses a policy aiming at non-declining volume flows and at a maximum 10 % increase of volume harvested in consecutive periods.

$$
\begin{aligned}
&H1-H2\le 0\\
&H2-H3\le 0\\
&H3-H4\le 0\\
&H2-1.1H1\le 0\\
&H3-1.1H2\le 0\\
&H4-1.1H3\le 0
\end{aligned}
\tag{6.6}
$$

(f) The set of area control constraints (6.7 - – analogous to Eq. 2.4 in Chap. 2). It expresses a policy aiming at maximum 10 % fluctuations of area harvested in consecutive periods.

$$
\begin{aligned}
&AH2-0.9AH1\ge 0\\
&AH2-1.1AH1\le 0\\
&AH3-0.9AH2\ge 0\\
&AH3-1.1AH2\le 0\\
&AH4-0.9AH3\ge 0\\
&AH4-1.1AH3\le 0
\end{aligned}
\tag{6.7}
$$

(g) The set of non-negativity constraints (6.8 – Eq. 2.6 in Chap. 2).

$$x_{11}, x_{12}, x_{13} \ldots, x_{164}, x_{165} \geq 0 \quad (6.8)$$

According to the solution by the LP model (Table 6.3), the optimal long-term management plan is associated with an objective function value Z equal to $2{,}041.3 \times 10^3$ €. It encompasses a periodic harvest of $4{,}089 \times 10\ m^3$. The area harvested ranged from 116.2 to 119.1 ha in periods 4 and 3, respectively. The solution thus demonstrates that the LP model may address efficiently and effectively both volume and area control objectives.

The plan proposed by the LP model reflects the stand age distribution. Younger stands are harvested later (Table 6.4) in periods 3 (stand 13) or 4 (stands 14–16). Conversely, older stands are harvested earlier in periods 1 (stands 5–8) or 2 (stands 1–4) in order to avoid opportunity costs associated with the delay of harvesting mature timber. Classical approaches might meet the regulation objectives and yet at a cost that might be higher than needed. The LP plan is efficient as it meets the regulation objectives while minimizing the opportunity costs of doing so.

Nevertheless the LP solution aims at providing further insight about the forest management planning problem. The distinction made by Geoffrion (1976) between the mathematical programming "ostensible purpose" – optimization of a particular problem, and its "true purpose" – generation of information to support decision making is illuminating. The value of the LP dual variables (shadow prices) conveys the impact of changing the independent term in a constraint in the value of the objective function (Chap. 2). Thus the LP solution may be used to check the impact of setting alternative area and volume control objectives.

The usefulness of this information may be illustrated by analyzing the value of the shadow prices associated to the area constraints (Eq. 6.3). They reflect the marginal value of each stand for the forest owner (Table 6.5). As expected the marginal value of younger stands (stands 13 –16) is lower as they are harvested later in the planning horizon. The highest marginal value of stand 6 is due both to its age and its productivity (Table 6.2). This information provides insight about the value structure of the current inventory as well as about the management planning problem. The forest owner may take advantage further of this information when making decisions on whether and how to expand the forest land base. The LP solution provides an estimate of the maximum amount he might pay when buying an additional hectare of each stand. Conversely, it conveys the minimum price he should consider when selling one hectare of each stand.

The LP solution also provides information about the opportunity costs associated with the selection of alternative plans. The reduced costs of decision variables measure the impact of selecting a non-optimal prescription on the value of the objective function (Chap. 2). For example, the solution highlights the costs of anticipating or delaying the timing of harvests (Table 6.6). It further shows that the option of not harvesting when available (stands 13–16) is associated with the highest costs. These result both from the loss of revenue that results from the harvests in

Table 6.3 Solution summaries

	Problem 1					Problem 2						Problem 3		
										PF				
Solution indicators	LP	MIP	GP	SA	PF	LP	MIP	GP	SA	Point A	Point B	MIP	GP	SA
H1 (m^3) ($*10^1$)	4,089.5	4,064.4	4,064.4	4,064.4	4,085.5	4,020.2	3,068.7	3,068.7	3,069.7	3,996.1	3,505.8	3,068.7	3,068.7	3,036.9
H2 (m^3) ($*10^1$)	4,089.5	4,108.4	4,108.4	4,108.4	4,085.5	4,020.2	3,223.9	3,192.4	3,223.9	3,996.1	3,505.8	3,200.5	3,200.6	3,229.4
H3 (m^3) ($*10^1$)	4,089.5	4,122.6	4,102.2	4,122.6	4,121.2	4,020.2	3,369.7	3,391.8	3,369.7	3,996.1	3,505.8	3,396.5	3,370.7	3,379.4
H4 (m^3) ($*10^1$)	4,089.5	4,204.5	4,085.5	4,204.5	4,399.3	4,020.2	3,507.5	3,523.6	3,507.5	3,996.1	3,505.8	3,507.5	3,537.1	3,537.1
Cstock1 (Mg) ($*10^1$)	2,785.5	2,787.8	2,787.8	2,787.8	2,785.9	2,799.1	3,042.0	3,042.0	3,042.0	2,805.3	2,930.4	3,042.0	3,042.0	3,020.1
Cstock2 (Mg) ($*10^1$)	2,320.2	2,322.4	2,322.4	2,322.4	2,321.2	2,519.2	2,880.3	2,888.3	2,880.3	2,524.7	2,662.5	2,886.2	2,886.2	2,889.5
Cstock3 (Mg) ($*10^1$)	1,868.7	2,194.8	1,867.4	2,194.8	1,906.9	2,267.3	2,657.1	2,662.0	2,657.1	2,272.3	2,396.3	2,658.1	2,664.6	2,662.6
Cstock4 (Mg) ($*10^1$)	1,496.1	1,459.8	1,494.6	1,459.8	2,129.0	2,040.6	2,436.8	2,435.0	2,436.8	2,055.4	2,183.1	2,435.0	2,436.1	2,433.1
Vol Ei (m^3) ($*10^1$)	5,080.4	4,875.0	5,076.4	4,875.0	5,027.0	5,500.0	8,335.5	8,325.3	8,335.5	5,597.6	7,497.9	8,329.5	8,325.5	8,316.0
AH1	118.7	117.9	117.9	117.9	118.5	116.6	89.0	89.0	89	115.9	101.7	89	89.0	87.9
AH2	119.1	119.7	119.7	119.7	119.0	115.0	89.7	88.8	89.7	114.1	99.2	88.7	88.7	88.7
AH3	118.2	119.3	118.6	119.3	118.6	111.2	89.4	89.3	89.4	110.5	97.0	90.4	89.5	90.6
AH4	116.2	115.4	116.1	115.4	116.1	106.8	88.1	89.1	88.1	106.1	92.1	88.1	89.0	89.0
NPVt (€) ($*10^3$)	2,041.3	2,021.0	2,041.0	2,021.0	2,012.4	1,973.7	1,867.1	1,866.1	1,867.1	1,971.5	1,929.1	1,866.6	1,866.1	1,815.0

Table 6.4 Harvest schedule: hectares harvested in each period

Harvest Plan		Problem 1						Problem 2						Problem 3		
												PF				
Stand	Period	Year	LP	MIP	GP	SA	PF	LP	MIP	GP	SA	Point A	Point B	MIP	GP	SA
1	**1**															
	2	13	31.0	31	31.0	31	31	31	31	31	31	31	24.3			
	3	23											6.7	31	31	31
	4															
	No harvest															
2	**1**	8	0.8													
	2	13	29.0	29.8	29.8	29.8	29.8	29.8	29.8		29.8	29.8	29.8	29.8	29.8	29.8
	3	23								29.8						
	4															
	No harvest															
3	**1**															
	2	13	30	30	30	30	30	23.8				22.4		30	30	
	3	23						6.2	30	30	30	7.6	30			30
	4															
	No harvest															
4	**1**	8					0.7									
	2	13	28.9	28.9	28.9	28.9	28.2	2.9		28.9		4.3	28.9			28.9
		18						26.0				24.6				
	3	23							28.9		28.9			28.9	28.9	
	4															
	No harvest															

5	**1**	3	29.2	29.2	29.2	29.2	29.2	27.9	29.2	29.2	29.2	27.2	29.2	29.2	29.2	29.2
	2	13						1.3				2.0				
	3															
	4															
	No harvest															
6	**1**	3	29.8	29.8	29.8	29.8	29.8	29.8	29.8	29.8	29.8	29.8	29.8	29.8	29.8	29.8
	2															
	3															
	4															
	No harvest															
7	**1**	3	28.9	28.9	28.9	28.9	28.9	28.9				28.9	12.7			28.9
	2	13							28.9	28.9	28.9		16.2	28.9	28.9	
	3															
	4															
	No harvest															
8	**1**	3	30	30	30	30	30	30	30	30	30	30	30	30	30	
	2	13														30
	3															
	4															
	No harvest															
9	**1**															
	2															
	3	23	29		29		21.9	20.4					29			
		28					7.1	8.6				29				
	4	33		29		29			29	29	29			29	29	29
	No harvest															

(continued)

Table 6.4 (continued)

Harvest Plan		Problem 1						Problem 2						Problem 3		
												PF				
Stand	Period	Year	LP	MIP	GP	SA	PF	LP	MIP	GP	SA	Point A	Point B	MIP	GP	SA
10	**1**															
	2															
	3	23	29.4	29.5	29.5	29.5	29.5			29.5		20.9	0.2			
		28						16.2				8.6				
	4	33	0.1					13.3	29.5		29.5		29.3	29.5	29.5	29.5
	No harvest															
11	**1**															
	2	18	0.3													
	3	23	29.3		29.6		29.6								29.6	29.6
		28		29.6		29.6		29.6				14.0	0.5			
	4	33							29.6	29.6	29.6	15.6	29.1	29.6		
	No harvest															
12	**1**															
	2															
	3	**23**	30.5	30.5	30.5	30.5	30.5	30.5	30.5		30.5	30.5	30.5	30.5		
	4	**33**								30.5					30.5	30.5
	No harvest															
13	**1**															
	2															
	3															
	4	33	28.5	28.5	28.5	28.5		17.7				2.9	4.8			
		38					28.5									
	No harvest	–						10.8	28.5	28.5	28.5	25.6	23.7	28.5	28.5	28.5

14	**1**															
	2															
	3															
	4	33	28.9	28.9	28.9	28.9							28.9			
		38					28.9	28.9				28.9				
	No harvest	–							28.9	28.9	28.9			28.9	28.9	28.9
15	**1**															
	2															
	3															
	4	33	29	29	29	29						14.2				
		38					29	17.0				14.8				
	No harvest	–						12.0	29	29	29		29	29	29	29
16	**1**															
	2															
	3	28		29.7		29.7										
	4	33	29.7		29.7											
		38					29.7	29.7				29.7				
	No harvest	–							29.7	29.7	29.7		29.7	29.7	29.7	29.7
	TOTAL		472.3	472.3	472.3	472.3	472.3	472.3	472.3	472.3	472.3	472.3	472.3	472.3	472.3	472.3

Table 6.5 Stand marginal value

Stand	Problem 1	Problem 2
1	138.6399	154.6636
2	138.7129	154.7451
3	135.5367	151.2018
4	134.5236	150.0716
5	147.4294	156.2438
6	170.8324	162.6433
7	146.7897	155.6337
8	151.4685	160.5945
9	107.6591	139.6389
10	110.0547	142.7462
11	108.8038	141.1238
12	111.5544	144.6913
13	95.69504	129.6633
14	96.56702	130.8448
15	97.84652	132.5785
16	98.75606	133.8109

those stands and from the loss of revenue that results from the need of adjusting the timing of harvests in other stands in order to meet the area and volume control targets.

However useful, LP is not able to convey the geographical location of forest activities. Yet this may be important even in a long-term planning framework. LP divisibility may lead to solutions where stands are split into non spatially referenced units to be managed differently. For example stand 10 was split into 2 units of 29.4 and 0.1 ha to be managed according to its prescriptions 2 and 4, respectively. Further, computational constraints lead frequently to the aggregation of stands into analysis areas thus compounding the impact of LP divisibility. Solutions to such highly aggregated models are very sensitive to changes in assumptions and aggregation schemes (Rose 1984). In this context, the information produced by the solution may be of little value to understand the management problem and to support effectively decision-making. Formulation F1 may then be changed within a Model I framework to further address concerns with locational specificity. The resulting MIP formulation of Problem 1 (F2) may be described as:

(a) Objective function Z. In this abbreviated form it displays the decision variables x_{kj}, i.e. whether stand *k is* assigned to prescription *j*, for stands 1, 2, 3 and 16:

$$\begin{aligned} MAX\ Z = {} & 155.6x_{11}+143.7x_{12}+131.5x_{13}+119.3x_{14}+107.6x_{15}+155.7x_{21} \\ & +143.8x_{22}+131.5x_{23}+119.4x_{24}+107.7x_{25}+152.1x_{31}+140.5x_{32} \\ & +128.5x_{33}+116.7x_{34}+105.2x_{35}+\cdots+85.9x_{161}+79.4x_{162} \\ & +72.6x_{163}+68.5x_{164} \end{aligned} \quad (6.9)$$

(Eq. 2.7 in Chap. 2)

Table 6.6 Prescription reduced cost

Stand	Prescription	Problem 1	Problem 2
1	**1**	0.00	−48.28
	2	0.00	0.00
	3	−12.65	0.00
	4	−14.53	0.00
	5	−25.99	−1.60
2	**1**	0.00	−48.31
	2	0.00	0.00
	3	−12.65	0.00
	4	−14.53	0.00
	5	−26.00	−1.60
3	**1**	0.00	−47.20
	2	0.00	0.00
	3	−12.36	0.00
	4	−14.20	0.00
	5	−25.41	−1.56
4	**1**	0.00	−46.85
	2	0.00	0.00
	3	−12.27	0.00
	4	−14.09	0.00
	5	−25.22	−1.55
5	**1**	16.68	0.00
	2	0.00	−53.45
	3	−1.69	0.00
	4		
	5		
6	**1**	0.00	0.00
	2	−17.36	−55.69
	3	−19.12	−0.05
	4		
	5		
7	**1**	16.61	0.00
	2	0.00	−53.27
	3	−1.68	−0.05
	4		
	5		
8	**1**	17.14	0.00
	2	0.00	−54.97
	3	−1.74	−0.05
	4		
	5		
9	**1**	0.00	−1.62
	2	0.00	0.00
	3	−8.53	0.00
	4	0.00	0.00
	5	−7.25	−1.49

(continued)

Table 6.6 (continued)

Stand	Prescription	Problem 1	Problem 2
10	1	0.00	−1.66
	2	0.00	0.00
	3	−8.72	0.00
	4	0.00	0.00
	5	−7.41	−1.53
11	1	0.00	−1.64
	2	0.00	0.00
	3	−8.62	0.00
	4	0.00	0.00
	5	−7.32	−1.51
12	1	0.00	−1.68
	2	0.00	0.00
	3	−8.83	0.00
	4	0.00	0.00
	5	−7.51	−1.55
13	1	−8.76	−1.53
	2	0.00	0.00
	3	−5.11	0.00
	4	−29.27	0.00
	5		
14	1	−8.84	−1.55
	2	0.00	0.00
	3	−5.16	0.00
	4	−29.54	0.00
	5		
15	1	−8.96	−1.57
	2	0.00	0.00
	3	−5.23	0.00
	4	−29.93	0.00
	5		
16	1	−9.05	−1.58
	2	0.00	0.00
	3	−5.28	0.00
	4	−30.21	0.00
	5		

In this model, the decision variables are integer to ensure that stands are not split by the assignment of prescriptions. Thus the coefficients of the decision variables c_{kj} correspond to the total net present value associated with prescription j for stand k. Again they include the value of the ending inventory. For example the net present value resulting from the assignment of stand 1 to its prescription 1 (c_{11}) is equal to 155.6×10^3 €. This is computed by multiplying the net present value per ha that results from this assignment – 5.0×10^3 €/ha – by the stand area – 31 ha (Table 6.1).

The maximization is subject to

(b) The set of area constraints (Eq. 2.8 in Chap. 2) (6.10)

$$
\begin{aligned}
&x_{11} + x_{12} + x_{13} + x_{14} + x_{15} = 1\\
&x_{21} + x_{22} + x_{23} + x_{24} + x_{25} = 1\\
&x_{31} + x_{32} + x_{33} + x_{34} + x_{35} = 1\\
&\ldots\\
&x_{161} + x_{162} + x_{163} + x_{164} = 1
\end{aligned}
\tag{6.10}
$$

(c) The set of accounting equations to determine the volume harvested in each period H_1 to H_4 (Eq. 2.9 in Chap. 2). Each Eq. (6.11) includes all decision variables that involve a harvest in the corresponding period.

$$
\begin{aligned}
&953.0x_{11}+953.5x_{21}+931.7x_{31}+\ldots+1027.5x_{81}+1106.4x_{82}-H1=0\\
&1040.5x_{12}+1120.4x_{13}+1041.2x_{22}+\ldots+933.2x_{91}+956.8x_{121}-H2=0\\
&1193.6x_{14}+1260.5x_{15}+1194.3x_{24}+\ldots+941.6x_{151}+950.3x_{161}-H3=0\\
&1156.5x_{94}+1221.2x_{95}+1182.2x_{104}+\ldots+1037.6x_{162}+1117.2x_{163}-H4=0
\end{aligned}
\tag{6.11}
$$

The coefficients of the decision variables v_{kjt} correspond to the volume harvested in stand k in period t when assigned to prescription j. For example, the volume harvested in period 1 from stand 1 if it is assigned to its prescription 1 (v_{111}) is approximately equal to 953×10 m^3. This was computed as the product of the stand area (31 ha) by the volume per hectare in period 1 that results from that assignment (307.4 m^3/ha (Table 6.2)).

(d) The set of accounting equations to determine the area harvested in each period AH_1 to AH_4 (analogous to Eq. 2.9 in Chap. 2). Each Eq. (6.12) includes all decision variables that involve a harvest in the corresponding period. Its coefficients correspond to the stand area.

$$
\begin{aligned}
&31x_{11} + 29.8x_{21} + 30x_{31} + \cdots + 30x_{81} + 30x_{82} - AH1 = 0\\
&31x_{12} + 31x_{13} + 29.8x_{22} + \cdots + 29.6x_{111} + 30.5x_{121} - AH2 = 0\\
&31x_{14} + 31x_{15} + 29.8x_{24} + \cdots + 29x_{151} + 29.7x_{161} - AH3 = 0\\
&29x_{94} + 29x_{95} + 29.5x_{104} + \cdots + 29.7x_{162} + 29.7x_{163} - AH4 = 0
\end{aligned}
\tag{6.12}
$$

(e) and (f) Volume and area control constraints may now be expressed just like in the case of the linear programming model (6.6 – Eq. 2.4 in Chap. 2 and 6.7 – analogous to Eq. 2.4 in Chap. 2).

(g) The set of constraints stating that the decision variables may take only the values 1 (if the prescription is assigned to the stand) or 0 (if the prescription is not assigned to the stand) (6.13 – Eq. 2.8 in Chap. 2)

$$
x_{11}, x_{12}, x_{13}, \ldots, x_{165} \text{ are binary} \tag{6.13}
$$

According to the solution by the IP model (Table 6.3), the optimal objective function value Z decreased to 2,021.0 × 10^3 €. Stands 2 and 10 are no longer split by different prescriptions (Table 6.4) and this thus cost 20.3 × 10^3 €. The harvest timings shifted in the case of three other stands in order to meet the new requirements. The requirement of locational specificity impacted the volumes and areas harvested in each period. They still meet the volume and area control constraints and yet are more uneven as management flexibility decreased (Table 6.3). In the case of larger forests with more stands the requirement of locational specificity may not have an impact as substantial on the evenness of areas and volumes harvested in each period. Yet it may have a higher impact on the computational cost.

Management Planning in Action 6.1: Volume Control in Long Term Industrial Forest Management Planning at Celbi in Portugal

Celbi is currently a factory of Altri, a leading Portuguese eucalypt pulp producer, with a capacity of up to 600 × 10^6 tonnes (http://en.altri.pt/aboutaltri/). Currently Altri manages about 84 × 10^3 ha of forest in Portugal all certified by the Forest Stewardship Council (FSC) and PEFC. Its wood self-sufficiency rate stands around 30 % and all its mills are entirely self-sufficient on power that is produced through the burning of wood components not suitable for pulp production (http://en.altri.pt/aboutaltri/).

In 2000, a priority of former Celbi owners (Stora Enso) was the assessment of the sustainability of pulpwood supply to this mill from its eucalypt land base extending over about 39 × 10^3 ha. This prompted the development of a linear programming (LP) model within a sustainability assessment project coordinated by the Forest Research Centre (http://www.isa.ulisboa.pt/cef/). The LP model included about 664 × 10^3 decision variables corresponding to prescriptions associated with 3,361 stands in Celbi eucalypt land base. The planning horizon included 31 one-year periods extending up to 2030. The model aimed at maximizing net present value. It included area and volume control constraints similar to the ones presented in 6.2. It further included constraints on the maximum area to be converted each year.

The project was successful as the LP model provided the information needed by the firm – e.g. the pulpwood potential supply ranging from 475 to 535 × 10^3 m^3, according to scenarios of productivity growth after a conversion as well of expansion of the eucalypt land base – to develop its strategic plan (Borges and Falcão 2000).

6.2.2 Heuristics

The locational specificity requirement may also be addressed by heuristic approaches. A heuristic may be defined as a technique that seeks good solutions at a reasonable computational cost without being able to guarantee optimality or even feasibility (Reeves 1993). The computational complexity of some industrial forest strategic management planning problems sometimes suggests the use of heuristics as these techniques may be more flexible and capable of addressing more complicated objective functions and constraints than exact algorithms. Moreover, given the uncertainties that derive from the large-scale attributes of the general forest management problem, good solutions may be adequate (Borges et al. 2002). As Gunn and Rai (1987) pointed out, solutions that are near optimal and near feasible may be adequate and even preferable if they can be produced with a greatly reduced solution effort, given the uncertainty about biological, technical and economic data in most forest management problems.

The design of heuristic approaches may often take advantage of the specific form of the forest management scheduling problem. Based on this design, specialized optimization solution processes can sometimes be evolved to address very large and complex problems (Borges et al. 2002). For example, Hoganson and Rose (1984) developed a specialized LP decomposition approach that may be used to solve this management planning problem thus circumventing the need to use MIP to address locational specificity requirements. Just like LP, this heuristic approach conveys information about the marginal values of resources and the volume flow targets thus contributing to the effectiveness of management planning.

The reader is referred to Borges et al. (2002) for a detailed review of the use of heuristics in multiple objective forest management. In this chapter we will consider for illustration purposes a meta-heuristic – simulated annealing (SA) – that has been widely used to address forest management planning problems (e.g. Lockwood and Moore 1993; Dahlin and Sallnas 1993; Murray and Church 1995; Tarp and Helles 1997; Boston and Bettinger 1999; Van Deusen 1999; Falcão and Borges 2002). In this illustration, the simulated annealing approach involved the conversion of the MIP formulation (Eqs. 6.6, 6.7, 6.9, 6.10, 6.11, 6.12, and 6.13) into a new objective (or evaluation) function. The resulting SA formulation of Problem 1 (F3) may be represented in abbreviated form as:

$$\begin{aligned} MAX\ Z = {} & 155.6x_{11} + 143.7x_{12} + 131.5x_{13} + 119.3x_{14} + 107.6x_{15} + \ldots \\ & + 85.9x_{161} + 79.4x_{162} + 72.6x_{163} + 68.5x_{164} - \theta \end{aligned} \tag{6.14}$$

Where θ stands for a global penalty function that decreases the value of Z if the constraints are violated. The literature reports several approaches to design a penalty function (e.g. Michalewicz 1996). In this chapter we will follow the approach proposed by Falcão and Borges (2001) so that the penalty function is more sensitive to large violations of the constraints and less responsive to small deviations. The separable penalty functions thus consisted of parabolas where deviations from the area constraints and the volume and area control objectives do contribute to decrease the value of Z. In all cases it was assumed that a 5 % deviation from feasibility was equivalent to a 1 % deviation from unconstrained *NPV* when estimating the parameter that characterizes the shape of the parabola for each constraint. Thus this parameter was computed as

$$A_{const\_j} = \Delta NPV / (\Delta D_{const\_j})^2 \tag{6.15}$$

Where ΔNPV stands for an unconstrained *NPV* reduction of 1 % and ΔD_{const_j} stands for a 5 % deviation from feasibility in constraint *j*. A_{const_j} is equal to zero if constraint *j* is met. In the case of our example problem, the penalty function will thus include three terms, one per Eqs. 6.6, 6.7 and 6.10. θ may thus be described in an abbreviated form as

$$\begin{aligned}\theta = {} & A_{const\_6\_10}\left[(x_{11}+x_{12}+x_{13}+x_{14}+x_{15}-1)^2+\ldots+(x_{161}+x_{162}+x_{163}+x_{164}-1)^2\right] \\ & + A_{const\_6\_6}\left[(H_1-H_2)^2+\ldots+(H_3-H_4)^2+(H_2-1.1H_1)^2+\ldots+(H_4-1.1H_3)^2\right] \\ & + A_{const\_6\_7}\left[(0.9AH_1-AH_2)^2+\ldots+(0.9AH_3-AH_4)^2+(AH_2-1.1AH_1)^2+\ldots\right. \\ & \left.+(AH_4-1.1AH_3)^2\right]\end{aligned} \tag{6.16}$$

with $x_{11}, x_{12}, x_{13}, \ldots, x_{165}$ as binary

In summary, the SA objective (or evaluation) function thus included as its first term the MIP objective function and as its second term a penalty function. The latter penalizes the violation of the area constraints (the first term in Eq. 6.16), of the volume control constraints (the second term in Eq. 6.16) and of the area control constraints (the third term in Eq. 6.16).

Afterwards, the SA approach involved an iterative process (Fig. 6.2) where solutions in each iteration were represented by a vector with 16 elements corresponding to the 16 stands. The value of each vector element consisted of a pointer to the prescription assigned to the corresponding stand. Solutions are thus integer.

The solution process started by selecting randomly a solution vector and by computing its objective function value (Fig. 6.2). Each SA iteration consisted of changing randomly the assignment of a prescription in 5 stands (5-opt approach) and computing its objective function value *Z2*. In order to avoid premature convergence to a local optimum, an inferior solution i.e. a solution associated with a lower objective function value might be accepted.

The SA solution strategy was implemented as described by Borges et al. (2002). It involved the design of a solution acceptance function and the definition of a stopping criterion. The former determined whether an inferior solution might be accepted. The probability of accepting inferior solutions increased with the temperature and it decreased with the magnitude of the inferior move (Fig. 6.2):

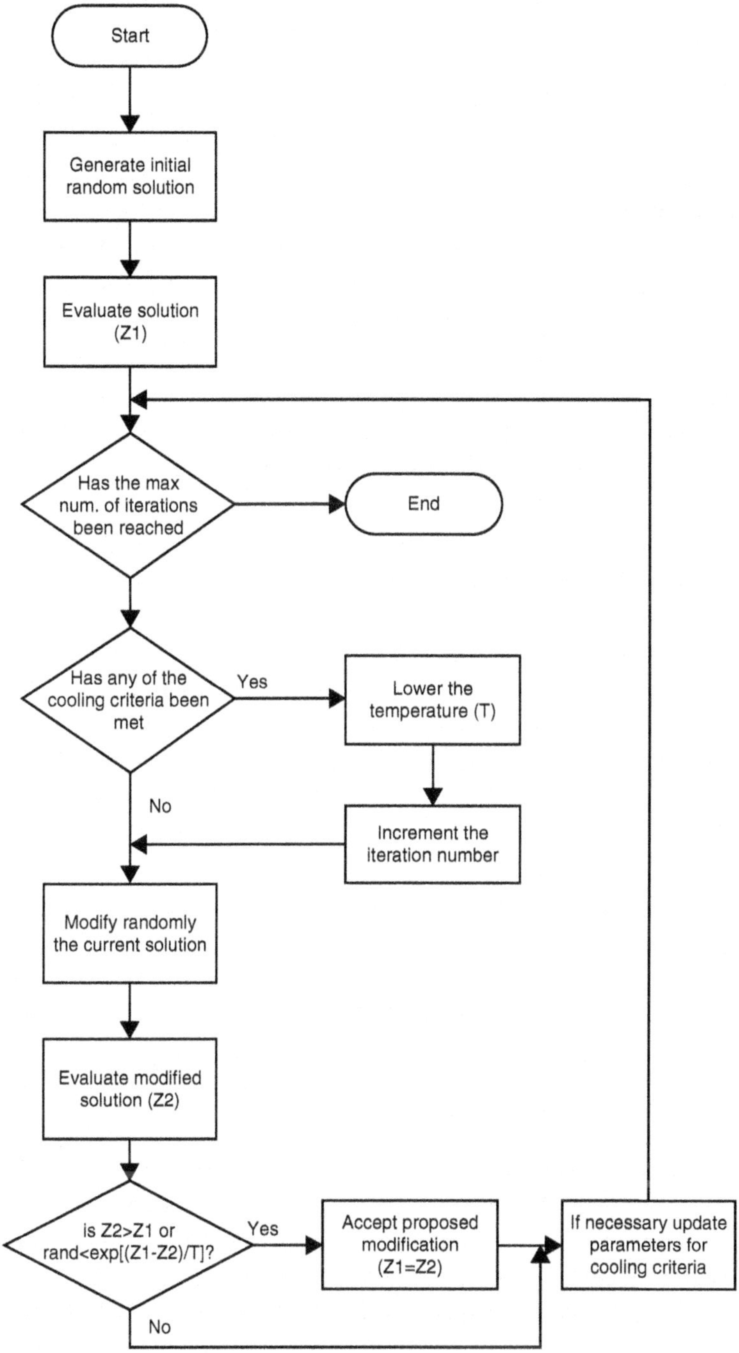

Fig. 6.2 Flowchart of the simulated annealing heuristic. *Z1* and *Z2* solution values before and after a proposed move, *T* system "temperature", *rand* random number between 0 and 1 (From Borges et al. 2002)

$$\exp\left[(Z_1 - Z_2)/T\right] > rand \tag{6.17}$$

where exp stands for the exponential function, Z_2 and Z_1 stand for the value of the SA objective (fitness evaluation) function (Eq. 6.14) after and before the 5-opt move, respectively, T stands for the control parameter (temperature) which decreases with the number of iterations according to the cooling schedule and *rand* stands for a random number in the interval [0 1]. In this example, the initial temperature was equal to 100,000. The temperature was decreased in each iteration according to a cooling rate equal to 0.9999. The stopping criteria was the number of iterations (100,000).

Meta-heuristic solutions are sensitive to parameter and penalty values. Typically these are problem specific and require pre-testing so that values are selected that may lead to approximate the optimal solutions. In this illustration, the parameterization did lead to the optimal MIP solution (2021.1×10^3 €) (Table 6.3). This is also a consequence of the problem small size.

6.2.3 Goal Programming and Pareto Frontier

Both LP and MIP have single criteria objective functions. Other objectives e.g. area and volume control were represented as constraints in the model. An alternative to this approach is to consider all criteria in the objective function. Goal programming is a technique that has been widely used for that purpose in strategic forest management planning. In the case of our example forest, the goal programming formulation of Problem 1 (F4) may be described as:

(a) Objective function Z

$$\begin{aligned}
\text{Min Z} = {} & 8.4746\, PdevAH1 + 8.4746\, NdevAH1 + 8.4746\, PdevAH2 \\
& + 8.4746\, NdevAH2 + 8.4746\, PdevAH3 + 8.4746\, NdevAH3 \\
& + 8.4746\, PdevAH4 + 8.4746\, NdevAH4 + 0.0244\, NdevH1 \\
& + 0.0244\, NdevH2 + 0.0244\, NdevH3 + 0.0244\, NdevH4 \\
& + 0.0049\, NdevNPVt
\end{aligned} \tag{6.18}$$

Where the decision variables consist of the deviations from the levels of the criteria set by the decision-maker. The variables *PdevAH1* to *PdevAH4* and *NdevAH1* to *NdevAH4* represent, respectively, the positive and the negative deviations from the target set to the area harvested in each planning period to reflect the area control regulation objectives. The variables *NdevH1* to *NdevH4* correspond to the deviations from the harvest levels targeted in each period. It was assumed that there was no constraint to harvesting higher volumes in each period and thus positive deviations were not minimized. Finally, the third criteria corresponds to the discounted net revenue from the forest e.g. the criteria listed in the LP and the MIP objective functions. The coefficients of the decision variables correspond to the product of 1,000 by the inverse of the criteria target levels measured in hectares, cubic meters and Euros, in the case of the area harvested, the volume harvested and the discounted revenues, respectively. The GP model was set to minimize the percent deviations to overcome dimensional constraints to the treatment of all three criteria. The discounted revenues criteria was provided with a weight equal to 10 while the remaining two were weighted equally as 1. For example, in the case of *NdevNPVt* the coefficient is computed as 10*1,000*1/2,041,298 = 0.0049 while in the case of *PdevAH1* the coefficient is computed as 1*1,000*1/118 = 8.4746.

(b) The equation that computes the discounted net revenue from the forest

$$\begin{aligned}\Big(&155.6x_{11} + 143.7x_{12} + 131.5x_{13} + 119.3x_{14} + 107.6x_{15} + 155.7x_{21}\\ &+ 143.8x_{22} + 131.5x_{23} + 119.4x_{24} + 107.7x_{25} + 152.1x_{31} + 140.5x_{32}\\ &+ 128.5x_{33} + 116.7x_{34} + 105.2x_{35} + \ldots 85.9x_{161} + 79.4x_{162} + 72.6x_{163}\\ &+ 70.2x_{164}\Big) \times 10^3 - NPVt = 0\end{aligned} \tag{6.19}$$

(c) The Eq. 6.20 that set the target for the area harvested in each period (118 ha) to reflect the area control objectives, where this area is defined by Eq. 6.12

$$\begin{aligned}AH1 - PdevAH1 + NdevAH1 &= 118\\ AH2 - PdevAH2 + NdevAH2 &= 118\\ AH3 - PdevAH3 + NdevAH3 &= 118\\ AH4 - PdevAH4 + NdevAH4 &= 118\end{aligned} \tag{6.20}$$

(d) The Eq. (6.21) that set the target for the volume harvested in each period (41×10^3 m^3) to reflect the volume control objectives, where this harvest level is defined by Eq. 6.11.

$$\begin{aligned}H1 - PdevH1 + NdevH1 &= 4100\\ H2 - PdevH2 + NdevH2 &= 4100\\ H3 - PdevH3 + NdevH3 &= 4100\\ H4 - PdevH4 + NdevH4 &= 4100\end{aligned} \tag{6.21}$$

(e) The Eq. 6.22 that sets the target for the discounted net revenue from the forest (2,041,298 €).

$$NPVt - PDevNPVt + NDevNPVt = 2{,}041.3 \tag{6.22}$$

(f) The set of constraints stating that the decision variables may take only the values 1 (if the prescription is assigned to the stand) or 0 (if the prescription is not assigned to the stand) (Eq. 6.13)

$$x_{11}, x_{12}, x_{13}, \ldots, x_{165} \text{ are binary}$$

In this illustration, the target values were set to emulate the area and volume control objectives and took advantage of the insights provided by the solutions by the other techniques. For example, the target for the criteria *NPV* was set as the value of the optimal LP solution. Generally, the targets are set based on *a priori* knowledge about the criteria space e.g. about the productive potential of the industrial forest.

The solution by the GP model (Table 6.3) shows that the *NPV* criteria target was almost achieved (2,040,967.45 €) while still guaranteeing that no stand was assigned to more than one prescription (Table 6.4). Yet this was at the cost of deviations from other targets. The MIP solution did point out that in order to meet the volume and area control objectives, while ensuring locational specificity of management options, the maximum net returns from the forest were equal to 2,021,000 €. For example, in the GP model the harvest levels in periods 1 and 2 did not meet the 41×10^3 m^3 targets. Further, there was a decline in harvest levels in periods 3 and 4.

This solution highlighted that often decision-makers lack the *a priori* knowledge about the productive potential of the industrial forest that is needed to specify coherent targets. It showed that setting a higher target level for the *NPV* criteria led to the underachievement of criteria emulating the area and volume control objectives. In fact, the information regarding the long-term impact of forest management options on objectives and conditions of interest is hardly ever perfect. The efficiency and the effectiveness of a multiple criteria approach to industrial forest strategic management planning calls for the use of models and methods as learning devices. The quality of decisions may be enhanced by a learning process that may provide additional insights about the resource capability model and the trade-offs between objectives (Borges et al. 2014a).

Most multiple criteria approaches reported in the forestry literature typically require the decision-maker to either specify the desired level of achievement or specify the preferences for the various objectives (Martins and Borges 2007). As often there is little information about what is possible to achieve (e.g. volume flows), defining *a priori* the goals and preferences may not be realistic and lead to poor management decisions (Tóth et al. 2006). Shortcomings of mechanistic approaches to the specification of the levels of achievement of various objectives as well as of the decision-makers preferences have been pointed out by Tóth and McDill (2009) and Romero (2004). In order to overcome them, Tóth and McDill (2009) demonstrated the possibility of developing and displaying a Pareto frontier e.g. of finding the non-dominated points in the feasible set in the criteria space (FSCS) in the case of problems with up to three forest management planning objectives. Romero (2004) discussed the use of several achievement functions and corresponding assumptions regarding decision-makers preferences.

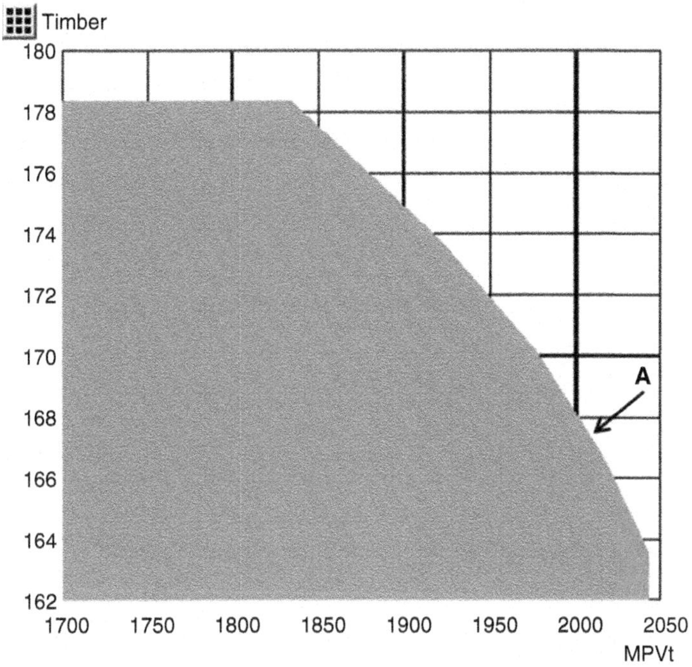

Fig. 6.3 Decision map showing the Edgeworth-Pareto Hull for a two criteria planning problem – *NPV* and *Timber* – subject to volume and area control constraints, considering minimum achievement levels min NPV = 1,700 × 10^3 € and min Timber = 162 × 10^3 m^3. Point A corresponds to the point in the Pareto frontier with *NPV* = 2,012.4 × 10^3 € and *Timber* = 166.9 × 10^3 m^3

Providing information about the set of efficient solutions can help the decision-maker understand the trade-offs between competing objectives. The analysis of these trade-offs may provide further insight about the forest management planning problem and help set adequate levels of achievement for various objectives (Borges et al. 2014a). In this section, we will apply an interactive modelling approach to generate the Pareto frontier of our industrial forest strategic management planning problem (Problem 1). The approach builds from the LP formulation of Problem 1 to display the trade-offs between discounted returns (*NPV*) and timber volume harvested (Fig. 6.3). The reader is referred to Borges et al. (2014a) for a detailed description of the modelling approach.

This trade-off information helps decision-makers set informed levels of achievement that reflect their preferences. It shows that harvesting over about 163 × 10^3 m^3 leads to a decrease of *NPV*. The LP solution did indeed highlight that it was not profitable to harvest more timber while meeting the volume and area control objectives. For illustration purposes lets assume that the decision-maker took advantage of this information to set as levels of achievement *NPV* = 2,012,415.5€ and *Timber* = 166,914.5 m^3 (Point A in Fig. 6.3). The modeling approach may then be used to retrieve the corresponding LP solution in the feasible set in the decision

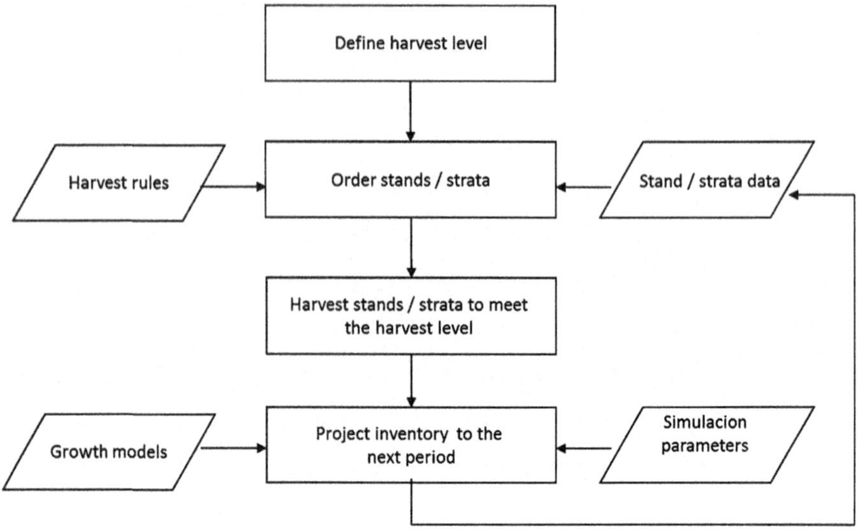

Fig. 6.4 Flowchart of simulation approach

space (Table 6.3). The solution highlights that the timber volume increase results from higher harvest levels in periods 3 and 4 at the cost of lower harvest values in earlier periods (Table 6.3). Harvesting later stands 9 and 14 (Table 6.4) leads to higher volumes yet this timber was financially mature earlier. As a consequence the value of *NPV* decreases.

6.2.4 Binary Search

Simulation is a technique that has also been widely used to address volume and area control objectives. In summary, binary search starts by setting the harvest level deemed as sustainable and by ordering the forest stands according to the priority for harvesting or harvest rule (Fig. 6.4).

The simulation approach proceeds by going down the list of ordered stands and harvesting them until the harvest level is met. Afterwards it projects the inventory to the next planning period. In this step, simulation parameters such as the area burned or new forest areas to include may be used to update the inventory. The stands are again ordered according to the harvest rule and this process is iterated successively to check whether the target volume level is sustainable or not. If, for example it finds that the initial level was too high, the estimate is decreased while if the inventory becomes too high the estimate is increased.

For illustration purposes, lets assume that the target level is set at 40,865 m^3 (the volume in the LP solution) and that stands are ordered according to the age. Results from the simulation approach show that the average age of the forest

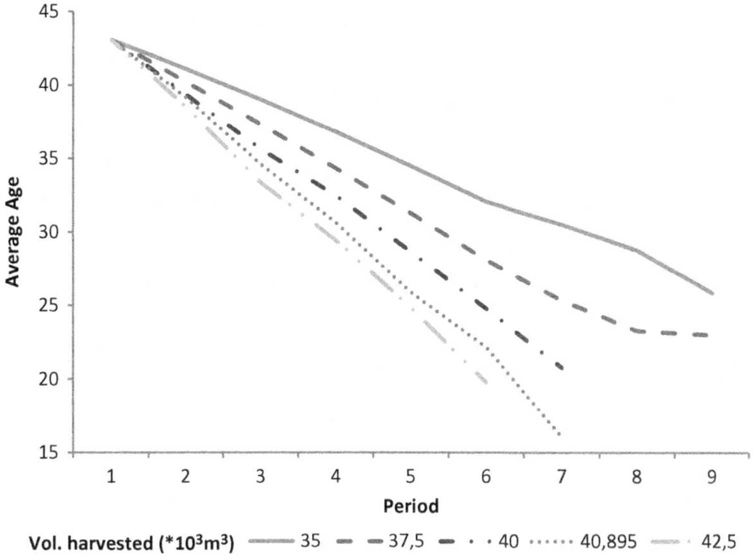

Fig. 6.5 Average stand age in the example forest as a function of periodic harvest levels

decreases over time from 43.7 years in the current inventory to 13.2 years in the inventory in period 7 when that harvest level may no longer be met. This suggests that the length of the planning horizon considered in Formulations 1–4 may have been too short. The harvest levels are not sustainable over longer periods. The simulation of other harvest levels help estimate the long term sustainable yield (Fig. 6.5).

Management Planning in Action 6.2: Assessing the Sustainability of National Pulpwood Supply: An Application in Portugal

Eucalypt (*Eucalyptus globulus* Labill) is the most important pulpwood producing species in Portugal. Eucalypt plantations extend over 647×10^3 ha – about 20.6 % of the total forest area in Portugal with a total yield of about 5.75×10^6 m3 per year. Nevertheless the land base owned or managed by vertically integrated pulp and paper companies provides at most 30 % of its pulpwood needs. In 2004 a priority of CELPA (http://www.celpa.pt), the Portuguese pulp and paper association, was the assessment of the sustainability of pulpwood supply from areas not owned or managed by the industry.

(continued)

(continued)

This prompted the development of a simulation approach similar to the one presented in 6.2 by the Large-Scale Informatics Systems Laboratory (http://lasige.di.fc.ul.pt/) and the Forest Research Centre (http://www.isa.ulisboa.pt/cef/). The eucalypt area not owned or managed by the industry was classified into analysis areas according to criteria such as age, coppice cycle, site index and location. Analysis areas were ranked for harvesting according to a weighted average of age, location and quadratic mean diameter. Simulation parameters included the probabilities of wildfire occurrence, of conversion i.e. of a clearcut and of abandonment. The possibility of expansion of the eucalypt area was also considered (Falcão 2006). This approach is being used by CELPA since 2004.

The mathematical programming and the simulated annealing approaches to address volume and area control in long term industrial forest management planning considered the objective of maximizing net present value. However, it is worthwhile to briefly note how the choice of objective function can influence the "personality" of the resultant model and the implications for how key constraints should be formulated. In a nutshell, the cost minimization model tends to be "lazy" and the profit maximization model tends to be "greedy."

In the case of the cost minimization model, the production constraints such as volume constraints must be specified as minimum targets (i.e., greater-than-or-equal-to constraints) or the optimal solution will be to produce nothing. This can be the most appropriate model when the plantation is owned by a vertically-integrated company that requires the forest to produce a certain volume of wood each year to meet the production requirements of a mill. In such models, a key concern may be the feasibility of meeting other constraints – such as sustainability or environmental constraints – while meeting the needs of the mill. Another possibility is that the mill requirements can be met too easily, and the forest is underutilized, which should make it easy to meet sustainability and environmental constraints but may not be in the best interest of the company.

With profit-maximization models, it is generally best to let the model determine the profit-maximizing level of production over time, but to add constraints that prevent the outputs (or inputs) of some products from fluctuating too wildly from one period to the next as illustrated in this section. If the model projects more production than is required by the mill this excess can potentially be sold to other companies, and if the projected production is less than is required, the company can either buy wood on the open market, buy more forestland, or develop more intensive management prescriptions that will produce higher yields.

Unless specifically required to by model constraints, profit maximization models often will not meet sustainability and environmental constraints. In particular, with a finite planning horizon, a profit-maximizing model will tend to harvest anything it can before the end of the planning horizon. This was highlighted by the binary approach to solve Problem 1. This is usually not a desirable outcome, so some kind of ending constraints must usually be imposed on profit-maximizing harvest scheduling models. As with any other aspect of harvest scheduling models, there are many ways to ensure that the model leaves the forest in a desirable condition at the end of the planning horizon. Furthermore, more than one approach can be incorporated into a model. One approach is to require the model to achieve a specific age-class distribution – such as a regulated forest – at the end of the planning horizon. While this approach may be useful in some circumstances, there are two disadvantages of this approach. First, if the planning horizon is not long enough it may not be possible to achieve the desired age-class distribution within that time frame and the model will be infeasible. Second, even if it is feasible, these constraints will tend to drive much of what the model does, especially in the final periods, and it leaves the model with very little flexibility to achieve any other management objectives. Three more promising ways to ensure that the model will leave the forest in a desirable ending condition are (1) to include a value of the ending forest in the objective function coefficients as in our case study, (2) to require the average age of the forest as a whole at the end of the planning horizon to be greater than or equal to some target, or (3) to require the total forest inventory at the end of the planning horizon to be greater than or equal to some target, as illustrated in the next section.

6.3 Long-Term Industrial Forest Management Planning to Address Both Multiple Product and Stock Control Objectives (Problem 2)

The focus of strategic management planning is on the assessment of the long-term sustainability of the industrial forest resource base. The implementation of volume and area control policies targets the provision of even or non-declining product flows from the industrial forest over the planning horizon. The solution of Problem 1 by several techniques (Sect. 6.2) did demonstrate the success of those policies. Nevertheless it highlighted that those policies, in particular if framed by revenue maximization objectives, do not guarantee the long-term sustainability of the industrial forest. Thus, Problem 1 was modified to address concerns with that sustainability. Specifically, for illustration purposes, the policy model was extended to include a condition on the value of the inventory at the end of the planning horizon.

Climate change concerns have led society and forest managers to focus on the potential of forests as carbon sinks. Accordingly, since 2000, carbon has emerged as an important product of industrial plantations (Chap. 14). The scale and scope of the carbon market has since expanded. Therefore it is increasingly important to assess the sustainability of carbon stock targets in strategic industrial forest management planning. The policy model of Problem 1 was thus further expanded to include conditions on the fluctuations of the carbon stock in the industrial forest.

In this section we will illustrate how to build mathematical programming models and how to design heuristic approaches to address jointly revenue optimization, area and volume control objectives as well as stock control concerns. The introduction of the new policy scenario defines our strategic management planning Problem 2. We will further interpret the solution of the new problem and assess the impact of constraints on the value of the ending inventory and on the average carbon stock on the harvest schedule and on the timber supply. Finally, we will illustrate how to assess the potential of specific solution techniques to address Problem 2.

6.3.1 Linear and Integer Programming

In the case of our example forest, the LP formulation of Problem 1 (F1) may be extended to represent Problem 2 (Formulation 5) by including (a) an accounting equation to compute the volume of the ending inventory, (b) a set of accounting equations to determine the average carbon stock in each planning period, (c) a constraint on the value of the ending inventory and (d) a set of constraints on the average carbon stock over the planning horizon.

(a) The accounting equation to compute the volume of the ending inventory. In this abbreviated form, it displays the decision variables x_{kj}, i.e. the area of stand k assigned to prescription j, for stands 1, 2 and 16:

$$18.2x_{11} + 14.0x_{12} + 9.7x_{13} + 5.4x_{14} + 2.1x_{15} + 18.9x_{21} + 14.6x_{22} + 10.1x_{23} + 5.7x_{24} + 2.1x_{25} + \cdots + 2.2x_{161} + 0.2x_{162} + 38.6x_{164} - VolEI = 0 \quad (6.23)$$

Where *VolEI* represents the volume of standing timber in the whole forest at the end of the planning horizon. The coefficients of the decision variables $VolEI_{kj}$ correspond to the volume per hectare in the ending inventory in stand k if it is managed according to prescription j. For example, if one hectare of stand 1 is managed according to prescription 1, the value of the ending inventory in that hectare will be $VolEI_{11} = 18.2 \times 10^3\ \text{m}^3$ (Table 6.2).

(b) The set of accounting equations to determine the average carbon stock in each planning period (Eq. 6.24):

$$6.1x_{11} + 9.0x_{12} + 9.0x_{13} + 9.0x_{14} + 9.0x_{15} + 6.4x_{21} + 9.4x_{22} + 9.4x_{23}$$
$$+ 9.4x_{24} + 9.4x_{25} + \ldots + 4.9x_{161} + 4.9x_{162} + 4.9x_{163} + 4.9x_{164} - CStock1$$
$$23.6x_{11} + 20.1x_{12} + 7.2x_{13} + 10.6x_{14} + 10.6x_{15} + 2.5x_{21} + 21.0x_{22} + 7.6x_{23}$$
$$+ 11.0x_{24} + 11.0x_{25} + \ldots + 7.4x_{161} + 7.4x_{162} + 7.4x_{163} + 7.4x_{164} - CStock2$$
$$20.5x_{11} + 9.2x_{12} + 23.6x_{13} + 22.9x_{14} + 8.2x_{15} + 21.4x_{21} + 9.6x_{22} + 2.5x_{23}$$
$$+ 23.8x_{24} + 8.5x_{25} + \ldots + 6.4x_{161} + 9.4x_{162} + 9.4x_{163} + 9.4x_{164} - CStock3$$
$$4.7x_{11} + 3.4x_{12} + 20.5x_{13} + 9.2x_{14} + 23.6x_{15} + 4.9x_{21} + 3.5x_{22} + 21.4x_{23}$$
$$+ 9.6x_{24} + 2.4x_{25} + \ldots + 24.5x_{161} + 20.9x_{162} + 7.5x_{163} + 11.0x_{164} - CStock4 \quad (6.24)$$

Where *CStockt* represents the average carbon stock in our example forest in period *t*. The coefficients of the decision variables CS_{kjt} correspond to the average carbon stock in period *t* in an hectare of stand *k* when managed according to prescription *j*. For example, if one hectare of stand 1 is managed according to prescription 1, the value of the average carbon stock in period 1 in that hectare will be $CS_{111} = 6.2 \times 10$ Mg C (Table 6.2).

(c) The constraint on the value of the ending inventory (Eq. 6.25)

$$VolEI \geq 5,500 \quad (6.25)$$

It is thus assumed that sustainability concerns may be addressed by setting the volume of the ending inventory criteria as 55,000 m^3.

(d) The set of constraints on the average carbon stock over the planning horizon (Eq. 6.26).

$$\begin{aligned} 0.9CStock1 - CStock2 &\leq 0 \\ 0.9CStock2 - CStock3 &\leq 0 \\ 0.9CStock3 - CStock4 &\leq 0 \\ CStock2 - 1.1CStock1 &\leq 0 \\ CStock3 - 1.1CStock2 &\leq 0 \\ CStock4 - 1.1CStock3 &\leq 0 \end{aligned} \quad (6.26)$$

No fluctuations over 10 % are thus allowed between the average carbon stocks in two consecutive periods.

According to the solution by the LP model (Table 6.3), the optimal long-term management plan is associated with an objective function value *Z* equal to $1,937.7 \times 10^3$ €. It encompasses a periodic harvest of $4,020.2 \times 10$ m^3. The area harvested ranged from 106.8 to 116.6 ha in periods 4 and 1, respectively. The average carbon stock decreased from $2,799.1 \times 10$ Mg C in period 1 to $2,040.6 \times 10$ Mg C in period 4. The solution thus demonstrates that the LP model may address too efficiently and effectively both volume and area control regulation objectives as well as concerns with stock control and the sustainability of the resource.

The LP solutions to Problems 1 and 2 also provide information about the opportunity cost associated with the new policy scenario. In fact, the *NPV* as measured by the objective function decreased by 5 % while the period harvest levels decreased by about 690 m^3. The constraint on the volume at the ending inventory is active and harvest levels must be lowered in order to meet the 55,000 m^3 target. The average carbon stock is always higher than in the solution to F1 yet it decreases substantially over time. This suggests that the current harvest levels may still not be sustainable and that the target for the volume of ending inventory may need to be adjusted. In fact, the simulation approach described in Sect. 6.2.4 did highlight that the $4{,}020.2 \times 10$ m^3 harvest level is not sustainable in the long term.

The plan proposed by the LP model reflects again the stand age distribution. Younger stands are proposed to be harvested later than in the case of F1 in periods 3 or 4 or are proposed not be harvested in order to meet the ending inventory constraint (Table 6.4). Conversely, just like in the case of F1, older stands are harvested earlier in periods 1 (stands 5–8) or 2 (stands 1–4) in order to avoid opportunity costs associated with the delay of harvesting mature timber. Nevertheless, some additional stands were split into more than one prescription in order to meet the stock control constraints.

The value of the shadow prices associated to the area constraints (Eq. 6.3) shows that in general the marginal value of each stand for the forest owner is higher in the case of F5 (Table 6.5). As expected the marginal value of younger stands (Stands 13–16) increases the most as they are instrumental to meet the stock control objectives. An additional hectare of one of these stands contributes directly to the net present value through the revenues resulting from its harvest. It contributes further indirectly to the net present value by relaxing the need to shift the harvest of older stands to later periods. Stand 6 is still the stand with the highest marginal value due both to its age and its productivity (Table 6.2). Nevertheless its value decreased as its harvest forces costly shifts of harvest timings in other stands. The LP solution thus adjusted the estimates of the maximum amount the forest owner might pay when buying an additional hectare of each stand when stock control objectives are considered. It further provided updated information about the opportunity costs associated with the selection of alternative plans e.g. the costs of anticipating or delaying the timing of harvests in each stand (Table 6.6). In particular it highlights the opportunity costs associated with the anticipation of harvest in older stands. These result from the loss of revenue that results from the need of adjusting the timing of harvests in other stands in order to regulate the harvest schedule and to meet the stock control objectives.

The strategic targets may turn out to be infeasible because of tactical and operational considerations that were left out from Problem 2. Feasible strategic targets may be approximated by enforcing locational specificity constraints. This may be even more critical in the case of Problem 2 as more stands had to be split between prescriptions to comply with the new policy scenario. It may thus be interesting to modify F5 within a Model I framework to further address concerns with locational specificity. The resulting MIP formulation of Problem 2 (F6) is an extended version of the MIP Formulation of Problem 1 (F2) to include further (a) an accounting equation to compute the volume of the ending inventory, (b) a set of accounting equations to determine the average carbon stock in each planning period, (c) a constraint on the value of the ending inventory and (d) a set of constraints on the average carbon stock over the planning horizon:

(a) The accounting equation to compute the volume of the ending inventory. In this abbreviated form, it displays the decision variables $x_{kj,}$ i.e. whether stand k is assigned to prescription j, for stands 1, 2 and 16:

$$\begin{aligned}&563.6x_{11}+434.1x_{12}+299.6x_{13}+168.3x_{14}+64.1x_{15}+563.8x_{21}+434.4x_{22}\\&\quad+299.7x_{23}+168.4x_{24}+64.1x_{25}+\cdots+63.9x_{161}+7x_{162}+1147.2x_{164}\\&\quad-\mathit{Vol\,EI}=0\end{aligned}\tag{6.27}$$

The coefficients of the decision variables $VolEI_{kj}$ correspond now to the volume in the ending inventory in stand k if it is managed according to prescription j. For example, if stand 1 is managed according to prescription 1, the value of the ending inventory in that stand will be $VolEI_{11} = 563.3 \times 10^3\ \text{m}^3$. This was computed as the product of the stand area (31 ha) by the volume per hectare at the end of the planning horizon that results from this assignment ($18.2 \times 10^3\ \text{m}^3$ (Table 6.2)).

(b) The set of accounting equations to determine the average carbon stock in each planning period (Eq. 6.28):

$$\begin{aligned}&190.1x_{11}+279.5x_{12}+279.5x_{13}+279.5x_{14}+279.5x_{15}+190.2x_{21}+279.6x_{22}\\&\quad+279.6x_{23}+279.6x_{24}+279.6x_{25}+\ldots+145.9x_{161}+145.9x_{162}+145.9x_{163}\\&\quad+145.9x_{164}-\mathit{CStock1}\\&73.2x_{11}+62.5x_{12}+225.1x_{13}+328x_{14}+328x_{15}+73.2x_{21}+624.8x_{22}\\&\quad+225.2x_{23}+328.2x_{24}+328.2x_{25}+\ldots+216.9x_{152}+216.9x_{153}+216.9x_{154}\\&\quad+219x_{161}+219x_{162}+219x_{163}+219x_{164}-\mathit{CStock2}\\&637.5x_{11}+286.2x_{12}+732x_{13}+710.3x_{14}+253.6x_{15}+637.8x_{21}+286.4x_{22}\\&\quad+73.2x_{23}+710.6x_{24}+253.8x_{25}+\ldots+189.5x_{161}+278.7x_{162}+278.7x_{163}\\&\quad+278.7x_{164}-\mathit{CStock3}\\&146.3x_{11}+105.2x_{12}+637.58x_{13}+286.3x_{14}+732x_{15}+146.4x_{21}+105.2x_{22}\\&\quad+637.8x_{23}+286.4x_{24}+73.2x_{25}+\ldots+729.9x_{161}+622.7x_{162}+224.4x_{163}\\&\quad+327.1x_{164}-\mathit{CStock4}\end{aligned}\tag{6.28}$$

The coefficients of the decision variables CS_{kjt} correspond now to the average carbon stock in period t in stand k when managed according to prescription j. For example, if one hectare of stand 1 is managed according to prescription 1, the value of the average carbon stock in period 1 in that stand will be $CS_{111} = 190.1 \times 10$ Mg C. This was computed as the product of the stand area (31 ha) by the average carbon per hectare in stand 1 that results from this (6.2×10 Mg C (Table 6.2)).

(c) and (d) The constraints on the value of the ending inventory and on the average carbon stock may now be expressed just like in the case of the LP model F5 (Eqs. 6.25 and 6.26, respectively)

According to the solution by the IP model (Table 6.3), the optimal objective function value Z decreased to $1{,}867.1 \times 10^3$ €. No stands are split between different prescriptions (Table 6.4) and this thus cost 106.6×10^3 €. The requirement of locational specificity in Problem 2 is thus five times more expensive than in Problem 1. The model proposed to delay the harvest of most stands. It further proposed a no harvest prescription in the case of the younger stands (stands 13–16). The stock control objectives when combined with the requirement of locational specificity thus did impact substantially the volumes and areas harvested in each period, which are considerably lower. The average carbon stock is thus higher. Moreover, the lack of management flexibility lead to a harvest plan that left a volume at the ending of the planning horizon that is much higher than required (83,355 m^3). Again, in the case of larger forests with more stands the requirement of locational specificity may not have an impact as substantial on the adjustment of harvest plans and on the criteria levels. Yet it may have an even higher impact on the computational cost.

6.3.2 Heuristics

The locational specificity requirement in Problem 2 may also be addressed by heuristic approaches such as simulated annealing (SA). In the case of our example forest, the SA approach involved the extension of the SA formulation for Problem 1 (F3) to address the new policy scenario. In summary, the SA formulation for Problem 2 (F7) takes the objective (evaluation) function of F3 (Eq. 6.14) and modifies its penalty function to include two further terms that penalize deviations from the target volume at the end of the planning horizon and from the average carbon stock constraints. In both cases, it was assumed that a 5 % deviation from feasibility was equivalent to a 1 % deviation from unconstrained *NPV* when estimating the parameter that characterizes the shape of the parabola for each of these two constraints. The penalty function may thus be represented in abbreviated form as:

$$\theta = A_{const\_6\_10}\left[(x_{11}+x_{12}+x_{13}+x_{14}+x_{15}-1)^2+\ldots+(x_{161}+x_{162}+x_{163}+x_{164}-1)^2\right]$$
$$+ A_{const\_6\_6}\left[(H_1-H_2)^2+\ldots+(H_3-H_4)^2+(H_2-1.1H_1)^2+\ldots+(H_4-1.1H_3)^2\right]$$
$$+ A_{const\_6\_7}\left[(.9AH_1-AH_2)^2+\cdots+(.9AH_3-AH_4)^2+(AH_2-1.1AH_1)^2+\ldots\right.$$
$$\left.+ (AH_4-1.1AH_3)^2\right]+A_{const\_6\_21}(VolEI-5500)^2$$
$$+ A_{const\_6\_22}\left[(.9CStock_1-CStock_2)^2+\cdots+(.9CStock_3-CStock_4)^2\right.$$
$$\left.+ (CStock_2-1.1CStock_1)^2+\cdots+(CStock_4-1.1CStock_3)^2\right] \quad (6.29)$$

Afterwards, the SA approach involved an iterative process identical to the process described in Sect. 6.2.2 (Fig. 6.2) where solutions in each iteration were represented by a vector with 16 elements corresponding to the 16 stands. Penalties and heuristic parameters were the same as in the case of Problem 1. In this illustration, the parameterization did lead again to the optimal MIP solution ($1{,}867.1 \times 10^3$ €) (Table 6.3). This is also a consequence of the problem small size.

Management Planning in Action 6.3: Impacts of Timber Management Scheduling on Multiple Product and Stock Control Objectives in Minnesota

Minnesota is located in the United States Upper Midwest and it extends over about 21×10^6 ha. Its forest area extends over about one third of the territory. In 1989, a citizen petition was submitted to the Minnesota Environmental Quality Board (EQB) to seek action to address a potential increase of nearly 3.62×10^6 m^3 in annual timber harvesting activity associated with a proposed 2.2×10^9 USD increase in the state's primary wood processing plant capacity (Kilgore 1992). As a consequence, the state of Minnesota decided to develop a Generic Environmental Impact Statement (GEIS) of timber harvesting.

For that purpose, prescriptions were associated to 13,536 USDA Forest Service's Forest Inventory and Analysis (FIA) plots, assumed to represent forest conditions in Minnesota. The most appropriate prescription for each stand/plot was selected by a scheduling model that matched demand for a product with the stand or forest area best able to supply that product and in consideration of mitigations and other constraints (Rose et al. 1993; Jaakko Consulting Inc 1994). The scheduling model was based on a Lagrangean relaxation of a typical forest management planning LP model such as the ones presented in 6.2 and 6.3 (Hoganson and Rose 1984). This model, developed at the University of Minnesota (http://www.forestry.umn.edu/), encompassed specialized techniques to search for the values of the dual variables of the LP

(continued)

(continued)

model as well as the design of maps of wood procurement zones for each market to overcome the combinatorial nature of integrating harvest timing decisions and wood shipping decisions (Hoganson and Kapple 1991).

The GEIS involved over 60 scientists and it represented one of the most extensive studies of timber harvesting and forest resources conducted in the United States (http://iic.gis.umn.edu/download/geis/documnts.html). It was influential to (1) determine the extent of industrial timber harvesting and related timber management activities in Minnesota; (2) identify and assess the environmental and related impacts of industrial timber harvesting; and (3) recommend strategies to mitigate adverse impacts where such were found to be significant (Kilgore and Ek 2007).

6.3.3 Goal Programming and Pareto Frontier Methods

In the case of our example forest, the goal programming formulation of Problem 2 (F8) consists of a modification of its formulation for Problem 1 (F4). The objective function includes new terms in order to minimize the deviations from targets regarding the average carbon stock in each planning period and the volume at the end of the planning horizon. The formulation may be described as

(a) Objective function Z

$$\begin{aligned} \text{Min } Z = {} & 8.4746\ PdevAH1 + 8.4746\ NdevAH1 + 8.4746\ PdevAH2 \\ & + 8.4746\ NdevAH2 + 8.4746\ PdevAH3 + 8.4746\ NdevAH3 \\ & + 8.4746\ PdevAH4 + 8.4746\ NdevAH4 + 0.0244\ NdevH1 \\ & + 0.0244\ NdevH2 + 0.0244\ NdevH3 + 0.0244\ NdevH4 \\ & + 0.0049\ NdevNPVt + 0.04\ NdevCStock1 + 0.04\ NdevCStock2 \\ & + 0.04\ NdevCStock3 + 0.04\ NdevCStock4 + 0.04\ PdevCStock1 \\ & + 0.04\ PdevCStock2 + 0.04\ PdevCStock3 + 0.04\ PdevCStock4 \\ & + 0.0182\ NdevVolEI \end{aligned} \tag{6.30}$$

Where the decision variables consist again of the deviations from the levels of the criteria set by the decision-maker. Two additional criteria were added, the average carbon stock *CStock* in period *t* and the volume of ending inventory *VolEI*. In the case of *CStock* the objective function penalizes both over and under achievements while in the case of *VolEI* it aims at minimizing the under achievement. The coefficients of the corresponding decision variables correspond to the product of 1,000 by the inverse of the criteria target levels measured in Mg and cubic meters, in the case of the average carbon stock and the volume of the ending inventory, respectively. Like before, the new GP model was set to minimize the percent deviations to overcome dimensional constraints to the treatment of all five criteria. The discounted revenues criteria was provided again with a weight equal to 10 while the remaining four were weighted equally as 1.

(b) The Eq. 6.20 setting a new target (89 ha) for the area harvested in each period to reflect the area control objectives, where this area is defined by Eq. 6.12.

(c) The Eq. 6.21 setting a new target for the volume harvested in each period (35×10^3 m^3) to reflect the volume control objectives, where this harvest level is defined by Eq. 6.11.

(d) The Eq. 6.22 setting a new target for the discounted net revenue from the forest (1,870,000€), where the discount net revenue is defined by Eq. 6.19

(e) The Eq. 6.31 setting a target for the volume of ending inventory, where this volume is defined by Eq. 6.27:

$$VolEi + PdevVolEi + NdevVolEi = 5,500 \tag{6.31}$$

(f) The Eq. 6.32 setting a target for the average carbon stock in each period (24×10^3 Mg C) to reflect the stock control objectives, where this average is defined by Eq. 6.28.

$$\begin{aligned} Cstock1 - PdevCstock1 + NdevCstock1 &= 2,400 \\ Cstock2 - PdevCstock2 + NdevCstock2 &= 2,400 \\ Cstock3 - PdevCstock3 + NdevCstock3 &= 2,400 \\ Cstock4 - PdevCstock4 + NdevCstock4 &= 2,400 \end{aligned} \tag{6.32}$$

(g) The set of constraints stating that the decision variables may take only the values 1 (if the prescription is assigned to the stand) or 0 (if the prescription is not assigned to the stand) (Eq. 6.13)

$$x_{11}, x_{12}, x_{13}, \ldots, x_{165} \text{ are binary}$$

In this illustration, the target values were set by taking advantage of the insights provided by the solutions by the other techniques. For example, the target for the criteria *NPV* was set as the value of the optimal MIP solution. The solution by the GP model (Table 6.3) shows that the *NPV* criteria target was almost achieved (1,866,130€) while still guaranteeing that no stand was assigned to more than one prescription (Table 6.4). Else the solution has a pattern similar to the MIP and SA solutions i.e. under achievement of harvest levels and over achievement of targets for both average carbon stock and volume in the ending inventory.

The Pareto frontier method may again be used to explore further the tradeoffs between the criteria and help set meaningful targets. The approach may build now from the LP formulation of Problem 2 (F5). Lets assume that the decision-maker wants to analyze the tradeoffs between four criteria e.g. *NPV*, *VolEI*, the average carbon stock over the whole planning horizon and the total volume harvested.

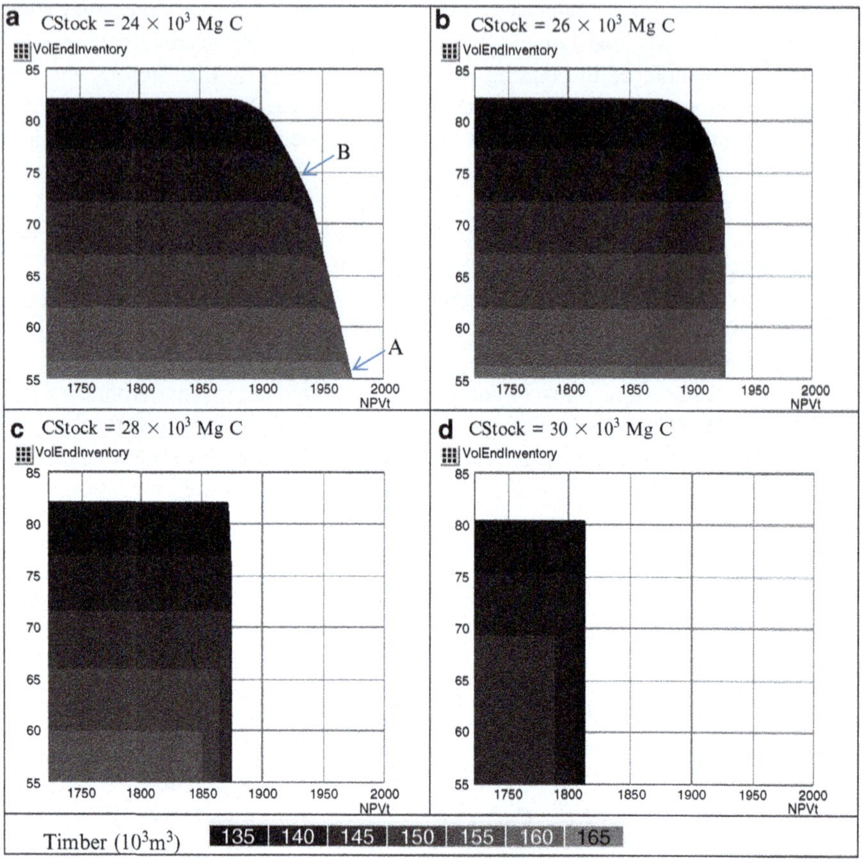

Fig. 6.6 Decision maps displaying the Edgeworth-Pareto Hull of a four criteria planning problem considering minimum achievement levels for the four criteria (min $NPV = 1{,}700 \times 10^3$ €, min $VolEI = 55 \times 10^3$ m^3, min $CStock = 24.0 \times 10^3$ Mg C and min $Timber = 135 \times 10^3$ m^3). Each set of decision maps corresponds to a specific average carbon stock (*CStock*) (**a**) 24.0×10^3 Mg C, (**b**) 26.0×10^3 Mg C . . . , d) 30.0×10^3 Mg C. Points *A* and *B* in (a) correspond to solutions selected by decision-makers from two decision maps associated with *VolEI* target levels of 55 and 75×10^3 m^3)

The Pareto frontier method may then be used to generate three-dimensional decision maps for which the values of the third, fourth,... criterion are fixed (e.g. Fig. 6.6 where the value of total timber harvested is fixed for each map and where the value of the average carbon stock is fixed for each set of decision maps). The maps are monotonic: a map contains all maps with better values of the third, fourth, . . . criterion (e.g. Fig. 6.6). When arranged in the form of horizontal series or even matrices, bi- or three-dimensional decision maps developed by this approach may provide information about the Pareto frontier in spaces up to five dimensions (Borges et al. 2014a).

Based on the tradeoff information provided by the Pareto frontier approach, lets assume that the decision-maker sets 24,000 Mg C as the level of achievement for the average carbon stock criteria (set of decision maps (a) in Fig. 6.6) and that he wants to explore solutions associated with an average carbon stock *CStock* = 24.0×10^3 Mg C (Figs. 6.6a) and alternative *VolEI* target levels (e.g. about 55 and 75×10^3 m^3 – points A and B in Fig. 6.6a, respectively).

The Pareto frontier approach may then be used to retrieve both solutions in the LP feasible set in the decision space (Table 6.3) and generate further insights about the management planning problem. As Borges et al. (2014a) pointed out, the usefulness of the Pareto frontier approach lies mostly in the preprocessing of the management planning problem to generate trade-off information so that the first specification of the levels of achievement by the decision-maker is more informed. Nevertheless, as this approach is based on the approximation of the Pareto frontier, the retrieval of a feasible solution requires the use of an achievement function to minimize the deviations between the solution selected in a decision map by the decision-maker and the feasible set in the criteria space, This is similar to the solution process by goal programming. Thus special attention should be also provided to the selection of an achievement function (Romero 2004).

Management Planning in Action 6.4: Long Term Cork Forest Management Scheduling to Address Multiple Product and Stock Control Objectives in Southern Portugal

In 2003, the Portuguese Regional Agriculture Office of Alentejo (DRAPAL) (http://www.drapal.min-agricultura.pt/) and the Extremadura Regional Government in Spain (http://www.gobex.es/consejerias/ceei_tecnologica.php) launched the project "Development of an information system for ecological-economic cork and holm oak ecosystem management". This cross-border cooperation initiative (http://ec.europa.eu/regional_policy/archive/interreg3/abc/voleta_west_en.htm) encompassed a regional sustainability assessment of cork flows to the forest industry.

This prompted the development and application of a modeling approach similar to LP model 6.3 by the Forest Research Centre (CFS) (http://www.isa.ulisboa.pt/cef/). The cork and holm oak land base extending over 1×10^6 ha in Alentejo was inventoried and firstly classified into 23,373 land units that were further aggregated into 84 analysis areas. The model included up to 8,400 prescriptions over a 5 ten-year periods planning horizon. Besides the area constraints, the LP model further included timber and cork non-declining flow constraints as well as carbon stock targets (Borges et al. 2009). The solution highlighted the constraining impact of the current inventory on potential cork supply in the first planning period and that both timber and cork non-declining flows policies may be sustained over the planning horizon (Borges et al. 2009).

More recently, in order to address DRAPAL concerns, the LP model was integrated by CFS and the Research Centre in Mathematics and Applications (http://www.cima.uevora.pt/) within a multi-criteria approach

(continued)

(continued)

targeting timber, cork, carbon, net present value and value of ending inventory achievement levels (Borges et al. 2014a). This approach is similar to the Pareto frontier approach described in Sect. 6.3.3

6.4 Designing the Industrial Forest Landscape in Long Term Management Planning

Financial efficiency usually dictates the concentration of activities of harvesting and infrastructure (e.g. road network) development. Conversely, broader economic and environmental goals often suggest its dispersion in time and space. For example, area and volume control objectives reflect concerns with the sustainability of timber supply and constrain the concentration of the timing of harvests in industrial plantations (Sect. 6.2). Nevertheless, the implementation of forest plans also requires spatially feasible prescriptions that may address simultaneously environmental concerns with the sustainability of the industrial land base and financial concerns with the dispersion of forest operations. These often encompass the definition of adjacency constraints: a maximum size of openings or a range of feasible opening sizes and a minimum exclusion period – the minimum time that must elapse between harvests on neighboring stands.

The spatial context of harvest scheduling decisions is typically addressed at tactical and operational planning scales (Chaps. 2 and 7). Nevertheless the possibility of balancing strategic and tactical goals may be facilitated if long-term management planning does provide a strategic design of the industrial forest landscape that may help accommodate requirements of tactical management planning. This may be achieved by extending Problem 2 and its concerns with timber harvesting regulation and with stock control objectives to include adjacency constraints.

Typically, this combinatorial optimization problem (Problem 3) involves very large numbers of integer variables and constraints. The literature reported several approaches to build and solve a forest management planning problem with adjacency constraints. For example, Borges et al. (2002) discussed thoroughly the use of heuristic approaches to address it, while Constantino et al. (2008) presented a range of mathematical programming representations of the adjacency problem. In fact, as Borges et al. (2002) pointed out, currently available optimization packages combine often the use of both heuristic and mathematical programming techniques. For example, heuristic techniques may utilized directly in optimization packages to help in key aspects of the solution process like in (1) finding initial feasible solutions, (2) finding initial bounds on optimal solutions, (3) selecting branches to search in a standard branch and bound technique and (4) selecting the specific mathematical programming strategy to use for solving a specific mixed integer problem (Borges et al. 2002).

In order to illustrate how to accommodate concerns with the design of our industrial forest landscape example, the policy model of Problem 2 was further expanded to include adjacency constraints. In this section we will illustrate how to build mathematical programming models and how to design heuristic approaches to address jointly revenue optimization, area and volume control objectives, stock control concerns and adjacency constraints. For illustration purposes we will consider the Path approach to build the adjacency constraints (McDill et al. 2002). The introduction of the new policy scenario defines our strategic management planning Problem 3. We will further interpret the solution of the new problem and assess the impact of adjacency constraints on the value of the ending inventory, on the average carbon stock, on the harvest schedule and on the timber supply. Finally, we will illustrate how to assess the potential of specific solution techniques to address Problem 3.

6.4.1 Mixed Integer Programming

The adjacency problem requires information about the location of management options. Thus LP may no longer be used to address Problem 3. In the case of our example forest, assuming that the maximum size of openings is 60 ha and that a 1-period exclusion is considered, the MIP formulation of Problem 2 (F6) should be expanded to include the path adjacency constraints (6.33 – Eq. 2.13 in Chap. 2) to define the MIP formulation of Problem 3 (F9):

$$\begin{aligned}
&x_{11} + x_{21} \leq 1\\
&x_{11} + x_{21} + x_{41} \leq 2\\
&x_{11} + x_{21} + x_{52} + x_{53} \leq 2\\
&x_{11} + x_{31} + x_{41} \leq 2\\
&x_{11} + x_{41} + x_{51} + x_{52} \leq 2\\
&x_{11} + x_{41} + x_{61} + x_{62} \leq 2\\
&x_{11} + x_{21} + x_{71} + x_{72} \leq 2\\
&x_{11} + x_{21} + x_{81} + x_{82} \leq 2\\
&x_{21} + x_{31} + x_{41} \leq 2\\
&x_{21} + x_{41} + x_{51} + x_{52} \leq 2\\
&x_{21} + x_{51} + x_{52} + x_{81} + x_{82} \leq 2\\
&x_{31} + x_{41} + x_{51} + x_{52} \leq 2\\
&x_{31} + x_{41} + x_{61} + x_{62} \leq 2\\
&x_{31} + x_{41} + x_{71} + x_{72} \leq 2
\end{aligned} \tag{6.33}$$

According to the solution by the MIP model (Table 6.3), the optimal objective function value Z decreased to $1{,}866.6 \times 10^3$ €. No stands were split between different prescriptions (Table 6.4) and all adjacency constraints were met. The solution was very similar to the MIP solution of Problem 2. This demonstrates that the 60 ha opening size limit is very easy to meet.

6.4.2 Heuristics

The adjacency constraints in Problem 3 may also be addressed by simulated annealing (SA). In the case of our example forest, the SA approach involved the extension of the SA formulation for Problem 2 (F7) to address the new policy scenario. In summary, the SA formulation for Problem 3 (F10) takes the objective (evaluation) function of F7 (Eq. 6.29) and modifies its penalty function to include one further term that penalizes deviations from adjacency constraints. These deviations were more penalized than the deviations from the other constraints. All other penalties and heuristic parameters were the same as in the case of Problem 2. The penalty function may thus be represented in abbreviated form as:

$$
\begin{aligned}
\theta = {} & A_{const\_6\_10}\left[(x_{11}+x_{12}+x_{13}+x_{14}+x_{15}-1)^2+\ldots+(x_{161}+x_{162}+x_{163}+x_{164}-1)^2\right] \\
& + A_{const\_6\_6}\left[(H_1-H_2)^2+\ldots+(H_3-H_4)^2+(H_2-1.1H_1)^2+\ldots+(H_4-1.1H_3)^2\right] \\
& + A_{const\_6\_7}\Big[(.9AH_1-AH_2)^2+\cdots+(.9AH_3-AH_4)^2+(AH_2-1.1AH_1)^2+\ldots \\
& \qquad + (AH_4-1.1AH_3)^2\Big] + A_{const\_6\_21}(VolEI-5500)^2 \\
& + A_{const\_6\_22}\Big[(.9CStock_1-CStock_2)^2+\cdots+(.9CStock_3-CStock_4)^2 \\
& \qquad + (CStock_2-1.1CStock_1)^2+\cdots+(CStock_4-1.1CStock_3)^2\Big] \\
& + A_{const\_6\_33}\left[(x_{11}+x_{21}-1)^2+\cdots+(x_{31}+x_{41}+x_{71}+x_{72}-2)^2\right]
\end{aligned}
\qquad (6.34)
$$

Afterwards, the SA approach involved an iterative process identical to the process described in Sect. 6.2.2 (Fig. 6.2) where solutions in each iteration were represented by a vector with 16 elements corresponding to the 16 stands. In this illustration, the parameterization lead to a suboptimal solution ($1{,}866.1 \times 10^3$ €) (Table 6.3). Nevertheless the best solution found by this random search heuristic did meet all constraints thus highlighting again that the 60 ha opening size is easy to meet. Computational costs were low as a consequence of the problem size.

6.4.3 Goal Programming

In the case of our example forest, the goal programming Formulation of Problem 3 (F11) consists of a modification of F8 to include the adjacency constraints (Eq. 6.33). No target was changed. The solution (Tables 6.3 and 6.4) again highlights that the adjacency constraints had little impact on the proposed plan.

Research is currently being conducted so that the Pareto frontier approach considered in earlier sections may address multi-criteria problems such as Problem 3, where the spatial context of stand-level decisions must be acknowledged. The integration of the Pareto frontier approach with combinatorial resource capability and policy models (e.g. mixed integer programming models) will provide the functionality needed to analyze tradeoffs between criteria before setting their levels of achievement.

Management Planning in Action 6.5: Designing the Forest Landscape: Applications in Portugal and in Minnesota

In 1998, Celbi, currently a factory of a leading Portuguese eucalypt pulp producer (Altri) (see Management Planning in Action 6.1), decided to lay out the harvest of one of its properties extending over about 640 ha in Central Portugal to minimize the environmental impacts of eucalypt stands clearcutting. All 144 stands in this property (Vale do Mouro) were mature. Yet it was decided to schedule the harvest over five 1-year periods so that no adjacent stands might be harvested in the same period. Decision-makers further wanted to minimize the opportunity cost of harvest delay.

This prompted the application of a dynamic programming approach to solve a simplified version of the MIP model in Sect. 6.3 (no product flow or stock control objectives were considered) within a project coordinated by the Forest Research Centre (http://www.isa.ulisboa.pt/cef/). For a detailed description of the technique the reader is referred to Hoganson and Borges (1998) and Borges et al. (1999a). The industry found that the adjacency constraints led to a decrease of about 3 % of the net present value (Borges et al. 1999b).

This technique was extended to further analyze trade-offs between timber production and environmental objectives in the early 2000s in a forest area comprising Minnesota DNR managed lands in Itasca and Cass County (http://www.dnr.state.mn.us/index.html) and USDA Forest Service lands within the Chippewa National Forest (http://www.fs.usda.gov/chippewa/). Approximately 302,000 ha distributed over 92,000 stands were modeled (Hoganson et al. 2004). This project coordinated by the University of Minnesota (http://www.forestry.umn.edu/) estimated the impact of landscape design, namely of the supply of interior space, on industrial timber management scheduling (USDA Forest Service 2004).

6.5 Integrating Road Building Schedules in Long Term Industrial Forest Management Planning

Harvest scheduling and road building problems have traditionally been addressed in two sequentially linked steps. In the first step, the optimal harvest scheduling model identifies the areas that will be harvested during the planning horizon, without accounting for road building considerations. In the second step, the minimal road cost program that provides road accessibility for the optimal harvest schedule is determined. During the 1960s and 1970s tactical forest planning was essentially designed using this sequential approach (Guignard et al. 1998). According to Weintraub and Navon (1976), a sequential approach leads to two main problems: the wrong set of stands (nodes) may be made accessible, and the selection of the access period for each stand may be not optimal. Kirby et al. (1986) suggest that since the integrated formulation is less constrained than the sequential formulation, the cost of implementing a management plan using an integrated approach will be always smaller than or equal to the cost of a management plan based on the equivalent sequential procedure. Thus, for example, a slight increase in harvest cost due to a change in harvest location could result in a significant reduction in road building cost, an improvement that could not be recognized in a sequential procedure. Guignard et al. (1998) found that if the objective is to maximize the net present value of the timber minus road building and transportation costs, then the integrated approach can generate up to 60 % greater profits than the traditional sequential approach.

Weintraub and Navon (1976) present one of the first integrated models reported in the literature, where silvicultural and transportation activities are simultaneously considered. The authors suggest a MIP approach to find the maximum value for discounted revenues earned from the sale of timber, where road, timber management, and transportation costs are included in the analysis. Kirby et al. (1986) introduce the Integrated Resource Planning Model (IRPM), which is a Goal MIP problem formulated as an assignment problem where management decisions are associated with treatment units explicitly identified on the ground. Decision variables are defined for every unit and management alternative, in terms of the proportion of the unit area (0–1) assigned to every management alternative. Constraints impose minimum requirements on water run-off, recreation usage, wildlife usage, erosion, timber yield, visual degradation, employment and revenue. The road network is formulated as a multi-commodity, multi-period, fixed charge, capacitated network problem with mutually exclusive road capacities. Their results produce up to a 21 % reduction in cost compared to the traditional sequential model.

Guignard et al. (1998) suggest additional improvements to the original Kirby et al. (1986) model. Lifting and branch and bound priorities based on double contraction considerations are used to reduce the computational time required to solve the problem to optimality. Different road standards are considered in the model, including dirt, gravel and paved alternatives along with the associated building and transportation costs and capacity and seasonality restrictions. Following the original strengthened model proposed by Guignard et al. (1998), Andalaft et al. (2003) combine road capacity reduction with triggers and lifting constraints using Lagrangian relaxation. The proposed model is applied to a Chilean pine plantation problem with planning horizons between 2 and 5 years. Operational seasonality (winter and summer) and road standards (dirt, gravel and paved roads) are included in the model. Adjacency constraints are formulated differently than in traditional methods, where a minimum harvest area (as opposed to a maximum harvest area) is imposed to guarantee acceptable operational fixed costs.

6.5.1 Potential Road Network

Forest road networks used as input for the model are defined considering potential and actual road segments meeting the minimum technical road requirements and providing full accessibility to every management unit in the forest. In the conceptual network, landings are represented by nodes and road segments are represented by arcs. In the initial step, a minimum number of landings is identified. Usually, landing locations are identified based on the machinery available for logging operations. A road exit node represents the connection point between the forest road network and the road that connects the forest with the final destination for the timber. Note that, even though forests can have multiple road exits, the problem can be easily be transformed into a one exit node problem by creating a dummy node that connects all the potential and existing exit nodes.

Based on a topographic analysis of the forest, a set of potential road segments is created such that every node is connected to the exit node. The set of potential road segments is identified by R_{ij}, representing the arc that connects node i with node j. f_{ij}^t represents the timber volume flow between node i and node j, and f_{ji}^t represents the timber volume flow between node j and node i. This differentiation is used to model road networks where cycles are allowed. Even though flow directions are differentiated by the flow variables f_{ij}^t and f_{ji}^t, road decision variables are undirected variables, and consequently r_{ij} represents both arcs (i,j) and (i,j) simultaneously.

6.5.2 The Integrated Harvest Road Model

The general MIP formulation for the harvest scheduling problem (Chap. 2) can be adapted to incorporate road building decisions using an integrated approach. Different formulations can be used for this purpose, and the strength of the resulting models can be appreciably different. Tight or strong formulations are important in practice, as computation times required to obtain acceptable solutions are directly related to this characteristic. Thus the integrated harvest road model can be modeled as follows:

Sets

T Set of periods in the planning horizon, such that $T = \{1, 2, \ldots, n_T\}$.
M Set of management units, such that $M = \{1, 2, \ldots, n_M\}$.
M^i Set of management units associated with transportation node i. In other words, the set of management units from which timber will be hauled through node i.
N Set of nodes in the road network.
A Set of the current and potential road segments where (i, j) describes an arc between nodes i and j such that $(i, j) \in A$
P^m Set of road segments that belong to the path between the node associated with management unit m and the exit node in the road network.

Decision Variables

x_m^t Binary variable which assumes a value of 1 if management unit m is harvested in period t, and 0 otherwise. Notice that $x_m^0 = 1$ represents that management unit m is not harvested at all during the planning horizon.
r_{ij}^t Binary variable which assumes a value of 1 if road segment (i, j) is built in period t, and 0 otherwise.
f_{ij}^t Total annual timber flow (mbf/year) from node i to node j in period t.

Constants

a_m Area of management unit m (acres).
v_m^t Volume per acre (mbf/acre) obtained from unit m if it is harvested in period t.
c_m^t Net discounted revenue per acre ($) if management unit m is harvested in period t. Assuming a discount rate r (expressed as decimal fraction), it is calculated as $c_m^t = c_m^{t*}/(1+r)^t$, where c_m^{t*} is the revenue per acre earned at period t if unit m is harvested in this period.
q_{ij} Timber flow capacity per time unit (mbf/year) of road segment (i, j).
R_{ij}^t Net discounted building cost ($) if the road segment (i, j) is built in period t. This cost includes road and bridge building costs. Assuming a discount rate expressed as decimal fraction, it is calculated as $R_{ij}^t = R_{ij}^{t*}/(1 + rate)^t$, where R_{ij}^{t*} is the road building cost incurred at period t if road segment (i, j) is built in this period.

Objective Function

$$Max\ Z = \sum_{m \in M} \sum_{t \in T \cup \{0\}} c_m^t a_m^t x_m^t - \sum_{(i,j) \in A} \sum_{t \in T} r_{ij}^t R_{ij}^t, \tag{6.35}$$

Subject to,

$$\sum_{m \in M^i} v_m^t a_m x_m^t + \sum_{(k,i) \in A} f_{ki}^t - \sum_{(i,l) \in A} f_{il}^t = 0, \quad i \in N \backslash \{exit\}, t \in T, \tag{6.36}$$

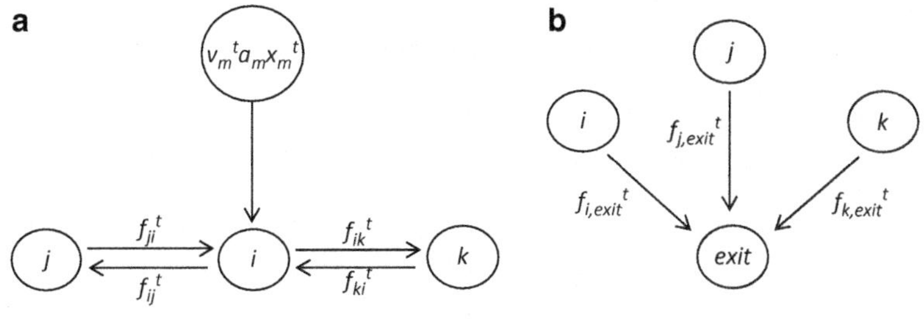

Fig. 6.7 Timber flow representation in a road network

$$\sum_{m \in M} v_m^t a_m x_m^t - \sum_{(k,exit) \in A} f_{k,exit}^t = 0, \qquad t \in T, \tag{6.37}$$

$$f_{ij}^t + f_{ji}^t - q_{ij} \sum_{t'=1}^{t} r_{ij}^{t'} \leq 0, \qquad (i,j) \in A, t \in T, \tag{6.38}$$

$$f_{ij}^t \geq 0, r_{ij}^t \in \{0,1\}, \qquad (i,j) \in A, t \in T. \tag{6.39}$$

Equation 6.35 represents the modified objective function for the integrated approach, where it is maximized the total discounted net revenue obtained from the forest. Note that in this formulation, road building costs are discounted from the net revenue obtained from the volume harvested during the planning horizon and the ending forest value. Equation set 6.36 defines the volume flow balance equations for non-exit nodes, where it is guaranteed that all flows entering a non-exit node will exit in the same period (Fig. 6.7a). Thus, in a given period *t*, this set of equations forces the total inflow volume coming from the associated management units and the connected nodes (f_{ji}^t,f_{ki}^t) to equal the total volume that leaves node i (f_{ij}^t,f_{ik}^t). This equation set differs from the constraints suggested by Kirby et al. (1986) and Guignard et al. (1998), since in this case it is assumed that all volume produced in the forest will be transported to external exit nodes. Equation set 6.37 is defined as the volume flow balance equations for the exit node at the end of each period *t*. In particular, it is assumed that all the volume produced in the forest will be transported to the exit node (Fig. 6.7b), from which it will be shipped to the final market destination. Constraint set 6.38, named as road capacity constraint set, defines the maximum timber flow allowed through a particular road segment. The upper flow bound is defined by the flow capacity q_{ij}, which is based on the technical road characteristics. Notice that if $f_{ij}^t \geq 0$ or $f_{ji}^t \geq 0$, this equation can only be satisfied when $\sum_{t'=1}^{t} r_{ij}^{t'} \geq 1$. A set of logical constraints ensuring that a road segment can only be built once during the planning horizon $\left(\sum_{t \in T} r_{ij}^t = 1\right)$ could be included, but such a constraint is unnecessary in the formulation. Finally, constraint set 6.39 describes the nature of the continuous and binary road variables used in the integrated formulation approach.

An advantage of this formulation lies in its flexibility to accommodate different road network structures, including road networks with and without cycles (spanning tree). However, as with the fixed charge network problem, if capacities are much larger than the actual flows, the LP relaxation of this model tends to produce weak bounds (Wolsey 1998). In this context, road capacity reduction has been proposed as a way to obtain tighter formulations (Weintraub et al. 2000). Even though, previous research has proposed reducing the capacity bounds as much as possible (Andalaft et al. 2003), it may be difficult to apply on road networks where cycles are allowed.

6.5.3 *Unitary Flow model (UF)*

The Unitary Flow model (UF) attempts to overcome the drawbacks observed with the traditional fixed charge network flow problem without losing the flexibility produced by accommodating different network structures. As mentioned, given that road capacities are usually much larger than the actual flows, capacity constraints mainly play a trigger function for binary variables associated with road-building decisions. Thus, in order to reduce the gap between road capacities and actual flows, unitary harvest volume can be assumed for road constraint modeling purposes. In this case, the potential volume that flows from a management unit to a given node (associated landing) can be assumed to always be unitary in the context of the road building constraints. Thus, an upper bound equal to the total number of management units in the forest can be used as the road capacity for any road segment in the network.

This formulation includes the objective function described in Eq. 6.35, constraints 6.39, and additional road constraints defined as follows:

$$\sum_{m \in M^i} x_m^t + \sum_{(k,i) \in A} f_{ki}^t - \sum_{(i,l) \in A} f_{il}^t = 0, \quad i \in N \backslash \{exit\}, t \in T, \tag{6.40}$$

$$\sum_{m \in M} x_m^t + \sum_{(k,exit) \in A} f_{k,exit}^t = 0, \qquad t \in T, \tag{6.41}$$

$$f_{ij}^t + f_{ji}^t - |M| \sum_{t'=1}^{t} r_{ij}^{t'} \leq 0, \qquad (i,j) \in A, t \in T. \tag{6.42}$$

Equation set 6.40, named as volume flow balance equation set for non-exit nodes, guarantees that no timber will be left at a non-exit node by requiring that all the traffic entering a particular node will exit in the same period. Inbound volumes from harvested management units are defined by $\sum_{m \in M^i} x_m^t$, assuming unitary harvest volume from management units. Equation set 6.41 guarantees that all volume harvested from the forest will exit through the exit node. Similar to equation set 6.40, it is assumed unitary volume from each management unit harvested. Constraint set 6.42 is defined to trigger the building of a particular road segment if there is a non-zero flow of timber on that road. For this set of constraints, the maximum flow for any road segment can be defined as the number of management units in the problem ($|M|$, cardinality of set M).

6.5.4 *Integer Constraint One-by-Unit Model (IC-OU)*

An alternative approach to the fixed charge network flow problem can be used when road networks follow a spanning tree structure. With a spanning tree structure, there is a unique path between every management unit and the exit node. This can be used to construct a set of trigger constraints to incorporate road building decisions into the model. In this case, when a management unit is harvested, all road variables associated with the path from that unit to the exit node must be built during or before the harvest period.

This formulation includes the objective function described in Eq. 6.35, constraints 6.39, and additional road constraints defined as follows:

$$|P^m| x_m^t - \sum_{t'=1}^{t} \sum_{(i,j) \in P^m}^{t} r_{ij}^{t'} \leq 0, \quad m \in M, t \in T, \tag{6.43}$$

$$\sum_{t \in T} r_{ij}^{t'} \leq 1, \quad (i,j) \in A. \tag{6.44}$$

Set of constraints 6.43 requires that if management unit m is harvested in period t, then the respective path P^m to the exit node must be built, either previously or in period t (Fig. 6.8). Notice that $|P^m|$ represents the cardinality of set P, i.e., the number of road segments in the path. Thus, when x_m^t equals 1, then $\sum_{t'=1}^{t} \sum_{(i,j) \in P^m} r_{ij}^{t'} \geq |P^m|$, otherwise the constraints will not be satisfied. In this context, we need the set of logical constraints defined by constraint set 6.44. Otherwise, a less expensive road segment in P^m could be built more than once to satisfy constraint set 6.43.

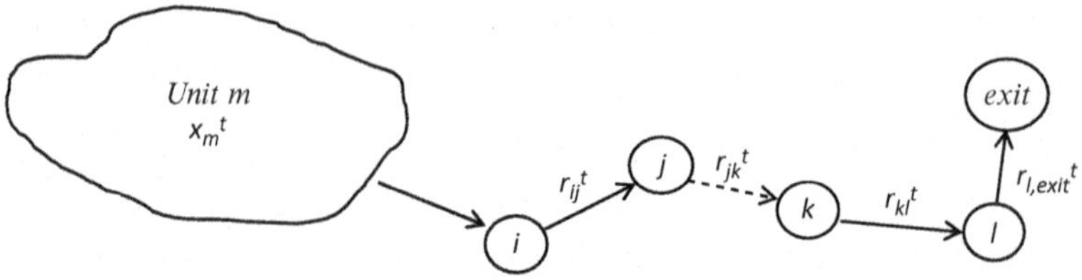

Fig. 6.8 Graphical representation of a path from a management unit to its respective exit node. In this case, all the road segments in the path must to be built to harvest management unit *m*

6.5.5 *Integer Constraint One-by-Road (IC-OR)*

An alternative way to incorporate road building decisions using an approach similar to IC-OU is to subdivide constraint set 6.43 into $|P^m|$ new sets of constraints, one for each road segment in P^m. This formulation includes the objective function described in 6.35, constraints 6.40 and 6.44, and additional road constraints defined as follows:

$$x_m^t - \sum_{t'=1}^{t} r_{ij}^{t'} \leq 0, \quad m \in M, (i,j) \in P^m, t \in T, \tag{6.45}$$

6.5.6 *Tightening Constraint (TC)*

Even though road constraints can be fully incorporated using any of the previous four formulations, additional superfluous constraints can be added to obtain tighter formulations. By including these constraints, integer-infeasible solutions can be eliminated from consideration, improving LP relaxation bounds. In particular, two sets of constraints that have been proposed in previous research are considered: road-to-road and project-to-road constraints, which are formulated as follows:

$$r_{ij}^t + \sum_{(i,l)\in A} \sum_{t'=1}^{t} r_{ij}^{t'} - \sum_{(k,j)\in A} \sum_{t'=1}^{t} r_{ij}^{t'} \leq 0, \quad (i,j) \in A, t \in T, \tag{6.46}$$

$$x_m^t - \sum_{(i,j)\in A} \sum_{t'=1}^{t} r_{ij}^{t'} \leq 0, \quad m \in M, t \in T, \tag{6.47}$$

Constraint set 6.46, named as road-to-road set of constraints, prevents isolated road segments from being built. As illustrated in Fig. 6.9, road segment (*i,j*) only should be built if node *i* or node *j* are already connected with some other nodes in

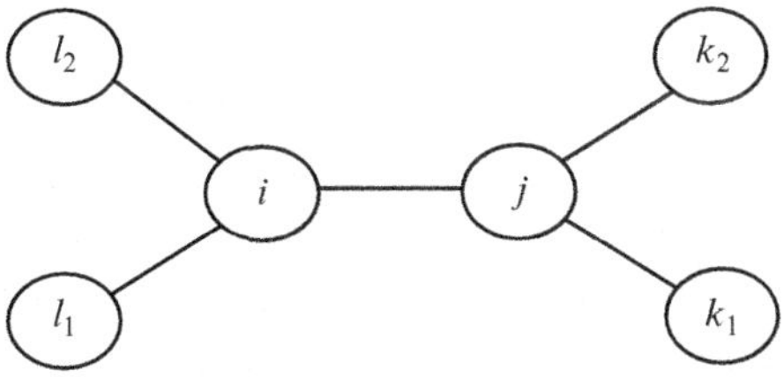

Fig. 6.9 Road-to-road constraint representation

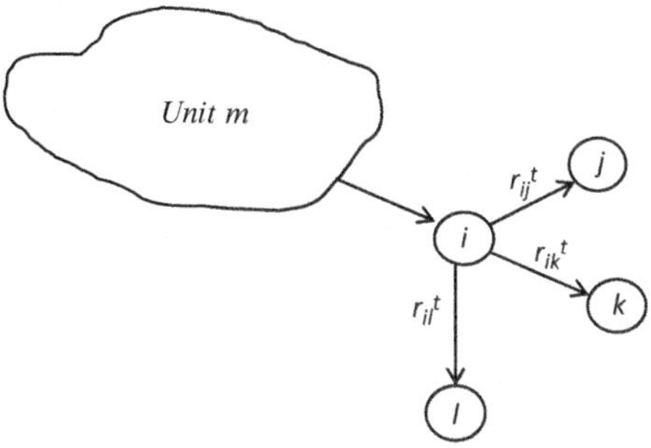

Fig. 6.10 Node accessibility representation, when the associated node to management unit *m* is potentially connected with the other three neighbor nodes

the network, in this case with nodes l_1, l_2, k_1 *or* k_2, otherwise an unconnected road segment would be built.

Constraint set 6.47, named as road-to-project set of constraints, requires that if a management unit is harvested, its associated node must be connected. In Fig. 6.10, management unit *m* can be harvested only if node *i* is already connected to one of its neighboring nodes *j*, *k*, or *l*.

Management Planning in Action 6.6: Integrating Road Building Schedules in Long-Term Forest Management Planning in Pennsylvania

To illustrate the integrated formulation, we consider from the PA Bureau of Forestry, the landscape named as Whitaker Hollow and the problem of road building scheduling over a five 10 years temporal horizon. This landscape is made up of even-aged stands in two forest types, Northern and Allegheny Hardwoods. Based on the topographic characteristics of the Whitaker Hollow landscape, a total of 113 nodes are identified on the map (Fig. 6.11a). Then, a road network following a spanning tree structure (without cycles) was designed, generating 113 road segments with a total distance of 20.85 miles. Considering the topographic requirements, 5 bridges were identified on this road network (spanning tree). The model assumes that a bridge will only be built if the respective road segment requiring a bridge is considered in the road building program.

(continued)

(continued)

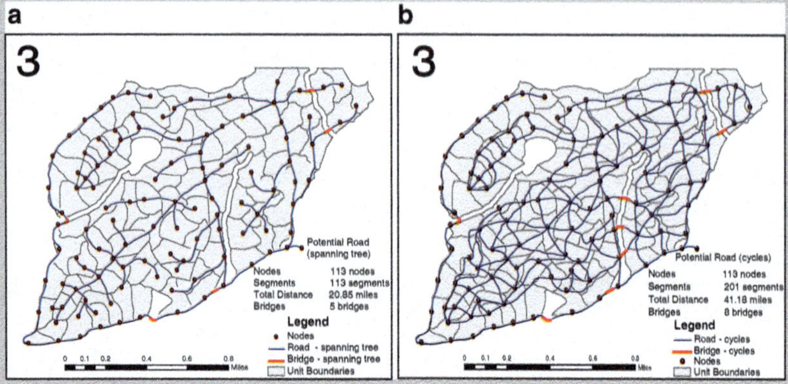

Fig. 6.11 (**a**) The potential spanning tree road network for the Whitaker Hollow landscape. (**b**) The potential road network allowing cycles for the Whitaker Hollow landscape

Similarly, a second road network allowing cycles was designed for the Whitaker Hollow landscape (Fig. 6.11b). In this case, 201 road segments were identified, almost twice the number in the spanning tree structure. Similarly, the potential road length increased from 20.85 to 41.18 miles when cycles were allowed. The network with cycles requires more variables and constraints in the model, however it provides a much larger set of potential solutions for the problem. Even if the initial road network allows cycles, given that transportation costs are not considered in the analysis, the road network built in the optimal solution will always follow a spanning tree structure. In fact, it can be shown that if a node is connected to more than one node on its way to the exit node, an improved solution can be obtained just by building the less expensive connected arc (road segment) that connects the arc with a path toward the exit node.

Figure 6.12a shows a map of the solution obtained using the traditional sequential approach for the Whitaker Hollow landscape. In this approach, the road building program is determined by finding the minimal cost road network satisfying the accessibility requirements imposed by the optimal harvest schedule, which had been determined irrespective of road building decisions. During the first half of the planning horizon (periods 1 and 2), harvest operations are mainly scheduled in the western part of the landscape, and during the last three periods (periods 3, 4 and 5), harvest operations are concentrated on the eastern part of the landscape. Similarly, Figure 8 shows a map of the optimal solution for the integrated model using a spanning tree road network structure. In this graphical representation, it is possible to verify that the sequential and integrated approach solutions harvest exactly the same management units and build the same road network.

(continued)

(continued)

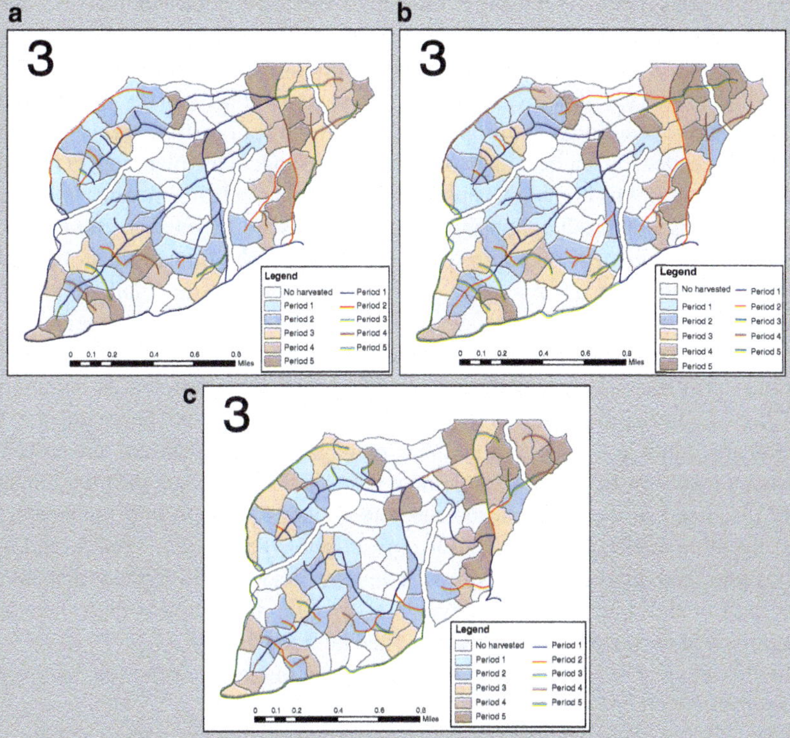

Fig. 6.12 (**a**) Solution for the traditional sequential approach. *Colors* indicate the periods in which management units are scheduled to be harvested and road segments are scheduled to be built. (**b**) Solution for the integrated model with a spanning tree road network. *Colors* indicate the periods in which management units are scheduled to be harvested and road segments are scheduled to be built. (**c**) Solution for the integrated model including a road network with cycles. *Colors* indicate the periods in which management units are scheduled to be harvested and road segments are scheduled to be built

The main difference between the solutions is the sequencing of the harvest decisions and the road building activities. Figure 6.12c shows a map of the optimal solution for the integrated model using a road network structure with cycles. The same management units are scheduled for harvest as in the solutions obtained via the traditional sequential approach and the integrated approach incorporating a spanning tree road network. However, the sequence of the harvest operations is different. In particular, the different harvest timings allow the model to take advantage of the flexibility offered by a road network with cycles. In this case, different road segments were built. Even though the potential road network allows cycles, the final solution follows a spanning tree structure.

6.6 Summary

Long term considerations typically frame all decision processes in industrial forests management planning. The sustainability of the resource and the stability of product supply can hardly be addressed at another temporal scale as a consequence of the biological processes at the core of the forest production system. Thus strategic management scheduling is at the apex of the hierarchy of planning levels. It provides the input needed by tactical and operational management planning that will be addressed later in this book in Chaps. 7 and 8. For example, it generates information about the range of product supply values that may be targeted in the short term by tactical management planning while safeguarding the future. Nevertheless, the adequacy of those estimates increases with the amount of detail of long-term analysis. The effectiveness of strategic management planning does depend in part on how it addressed tactical and operational constraints.

Long-term management planning of industrial forests unfolds as summarized in Chap. 2. It requires data and information about the land base and the vegetation dynamics as well as about the management options and the costs and revenues resulting from their implementation. These data and information items were discussed in detail in earlier chapters. Data acquisition and management – e.g. inventory and land classification, prescription generation, growth and yield projections and market analysis – is a lengthy and costly process that takes a huge part of the time needed to develop a strategic management plan. It requires computerized tools that will be characterized later in Chap. 9.

In this chapter we have built from the state of the art reported by over 94 researchers from all over the world (Borges et al. 2014b) to characterize the dimensions – e.g. spatial scale, number of objectives, type of goods and services, spatial context – of strategic forest management planning of industrial plantations. We further provided a brief historical overview of how the relative importance of these dimensions evolved in recent decades. This led to the classification of long term problems into four main clusters. The first cluster was characterized by objectives (criteria) such as revenue maximization e.g. maximization of the value of the industrial forest, cost minimization and the regulation of the forest over the planning horizon. The second cluster included additional stock control objectives. The third cluster included further landscape design objectives. The forth cluster addressed strategic road building objectives within the framework proposed by the first cluster.

For illustration purposes, in the case of the first cluster we considered concerns with the regularity of the volume and the area harvested over the planning horizon. This is often a concern of managers of industrial plantations: how much to harvest so that there is no rupture in raw material supply. In the case of the second cluster we considered further both carbon sequestration objectives and concerns with the sustainability of the industrial forest by including a constraint on the inventory at the end of the planning horizon. In the case of the third cluster we included adjacency constraints. All these three clusters encompassed only harvesting decisions. The fourth cluster included further road building decisions. Other criteria might have been included in each cluster yet we think that we have considered the most representative for illustrating the processes of model building and for analysing the potential of operations research techniques to address strategic forest management scheduling as well as for interpreting its solutions.

The forestry literature reports several alternative modelling approaches to address strategic management planning problems within each cluster (e.g. Martins and Borges 2007; Diaz-Balteiro and Romero 2008). In this chapter we have classified them into mathematical programming and heuristic approaches. We have differentiated further between approaches that address explicitly one or else several objectives. Moreover we have referred to the potential of *a posteriori* preference modeling approaches. This was influential to highlight the potential of each broad class of techniques. For illustration purposes we have selected from each group, modeling approaches such linear programming, mixed integer programming, goal programming, simulation, simulated annealing and Pareto frontier methods. Again, other techniques are available that may be used to address strategic management planning. Nevertheless we think that the selection made does provide the information needed to help develop both model building and model solving to address the most representative problems.

This chapter builds from the formal presentation of management planning models in Chap. 2 to illustrate how each technique may be used to address each problem. For that purpose we have introduced an example forest with 16 stands. This academic example does provide the data and the information needed for a detailed numerical illustration of model building. It facilitates further the interpretation of results. A list of problems at the end of the chapter builds from the same example forest to support model building, model solving and interpretation of results by readers

6.7 Problems

1. Discuss the potential of mathematical programming to address strategic industrial forest management scheduling.
2. The prescriptions available to each of the 16 stands are summarized in Table 6.2. It was assumed that timber prices were constant and independent of the stand age. Often this is not the case. Consider the information in Table 6.7 and build a new prescription table to reflect the relation between stand age and timber price.
3. The solutions by linear programming (LP) to Problems 1 (LP Run #1) and 2 (LP Run #2), assuming constant prices, are reported in Tables 6.3, 6.4, 6.5, and 6.6. Use LP to optimize the objective function of Problems 1 and 2 subject only to the stand area constraints (Eq. 6.3) and assuming that prices are either constant or as reported in Table 6.7. Summarize the solutions of these runs (LP Run #3 and LP Run #4, respectively) in Tables similar to 6.3, 6.4, 6.5,

Table 6.7 Timber prices

Interest rate (i)	3 %
Timber age (years)	Stumpage price (€/m^3)
0–25	15.5
25–45	24
45–50	30
>50	45.5

and 6.6. Explain how you could determine the optimal schedule without even using linear programming. How could one derive the dual prices for each stand?

4. Use LP to solve Problems 1 and 2, assuming that prices are as reported in Table 6.7. Summarize the solutions of these runs (LP Run #5 and LP Run #6, respectively) in Tables similar to 6.3, 6.4, 6.5, and 6.6.
5. The binary search check demonstrated that the inclusion of the value of ending inventory value in the objective function coefficients (Problems 1 and 2) as well as the inclusion of a 55,000 m^3 constraint on the volume of ending inventory (Problem 2) did not prevent the models from prescribing harvest levels that are not sustainable in the long term. Change Problem 2 to include a 75,000 m^3 constraint on the volume of the ending inventory. Consider prices reported in Table 6.7 and use LP to solve this problem and summarize the solution of this run (LP Run #7) in Tables similar to 6.3, 6.4, 6.5, and 6.6.
6. Compare Run #1 to #7 in terms of the differences in volumes harvested in each planning period and volume of ending inventory. Based on the characteristics of each LP model, explain the differences, if any.
7. Use LP to solve Problem 2 (considering a 75,000 m^3 constraint on the volume of the ending inventory), assuming that prices are as reported in Table 6.7, and modified so that no area, volume or carbon stock control constraints are considered. Summarize the solution of this run (LP Run #8) in Tables similar to 6.3, 6.4, 6.5, and 6.6.
8. Compare Run #4 and Run #8 in terms of the differences in the stands that are not harvested over the planning horizon. Based on the characteristics of the stands involved, explain the differences. i.e. why did each run select not to harvest the stands that it did?
9. Consider the reduced costs for stand # 1 in the case of Runs #4 and #8. Try to explain why the reduced costs changed as they did between these runs. Be as specific as you can. Why did they change more for some prescriptions than for others?
10. Determine how much the marginal value changed between Runs #4 and #5 for each stand. Explain these changes the best you can i.e. why are some changes larger than others? Why do some increase and others decrease?
11. Consider the changes in reduced costs for prescriptions for stands #1, #5, and #11 between Run #4 and #5. Explain the changes (direction and relative magnitude) the best you can in terms of the characteristics of the stands and the characteristics of the management planning problems.

12. Using Run #4 as the basis of comparison, the simple "with and without principle" of economics, and the value of the objective function found for Runs #4 through #8 determine on a per hectare basis:

    (a) The cost of area and volume control constraints
    (b) The combined cost of area and volume control constraints + the constraint on the carbon stock + the 55,000 m^3 ending inventory constraint
    (c) The cost of the 75,000 m^3 ending inventory constraint
    (d) The combined cost of area and volume control constraints + the constraint on the carbon stock + the 75,000 m^3 ending inventory constraint

13. From your understanding of the original problem try to explain why you think the costs in question 12 are of the magnitude that they are. Do you think these costs are significant? Why?
14. If you were asked to estimate the cost of the constraint on the carbon stock in Problems 1 and 2 how would you proceed? Do you expect that cost to be significant? Why?
15. Assume that the industrial forest owner learns that timber demand in each period is equal to 10,000 m^3. Modify the model for Run #4 to include the new constraints. Summarize the solution of this run (LP Run #9) in Tables similar to 6.3, 6.4, 6.5, and 6.6.
16. In the case of Run #9 very few stands are scheduled for harvest during the planning horizon. Based on stand characteristics, explain why the model picked to harvest the stands that it did.
17. Consider Runs #4 and #9. Explain the marginal values of the stands in the case of Run #9 in terms of values associated to Run #4. Be as specific as you can.
18. One might say that with forest management scheduling models "the marginal value of valuable stands likely decreases and the marginal value of marginal stands likely increases." Do you agree with this statement? Explain why or why not.
19. Replicate Runs #4 to #9 with the requirement that no stands are fragmented:

    (a) Do you need to use Mixed Integer Programming (MIP) or Simulated Annealing (SA) to get the integer solutions to Runs #3 and #4? Why?
    (b) What is the cost of the locational specificity requirement in the case of Runs #4 to #9. Explain the differences based on the characteristics of the management planning problems.

20. Build and solve a goal programming (GP) model representing the problem that corresponds to Run #8. When would you rather use GP than pure LP to address an industrial forest management planning problem?
21. Discuss the potential of Pareto frontier approaches to address industrial forest management planning problems.
22. Develop your own heuristic to solve Problem 3, assuming that prices are as reported in Table 6.7 and a 75,000 m^3 ending inventory constraint. Summarize the solution of this run in Tables similar to 6.3, 6.4, 6.5, and 6.6.

(a) Compare your solution to the Mixed Integer Programming (MIP) solution to this problem.
(b) What is the cost of the adjacency constraints? Do you think they are significant?

23. Discuss the potential of heuristics to address industrial forest management planning problems.

Acknowledgments The development of this chapter was partially supported by funding from the European Union's Seventh Programme for research, technological development and demonstration under grant agreements: n° 282887 (INTEGRAL – Future-oriented integrated management of European forest landscapes) and n° PIRSES-GA-2010-269257 (ForEAdapt – Knowledge exchange between Europe and America on forest growth models and optimization for adaptive forestry) and from the EMMC MEDfOR – Mediterranean Forestry and Natural Resources Management as well as by the Grant SFRH/BD/62847/2009' funded by Portuguese Science Foundation.

References

Alves AAM (1984) Gestão da empresa florestal; apontamentos para estudo. Departamento de Engenharia Florestal, ISA, Lisboa

Andalaft N, Andalaft P, Guignard M, Magendzo A, Wainer A, Weintraub A (2003) A problem of forest harvesting and road building solved through model strengthening and Lagrangean relaxation. Oper Res 51(4):613–628

Borges JG, Falcão A (2000) Simulação expedita de planos estratégicos para a área de eucaliptal da Stora Celbi, Celulose Beira Industrial, SA, Grupo de Economia e Planeamento em Recursos Florestais. Departamento de Engenharia Florestal, ISA, Lisboa, 490 p

Borges JG, Hoganson HM (1999) Assessing the impact of management unit design and adjacency constraints on forest wide spatial conditions and timber revenues. Can J For Res 29:1764–1774

Borges JG, Hoganson HM, Rose DW (1999a) Combining a decomposition strategy with dynamic programming to solve spatially constrained forest management scheduling problems. For Sci 45:201–212

Borges JG, Gomes JA, Falcão e AH, Hoganson M (1999b) Heurística baseada em programação dinâmica e restrições de adjacência em gestão florestal. Exemplo em montado de sobro e aplicação em eucaliptal. Revista Florestal XII(1/2):99–107

Borges JG, Hoganson HM, Falcão AO (2002) Heuristics in multi-objective forest management. In: Pukkala T (ed) Multi-objective forest planning, vol 5, Managing forest ecosystems. Springer, Netherlands, pp 119–152

Borges P, Marques S, Borges JG, Tomé M (2009) Scenario analysis applied to cork and holm oak forest ecosystems in Southern Portugal. In: Palahi M, Birot Y, Bravo F, Gorris E (eds) Modelling, valuing and managing Mediterranean Forest ecosystem for non-timber goods and services, EFI proceedings, vol 57, pp 49–56

Borges JG, Garcia-Gonzalo J, Bushenkov VA, McDill ME, Marques S, Oliveira MM (2014a) Addressing multi-criteria forest management with Pareto Frontier methods: an application in Portugal. For Sci 71:187–199. doi:http://dx.doi.org/10.5849/forsci.12-100

Borges JG, Nordström EM,Garcia-Gonzalo J, Hujala T, Trasobares A (eds) (2014b) Computer-based tools for supporting forest management. The experience and the expertise world-wide. SLU University Press

Boston K, Bettinger P (1999) An analysis of Monte Carlo integer programming, simulated annealing and tabu search for solving spatial harvest scheduling problems. For Sci 45:292–301

Constantino M, Martins I, Borges JG (2008) A new mixed integer programming model for harvest scheduling subject to maximum area restrictions. Oper Res 56:542–551

Dahlin B, Sallnas O (1993) Harvest scheduling under adjacency constraints – a case study from the Swedish sub-alpine region. Scand J For Res 8:281–290

Diaz-Balteiro L, Romero C (2008) Making forestry decisions with multiple criteria: a review and an assessment. For Ecol Manag 255:3222–3241

Falcão A (2006) CELPASIM 2.0. Manual do Utilizador. Departamento de Informática – Faculdade de Ciências da Universidade de Lisboa, 39 p

Falcão AO, Borges JG (2001) Designing an evolution program for solving integer forest management scheduling models: an application in Portugal. For Sci 47:158–168

Falcão AO, Borges JG (2002) Combining random and systematic search heuristic procedures for solving spatially constrained forest management scheduling problems. For Sci 48:608–621

Garcia-Gonzalo J, Palma J, Freire J, Tomé M, Mateus R, Rodriguez LCE, Bushenkov V, Borges JG (2013) A decision support system for a multi stakeholder's decision process in a Portuguese national forest. For Syst 22:359–373

Geoffrion AM (1976) The purpose of mathematical programming is insight, not numbers. Interfaces 7(1):81–92

Guignard M, Ryu C, Spielberg K (1998) Modeling tightening for integrating timber harvest and transportation planning. Eur J Oper Res 111:448–460

Gunn EA, Rai AK (1987) Modeling and decomposition for planning long-term forest harvesting in an integrated industry structure. Can J For Res 17:1507–1518

Hoganson HM, Borges JG (1998) Using dynamic programming and overlapping subproblems to address adjacency in large harvest scheduling problems. For Sci 44:526–538

Hoganson HM, Kapple D (1991) DTRAN 1.0: a multi-market timber supply model. College of Natural Resources and Agricultural Experiment Station, Department of Forest Resources Staff Paper Series Report No. 82, University of Minnesota, St. Paul, 66 p

Hoganson HM, Rose DW (1984) A simulation approach for optimal timber management scheduling. For Sci 30:220–238

Hoganson HM, Bixby J, Bergmann S, Borges JG (2004). Large-scale planning to address interior space production: three case studies from northern Minnesota. *Silva Lusitana* N° Especial: 35–48

Jaakko Consulting Inc, (1994) Final generic environmental impact statement study on timber harvesting and forest management in Minnesota. Minnesota Environmental Quality Board, St. Paul, 504 p

Kilgore MA (1992) Government response to sensitive timber harvesting issues: Minnesota generic environmental impact statement. In: Conference proceedings: forestry and the environment: an international perspective. University of Alberta, Edmonton, AB, Canada

Kilgore MA, Ek AR (2007) Cumulative forestry impact assessments: lessons learned and planning for states. Staff paper series no. 193, Department of Forest Resources, College of Food, Agricultural and Natural Resource Sciences, University of Minnesota, St. Paul, MN

Kirby MW, Hager WA, Wong P (1986) Simultaneous planning of wildland transportation alternatives. TIMS Stud Manag Sci 21:371–387

Lockwood C, Moore T (1993) Harvest scheduling with spatial constraints: a simulated annealing approach. Can J Res 23:468–478

Martins H, Borges JG (2007) Addressing collaborative planning methods and tools in forest management. For Ecol Manag 248:107–118

McDill ME, Rebain SA, Braze J (2002) Harvest scheduling with area-based adjacency constraints. For Sci 48(4):631–642

Michalewicz Z (1996) Genetic algorithms + data structures = evolution programs, 3rd edn. Springer, Berlin, 387 p

Murray A, Church R (1995) Heuristic solution approaches to operational forest planning problems. OR [Oper Res] Spektrum 17:193–203

Reeves CR (1993) Modern heuristic techniques for combinatorial problems, 1st edn. Wiley, New York, 320 p

Rodriguez LCE, Nobre S (2013) The use of forest Decision Support Systems in Brazil. In: Borges JG, Nordström EM, Garcia-Gonzalo J, Hujala T, Trasobares A (eds) (2014b) Computer-based tools for supporting forest management. The experience and the expertise world-wide. SLU University Press

Romero C (2004) A general structure of achievement function for a goal programming model. Eur J Oper Res 153:675–686

Rose DW (1984) A critical review of mathematical programming in forest resource planning. In: Proceedings of the IUFRO symposium on forest management planning and managerial economics, Tokyo, pp 306–317

Rose D, McDill M, Hoganson H (1993) Development of an environmental impact statement of forestry programs: a Minnesota case study. The Compiler 10:18–27

Tarp P, Helles F (1997) Spatial optimization by simulated annealing and linear programming. Scand J For Res 12:390–402

Tomé M, Borges JG, Palma J, Garcia-Gonzalo J, Freire J (2010) Relatório – simulação de planos estratégicos para a Mata Nacional de Leiria. FORCHANGE/Centro de Estudos Florestais, Departamento de Engenharia Florestal, Instituto Superior de Agronomia, 32 p

Tóth SF, McDill ME (2009) Finding efficient harvest schedules under three conflicting objectives. For Sci 55(2):117–131

Tóth SF, McDill ME, Rebain S (2006) Finding the efficient frontier of a bi-criteria, spatially-explicit harvest scheduling problem. For Sci 52(1):93–107

USDA Forest Service (2004) Final environmental impact statement: forest plan revision: Chippewa and Superior National Forests. Eastern Region, Milwaukee, 1732 p

Van Deusen P (1999) Multiple solution harvest scheduling. Silva Fenn 33:207–216

Weintraub A, Navon D (1976) A forest management planning model integrating silvicultural and transportation activities. Manag Sci 22(12):1299–1309

Weintraub A, Church RL, Murray AT, Guignard M (2000) Forest management models and combinatorial algorithms: analysis of state of the art. Ann Oper Res 96(2000):271–285

Wolsey LA (1998) Integer programming. Willey-Interscience, New York, 264 p

Chapter 7
Tactical and Operational Harvest Planning

Alexandra S. Marques, Jean François Audy, Sophie D'Amours, and Mikael Rönnqvist

7.1 Introduction

Forest harvesting planning establishes the plans for the operations related with cutting down the trees at the extraction sites and forward them to pick-up points close to logging roads. Forest harvesting includes tree felling (final felling or thinning operations) and bucking of trees into logs using manual chainsaw operations or different type of mechanized harvesting machine systems depending on the characteristics of the stand. Forwarding operations can also be done with different systems including mechanized forwarders and tower hauling.

There are many important aspects to consider in the tactical and operational harvest planning depending on the planning horizon and the spatial scale to be considered. The plans may include aggregated operations to be performed in the next 1–3 years or detailed operations for the next month, week or even on a daily basis. The decisions may focus on an entire forest region or in a single stand. The temporal-spatial planning dimensions determine the level of aggregation of the required information. In general terms, The stand information includes for example area, volume of different assortments, average tree size, average forwarding distance and which time periods harvesting is possible. The assortments can be defined by species, dimensions (length and diameter) and use (saw log, pulp log or energy in the form of forest biomass). The machine systems and teams have specific capacities

A.S. Marques (✉)
INESC Tec, Porto, Portugal
e-mail: alexandra.s.marques@inesctec.pt

J.F. Audy
Université Laval, Québec, QC, Canada

S. D'Amours
Department of Mechanical Engineering, Laval University, Laval, Canada

M. Rönnqvist
FORAC, Université Laval, Québec, QC, Canada

J.G. Borges et al. (eds.), *The Management of Industrial Forest Plantations: Theoretical Foundations and Applications*, Managing Forest Ecosystems 33,
DOI 10.1007/978-94-017-8899-1_7,

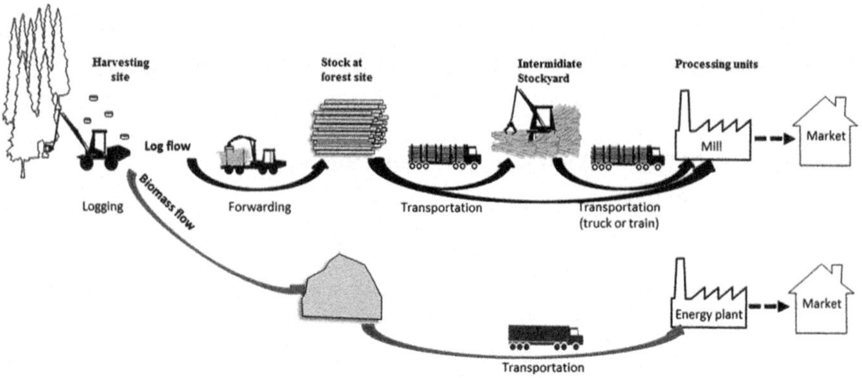

Fig. 7.1 Generic representation of the procurement network of forest products

regarding the stand characteristics which need to be considered. Moreover, the location of the exploitation sites, the costs of the forest operations and the prices of the assortments influences the planning objectives.

Tactical and operational harvest planning in case of forest plantations is often driven by the demand of wood at the processing facilities. Thus, the planning objective is generally to supply a given volume of a set of assortments making the best use of the available machinery and teams, with lowest possible cost. Shortage in volume and/or in harvesting capacities may not allow to entirely supply all the demand. Variable cost penalties for unfilled volume will usually be used to manage different priority level in the fulfillment of the demand. In these cases, the planning process also considers the stand-to-industry allocation for each assortment and transportation to the processing facilities, where it will ultimately lead to the production of finished wood products sold to the customers. Hence, we often search for integrated harvest and transportation planning models. However, it is important to distinguish between what decisions that are implemented in practice (often called business decisions) and decisions that are used to model the impact of decisions (anticipation decisions). In this section, we focus on harvesting as the business decisions although we often use transportation decisions as anticipation decisions. It some cases, the operations further include the intermediate storing of the wood in stockyards before being delivered at the mills. Also, there is a possibility to move full length trees in the system. The sequence of operations performed in these locations material flows (trees, logs and biomass) across the locations is often called the procurement network, see Fig. 7.1.

What You Will Learn in This Chapter

- Distinguish between strategic and tactical/operational harvest planning
- Good understanding of the main harvest tactical and operational planning problems related with the business management decisions undertaken by the players involved in the management of industrial plantations

- Comprehend the Operations Research (OR) models and methods that may be used to approach harvest tactical and operational planning problems
- Distinguish between business decision variables and anticipation decision variables
- Viewed examples of integrated harvest tactical and operational planning problems encountered in the literature
- Have knowledge on open research topics that may inspire future work in this domain

The chapter unfolds as follows. Section 7.2 discusses the specificities of tactical and operational harvest planning in comparison with strategic planning. Section 7.3 proposes a classification for the tactical and operational forest harvesting planning problems, driven by the distinct decisions related with managing industrial plantations, undertaken by the diverse types of decision-makers. Section 7.4 describes the tactical harvest planning problem and discuss decision problems arising during this process. Section 7.5 follows a similar structure but for the forest operational planning process in industrial plantations. Section 7.6 discusses one example of a fully integrated forest tactical and operational planning problem. The chapter ends with a discussion of open questions and future research guidelines. Some problems and literature references for further reading are also presented.

7.2 Relation Between Strategic, Tactical and Operational Harvesting Planning

Forest harvesting planning is commonly managed according to three temporal dimensions: strategic, tactical and operational. As discussed in a previous chapter of this book, strategic harvest management planning sets long-term policies, therefore often relying in aggregated information about the predictions of wood availability and demand across the planning horizon that can go up to several decades. It generically states *when* to perform the harvest operations as a whole without taking into account any technicalities related with the way the operations should be conducted.

On the contrary, tactical and operational planning deal with medium to short term decisions at forest and stand level, respectively. The tactical/operational plans state *when* and *how* the operations are performed (*the means and the timing*). For this purpose, the technical issues related with the forest operations and the characteristics and limitations of the machine systems are of great importance. The plans further rely on detailed information about the site conditions and the machinery, workers and trucks available locally.

The distinction between strategic and tactical/operational is somewhat consensual in the literature. However, the distinction between tactical and operational planning is narrow and greatly case-specific. For example, some studies particularly in the context of pulp and paper industries do not use the term tactical (e.g. Murray and Church 1995; Epstein et al. 1999). They designate by operational planning the process of scheduling forest operations to be executed in a monthly basis, for the purpose of preparing the company's annual budget and negotiating the outsourcing contracts.

In general terms, it is acknowledged that tactical harvest planning address problems related with allocating available resources such as harvesting machinery systems and teams as well as scheduling the execution of the harvesting operations. Tactical plans encompass monthly to yearly planning periods and time horizons that may vary from 1 to 3 years, depending on the complexity of the problem and the species composition. The operational planning problems relate to detailed scheduling decisions that precede and determine the real-world operations (D'Amours et al. 2008). The time frames can go from minutes to hourly planning periods spread across time horizons that can go from few weeks to months. Another comparison is to say that tactical decisions regard the time period when harvesting starts and operational address decisions related with a stand or harvesting unit together with machine systems. The length of the time periods is generally such that in tactical planning several stands can be harvested in the same time period whereas in operational the harvesting of a stand cover several time periods. Another difference is that tactical planning often uses an aggregated demand information on assortments without location information whereas operational planning include detailed description of the assortments demanded at each industrial site as well as the availability at each stand.

These temporal dimensions of harvest planning are often hierarchically integrated. Examples of top-down hierarchical planning may be found in the forestry literature (e.g. Weintraub and Cholaky (1991) and in some pulp and paper companies. The strategic plan often sets budget constraints to the tactical plans. The stands selected for harvesting in the first period (often 3–5 years) of the strategic plan will be the focus of the tactical harvesting plans. The tactical plan for a larger forest region will in turn constrain the decisions taken in the operational plans (often called forest project) for each smaller scaled harvesting unit in that region.

Complementary, bottom-up hierarchical planning approaches are of importance in cases of dynamic planning and when large amount of information on the execution updates is collected in real-time from the forest sites, and is available for re-planning. In these cases, the control processes continuously check the compliance with the production targets set by the operational plans. Non-compliance situations may lead to updates in operational plans and revisions in the tactical plans. The information provided about the forest and the industry status will be input for new strategic plans (Fig. 7.2).

7.2.1 *Specificities of Tactical and Operational Harvest Planning*

There are some specific features encountered in harvest tactical and operational planning that distinguishes these problems from strategic planning. The first relates

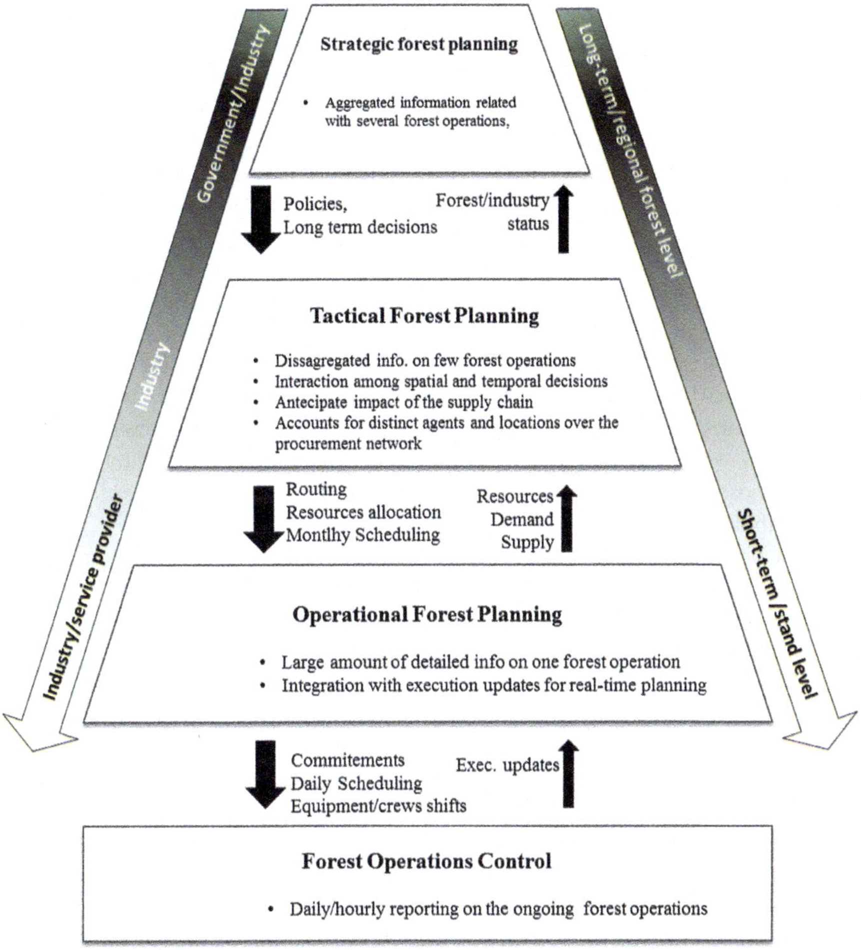

Fig. 7.2 Forest planning hierarchical approach (Marques 2012)

with combining decisions about a large number of operations, each with possible variants that depend on the forest site specific conditions. Felling operations are performed to cut down the tree usually with chainsaws or mechanized harvesters. There are several ways to transport the tree from the extraction site to the landing site or roadside. Forwarders carry the trees out from the forest while skidders simply drag the tree in the terrain. Logging roads are often built for transporting the logs from the harvest sites to the roadside. The quality and accessibility seasonal conditions of these roads impact on the timing of harvesting operations. Alternatively cables and yarders can be used to lift the tree off the ground and transport it by air to cover large distances particularly in climbing regions. Log processing and sorting can occur at the harvest site or at the landing site. It involves removing the limbs and the tops of the trees and bucking them into merchantable log lengths. The logs are sorted into assortments according to grade, dimension and species. The assortment piles are placed near to the roadside. From there, the logs are transported by logging trucks to the stockyards or wood processing facilities.

Contrary to strategic planning the *seasonality* of the harvest operations plays a key role. The soil conditions during part of the year may delay the persecution of the harvesting operations. Also the conditions of the road network may preclude the access to stands at sensitive parts of the network. Specifically, harvesting in Nordic countries prevails during the winter when the ground is frozen, thus reducing the risk of soil erosion while moving the logs out of the forest (D'Amours et al. 2008). While in Mediterranean countries harvesting and transportation is forced to occur mainly during the summer. In Chile the autumn is problematic due to lots of rain that has a negative impact on the road network. Moreover, in some countries, the saw mills or harvest operations are closed during summer holidays whereas the pulp and paper mills work continuously during the year. This impacts the inventory planning of the assortments and lead to significant fluctuation in inventory level along the wood supply chain.

In tactical and operational planning, the organizational context where harvest planning takes place is of great importance. The number and role of the players involved in the procurement network, the relationship among their, the impact of their decisions on the other players (i.e. interaction network) needs to be acknowledged in the forest harvesting models and further sets requirements for new technological solutions for data sharing and collaboration. There are significant differences in the organizational context of planning between countries and even between the distinct forest-based supply chain that may be found in each country. In the case of industrial plantations managed by vertically integrated pulp and paper companies in Southern Europe and Southern America (e.g. Marques et al. 2011), harvest planning takes place at central and regional forest offices and involves mainly the industry's harvest managers, although the operations are outsourced to companies specialised in logging and transportation. The industry's harvest scheduling and wood delivery plans rely on aggregated information about the total availability of harvesting crews and machine systems in the region. The outsourced companies rely in the industry's plan to produce detailed plans concerning the allocation and schedule of their crews and equipment. The wood carriers may plan collaboratively, in order to reduce the total transportation costs (e.g. Audy 2013). The operational plans related with the extract the logs from the harvest site often involve negotiation among the harvest managers, the harvesting and the wood transportation companies.

Harvest planning decisions further impact in the decisions related with the remaining operations of the supply chain. Studies (e.g. Weintraub et al. 1995) show that the cost of the harvesting operations taking place at the procurement stage may impact on 30–50 % on the total wood procurement cost, therefore affecting the cost of producing and distributing the finished wood-based products, ultimately increasing the price paid by the costumers at the end of the supply chain.

The reverse is also truth as the production levels set according to sales forecasts will determine the monthly demand levels of the wood assortments in the mills, ultimately impacting and impacted by the harvest planning decisions.

The complexity of the tactical and operational planning processes and the economic importance of the procurement activities motivated research since the 1990s. Operations Research (OR) techniques are now commonly used to approach harvest planning decisions (e.g. Weintraub et al. 2007). The literature reports the use of linear, integer, mixed-integer and nonlinear models (Chap. 2). The solution methods in use depend on the required solution time and may include dynamic programming, LP methods, branch & bound methods, heuristics, column generation and Multi-Criteria Decision Making approaches (Rönnqvist 2003). Such techniques are often implemented within wider decision support systems for forest planning (Chap. 10). These computerized-tools manage the information required for running the models, execute the OR methods and display the optimal plans using graphical user interfaces and spatial maps.

7.3 Tactical and Operational Forest Harvesting Planning Problems

Recent research (Marques et al. 2011; Marques 2012) describes the decisions related with managing industrial plantations taken during the wood procurement stage, in the perspective of distinct decision-makers. These decisions are common to the majority of the forest systems and countries and have been approached with mathematical programming models with specific objective functions and constraints. Each of these elementary decisions corresponds here to a Forest Tactical and Operational Planning Problem (FTOPP).

The FTOPP may be classified in a planning matrix according to the temporal dimension of the elementary decisions and the harvesting operation that relates to those decisions (Table 7.1). Please note that this table includes also the transportation and storage planning that will be described in Chap. 8. However, the main business decisions we consider is related to the harvesting decisions.

In many cases, the harvest planning deals simultaneously with many of this decisions. Yet, it is important to distinguish between fully integrated planning problem and decoupled planning problem with the anticipation of related decisions. The first type of problem merges the single FTOPP into a unique decision problem where the decisions are all *business decision variables*. This means that obtaining the problem result ends the decision-making process. The choices made in respect to each of the single FTOPP will then be implemented in the course of processes that are often conducted separately. Even if this model is tractable, it is not often that all decisions are implemented in practice.

The second type of problem focus on business decisions of a primary FTOPP (the decision variables), but include other *anticipation variables* in order to anticipate the

Table 7.1 Planning matrix for tactical and operational harvest planning (adapted from Marques et al. 2011)

	Logging	Transportation	Storing at the stockyard	Delivering and the mill
Tactical	Harvest scheduling	Harvest unit assignment	Terminal layout	Wood supplies negotiations
	Roads network planning	Wood assortments	Inventory planning	Wood reception planning at the mill
	Machine system/team assignment	Transportation adjudication	Yard equipment/team assignment	
	Harvest sequencing	Trucks/drivers assignment	Wood reception planning at the stockyard	
	Harvest service adjudication	Fleet management		
Operational	Extraction of logs	Truck routing and scheduling	Wood reception planning at the stockyard	Wood reception planning at the mill
	Bucking & sorting strategies		Yard equipment/team scheduling	
	Harvest & transport. Synchronization			
	Machine system/team scheduling			

impact on/from other related secondary FTOPP. The anticipation variables improve the quality of the results (e.g. feasibility, robustness) of the primary FTOPP problem as their impact can be modeled. The outcome of such problems ends the decision-making process but only for the primary FTOPP. A new decision-process will be conducted for the secondary FTOPP, which will then provide the best choice to be implemented.

The distinction between these situations is done from the analysis of the set of model constraints and it is not usually encountered in the literature. As mentioned earlier, a typical example is combined planning of harvesting and transportation activities but with harvesting as business decisions and transportation as anticipation variables.

The remaining chapter discusses the FTOPP listed in the planning matrix for tactical and operational harvest planning. Whenever relevant, discusses the related decisions that give rise to elementary planning problem with anticipation. The description of the problems found in the forestry literature include a review of the frequently used OR techniques. However, the topics related with wood transportation will be briefly addressed because one of the next chapters of this book is dedicated to this theme.

7.4 Tactical Harvest Planning

The tactical harvest planning process consists in scheduling harvesting operations across the planning horizon that can go from 1 to 3 years. For this purpose, harvest units are usually settled by grouping the forest stands to be felled each time period (often 1 month). Then, the available resources (crews and equipment) are assigned to the harvesting units. If no roads are available, it further determines the investments in forest roads used for transporting the logs from the harvest sites to the mills.

In the case of harvest planning in industrial plantations, this process typically takes place once a year, contributing for the definition of the company's forest annual budget for the next year (Fig. 7.3). The tactical plan is built by the company's harvest managers upon information about target wood demand levels and the set of forest stands that were previously selected for harvesting in that period according to the strategic plan. The tactical plan is constrained by the characteristics and crews and equipment available at the outsourcing companies working in the forest regions.

The harvest schedule and the corresponding budget are subjected to negotiation among the harvest managers and the forest practitioners involved in managing the forest regions. This process often requires surveys (aerial or on-site) to the selected harvest stands for yield evaluation and/or assessing the conditions of the existing forest roads.

When consensus is reached, the tactical harvest plan is approved by the company's board of directors and/or government organisations. The approved tactical harvest plan will set the basis for negotiating the outsourcing contracts with the harvest entrepreneurs, road construction companies, wood carriers and market wood suppliers. The industry opens calls for adjudicating the annual harvesting and road construction work, often split into work-orders. Similarly, the industry sets delivery requests based on the best assortment and assignment of the wood to the mills. The outsourcing companies are then selected based on historical and empirical data about the operations costs and equipment productivity as well as in the best match between the companies' machinery and the requirements of the work-orders. The outsourcing companies can in many cases be, like in Sweden, individual machine teams with a harvester and a forwarder.

Following contractualisation, there are two options for further re-planning. Either the company still manage and is responsible for the rolling horizon planning or the outsourcing companies will individually plan the allocation and scheduling of their equipment and crew. The wood carriers will further establish the routes and schedules for the wood trucks fleet.

The main decision problems rising in the forest tactical harvest planning process include harvest scheduling, machine system/team assignment, harvest sequencing and harvest service adjudication. Some of these FTOPP, like harvest scheduling, have been widely studied. The elementary decisions, existing mathematical formulations and computerized-tools may be driven from a thorough the literature. Other

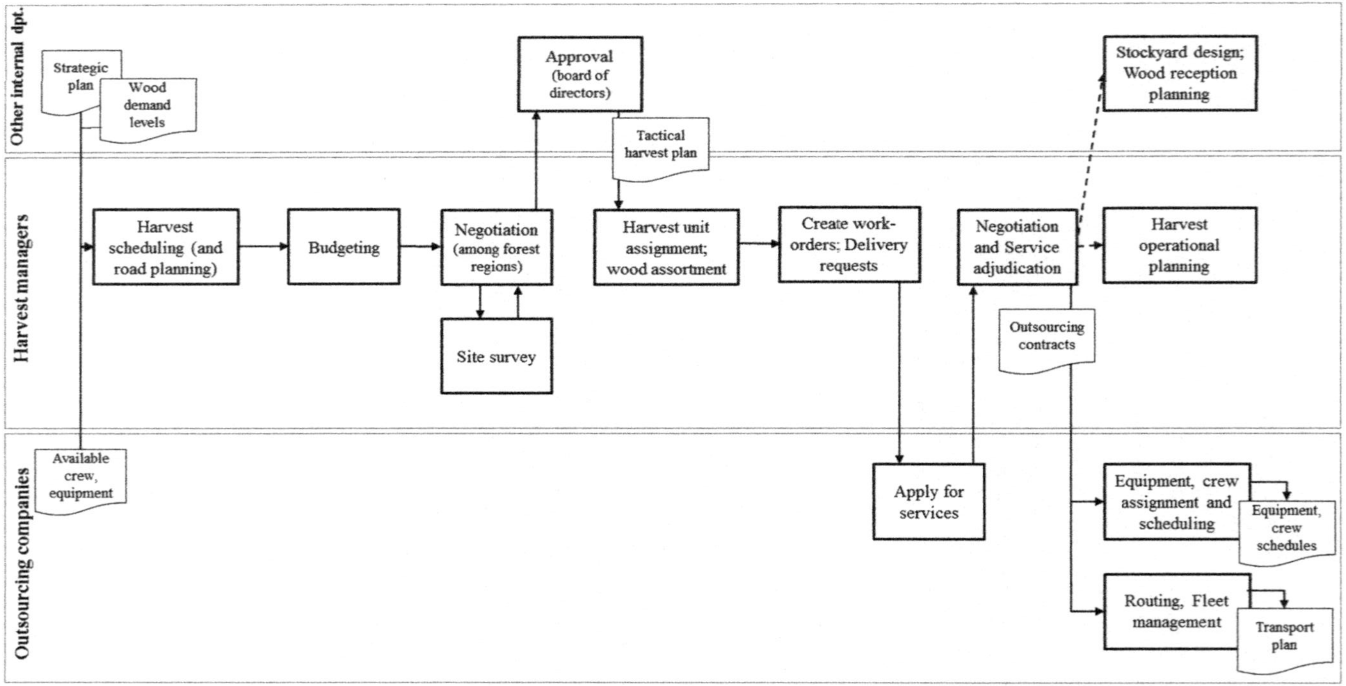

Fig. 7.3 Representation of the generic processes related with tactical harvest planning process in industrial plantations

FTOPPs, like harvesting adjudication, are seldom addressed. The related elementary decisions come from collaborative work with the forest practitioners and decision-makers (e.g. Marques et al. 2013a, b).

7.4.1 Harvest Scheduling

The forestry literature presents several solution methods and computerized-tools developed to approach this decision problem. Examples are found in Bredström et al. (2010), Karlsson et al. (2004), Andalaft et al. (2003), Murray and Church (1995), and Weintraub and Murray (2006). Yet, few authors provide a detailed description of the harvest scheduling process. The discussion about the factors impacting in the harvest scheduling elementary decisions is rarely found.

The tactical harvest scheduling problem states where to harvest and how much to cut each planning period (usually a month). It relies in the list of stands selected for harvesting in the strategic forest plan; in the estimates of wood availability (in the market and self-owned forests) and in yearly demand levels set by the mills. Suppose we use monthly time periods. Then, the decision problem consists in establishing the *best month for starting the harvesting operations in each of the management units* previously selected for harvest in that year. These business decisions are modeled with binary decision variables that take the value 1 if the management unit is started to be harvested in the month and zero otherwise. Tactical harvest scheduling problems are often modeled with Integer Programming (IP) or Mixed Integer Programming (MIP) formulations. These variables resemble those used for strategic planning although using monthly planning periods instead of yearly.

The business decisions related with tactical harvest scheduling depend upon a series of factors that give rise to new decision variables and problem constraints, including the fulfillment of the mills wood demand levels, the seasonality, the availability of harvest resources, the company's silvicultural best practices and national regulations.

To illustrate a tactical harvest scheduling model, consider the following variables and coefficients:

x_{mt} = a binary variable whose value is 1 if management unit m is selected to be harvested in period t for $t = 1, 2, \ldots T$; when $t = 0$, the value of the variable gives the area from management unit m not scheduled for harvest during the planning horizon (hectares),
c_{mt} = unit harvesting cost of management unit m in period t
v_{mt} = the per-hectare volume harvested from management unit m if it is harvested in period t (m3/ha),
w_m = amount of work required for harvesting management unit m (hours)
a_m = the area of management unit m (hectares),
d_{jt} = volume demand from mill j in period t (m^3/ha),
e_t = amount of harvesting work that can be conducted with the available resources in period t (hours).

The model formulation is as follows:

[P1]

$$\min z = \sum\nolimits_{m=1}^{M} \sum\nolimits_{t=1}^{T} c_{mt}.a_{mt}.x_{mt} \tag{7.1}$$

Subject to

$$\sum\nolimits_{t} x_{mt} \leq 1 \quad \text{for } m = 1 \text{ } to \text{ } M \tag{7.2}$$

$$\sum\nolimits_{m} v_{mt}.a_{m}.x_{mt} \geq \sum\nolimits_{j} d_{jt} \quad \text{for } t = 1 \text{ } to \text{ } T \tag{7.3}$$

$$\sum\nolimits_{m} w_{m}.x_{mt} \leq e_{t} \quad \text{for } t = 1 \text{ } to \text{ } T \tag{7.4}$$

$$x_{mt} \in \{0, 1\} \quad \text{for } m = 1 \text{ } to \text{ } M \text{ and for } t = 1 \text{ } to \text{ } T \tag{7.5}$$

The objective function defines the objective of minimizing the total harvest tactical planning costs (Eq (7.1)). The set of constraints typically force all the management units to be harvested at most once in the planning period (Constraint set (7.2)), oblige the fulfillment of the aggregated demand from the mills (Constraint set (7.3)) and limit the work conducted each period to the availability of the harvesting resources (Constraint set (7.4)). Constraint set (7.5) set the nature of the decision variables.

The solution approaches found in the literature include both exact method and heuristics such as simulated annealing, tabu search and genetic algorithms.

This is a base model that can be extended to include demand levels by assortment type, budget constraints and spatial and temporal constraints. One concern of the harvest manager responsible for the tactical plan is to assure the compliance with the national and regional regulation. The plan should also be consistent with silvicultural best practices applied to harvesting operations. As an example, current practices often acknowledge the importance of the preserving the forest services, producing non-wood products (e.g. wildlife, biodiversity, soil, and water) and look for mitigating the environmental impact of harvesting in terms of erosion and landscape aesthetics. Consequently, these practices suggest the use of spatial adjacency constraints for bounding the size of the clear-cut opening area and of the patches, corresponding to the group of stands that emerge together as a consequence of clear-cut operations (e.g. Martins et al. 2005). There are several ways to model adjacency concerns and readers are directed to Chaps. 2 and 6 for an in-depth discussion about this topic. As an example, adjacency constraints (7.6) ensure that adjacent stands are not harvested in the same period.

$$x_{mt} + x_{st} \leq 1 \quad \text{for all adjacent areas } m, s \text{ and } t = 1, 2, \ldots T \qquad (7.6)$$

Moreover, seasonality of the weather conditions may compel the concentration of the harvesting operations in some periods of the year when the forest roads are easily accessible and the terrain conditions are favorable to the use of harvesting machinery. The time window when harvesting can occur in each management unit increases the complexity of the decision-making process. The harvest manager may handle this factor during the definition of the binary variables of the harvest scheduling problem. The undesirable harvest timing for the management units due to inaccessibility of the forest roads should be identified beforehand and the correspondent decision variable must be left out from the model.

Model [P1] may also be improved to account for minimum wood demand levels set by the mills specified in terms of assortments (species, quality grade, length interval, diameter class) and delivery month. Pulp mills usually operate in continuum the whole year. The fluctuations in the total amount of wood and wood assortments demanded each month is the result of the wood consumption estimates associated to the mills monthly production plans. Consequently, the harvest manager needs to balance the wood demand with the production in a monthly basis. Thus avoiding wood stocks whenever possible and balancing the work of the crews and equipment along the year.

For this purpose, the harvest scheduling problem may be combined with wood transportation problem. Wood flow decision variables y_{mtj} are added to the decision problem [P2], setting the amount of wood transported from a management unit to the mill in a given period. Note that these are anticipation variables and will not be implemented in the harvest scheduling plan. The growing stock overall production for each management unit is estimated with computerized-tools and empirical knowledge. Information about bucking and sorting strategies enable a rough initial estimate of the production split into the required product assortments. The consistency of this MIP formulation is achieved by adding new connection constraints that relate both types of decision variables, assuring that wood transportation from a management unit cannot occur before the period selected for harvesting.

To illustrate the important extension with anticipation in flow variables we define the following additional variables and parameters:

y_{mtj} = a variable representing the flow from management unit m to mill j in period t
t_{mtj} = unit transportation cost between management unit m and mil j in period t

The modified model formulation is as follows:

[P2]

$$\min z = \sum_{m=1}^{M} \sum_{t=1}^{T} c_{mt} . a_{mt} . x_{mt} + \sum_{m=1}^{M} \sum_{t=1}^{T} \sum_{j=1}^{J} t_{mtj} . y_{mtj} \qquad (7.7)$$

Subject to

$$\sum_t x_{mt} \leq 1 \quad \text{for } m = 1 \text{ } to \text{ } M \tag{7.2}$$

$$\sum_j y_{mtj} \leq v_{mt}.a_m.x_{mt} \quad \text{for } m = 1 \text{ } to \text{ } M \text{ and } t = 1 \text{ } to \text{ } T \tag{7.8}$$

$$\sum_m y_{mtj} \geq d_{tj} \quad \text{for } j = 1 \text{ } to \text{ } J \text{ and } t = 1 \text{ } to \text{ } T \tag{7.9}$$

$$\sum_m w_m.x_{mt} \leq e_t \quad \text{for } t = 1 \text{ } to \text{ } T \tag{7.4}$$

$$x_{mt} \in \{0, 1\} \quad \text{for } m = 1 \text{ } to \text{ } M \text{ and for } t = 1 \text{ } to \text{ } T \tag{7.5}$$

$$y_{mtj} \geq 0 \quad \begin{array}{l} \text{for } m = 1 \text{ } to \text{ } M, \text{ for } t = 1 \text{ } to \text{ } T \text{ and} \\ \text{for } j = 1 \text{ } to \text{ } J \end{array} \tag{7.10}$$

Here we comment the new parts and constraints included. The objective function (Eq. (7.7)) is extended with the transportation cost. Constraint set (7.8) limits the flow from the stands and Constraint set (7.9) oblige the fulfillment of the demand from the mills. Constraint sets (7.10) set the restrictions on the continuous variables.

Further improvements in model [P2] may lead to tactical harvest scheduling plans that explicitly acknowledge forest management rules imposed by the forest company. Examples of such rules include: obliging harvesting of all the management units within the planning horizon, limiting the budget spent each planning period, setting minimum levels of certain product assortments per period (e.g. barked logs).

Harvest scheduling decisions are frequently combined with other investment decisions such as **road network design and planning** in the course of the establishment of the forest annual budget. Combining both decision problems thereby assures that the conditions of the road network enable the extraction of the logs from the harvest sites, and may further set bounds for the forest budget spent in each period. The modeling approaches often found in the literature for this fully integrated problem include the binary variables and constraints described for the harvest scheduling problem, additional decision variables for modeling the road upgrade and construction decisions and new constraints that affect road network planning. New connection constraints are further added to the model in order to assure that new forest roads are built or existing roads are upgraded before the planning period when harvesting is due to occur. The single integrated plan will set the schedules for both harvesting and road construction operations that will be independently executed (further details about integration issues are described in Sect. 7.6).

7.4.2 Machine System/Team Assignment

Sophisticated harvest scheduling planning processes may take into account that the duration and performance of the harvesting operations in the management units are actually affected by the characteristics of the harvesting equipment and crew. These processes combine the harvest scheduling problem with detailed description of the crew and equipment allocation and scheduling leading to extended formulations where the basic binary decision variables are changed to account for the additional index related with the crew/equipment identification. New constraints are added to force the allocation of the crews to the management units (further details in Sect. 7.6). The combined problem relies in detailed information about the site conditions, including the terrain physiography and stands characteristics, such as trees density, height and stand composition. The problem further requires updated information about the characteristics of the available resources and the experts' knowledge on which are the desirable resources characteristics for specific site conditions. Such detailed information about the available resources may not be easily obtain before the adjudication of the harvesting services to the outsourcing companies that occur later in time. For that reason, the harvest scheduling problem with team and machine system allocation and scheduling is expected to occur when the equipment and crew are managed by the harvest manager or when there are larger outsourcing companies with multiannual harvesting contracts set before the elaboration of the tactical plan.

7.4.3 Harvest Sequencing

So far we have addressed the harvesting and transportation cost. A third important cost component is the cost of moving machine systems between harvest units. When the felling operations are finished the harvesters and other harvesting equipment are moved to the next stand to be harvested. Equipment transportation is a non-profit operation that further contributes to the increase of harvesting costs whenever there is the need to hire specific equipment movers for travelling long distances between harvesting units. Such practices may also be addressed by adjacency constraints that force harvesting to occur in consecutive stands located in the neighborhood of previously harvested stands.

The minimization of equipment transportation between harvesting units is seldom addressed in the forestry literature. One of the few examples is provided by Bredström et al. (2010) that propose a two phased heuristic to tackle all three cost components. One reason for few articles may be fact that these constraints motivated by operations efficiency may be conflicting with other spatial constraints related with the maximum clear-cut opening area due to environmental goals leading to complex and unfeasible problems.

7.4.4 Harvest Service Adjudication

The harvest service adjudication problem consists in finding the best outsourcing contracts for executing the forest operations foreseen in the tactical plan. This decision problem occurs in cases where the industry is responsible for managing forest plantations (self-owned and rented) but relies on a large number of small scaled outsourcing companies for executing the harvesting operations with their own equipment and crews hired locally for the harvesting season. This decision problem has not yet addressed in the forestry literature. The reasoning may be the fact that the organizational context where the forestry decisions take place is often overlooked in the studies about forest planning.

The harvest managers involved in harvest service adjudication face two main types of business decisions. First they need to find the best way to *cluster the harvesting operations scheduled in the taking plan into work orders* (or work packages). There are several criteria that may be applied for clustering the operations, including the geographical proximity among management units, the expected harvesting date, specific characteristics of the crew and equipment required for harvesting. The work orders may have a balanced amount of work per month, minimize the crews movements (e.g. daily back and forth travelling from home base to harvest area) and equipment movements and assure continuous work levels across the entire planning period. Then, the harvest manager needs to handle the *assignment of the work orders to the local entrepreneurs*. The problem is focused on the characteristics of the harvesting equipment and crew, and consists in finding the best match between the requirements and the availability of these harvesting resources, while minimizing the total cost of outsourcing harvesting operations. There may be work orders mandatorily assigned to specific entrepreneurs due to business or technical reasons. The problem may further acknowledge priorities to some entrepreneurs, therefore reflecting the company appreciation about the entrepreneurs' individual performance in previous contracts.

Both decision problems are often lagged in time. The second problem relies in the proposals of the local entrepreneurs to the work-orders previously put out to tender. Nevertheless, integrated approaches may be foreseen for future harvest adjudication processes, fostering the best use of the available harvesting resources. Regardless, the nature of the business decisions suggests the development of IP formulations, based on other clustering and assigning models found in the OR literature.

Management Planning in Action 7.1: Harvest Planning Including Crew Allocation and Routing

One planning problem at one of the larger Swedish forest companies is to make an annual plan for all its contracted teams. The reason is to make sure that there is a match between harvesting areas and capacity at harvest

(continued)

(continued)

teams. The supply is given as a volume for each assortment at each of a large set of harvest areas. In order to compute the harvest scheduling additional information is required, such as for forwarding distance, average tree size and area of each harvesting stand. Each team consists of a harvester and a forwarder and there is detailed information available, for example, machine sizes (both harvester and forwarder), home base location and working hour capacity. There is also information on particular skill like ability to harvest in steep terrain, efficiency measure as compared to an average team, and if the team is used for thinning or final felling operations. The objective is to establish a feasible plan for harvest scheduling and machine systems/team assignment with smallest possible cost. The total cost has three components. The first is the production cost which is based on the hourly rate times the number of hours a team is working. The working time needed for each team is computed using the so-called performance functions. These functions take harvest area, tree size, forwarding distance, machine type and size as arguments and use measured (through time measurements) functions to compute the time. The second cost component is the travel cost associated with the daily travel back and forward to the home base. The third is the cost to move the teams between harvest areas. This cost depends on the distance and the possible need to use a trailer to move over longer distances.

The model is an integrated allocation and routing problem. Such problem is very difficult to solve and the solution approach used is based on a two-phase heuristic (Bredström et al. 2010). In the first phase, a general assignment problem is solved which allocates teams to harvest areas together with flows between areas and industries. The second stage makes the routing for each of the contracted teams; this is a traveling salesman problem. In order to have some coordination mechanism between the two problems, the solution approach includes a penalty in the first phase in order to allocate areas such that the teams in the second problem have a good set of harvest areas.

The case study at the company involves 46 machines and 968 harvest areas representing a log volume of 1.33 million cubic meters. In the case study, the managers made many what if scenarios to get the answers on many practical questions. Typical examples are what happens if the demand increases by 5 % at mill x in period t? In what area should we contract another harvest team? What new machine should team m purchase? All data and solution is embedded in an Excel framework and using AMPL and CPLEX as the solver. As results, there is a set of maps generated with the allocation of areas to each team (see example below) and many tables providing detailed information on the flows, allocation, and harvesting schedules. In a further development the

(continued)

(continued)

demand with lower levels, upper levels and target values at specific industries were added. Here anticipation variables for the flows were also added to the model (Fig. 7.4).

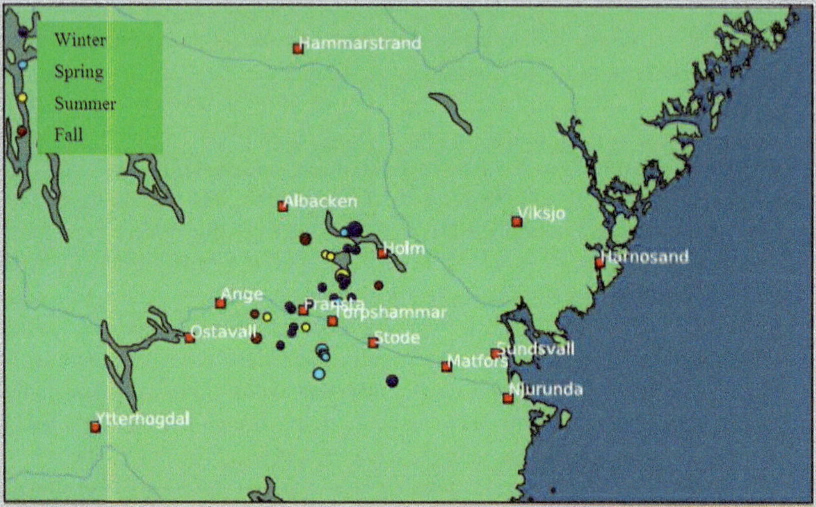

Fig. 7.4 Illustration of the assignment of areas to one team. *Red squares* are all home bases and *circles* harvest areas harvested in different seasons. In this particular example, the team has its home base in the middle of the set of areas

7.5 Operational Harvest Planning

The operational harvest planning process consists in planning the extraction of the wood in each harvest unit. This process starts by charactering the harvest unit in respect to physiography, stand composition and age and estimates of wood production. The areas with conservation interest (e.g. with protected species, buffers around water courses) in the harvest unit are also described. For this purpose, the harvest manager relies on both historic data stored in the company's information systems and in stand information collected in-site during a survey. Then, the planning process addresses the best way to extract the logs from the harvest site, decides how and where to perform the logs processing and sorting operations. Finally, the process specifies the best type of machine system to be used and the required skills of the team. Consequently, target values for the main operations indicators may be established (including expected productivity of the equipment and crew, amount of wood extracted per day), therefore leading to an estimate of the duration of the harvesting operations. The harvesting operational plan is discussed and approved by the outsourcing harvesting company and will set the basis for scheduling the work of the harvesting team and routing/scheduling the wood transportation. During the execution of the harvesting operations, the indicators established in the plan may be monitored in order to provide valuable information for the industry and the outsourcing companies (Fig. 7.5).

The main decision problems related to harvest operational planning relate are planning the extraction of logs and establishing bucking and sorting strategies.

7.5.1 Extraction of the Logs

When a harvest area is to be harvested it is important to plan the routes used by the harvester and forwarder as this impact the overall harvesting time. Here, we discuss it from the view of a team of one harvester and one forwarder. First the planner decides on the basic extraction road where the forwarder will run many times (see Fig. 7.6). This road is decided to avoid soil damage and at the same time minimize the driving time. There are also rules for when to enforce the soil, avoidance of steep terrain or when to construct bridges. Second, the small roads are determined. These are roads where the harvester will drive once and put the logs after bucking in assortment piles and the forwarder as few as possible. These routes need to be planned in a way that all trees are reachable by the crane of the harvester.

Once the harvester has put the logs in piles, the forwarder will use the harvester routes to load the piles and move them to pick-up points adjacent to the forest roads. These are temporary stocking locations in the forest sites, from where the wood is loaded to logging trucks and moved to the mills.

The establishment of the forwarding and harvesting routes may be made manually by the harvester and forwarder driver. However, different studies shows there is a large potential of savings, up to 10 %, if some planning is done. The article by Flisberg et al. (2007) describes a case where GPS data from the harvester is used to define the log extraction network and production files from the bucking process are used to define the location of piles of the specific assortments (including both assortment type and volume) and then optimize the routing for the forwarder. Once the network is defined and the piles are known, the routing problem is a Vehicle Routing Problem (VRP) (This is further studied in Chap. 9). A common complication is that several loading and unloading principles can be used if several assortments are mixed. As a solution method several VRP methods can be used. In the article the best tested is based on a repeated matching algorithm.

The challenge today is to make a pre-plan for both the harvester and forwarder. This requires 3D information of the ground as well as information about the tree location. Using aerial scanning this information is available. The forwarding route planning problem is a general network design problem.

The left part of Fig. 7.7 gives an aerial photo (Google Earth) of an area in middle Sweden and the right part the aerial recognition of the individual trees. Figure 7.8 shows a proposed network design of the basic routes (yellow) and the small roads

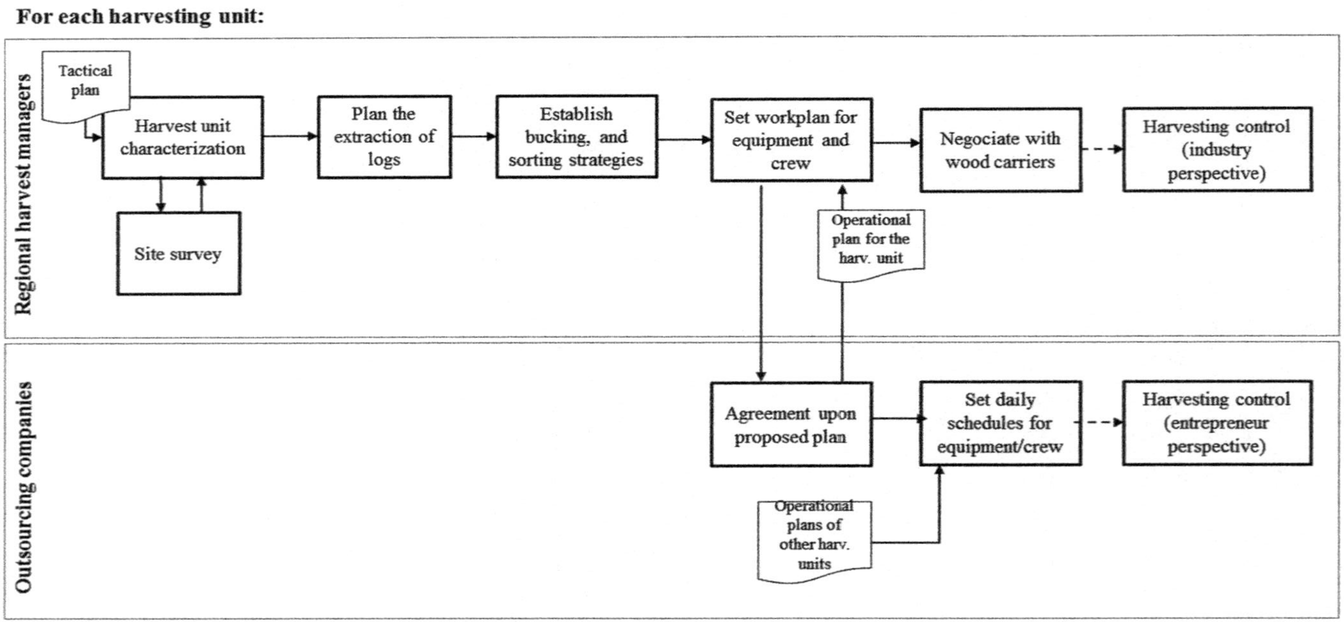

Fig. 7.5 Representation of the generic processes related with tactical harvest planning process in industrial plantations

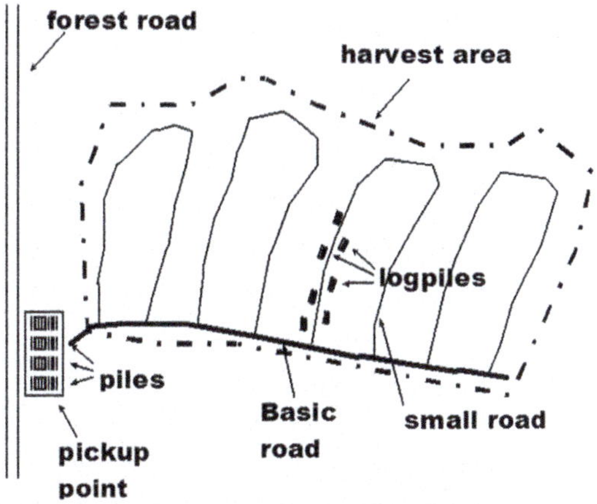

Fig. 7.6 Representation of the basic forwarding routes and small roads for the harvester (from Flisberg et al. 2007)

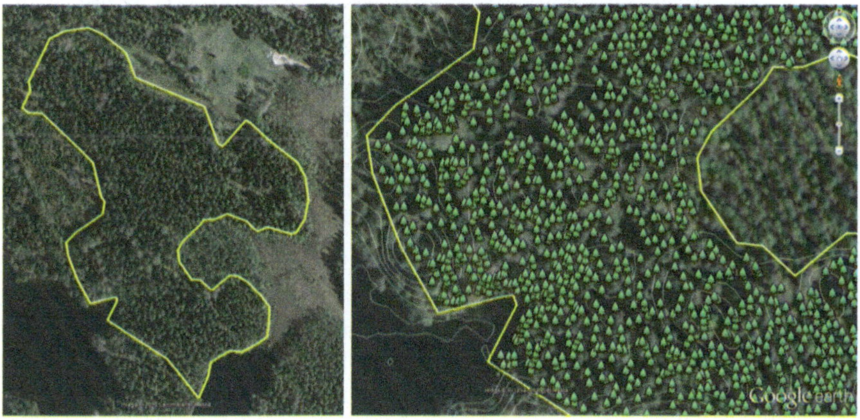

Fig. 7.7 Aerial photo of the harvest area (*left*) and identification of individual trees of a part of the area based on aerial laser scanning (*right*)

(red and pink). In this solution, the objective was to minimize the soil damage which is based on information on soil type and relief maps. Also, the height information is used to identify parts where no routing can be done due to too large angles in sideways for a loaded forwarder. In this particular example in the figure, the discretization level of the road network is 1.5 m (i.e. the surface is divide into 1.5 m squares between which the arcs defined). Given a solution, it is possible for the planner to make an early assessment of the total harvesting time without making a stand visit. For low value biomass such as branch and tops, a given solution also allow an early assessment of where is located in the harvest stand the biomass volume that are economically profitable to extract. Also, the basic roads can be used directly as suggestions for the operations as this is normally something the manager provides on printed maps.

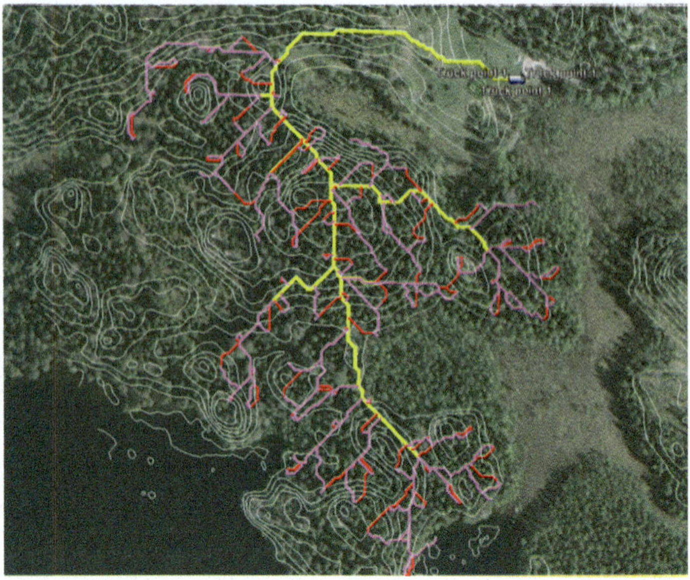

Fig. 7.8 Example of a network of basic routes (*yellow*) and small roads (*red* and *pink*)

7.5.2 *Bucking and Sorting Strategies*

This planning problem that is handled by the harvesting managers responsible for managing the industrial plantations, aims at establishing which bucking instructions and sorting rules to be followed by each harvesting team. The bucking instructions set how a felled tree must be process to obtain one (stem in full-tree method) or a set of products (logs in cut-to-length method) while the sorting rules set how each of the expected products must be separated in distinct inventory piles at roadside. The template of the bucking instructions will depend according the utilization or not of a mechanised processing head for the bucking.

When a mechanised processing head is used, the bucking instructions are provide through a prices list of products specified in a two dimensions matrix of a diameter class and a length interval, see Fig. 7.9. The higher the price of a product is the higher is preferable to produce it while a product with a null price value indicates that its production is not allowed. The matrix is use by the computer of the processing head to optimize the cross-cutting (i.e. bucking) of the felled tree in products.

The price used can either be based on the commercial price or a price that is used to produce a particular distribution of the length diameters. A matrix is provided for each commercial specie (or a group of species managed jointly) on the standing timber. Moreover, for some products (i.e. here, a combination of a diameter class and a length), additional instructions regarding the specifications of a product could

diameter (cm) \ length (dm)	48	49	50	51	52	53	54	55	56
12	9	11	219	58	5	1	2	8	151
14	27	37	505	94	31	8	33	252	361
16	46	53	518	62	102	51	71	627	515
18	44	134	551	66	165	60	115	1189	616
20	67	188	1241	131	164	130	143	1646	1564
22	30	86	765	103	155	63	151	1692	1530
24	19	72	443	73	156	64	175	1914	1056
26	65	99	605	63	152	84	126	1739	926
28	23	129	525	42	130	88	125	1630	442
30	6	18	244	23	86	57	80	1364	517
32	4	18	273	43	73	79	73	1122	322
34	4	7	6	10	101	35	53	957	191

Fig. 7.9 Example of the output given in number of logs of different diameter and length when a particular price list has been used

be provided to the operator of the processing head. Such instructions are external criteria that could be fairly easily observed by the operation in respect to the felled tree, such as external defects. The operator is then responsible to validate or change the optimization bucking solution proposed by the computer before the processing head execute the bucking.

The bucking problem is formulated as a longest path problem and solved very efficiently with dynamic programming, see Näsberg (1985). In the longest path problem, the tree is discretized into, say 1 cm, slices. Each slice denote a node and arcs representing each possible length start in the node and end in a node further ahead corresponding to the length of the log. For positions where the diameter or any quality requirement is not satisfied, the arc is removed. For the remaining arcs, a longest path problem which essentially maximizes the value along a feasible cutting path is solved.

Where no mechanised processing head is used, the bucking instructions are usually called cutting list. Cutting list consist in an ordered listing of products that can be produced. Each product is defined by a set of specifications such as specie (or group of species managed jointly), length interval, diameter classes (big and small end) and quality grade (that is in turn defined by external criteria that could be fairly easily observed on the felled tree). Thus, starting from the product at the top of the cutting list, the first product that meets its requirements must be produce and so on. Cutting lists are used in different configuration of operator and location along the wood supply chain (e.g. a chainsaw operator at the stump, a loader and pull-through delimber operator at landing site). The requirement of a log can be very advanced with rules on for example length, diameter, knot sizes, whorls, sweep and machine damage.

According to the characteristic of the standing timber, the type of treatment (e.g. clear-cut or thinning) and the type of equipment of the harvesting team, a set of preferable bucking instructions and sorting rules is defined at the tactical level. At the operational level, the choice of one of these preferable bucking instructions and sorting rules is usually driven by the demands.

When a mechanised processing head is used for bucking, two bucking problems are discussed by Marshall (2007): the individual tree optimization problem (buck-to-value problem) and the multiple trees with demand constraints problem (buck-to-order problem). The objective of the buck-to-value problem is to obtain from each individual stem the output in products that provide the maximum value according to the bucking prices (or values) list. The objective of the buck-to-order problem is to obtain from a block (or a set of blocks) the output in products that provide the maximum value according to the bucking prices list while meeting the demand and the available input in trees. The latter will need some continuously updated prices. The principle is the same as in for example Todoroki and Rönnqvist (2002) which study dynamic price settings for sawmill production.

7.6 Examples of Fully Integrated Forest Tactical and Operational Planning Problems

7.6.1 Integrating Harvest Operations with Road Building, Upgrade and Maintenance

The forest industry is highly dependent on an efficient road network, since most of the wood is transported some distance by trucks. To access new harvest areas, there is a need to build new roads. Even though the decisions are related to the road building, it is critical to integrate it with harvest decisions. The objective is typically to minimize the combined cost of road building and harvesting. In such cases, the road investments are business decisions and harvesting decisions anticipation decisions. Decisions about road building are typically included in MIP models supporting tactical harvest planning on time horizons of one to five decades. The first models were presented in Weintraub and Navon (1976) and Kirby et al. (1986). Several methods were then proposed for strengthening the MIP formulations by adding various valid inequalities. Andalaft et al. (2003) presents a solution approach for a road building problem including typical operational harvest planning decisions, as well as the roads to build or upgrade for access, considering different road standards (accessible Summer/Winter). The paper by Cea and Jofre (2000) presents a two-level model and optimization algorithm to assist combined strategic and tactical planning. The tactical planning problem includes decisions about harvesting, transportation, road construction and upgrading considering different road standards.

In many other applications, the roads exist but the quality may be too low for transports for certain conditions. The weather conditions in many countries vary during the year (cold winters, warm summers), due to the climate. Every year, there is a period of thaw in the spring and typically a period of heavy rain in the fall. During these periods there is uncertain accessibility to parts of the road network due to unfirm ground which causes in the Nordic countries considerable forestry costs. Pulp-, paper- and sawmills often require a continuous wood supply all the year round. In order to satisfy this requirement, forest companies use large reserve stocks of raw material located at roads that are accessible all year. The outdoor storage of raw material causes considerable costs due to extra handling and deterioration in quality any special wood inventory care (e.g. water sprinkler) and, of course, capital cost of the inventory. Sometimes impassable roads can be avoided by alternative routes though the increased haulage distance causes a significant extra transportation cost. Road upgrading is done to provide prerequisites for all year round accessible harvest areas. The objective is to minimize costs for transportation, upgrading and storage subject to constraints on continuous supply and requirements on the lowest road standard used for transportation. This implies upgrading of roads or alternatively, detours (longer routes) or storage to achieve a feasible flow. The trade-off between upgrade cost and storage cost can be evaluated by comparing plans for different levels of required available volumes. A larger available volume along all year around accessible roads means a decreased need for reserve storage.

Mathematical models and solution methods for integrated harvesting and road network design may be found in the forestry literature. As an example, Frisk et al. (2006) propose a decision support system that is based on the detailed information from Swedish national road database (NVDB). This includes information about all road segments in Sweden, such as their length, accessibility class and maximum allowed speed. The model used was however difficult to solve and some modified models were suggested in Henningsson et al. (2007).

Today there are new challenges. Since the planning horizon is as long as 10 years, there is a need of detailed description of the forest growth and the related net present values. For this reason, it is important to integrate with strategic forest management software with growing functions. There are new equipped with central tire inflation (CTI) system, which can pass on roads of different quality compared to standard trucks. Also, using aggregated road databases creates difficulties in data handling, data quality, conflict data and there is a need to make aggregations for more practical use.

The management in action below illustrates the need for integration of harvesting, network design and transportation.

Management Planning in Action 7.2: Logistic Planning After a Storm

Early January 2005, the storm Gudrun hit an area in southern Sweden with hurricane force winds. During these days, an estimated 70 million cubic meter of forest was wind felled. This volume amounts to one annual harvest

(continued)

(continued)

volume for the whole of Sweden. Most forest damage was concentrated to a region close to the west coast. The monetary value of the wood alone was approximately 3.2 billion Euros. Besides the forest damage, considerable damage also occurred to the infrastructure. All plans and logistic network were immediately obsolete. Many questions appeared. How much volume was wind felled and where? Was the available harvesting and transportation capacity enough to recover felled volume? How to supply existing customers? Where to store the extra wood? What are the immediate effects on the wood market?

Due to the risk that the quality of the timber would fall, it was important to deal with the wind felled volumes as soon as possible. Given that there was a need to harvest and to clear the damaged areas as soon as possible, a critical question was where all the logs would be stored. A large number of storage terminals had to be established and brought into operation as most customers (saw, pulp and paper mills and heating plants) did not have enough storage space. A number of storage alternatives were drawn, including storage in the forest, at road sides, in lakes, at airfields and on farm land. Furthermore, many transportation modes were possible, including trucks, trains and vessels. At the same time, it was critical to plan in which damaged areas the harvest teams were to work depending on their machine system size.

The company Sveaskog developed new logistic plans after the storm for tackling 3.1 million cubic meters wind felled. This Swedish company adopted a decision support system FlowOpt (Forsberg et al. 2005). The optimization provided plans on storage locations, detailed harvest plans and all transportation flows using trucks, train systems and vessels. The overall strategy adopted was to find as many customers as possible in the storm felled area and to ensure volumes on trains and trucks for prioritized customers. In total there were up to 200 harvest teams, 122 trucks, 21 train systems and 110 potential vessel routes available to move logs for 22 different assortments. The actual operation followed very well the plans provided by the DSS.

In the project, Sveaskog were successfully able to develop a support tool to assist in its logistics planning from both strategic and tactical perspectives (Broman et al. 2009). This was possible, despite short delays, due to the existing system FlowOpt and through making use of the Swedish road database. The model developed for the operations describes the situation accurately and the results provided new and valuable insights for the logistics managers at Sveaskog. The analyses helped Sveaskog's managers to find priority customers and support them in negotiations. Basic data has been used in many "what if" scenarios in order to evaluate alternatives.

7.7 Open Questions and Future Research Guidelines

Existing approaches for tactical and operational harvest planning are mainly based for deterministic planning environments. Even though there exist many methods to forecast the volumes available at harvest areas there is a large degree of uncertainty. This is true for, for example, the volumes available of different assortments at stands, the harvesting times for different machine systems, the weather conditions etc. One approach to take uncertainty into account is to use a rolling horizon planning where a new problem is re-solved a soon as new information is available. Another is to use methods that model uncertainty. Examples of methods are robust optimization and stochastic programming.

Another issue is that problems become very large for typical sized industrial cases, leading to long computational times for reaching a solution. Consequently, further research is need to keep-up developing efficient methods which are capable of solving very large problems and at the same time generate near-optimal solutions.

Many of the solution methods have been implemented within Decision Support Systems for Forest Management. Such computerized-tools store the information required for planning and enable many features useful for decision-support, such as the generation of "what if" scenarios and analysis of the optimization results making use of tables and maps. Yet, further research is needed in developing Information and Communication solutions for collecting information relevant for planning in real-time and displaying it to the decision makers in a way that it may contribute for flexible and reactive planning processes.

The topics of collaboration between the agents of the forest-based supply chain are still seldom addressed. New collaboration strategies and collaborative business models may introduce new objectives and constraints to the existing tactical and operational harvest planning problems.

Further research is needed to cover some of the management decisions loosely addressed in the forestry literature, such as harvest sequencing and harvest service adjudication. New studies on integration of management decisions may pave the way to new sophisticated holistic planning processes.

7.8 Summary

Tactical and operational harvesting is a key component in the wood supply chain. It corresponds to a large part of the overall cost and it is hence critical that it is well planned. Although related with strategic harvest planning (Chap. 7), medium- and short term planning has specificities that need to be acknowledged. The relations with transportation planning (Chap. 9) are also of great importance.

The chapter presents general forest tactical and operational planning processes and focus on the forest management decisions undertaken by managers of industrial plantations concerning medium- and short-term planning. Specific sections cover each of the typical tactical and operational harvest planning problems, including harvest scheduling, machine system/team assignment, harvest sequencing, harvest service adjudication, extraction of logs, bucking and sorting strategies. These decision problems are described in terms of the core objectives and constraints, modelling approaches and solution methods frequently found in the forestry literature.

The integration of elementary management decisions will be further discussed and exemplified. Special attention goes for distinguishing between situations where the elementary planning problem is extended to accommodate the impact on/from other planning problems and situations where the decisions are actually combined and aiming simultaneous execution. The modelling approaches and solution methods proposed in the literature for solving these problems are also described.

7.9 Problems

1. List the specific characteristics of tactical and operational harvest planning, when compared with strategic harvest planning and transportation planning.
2. Differentiate between business decisions and anticipation decisions. Provide examples of problems where anticipation decisions are often used and other examples of fully integrated problems with more than one type of business decisions.
3. Consider the formulations proposed for harvest scheduling [P1] and [P2] and add new constraints related with: (a) budget restrictions per period; (b) production of minimum levels of certain product assortments per period; (c) spatial adjacency constraints.
4. Propose a model formulation for the harvest service adjudication problem.

References

Andalaft N, Andalaft P, Guinard M, Magendzo A, Wainer A, Weintraub A (2003) A problem of forest harvesting and road building solved through model strengthening and Lagrangean relaxation. Oper Res 51(4):613–628

Audy JF (2013) Inter-firm collaboration in transportation. PhD thesis, Université Laval, Canada, 224 pp

Bredström D, Jönsson P, Rönnqvist M (2010) Annual planning of harvesting resources in the forest industry. Int Trans Oper Res 17(2):155–177

Broman H, Frisk M, Rönnqvist M (2009) Supply chain planning of harvest operations and transportation after the storm Gudrun. INFOR 47(3):235–245

Cea C, Jofre A (2000) Linking strategic and tactical forestry planning decisions. Ann Oper Res 95(1–4):131–158

D'Amours S, Ronnqvist M, Weintraub A (2008) Using operational research for supply chain planning in the forest products industry. INFOR 46(4):265–281

Epstein R, Nieto E, Weintraub A, Chevalier P, Gabarró J (1999) A system for the design of short term harvesting strategy. Eur J Oper Res 119:427–439

Flisberg P, Forsberg M, Rönnqvist M (2007) Optimization based planning tools for routing of forwarders at harvest areas. Can J For Res 37:2153–2163
Forsberg M, Frisk M, Rönnqvist M (2005) FlowOpt – a decision support tool for strategic and tactical transportation planning in forestry. Int J For Eng 16(2):101–114
Frisk M, Karlsson J, Rönnqvist M (2006) RoadOpt – a decision support system for road upgrading in forestry. Scand J For Res 21(7):5–15
Henningsson M, Karlsson J, Rönnqvist M (2007) Optimization models for forest road upgrade planning. J Math Model Algorithms 6(1):3–23
Karlsson J, Ronnqvist M, Bergstrom J (2004) An optimization model for annual harvest planning. Can J For Res Rev Canadienne De Recherche Forestiere 34(8):1747–1754
Kirby MW, Hager WA, Wong P (1986) Simultaneous planning of wildland management and transportation alterations. In: TIMS studies in the management sciences 21. Elsevier, North Holland, pp 371–387
Lehoux N, D'Amours S, Rönnqvist M (2012) A framework for an efficient implementation of logistics collaborations. Int Trans Oper Res 19(5):633–657
Marques AF (2012) New decision support tools for forest tactical and operational planning. PhD thesis, Instituto Superior de Agronomia, Universidade Técnica de Lisboa, 69 pp
Marques AF, Borges JG, Sousa P, Pinho A (2011) An enterprise architecture approach to forest management support systems design. An application to pulpwood supply management in Portugal. Eur J For Res 130(6):935–948. doi:10.1007/s10342-011-0482-8 (WOS:000296063100004)
Marques AF, Rosset C, Rasinmaki J, Varik A, Nobre S (2013a) Collaborative development of a semantic wiki on forest management decision support. Scand J For Res. doi: 10.1080/02827581.2013.832791
Marques AF, Borges JG, Pina JP, Lucas B, Garcia J (2013b) A participatory approach to design a regional forest management planning decision support toolbox. For Syst 22(2):240–258. doi:10.5424/10.5424/fs/2013222-03120
Marshall H (2007) Chapter 19: Log merchandizing model used in mechanical harvesting. In: Weintraub A, Romero C, Bjørndal T, Epstein R (eds) Handbook on operations research in natural resources. Kluwer Academic Publishers, New York
Martins I, Constantino M, Borges JG (2005) A column generation approach for solving a non-temporal forest harvest model with spatial structure constraints. Eur J Oper Res 161:478–498
Murray AT, Church RL (1995) Heuristic solution approaches to operational forest planning problems. Or Spektrum 17(2–3):193–203
Näsberg M (1985) Mathematical programming models for optimal log bucking. Linköping Studies in Science and Technology, Dissertation 132, Department of Mathematics, Linköping University, Sweden, 198 p
Ronnqvist M (2003) Optimization in forestry. Math Program 97(1–2):267–284
Todoroki CL, Rönnqvist M (2002) Dynamic control of timber production at a sawmill with log sawing optimization. Scand J For Res 17(1):79–89
Weintraub A, Cholaky A (1991) A hierarchical approach to forest planning. For Sci 37:439–460
Weintraub A, Murray AT (2006) Review of combinatorial problems induced by spatial forest harvesting planning. Discret Appl Math 154(5):867–879
Weintraub A, Navon D (1976) A forest management planning model integrating silvicultural and transportation activities. Manag Sci 22(12):1299–1309
Weintraub A, Jones G, Maecham M, Magendzo AA, Malchuk D (1995) Heuristic procedures for solving mixed-integer harvest scheduling – transportation-planning models. Can J For Res Rev Canadienne De Recherche Forestiere 25(10):1618–1626
Weintraub A, Romero C, Bjorndal T, Epstein R, Miranda J (eds) (2007) Handbook of operations research in natural resources, International series on operations research & management science. Springer, New York

Chapter 8
Transportation and Routing

Jean François Audy, Alexandra S. Marques, Mikael Rönnqvist, Rafael Epstein, and Andres Weintraub

8.1 Introduction

Transportation plays a vital role in forest operations. As a fraction of total procurement costs, transportation can be as high as 45 %, which is the case in Chile. Corresponding numbers in other countries are Canada 36 %, Australia 18–25 %, Southern US 25–35 %, Sweden 30–40 % and New Zealand 40 % (Audy et al. 2012a). The characteristics of transportation depend significantly on the type of forest and managements as well forest ownership state. For example, in US and Canada a significant percentage of forests are native, where road building and maintenance plays an important role, and operational transportation costs are less important. In plantations, which are usually closer to existing roads, transportation costs play a larger role. Such is the case in countries like New Zealand, Chile, South Africa and Southern US. In Sweden, the forest ownership characteristics play a role. The existence of many small owners leads to a different form of transportation, as will be shown.

The main element in transportation is to move timber (or logs, round-wood) from the forest to first destination, usually a mill: pulp, paper, sawmill, or energy. In

J.F. Audy (✉)
Université Laval, Québec, QC, Canada
e-mail: Jean-Francois.Audy@cirrelt.ca

A.S. Marques
INESC Tec, Porto, Portugal
e-mail: alexandra.s.marques@inesctec.pt

M. Rönnqvist
FORAC, Université Laval, Québec, QC, Canada
e-mail: mikael.ronnqvist@gmc.ulaval.ca

R. Epstein • A. Weintraub
Department of Industrial Engineering, University of Chile, Santiago, Chile
e-mail: repstein@dii.uchile.cl; aweintra@dii.uchile.cl

J.G. Borges et al. (eds.), *The Management of Industrial Forest Plantations: Theoretical Foundations and Applications*, Managing Forest Ecosystems 33,
DOI 10.1007/978-94-017-8899-1_8,

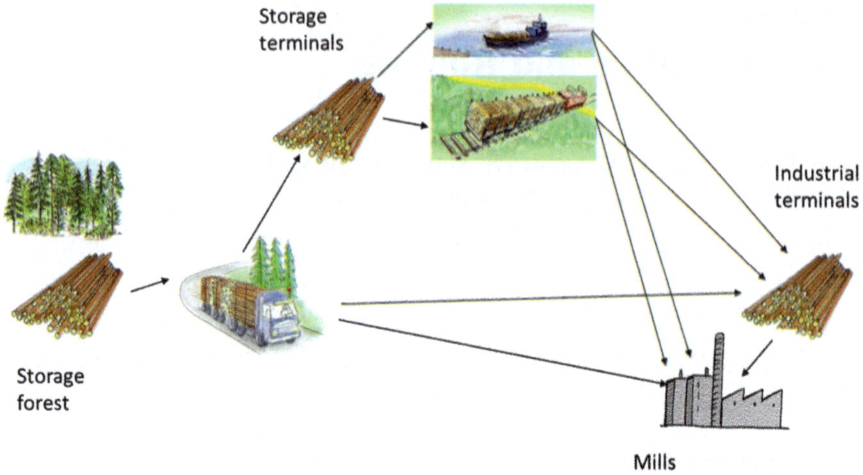

Fig. 8.1 Illustration of the main transport of timber from forest to storage facilities or mills using trucks, trains and vessels (from Rönnqvist (2012))

other cases it can be a port for direct export, or a terminal, where timber will be stored for later use. Moreover, there may also be storage possibilities close to the mills. The main transportation is by different truck types but train multimodal systems and vessels are also important transport modes, in particular for longer distances and large volumes. When no transportation infrastructure exist for such multimodal systems, truck multimodal systems combining heavy-load trucks (e.g. up to 165 ton payload in Canada) for upstream transshipment point flow and standard ones for downstream flow can take place. Most systems developed deal with this type of transportation which is illustrated in Fig. 8.1. But we also need to consider transportation from primary destination to secondary ones, for example, from sawmill to clients or to secondary manufacturing plants.

We consider different levels of planning. At strategic level decisions are made over up to 5–10 years on the transportation mode to be used and its network infrastructure system. Typical decision involves location of terminals, primary road network and selection of transportation mode (including multimodal system). At tactical level decisions involve planning for a year, a season or a month. For the longer horizon, for example one year, the number of trucks per truck type can be considered a tactical decision. How many trucks will be contracted for the next year is a typical example in Chile. Here truck assignment models are used to estimate the need for trucks. On the shorter horizon, e.g. one month, the truck fleet is assumed fixed and there is a given industrial demand to satisfy each time period. Several models developed in this area have been carried out in the Scandinavian countries. The models are used to determine the allocation from supply areas to industries taking into account backhauling flows. Where OR models have been most prominent is at operational level, where the actual routing of vehicles is carried out. On this level, any use of train system and vessels are considered to be given supply and demand points in the network. A fourth level is real-time decisions and essentially concern truck dispatching, i.e. the assignment of the next load (or more) to a truck as the transportation operations occur.

Most models developed show the use of trucks in carrying timber from origins at the forests to destinations. Decisions here involve which origins are served by which trucks and the routing of the vehicles. We will show successful applications in Chile, Brazil, South Africa, New Zealand, Sweden, where systems have been developed to consider these problems. The approaches have included heuristics, simulation and exact mixed integer formulations, solved approximately in most cases.

Ideally transportation decisions should be taken jointly with harvest operation decisions and integrated into the whole value chain from forests to final customers. This important issue has been discussed earlier in Chap. 7 and we refer to Table 7.1 for different integration problems. We will discuss this integration in this chapter. There is considerable interest worldwide in finding ways toward cost-savings opportunities in transportation (Murphy 2003), including making transportation operations more efficient with better planning. Fuel cost representing a significant proportion of transportation costs, volatile crude oil world markets and growing environmental concerns (i.e. greenhouse gases emissions reduction) are also drivers to improve transportation efficiency. In addition, when using many trucks, there is a desire to balance the workload such that all truck drivers receive a fair share of the overall work load. This means that there often is a multi-objective goal to consider in the planning process.

What You Will Learn in This Chapter

- Distinguish between strategic and tactical/operational transportation/routing
- Understand strategic transportation models and related solution methods.
- Understand tactical transportation models and related solution methods.
- Understand routing and operational models and related solution methods.
- Provide examples of transportation and routing applications encountered in the literature
- Have knowledge about open research topics that may inspire future work in this domain

The remaining chapter unfolds as follows. Section 8.2 describes the network and information needed for all transportation. Sections 8.3 and 8.4 focus on strategic and tactical transportation problems, respectively. Section 8.5 describes routing models and methods, and Sect. 8.6 focus on the wood reception planning. The chapter ends with a discussion of open questions and future research directions in Sect. 8.7, a summary in Sect. 8.8, and finally, some problems are presented in Sect. 8.9.

8.2 Network Structure

We present the problem in the form of the value chain that spans from harvest areas to final destinations with deliveries of final products to customers. The value chain can be viewed in a network representation where the nodes represent physical

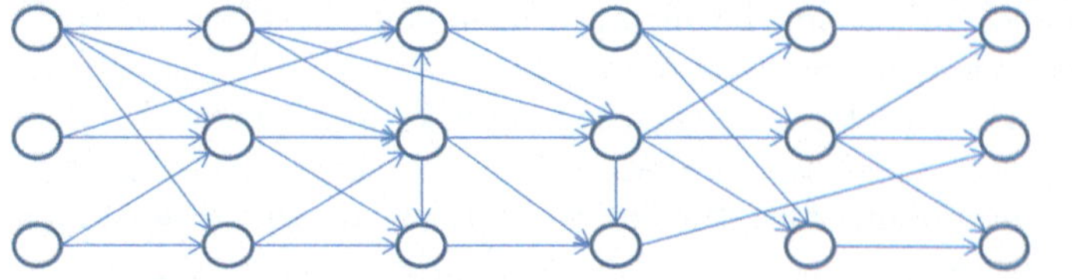

Fig. 8.2 Illustration of a general network representing the value chain

entities where activities are carried out, and the arcs represent transportation in different forms. Figure 8.2 shows a schematic form of a value chain that includes forests, where harvesting will be carried out, terminal and storing areas, where timber will be stored for a period of time, different plants, including pulp and paper plants which are major investments and play a large role in the industry, sawmills, secondary plants like panel mills, energy plants which receive residues as well as low quality timber in some cases, ports or other embarkation sites, and final customers of the different products. We will concentrate in particular on the arcs linking the nodes, which will represent the different forms of transportation. According to the type of decisions made the network will take different forms. At strategic level the types of nodes (plants for example) and arcs (transportation mode) are part of the decision process. At tactical level the network is basically fixed, and decisions involve the flows along the network. At operational level, it is more routes and sequences of arc use that are the decision variables. As mentioned, at all levels there is important interaction between the nodes (production) and arcs (transportation).

In the next sections, we describe a number of decisions at strategic, tactical and operational level. We will see how the basic structure of the network in Fig. 8.2 is preserved and how at different planning levels there are significant variations on the relevant part of the network depending on the decisions to be taken, and the information that will be needed. In these sections, we present different models that have been proposed and used, as well as what data is needed. For the network, it is important to have necessary information regarding the transportation infrastructure. For instance, on the roads network this includes road distance, road quality, road width, road slopes, speed limits etc. All this is needed in order to compute accurate distance or travel time between nodes in the network (Flisberg et al. 2012) and to compute upgrading costs to improve the road quality (Frisk et al. 2006). Here, Geographical Information Systems (GIS) are very important. At the same time, it can be costly and highly resource consuming to both collect and maintain this information with a high quality.

8.3 Strategic Transportation Planning

Both at strategic and tactical planning, forest transportation can be either separated or integrated with harvesting. If we study Fig. 8.2 showing the value chain, we can identify the strategic decisions related to harvesting and timber transportation activities. At the strategic level, the problem is not defined in high spatial detail and data are highly aggregated. The first nodes correspond to harvesting in the forest. At strategic level main decisions involve long range level of harvesting and silvicultural practices, replanting with different options, and also purchase of timber lands. At this level, the nodes correspond typically to 'macro stands', that is areas which are silviculturaly similar and geographically close and thus can be grouped together for management treatment (but not geographically identified). Spatial identification is carried out at tactical level.

The second set of nodes corresponds to plants at primary level. At strategic level we will be concerned with decisions involving major investments, in particular pulp and paper plants, but also sawmills and other plants. Pulp and paper plants need to be identified spatially, whereas smaller scale investments can be defined with less spatial detail. The third set of nodes indicates secondary plants, like panel mills, while the fourth corresponds to final customers. The arcs of the network correspond to the transportation part of the value chain. While construction of major roads can be defined explicitly, most road building need not be defined in spatial detail at strategic level. As described in Weintraub and Cholacky (1991), at this level of decisions, the plans for road building can be described in a more general way, indicating a level of investment and connectivity. Collection of information and data is not trivial at strategic level, given the long range horizons considered and the large amount of data. Uncertainty plays a major role, in particular in long range market conditions. Obtaining robust solutions is important and this is discussed later in this section. We present a simplified deterministic strategic planning model. Note that this model is one of many possible versions. Given the investments involved, the length of a rotation (in Chile) might be a minimum span of 40 years or more.

As described earlier, 'macro stands' will be defined for the model. Each macro stand will be composed of multiple basic stands, forming a non-compact area, but in a geographically close neighborhood to consider transportation costs. To describe the timber production, growth models are used separately to determine the volumes of timber obtained per hectare and species when harvesting in any period. At this level no detailed description of timber type, such as lengths and diameters is normally used. Both investments costs and operating costs are considered. These involve the management of timber lands, the different types of plants, where investment and processing costs need be defined, transportation where investments in road building or upgrading, fleet acquisition of trucks, (or other means of transportation) and hauling costs need be considered. The objective has revenues due to sales of timber and products. The production at plants, capacities, processes and costs at each plant need to be defined. To simplify the model, we make several assumptions. We consider only one level on plants, transportation between macro stands and plants can be defined through one road that either exists or needs to be built. We neither consider land acquisition nor complex silvicultural separate alternatives, to concentrate in the transportation part of the problem. We consider plants decisions as already taken. In reality, given the importance of location to consider road building and transportation costs decisions on investment in plants, land acquisition and major silvicultural policies should be considered integrated to strategic transportation planning. We do not include in the model considerations on environmental preservation and sustainability, which are clearly major issues (Weintraub et al. 2000). We consider one type of timber but several wood products.

A deterministic model assumes the data is correct and no uncertainty takes place. In long range models this assumption will clearly not be true. Not much has been developed to obtain in a quantified way robust solutions at strategic level. Simple ways of analysis include sensitivity and parametric analysis, varying the uncertain parameters. In particular market conditions will be subject to uncertainty, prizes and levels of demand.

To illustrate a strategic transportation model (simplified), consider the following variables and coefficients:

Variables

x_{it} = Hectare of area harvested in macro-stand *i*, period *t*
z_{jbt} = Volume (m^3) of product *b*, delivered at market *j*, period *t*
y_{irt} = Volume (m^3) of timber transported between macro-stand *i*, and plant *r*, period *t*
w_{rjbt} = Volume of product *b*, sent from plant *r*, to market *j*, in period *t*
h_{irt} = 1, if road-building (route) between macro-stand *i*, and plant *r*, period *t*, is built; 0 otherwise

Note: Some roads between macro-stand *i* and plant *r* already exist.

u_{rbt} = Volume (m^3) produced in plant *r*, of product *b*, period *t*

Parameters

c_{it} = Cost per m^3 of harvesting in macro-stand *i*, period *t*
p_{rbt} = Cost per $m^3$of producing product *b*, in plant *r*, period *t*
e_{jbt} = Revenue per m^3 of delivery product *b*, in market *j*, period *t*
f_{irt} = Cost per m^3 of transporting timber from macro-stand *i*, to plant *r*, in period *t*
g_{rjbt} = Cost per m^3 of transporting product *b*, from plant *r*, to market *j*, period *t*
q_{irt} = Cost of building road from macro-stand *i*, to plant *r*, in period *t*
d_{jbt} = Demand (m^3) for product *b*, in market *j*, period *t*
a_i = Total area (in hectare) of macro-stand *i*
m_{ir} = Capacity of road from macro-stand *i* to plant *r*
b_{rb} = Conversion factor: number m^3 timber to produce one m^3 of product *b* at plant *r*
k_{it} = Volume (m^3) per hectare of area harvested in macro-stand *i*, in period *t*

Objectives Max: Revenue – Costs

Revenue by sales $\sum_t \sum_j \sum_b e_{jbt} z_{jbt}$

Cost of road building $\sum_t \sum_i \sum_r q_{irt} h_{irt}$

Cost of harvesting $\sum_t \sum_i c_{it} k_{it} x_{it}$

Cost of transporting to plants $\sum_e \sum_i \sum_r f_{irt} y_{irt}$

Cost of production of plants $\sum_t \sum_r \sum_b p_{rbt} u_{rbt}$

Cost of transport to markets $\sum_t \sum_r \sum_j \sum_b g_{rjbt} w_{rjbt}$

Constraints

1. Timber production in macro-stand i
   $\sum_t x_{it} \leq a_i$ (Total area harvested in macro-stand i cannot exceed a_i)
2. Timber sent from macro-stand i to plant r through roads.
   $y_{irt} \leq m_{ir} \sum_{\theta=1}^{t} h_{ir\theta}$ For non-existing roads (ir) that needs to be built.
   $y_{irt} \leq m_{ir}$ For existing roads (ir)
3. All timber produced at macro-stand i in period t is sent to plants
   $\sum_r y_{irt} = x_{it} k_{it}$
4. Production at plant r in period t
   $\sum_i y_{irt} = \sum_b b_{rb} u_{rbt}$ (All timber is used to produce product(s))
5. Possible product b is sent from plant r to market j in period t
   $u_{rbt} b_{rb} = \sum_j w_{rjbt}$
   $z_{jbt} = \sum_r w_{rjbt}$ (Products arrives to markets j from any plant r)
6. Demand is bounded in all markets j for all products b in all time periods t
   $z_{jbt} \leq d_{jbt}$
7. All variables are non-negative, $h_{irt} \in \{0, 1\}$

The summation is done over all members of a set for each index used.

Management Planning in Action 8.1: Network Design Using Train Systems

Sveaskog is Sweden's largest forest owner and leading supplier of timber, pulpwood and biofuel. Some years ago, essentially all transports were carried out by logging trucks. At this time, there was a discussion whether the company should change the network to also use a train system taking pulp logs

(continued)

(continued)

from southern Sweden to some large mills in middle Sweden and then move sawlogs from middle Sweden to several large sawmills in southern Sweden. Beside the decision to use a train system, decisions must be taken where to locate the terminals for loading and unloading. These decisions are very strategic but how is it possible to evaluate the possible decisions in a model?

It was decided to set up a tactical model where a set of potential terminals and train systems were used to define a number of scenarios. First, Sveaskog wanted to include binary decisions directly into the model. However, it turned out that to define the cost coefficients for the location of terminals and use of train systems required several weeks of negotiations, for each possible configuration (e.g. capacity). At this stage, it was decided to set up five main scenarios with given train system and configuration of terminals. Also, the actual costs of these scenarios were not known. However, each scenario had a fixed network so it would be possible to compute the annual cost of the transportation using a tactical flow model.

Map describing the area of the operations where the train system could make a difference. Red squares are potential locations of terminals, the main pulp-, paper- and saw-mills are illustrated with small factories and green arrows denote the forest areas. The five scenarios each included a set of terminals together with different configuration of the train system. This could for example be defined on the frequency of the train, its capacity and the number of locomotives in each system.

To provide information for the tactical model, information on all transports done the year before was available, and hence there was information on supply and demand. In total the case study had 1,500 supply points, 220 industries, 12 assortment, 5 assortment groups, 5 train systems and 10 potential terminals. The number of constraints was no more than 3,000 but the number of variables is 30 million. The reason for this is the fact that backhauling flows was allowed and they increases the number of variables drastically. The solution time for the scenarios ranged between a few minutes and several hours depending on the tolerance requirement.

(continued)

(continued)

In the scenarios, it was clear that the transport by trucks could be reduced by 35 % and the overall energy consumption by 20 %. The latter also correspond to the decreased emissions of CO_2. Each of the scenarios provided the total cost beside the fixed cost for trains and terminals. However, when Sveaskog were doing the negotiations to establish these costs, it was very clear which alternative that would be the best scenario. However, in the end the total improvement in terms of cost was relatively small. Also, there were other concerns to consider. For example, if the train system was selected, it meant that 35 % of the truck drivers would not have any job and this in areas where the unemployment is high from start. The final decision was to start using the train system using one of the design scenarios.

8.4 Tactical Transportation Planning

At this level of decisions, the basic elements of the value chain, consisting of forests, primary and secondary plants, and roads are in place. Planning horizons ranges from around 1 month, a season to 1 year. The spatial configuration plays an important role, as locations and travel distances are important elements for the decision process.

The basic elements of decisions involve detailed definition of fleets, aggregated capacity level of trucks and other transportation means. This implies definition on how supply of timber at origins at the forest will be coordinated with demands at plants for timber, and processed products down the chain. Different transportation mode are used between forests and first destinations, plants, ports, stocking areas, and then between primary and secondary destinations. The main costs typically arise in the first transportation, between forests and primary destinations, so we will concentrate on this aspect.

A basic model (Epstein et al. 2007) to distribute timber is a simple transportation model, where at the origins supply is given by the expected harvest at stands and demands are given at destinations. Costs on each arc are defined by the unit costs of a back-and-forth transportation between each origin and destination. This is a simple LP model which gives an approximate solution to the distribution problem, an allocation of stands or catchment areas to specific destinations, and indirectly the fleet needed. If different assortments are defined, the problem can be transformed into a multi-commodity model and product substitution can occur by defining demand on assortment groups that could be satisfied by more than one assortment. Note the approximation taken in not considering explicitly the empty truck returns or queuing. Transportation costs are typically based on a full load from forest to mill, and then empty back to the forest. One main reason for this is that the volumes

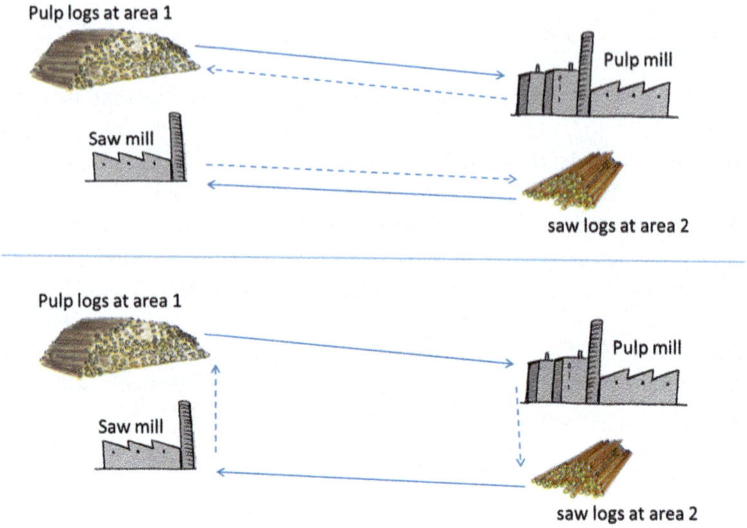

Fig. 8.3 Illustration traditional back and forward transportation (*top*) and backhauling (*bottom*)

available at harvest areas in forests are generally large and require several truck loads before it is empty. This implies that the loaded proportion is up to 50 %. Since most trucks need to drive from and to a home base in the beginning and end of the day, the loaded proportion is even more reduced. However, it is possible to generate backhauling possibilities where loads from two harvest areas are combined. This is illustrated in Fig. 8.3. In the basic version (top) the truck drives back and forward, and using backhauling (bottom) the truck drive the backhaul route: area 1 → pulp mill → area 2 → sawmill → area 1. Suppose the distance between areas and mills are 100 km and between pulp mill and area 2 (and sawmill to area 1) is 20 km. Then, the distance saved is 160 km which correspond to 160/400 = 40 %. It is evident that backhauling can improve the transportation efficiency. A model and solution approach based on column generation dealing with backhauling can be found in Carlsson and Rönnqvist (2007).

At his level of decision, more detailed information is required than at strategic level. Note here that spatial location of the timber production is relevant, as well as a clear definition of different types of timber and what volumes are produced. There is a need to have access to harvesting costs and transportation costs between stands and mills. In case the demand is not fixed, we need the market value of each product or assortment delivered to plants and/or customers. As inventory is important, we also need costs and capacities for holding timber at stocking yards or terminals and also wood freshness guideline.

Typical transportation situations have seasons when transportation is difficult. For example, in Chile, roads are made of gravel, which can be used year round and cheaper dirt roads, which cannot be used during the winter season due to rains. To make better use of the system, stocking areas, closer to plants and connected to them via gravel roads are used to stock timber in summer for use in winter, as there is a need of a steady supply of timber to plants throughout the year. In a similar way, in Nordic countries such as Sweden and Canada, transportation is easier in winter when roads are frozen and more difficult in spring due to thawing roads. In this form, the use of inventories plays an important role in the yearly planning. We present a basic model which considers supply of timber known, demands as given, as well as transportation capacities and costs in all seasons, stocking yards costs and capacities. For the stocking yards, we consider the case of Chile, where timber is sent to stocking yards in the summer, and from the stocking yards to plants in the Winter. Since timber deteriorates, inventories at stocking yards must be sent completely to plants in the summer season. Periods will be defined by season, summer and winter.

For the sake of simplicity, we assume only one type of timber. In reality, multiple types of timber or assortments are considered, which can be exchangeable if allowed to downgrade. That is, wide diameter high value logs, normally sent to sawmill can be sent to the pulp mill, or even the energy mills. Opticort, a model representing this more complex reality is presented in Epstein et al. (1999a, b). Also for the sake of simplicity, we assume no capacities on roads. It is however easy to include such constraints. To illustrate a tactical model, consider the following variables and coefficients:

Variables

$x_{it} = 1$, if stand i is harvested in period t; 0 otherwise
$y_{ijt} =$ Volume (m^3) of timber transported from stand i, to plant j, in period t
$z_{irt} =$ Volume (m^3) of timber transported from stand i, to stocking yard r, in period t
$u_{rjt} =$ Volume (m^3) of timber transported from stocking yard r, to plant j, in period t
$v_{rt} =$ Volume (m^3) in inventory in stocking yard r, in end of period t

Parameters

$c_{it} =$ Cost of harvesting stand i, in period t
$e_{jt} =$ Revenue per m^3 at plant j, in period t
$f_{ijt} =$ Cost of transportation per m^3 between stand i, and plant j, in period t
$g_{irt} =$ Cost of transportation per m^3 between stand i, and stocking yard r, in period t
$h_{rjt} =$ Cost per m^3 of transportation between stocking yard r, and plant j, in period t
$b_{rt} =$ Capacity (m^3) of stocking yard r, in period t
$d_{jt} =$ Demand (m^3) of plant j, in period t
$a_{it} =$ Volume (m^3) in stand i, in period t
$l_{rt} =$ Cost for holding inventory in stocking yard r, in period t (per m^3)

Objectives Function: Max Revenue – Costs

Revenue: $\sum_t \sum_j e_{jt} \left(\sum_i y_{ijt} + \sum_r u_{rjt} \right)$

– Cost of harvesting: $\sum_t \sum_i c_{it} x_{it}$

- Cost of transportation from stands to plants: $\sum_t \sum_i \sum_j f_{ijt} y_{ijt}$
- Cost of transportation from stands to stocking yards: $\sum_t \sum_i \sum_r g_{irt} z_{irt}$
- Cost of transportation from stocking yards to plants: $\sum_t \sum_r \sum_j h_{rjt} u_{rjt}$
- Cost of inventory at stocking yards: $\sum_t \sum_r l_{rt} v_{rt}$

Constraints

1. Harvesting at stand i only once
   $\sum_t x_{it} \leq 1$
2. All timber harvested is transported to plants j or stocking yard r in period t
   $a_{it} x_{it} = \sum_j y_{ijt} + \sum_r z_{irt}$
3. Inventory at stocking yard r in period t
   $v_{rt} = v_{r,t-1} + \sum_i z_{irt} - \sum_j u_{rjt}$
4. Capacity of stocking yard r in period t
   $v_{rt} \leq b_{rt}$
5. Volume delivered at plant j must satisfy demand each period t
   $\sum_i y_{ijt} + \sum_r z_{irt} = d_{jt}$
6. $x_{it} \in \{0, 1\}$
   all other variables non-negative

8.4.1 Road Network Upgrading

In the strategic model, there were binary variables associated to use a particular link between forest and plant. This can be used to model both transportation mode and full roads availability. This is true when there is a need of new roads to access harvest areas. However, in many cases the roads do exists but may need to be upgraded to a higher quality. For example, a road may exist and it is possible to drive on the road during seasons of good weather conditions. But when the conditions are bad, the roads are not usable for transport. This is a typical situation in Sweden and Canada in the thawing period in the spring. The roads are then closed for a number of weeks. In order to cope with this situation, there are a few main alternatives. The first is to move logs to roads sites or terminals from where the logs can be transported without any problem. This gives rise to additional loading and unloading and hence extra cost and timber handling damage. The second is to use trucks equipped with central tire inflation (CTI) system where the air pressure can be controlled (e.g. lower pressure to increase the area of contact between the tire and the ground) and hence the trucks can drive on roads impassable for trucks without CTI system. This is restricted to a few types of roads. A third alternative is to upgrade the roads so that they are not (or very limited) affected by the thawing.

Most of the models dealing with road upgrading, as well as building, are integrated with harvesting decisions (e.g. Andalaft et al. (2003)). These models have a medium term horizon but here we will focus on the decisions for road upgrading and harvest planning for the next year. However, in order to make the business decisions to be implemented, we need a methodology to analyze the impact of these business decisions. This can be done through anticipative planning where we include the behavior of other activities given the business decisions. For this reason, we include harvesting decisions over 5 or 10 years as anticipation variables (Frisk et al. 2006). With this time frame, we can guarantee more long term effects like keeping equal average distance from stands to mills, even harvest levels, etc. Moreover, we include industrial demand and the associated transportation flows as anticipation variables. As there are different requirements and restrictions over a year, we also divide each year into seasonal periods. None of the anticipation variables will be implemented in practice as they are only used to measure the impact of the business decisions. For example, transportation flows will be implemented in short term flow models on a monthly basis.

The upgrading problem requires detailed information on the roads and a GIS is required. Often there are many candidate roads or road segment that can be upgraded and the many roads give rise to a very large problem. This is particular true when the network flows are represented by nodes and arcs. This has been identified by e.g. Henningsson et al. (2007). In order to cope with the dimension, it is possible to generate a limited number of origin–destination routes. Thus, instead of using nodes and arcs, we can now use flows on routes which drastically can reduce the model size.

Management Planning in Action 8.2: Use of Backhauling in Decentralized Planning at Holmen Skog

Holmen Skog is a large Swedish forest company that manage more than 1.2 million hectares of productive forest land. The annual harvesting carried out within Holmen Skog is about 2.9 million cubic meters. Besides harvesting its own forest it also trade logs with private forest owners, forest associations and sawmills. Holmen Skog also co-operate with other forest companies. In 1999, Holmen Skog started to develop a web-based system Åkarweb [combination of Swedish word 'åkare' for truck driver/owner and 'web'] to support the transportation planning carried out within the company (Eriksson and Rönnqvist 2003). The transportation work is carried out by a number of independent transport companies and organisations. The development

(continued)

(continued)

focused on a number of issues. The purposes with the system were to enable efficient but decentralized planning that involves several transporters and companies. The actual information on supply, demand and inventories is given online. It also had a map support to show inventories and contact information on any transport order. A very important part was to identify backhauling opportunities between transporters even if it is the transporters themselves that make the decisions.

Backhauling is essentially improvement of two direct flows with different assortments from supply to demand points as illustrated in Fig. 8.3. The problem with backhauling is that the number of potential backhauls becomes very large. As a solution method it is critical to use column generation (Carlsson and Rönnqvist 2007).

During operation about 50 transporters associated with Holmen using the system. In addition there are 10 transporters associated with another forest company using it. In total there are about 180 logging trucks connected (direct or indirect through their transport planners). Once all information is collected each morning, an optimisation system solves a model to find the best possible backhauling trips. These trips are on a daily basis viewed by the users as a support for further planning. The system can be used both as a centralised and decentralised system, currently the latter is used. Given the potential backhauling trips it is then up to the transport planners to use them. This is a decentralised way to coordinate the transportation planning.

There is a number of potential savings using Åkarweb. There is considerably less administration with transportation management and the efficiency in the transportation increased. The experience show that the empty trucking distance for Holmen Skog (in the studied area) is in average 80 km per load. If the transports are optimized with respect to backhauling the empty trucking distance will be reduced with 15 %. The overall cost savings have been computed as 4 % in average. In the same study, interviews with hauliers showed that over 80 % of the users enjoyed working with Åkarweb. They believe Åkarweb simplifies and improves the transport planning and that it gives a good overview of the landings and helps navigation in unknown areas.

8.5 Operational Transportation Planning

Operational transportation decisions involve actual dispatching and scheduling vehicles to perform tasks. While other transportation modes exist, such as trains and barges, trucks constitute the main mode of transport by far, and models developed have centered on them. The tasks are carried out by trucks of different types, differentiated by capacity, power of engine to have flexibility to carry logs of different dimensions and handle different types of difficulty in roads. Different types of trucks are used at different levels in the value chain. It is quite different to transport logs from forests to plants, then products from plants to customers.

As in the tactical model, we will concentrate on the first transportation leg, from forests to first destinations, which is where most of the effort in coordinating and gaining efficiency has concentrated. The most typical problem involves moving timber from origins at forest to destinations. This implies assigning trucks, which may be of different types to these tasks. There are basically two modes of hauling. In one case, trucks go back and forth from origins at forests where they are loaded in one origin with timber which need to be delivered at one destination. This is the case of forests in Chile, South Africa and other countries, where large companies own forests, so the volume at origins is enough to fill trucks. Alternatively, trucks follow a route, collecting timber at different origins and delivering at one or more destinations. This is a typical case of Sweden for example, where timber is produced by small owners, who deliver high value, smaller volumes of timber, and several owners need to be visited to fill trucks. Different models have been proposed for each case. Note that in the case of back and forth trips, the whole day travel of a truck can be defined as a route, alternating between pick up points and deliveries at destinations.

The Vehicle Routing Problem (VRP) in timber transportation is quite different from standard VRP problems found in the literature. Audy et al. (2012a) report a number of attributes that distinguish a PDP in timber transportation from a more general PDP. VRP in timber transportation is a variant of the *pick-up and delivery vehicle routing problem*, more commonly designated a *pick-up and delivery problem* (PDP). Mills and stands may have particular times available for delivery and pick-up and we have so-called time windows. When time windows are used, the problem is called a VRP with Time Windows (VRPTW).

In their classification scheme for PDPs, Berbeglia et al. (2007) differentiate three *structures* to describe the number of origins and destinations of the commodities involved in the PDP. Two of them can be seen in timber transportation depending on whether the supply and demand points are paired (i.e. *one-to-one*) or unpaired (i.e. *many-to-many*). The structure *many-to-many,* in which any site can serve as a source or as a destination for any commodity and the structure *one-to-one,* in which a commodity has a given origin and a given destination. This means that in the *many-to-many* structure, the PDP includes allocation decisions (i.e. which supply points satisfy which demand points in what volume of a given product) in addition to the truck routing decisions. With reference to both structures, we provide a general description of the main attributes defining PDP in timber transportation. In a PDP in timber transportation, a set of vehicle routes must be generated in order to deliver a set of *requests* (one-to-one structure) or to satisfy a set of *demand points* (many-to-many structure) according to a given objective (e.g. total minimum cost and/or total minimum empty driving distance) and subject to a set of constraints. A *request* specifies a volume, an assortment, the site where it is to be picked up (origin) and the site where it is to be delivered (destination). Time constraint(s) can be added onto a request (e.g. a latest delivery time or a time window when the pick-up must be made). A *demand point* is a location requiring specific volume in an assortment group (defined by one or several assortments). To satisfy the set of demand points, a set of *supply points* is available; each supply point is a location that can provide specific volume in an assortment. Both the origin/supply and destination/demand sites can be visited more than once. This is the typical situation as the volume available usually exceeds one truckload. On a planning horizon over a day, the entire demand can be divided into daily minimum and maximum accumulated volume, while the entire supply can be released into daily volume. This allows spreading out the deliveries/pickups at a demand/supply site over the whole planning horizon and, for the latter, representing a daily production (e.g. by a harvest team) at supply site. Transportation priority can be put on e.g. certain urgent requests to deliver or critical supply/demand points to empty/fulfil.

To execute the transportation, a fleet of vehicles is available. This fleet of vehicles may consist of the same (homogenous) or different (heterogeneous) vehicle types, each with a unique set of transportation-relevant characteristics (e.g. capacity, set of assortments allowed to haul, fuel consumption, trucks with or without a crane, set of sites not allowed or impossible to visit). The vehicles are spread throughout a set of sites (multi-depot) or based in only one site (single depot). A route usually starts and ends at the vehicle's depot. For a planning horizon exceeding 1 day, the vehicle may be allowed to come back to the depot (or home base) not fully unloaded (i.e. stay loaded overnight), in which case the delivery must be performed the following day. Multiple pickups may be necessary before the truck is full, which is the typical situation when the harvesting is finished and there is a need to clean off all piles, including some with less-than-truckload size. To fill-up the truck, some piles are subject to a partial pick-up and this complicates the planning process. Different approaches are used to deal with this; most are heuristic based.

A route must respect different time constraints such as vehicle's working hours availability (e.g. to disallow working at night), length of driver's work shift, time windows at supply/demand points, etc. More than one driver's work shift could be scheduled on a vehicle. The change of driver can be performed from among a set of predefined changes over sites or only at the truck's depot. Time windows at supply/demand points consist mainly of two forms: opening hours and on-site loader(s) operation hours. The first specifies the site's opening hours in which a vehicle can perform a pickup/delivery, while the second specifies the hours in which on-site loader(s) are available for (un)loading operations. Vehicle types without crane must be scheduled inside both time windows at any site while usually, vehicle type with a crane (i.e. self-loading) must be scheduled inside both time windows only at delivery site. Waiting time is generally allowed when a vehicle arrives before the beginning of a mandatory time window and waiting time for vehicle queuing can also be computed (e.g. when a vehicle waits for a loader already in use by another vehicle). Rather than specify predefined time windows for on-site loader(s) operation hours, the PDP can also include the scheduling of on-site loader(s). We refer to *multiple time windows* (as in Xu et al. 2003) to designate e.g. the site's daily opening hours that can change according to the day of the week. It is also possible to address queuing of trucks at mill gates, which is typical for large industries with several specialized production lines. In such a case, it is necessary to come up with a good queuing strategy in order to minimise the waiting time in the industry's yard as well as to minimise additional movements in the yard transportation from log-piles to productions lines. An approach based on revenue management principles has been tested in a Portuguese pulp and paper mill by Marques et al. (2012) and this is described in detail below in Management Planning in Action 8.3.

To solve PDP in timber transportation, several planning methods have been proposed in the literature. In the next section, we review a number of solution methods and refer to Audy et al. (2012a) for a exhaustive review.

Linnainmaa et al. (1995) propose a three-phase approach using exact mathematical programming methods and heuristics to generate a weekly truck schedule. Weintraub et al. (1996) propose a simulation-based method described in Management Planning in Action 8.3. A column generation method in which each column corresponds to one feasible route is proposed by Palmgren et al. (2003, 2004) and Rey et al. (2009) to generate a daily truck schedule. McDonald et al. (2010) propose a simulated annealing method in which each new solution (daily routes schedule) generated is evaluated according to four performance metrics. Gronalt and Hirsch (2007) and the third phase of Hirsch (2011) propose a Tabu Search (TS) method to generate a daily route schedule to deliver a set of requests. Flisberg et al. (2009) also propose a TS in the second phase of their method generating routes schedules for up to 5 days. With Rummukainen et al. (2009), Flisberg et al. (2009) propose the two planning methods that support the consolidation of less-than-truckload (LTL) size requests in full (or nearly) truckload-size request. Rummukainen et al. (2009) propose a three-phase method embedding a mixed integer programming (MIP) model, a dynamic programming algorithm and two TS heuristics. El Hachemi et al. (2009) and El Hachemi et al. (2011) propose a two-phase method to solve consecutive daily PDPs from an initial weekly PDP. The first method embeds local search algorithms enhanced with a tabu component and a greedy heuristic. The second method embeds an MIP model and a constraint-based local search model with two solving approaches: an iterated local search algorithm and a hybrid algorithm combining previous iterated local search algorithm and constraint programming.

The availability of information at planning time is an important dimension present in PDP. In *static* problems, all information is assumed to be known a priori, while in *dynamic* problems, information is revealed gradually and/or subject to change over time. Nearly all papers in the literature on VRP in timber transportation address static PDPs, while Rönnqvist and Ryan (1995) and Rönnqvist et al. (1998) are exceptions by proposing a truck dispatching solution method for a dynamic PDP.

However, it is critical to be able to anticipate the future and make a full day plan that can be changed later. A key component for such a system is to be able to re-optimise given a current partial solution. Also, the information handed back to each truck's driver is usually only information about their next trip, as their expected planned route may change after a re-optimisation. To allow decision maker in timber transportation to use computer-based routing methods, decision support systems (DSSs) embedding the planning methods have been developed and deployed in the industry. Audy et al. (2012a) discuss a number of DSSs in timber transportation, including ASICAM discussed below, RuttOpt in Sweden (Andersson et al. 2008) and VTM in Canada (Audy et al. 2013).

Management Planning in Action 8.3: ASICAM, a Forest Truck Scheduling System

ASICAM is a simulation based computational system used successfully by forest firms in Chile and other countries. Traditionally schedules were generated manually, where to simplify the problem, each truck always traveled between one origin and one destination, with no timing considerations. In addition to the loss in optimality due to the manual scheduling, since it took a long time to develop a schedule, it was used through a week or longer. Obviously, since supply and demand for specific logs change every day, a rigid schedule could not lead to good solutions in different days. The consequences of the manual planning were deficient schedules. Demand for some logs could not be satisfied, there was friction among the drivers who competed to do more trips, there was poor coordination with downstream operations and, mainly, the transportation costs increased specially as trucks had to queue for long times due to congestion both at origins and destinations. In 1989, the firms, who had organized a center for innovation jointly through a consortium, asked us to develop a system to improve the truck scheduling. The basis for the system was organizational, a central office planned all trips (origin, destination, time of departure, and since the travel time is known, time of arrival to destination), and the use of a system to schedule the trips. It was based on simulation with heuristics. The model works by time increments (see below Figure). It starts at 6.00 am when the first trucks are being loaded at origins, at 6:20 am the trucks leave for their destinations, to which they will arrive at 6:20 plus the travel time (TT) required. At 6:20 plus TT plus 20 min the trucks are empty ready to travel to a new origin. At 6:20 a second batch of trucks starts being loaded at origins, and leave loaded at 6:40 am. In this form the model advances through the day, traveling between origins and destinations. Note that if a truck reaches an origin or destination with trucks ahead of it, it has to queue. The planning horizon ends at 6 pm, when the transportation system is stopped, or longer if there is overtime. Periods

(continued)

(continued)

were defined of 20 min each, corresponding to loading, unloading time. Travel times were approximated to these 20 min periods.

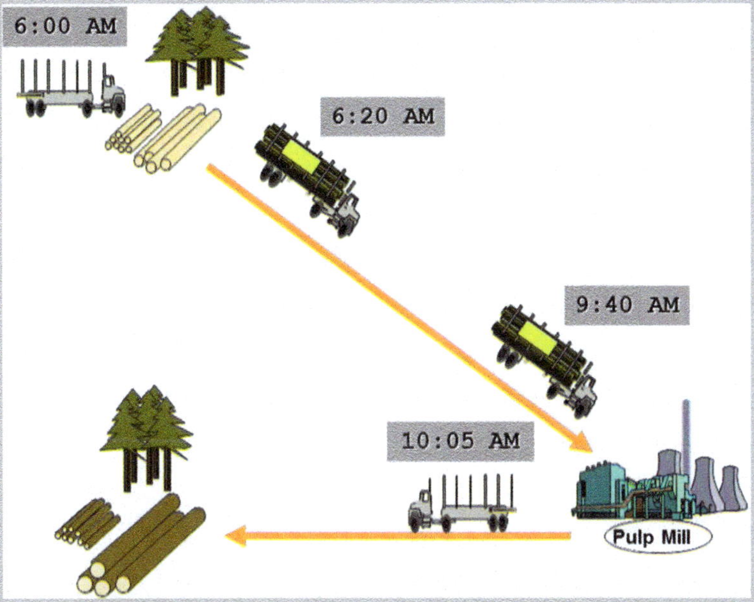

The simulation needs to take into consideration: demand and supply for each type of log at each node, availability of trucks and cranes for loading, unloading times and costs for all activities. There are multiple constraints where the main ones are: demand for each log type at each node, arrival of logs at destinations should be steady to make downstream operation easy, a number of organizational issues must be included, such as lunch breaks, drivers starting and ending their day near their homes, have similar incomes for all trucks in each week, etc.

The main challenge was to find heuristic rules to assign trucks once they become empty. The rules need to consider the constraints as well as minimizing costs. The model considered a horizon of 1 h, to be able to schedule multiple trips jointly, but only implemented the first period of 20 min. A desirability index for each trip was calculated as the real cost of the trip (including queuing) plus a congestion cost, derived by queuing caused to other trucks. Priorities of trips were defined, the main one to fulfill the demand for each log type at each node. Typical forest operations carry out between 100 and 700 trips per day and need between 40 and 250 trucks. The system runs in a few minutes on a regular PC. It is typically run early in the afternoon and the instructions are sent to the drivers by email. There are unexpected events

(continued)

(continued)

during the day, such as breakdowns of cranes, which are handled manually, via radio, by the operators. The system was implemented by all major forest firms in Chile, very successfully. Costs were reduced between 15 and 25 % and the number of truck even more. Average queueing was reduced from 4.5 to 0.5 h. Below the Figure shows how the arrival of trucks to destinations became much more steady.

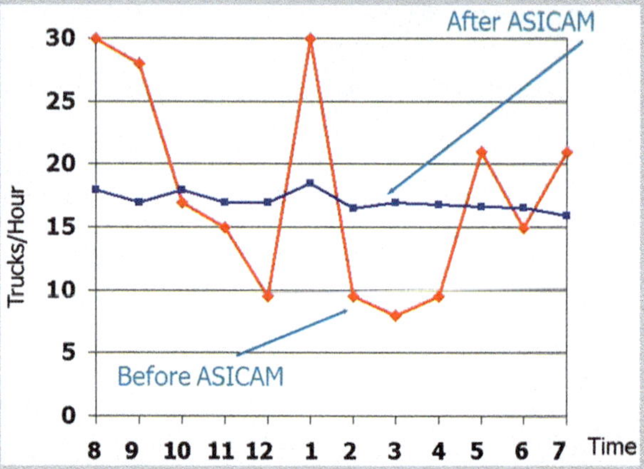

The system was also used by forest firms in South Africa (MONDI won the South African Logistics prize 1996 using ASICAM), Brazil, Uruguay, Venezuela, Argentina. ASICAM, with upgrades is being widely used. ASICAM, with other operational systems, won the INFORMS Edelman Prize Competition in 1998 (Weintraub et al. 1996).

8.6 Real Time Operations – Wood Reception Planning

The timber harvested in industrial plantations is transported by truck directly to the mills or to intermediate stockyards where it may be temporary stored and then transported to the mills. The reception of the wood at the mills and stockyards has been poorly addressed in the literature. Transportation planning is often conducted without taking into account the trucks arrival and entrance at the mill, unloading and internal movements until exiting the facilities. However, during the delivery day, the wood trucks are confronted with congestion and queuing for loading at the harvesting sites and for unloading at the mills. There may be hundreds of trucks arriving at the mill each day, coming from neighboring harvesting sites, with similar schedules and travelling identical routes to the mill. In some cases unloading at the mill during certain rush hours (close to lunch time and at the end of the day) may take up to 4 h when the estimate duration is 40–50 min in non-congestion conditions (Marques et al. 2012). Congestion and queuing particularly at the mill's gate increases the duration and costs of the transportation services and decreases the efficiency of the reception process. Increasing concerns about truck waiting time can be noted in the recent literature with the development of a few routing methods aiming a better coordination with the scheduling of the loading equipment.

In most of the mills, the main decisions related with wood reception planning are taken by the mill receptionist that deals with sequencing the trucks when they actually arrive at the mill's gate and decides its best unloading location. The arrival of the trucks is not known beforehand. Wood reception is often conducted independently from mills production processes. Consequently, the trucks enter the mill on a First-In-First-Out (FIFO) basis according to their arrival time and mainly unload at an active cell of the stockyard. Consequently, other transporters are needed to move the stock to the production lines assuring its continuous operation.

More sophisticated wood reception planning processes may occur when the set of trucks arriving at the mill each day is known beforehand (Marques et al. 2012). The harvest manager may provide information about the trucks expected to departure from each harvest site and destined for the mill. In this case, wood reception planning starts during tactical transportation planning for establishing optimal schedules for the planned trucks in order to avoid congestion.

This decision problem relies in the discretization of the mills opening hours into time slots and consists in assigning the planned trucks to the available time slots, assuring that the arrivals are evenly distributed along the day. The assignment problem may take into account a priority index that reflects the relative importance of each truck. This index may be built upon the historical behavior of the carrier/truck over the last deliveries, its number of next scheduled trips for the same day and/or the freight/truck specific characteristics. The minimum cost assignment problem may foresee a set of time slots that should be used by the unplanned trucks that will arrive at the mill each day carrying market wood.

The time slots assigned to each planned truck should be communicated to the carrier. The transportation contract should foresee benefits for arriving within the reserved slot, which may include reduced waiting time. Such sophisticated wood reception planning processes may further improve the decisions of the mills receptionist during the delivery day. Therefore, the sequence of the trucks may be set according to the previously established priority index and the compliance with the planned schedule.

The wood reception planning process may be integrated with the wood supply to the production lines. The trucks may preferentially unload directly at the production lines, therefore reducing the number of freights coming from the stockyard. Thus, match between the wood assortment carried by the truck and the assortment consumed at the production line is of great importance during the assignment process.

The wood reception problem was firstly presented in Marques et al. (2012). The authors propose to approach this problem with Revenue Management (RM) techniques (Phillips 2005; Quante et al. 2009; Talluri and van Ryzin 2004). The proposed technique consists in classifying the trucks/carriers into priority segments and progressively assigning them to time slots at the gate and at the unloading location, currently available for its segment. This approach was implemented in a three-phase method that runs in distinct time frames.

Before the delivery day, Time Slot Allocation Planning (Phase 1) finds the best time slot/unloading docks for the planned deliveries, minimizing the overall daily reception extra-cost. It acknowledges both the cost of materials handling at the stockyard (i.e. related with the internal stock movements from there to the unloading dock at the line) and the cost of having the trucks stationary inside the mill. Consequently, the direct unloading at the production lines is prioritized.

During the delivery day, the time slots reserved are kept available as long as possible. Each time a truck arrives (planned and unplanned) the Time Slot Order Promising (Phase 2) assigns the truck to the best time slot/unloading dock that are available for its segment. Planned trucks arriving on-time have direct access to the mill. Delayed trucks in higher segments will have a large range of slots available, therefore will have lower waiting time.

Nevertheless, just before starting the time slot at the line, Time Slot Order Allocation (Phase 3) selects the best truck in queue according to the minimum cost criteria. Consequently, if the slot was still reserved for a higher priority truck, it will be assigned to another truck waiting in the queue. The daily delivery schedules obtained with the three-phase solution method were compared with the results of the FIFO approach used today. The daily wood reception costs were reduced in 55 %. Significant reductions were also reported for the maximum waiting time to unload, particularly for the trucks within higher priority segments.

Management Planning in Action 8.4: Wood Reception Planning

The pulpwood reception at a pulp mill was studied by Marques et al. (2012). The case study was based on data from the pulp and paper mill Europac Kraft Viana at Viana do Castelo from the EuroPac Group, located at the northern region of Portugal. The mill produces since the 1980s Kaftliner paper for packaging, using Pinus pinaster and Eucalyptus globulus pulpwood as well as recovered paper. Its annual production rounds up 350 thousand metric tons of paper, consuming about 700 thousand metric tons of maritime pine and 180 thousand metric tons of eucalyptus pulpwood. The majority of the pulpwood comes from the national market. Only a small fraction is produced in self-managed forests. The wood reception at the mill is often conducted without planning in advance. Yet, an average sized pulp mill can receive more than

(continued)

(continued)

120 trucks per day. Total waiting time of over 4 h was reported for trucks arriving during rush hours (9–11 am, 11 am–01 pm and 7 pm–9 pm).

The problem instance included 120 trucks (72 planned and 48 unplanned). Each truck carried one log assortment, consisting in barked or unbarked logs of different volume and density. Their maximum tonnage is limited by national regulation to 40 ton or 60 ton (about 28 or 48 ton of payload). Three distinct wood assortments and six possible unloading docks inside the mill were considered. The time was divided into 496 time slots.

The wood reception process at the mill may be represented by the wood flows over the mill layout. The truck arriving at the mill is placed at the service queue, where it may stay for several minutes or even hours. The operations at the service gate take less than 10 min. The load is weighted and the reception manager decides the unloading location. It can be the stockyard or directly at the production lines. Unloading operations often take only take 10–15 min using a stationary electric crane. When empty, the truck exits on the same gate where a new weighting estimate is recorded. The supply stage assures the continuum operation of the production lines during 10–14 h per day. The line includes an unloading table connected to a rolling runway that forward the pulpwood to the log feeders, where the wood will be mechanically barked and chipped.

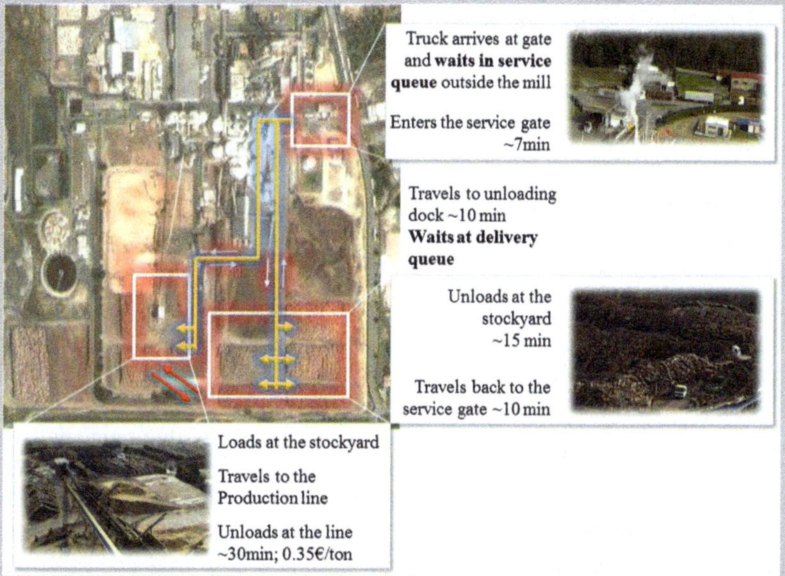

(continued)

(continued)

The proposed solution approach based on Revenue Management principles reduced the queuing effects and led to a 55 % reduction on the daily reception extra-cost for one delivery day at the Portuguese pulp mill, when compared with the FIFO procedure used today. Both the total waiting time to complete the reception process and the materials handling cost were reduced. This method further accomplished the reduction of the waiting time for both the planned deliveries arriving on-time and the trucks with higher priority index.

The implementation of this solution approach in the Wood Delivery System prototype enabled the generation of several "what-if" scenarios, those comparative analysis provided valuable information to design ad plan the reception process under both the carrier's and the mill's perspectives.

8.7 Open Questions and Future Research Guidelines

One key issue for transportation planning is the uncertainty. Even if most models are based on deterministic models there is a need to take into account uncertainty. On a strategic level, the volumes available in the stands are uncertain depending on the growth model and their uncertainty. However, at this aggregated level, it will not impact feasibility as this can be dealt in later tactical and operational models. At strategic level there is also uncertainty about future economic and market variables which may impact decisions taken. For example, sharp increase in oil price or new environmental constraints which compel to change structural options. On a tactical level, the uncertainty can be dealt with keeping a safety stock available either at stands or terminals. Also, it is possible to make sure that the average distance is kept throughout the planning period in order to enable that the transportation capacity is not limited in some periods. On an operational level, it is possible to include also safety stocks or to include extra times to consider uncertainty when deterministic travel times are used. Even if there are ways to cope with uncertainty it will be one of the challenges in the future to improve the planning capability using either stochastic programming and/or robust optimization.

There is a lack of methods to deal with real time operations. Such models will require more information that is updated in real time. Furthermore, the models must include synchronization with loading and unloading as well as considering the queuing issues. Finally, the topics of collaboration between the agents of the wood supply chain are still seldom addressed in the literature. New collaboration strategies and collaborative business models may introduce new objectives and constraints to the existing transportation planning problems (e.g. Audy et al. 2012b).

8.8 Summary

In this chapter, we have described the differences between strategic, tactical and operational transportation planning. As described in Chapter 7, transportation planning is often used to enhance other models. The strategic models often deal with designing the network and transportation mode selection. Hence, such models naturally involve many binary variables. The tactical models are more concerned with allocation rules i.e. how to connect supply and demand points. Important for such models are also seasonality which implies that inventory management is an important aspect. The operational planning typically involves routing and the size of the models increases dramatically. Even more difficult with data and optimality is real time planning as e.g. described for the wood reception problem. In the strategic and tactical models it is often possible to solve them to optimality. This is due to the fact that they are often aggregated and limited in size and the fact that the solution time does not need to be very quick. For the operational models, it is often necessary to develop various heuristic approaches. It is also important to note that these models are often different based on the characteristic of the company and or country.

8.9 Problems

1. Describe how backhauling can be included in tactical transportation planning. List different alternatives in how to generate backhauling routes.
2. Describe which characteristics that will provide different routing problems for forest transportation.
3. Describe and suggest a model when truck transportation and train systems are integrated. In particular suggest how to model the train system with its possible characteristics.
4. Propose a model formulation for including queuing in the routing problem.
5. Describe how different objectives can be included in strategic, tactical and operational planning.

References

Andalaft N, Andalaft P, Guignard M, Magendzo A, Wainer A, Weintraub A (2003) A problem of forest harvesting and road building solved through model strengthening and Lagrangean relaxation. Oper Res 51(4):613–628

Andersson G, Flisberg P, Lidén B, Rönnqvist M (2008) RuttOpt: a decision support system for routing of logging trucks. Can J Forest Res 38(7):1784–1796

Audy J-F, D'Amours S, Rönnqvist M (2012a) Planning methods and decision support systems in vehicle routing problems for timber transportation: a review, CIRRELT research paper 2012-38. CIRRELT, Canada. Montreal, Canada

Audy J-F, Lehoux N, D'Amours S, Rönnqvist M (2012b) A framework for an efficient implementation of logistics collaborations. Int Trans Oper Res 19(5):633–657

Audy J-F, D'Amours S, Rousseau L-M, Favreau J (2013) Virtual transportation manager: a decision support system for collaborative forest transportation, CIRRELT research paper 2013-34. CIRRELT, Montreal

Berbeglia G, Cordeau J-F, Gribkovskaia I, Laporte G (2007) Static pickup and delivery problems: a classification scheme and survey. TOP 15(1):1–31

Carlsson C, Rönnqvist M (2007) Backhauling in forest transportation – models, methods and practical usage. Can J For Res 37:2612–2623

El Hachemi N, Gendreau M, Rousseau LM (2009) A heuristic to solve the weekly log-truck scheduling problem. In: International Conference on Industrial Engineering and Systems Management, Montreal, Canada, 13–15 May

El Hachemi N, Gendreau M, Rousseau LM (2011) A heuristic to solve the weekly log-truck scheduling problem. Comput Oper Res. doi:10.1016/j.cor.2011.02.002 (Forthcoming)

Epstein R, Nieto E, Weintraub A, Chevalier P, Gabarro J (1999a) A system for the design of short term harvesting strategy. Eur J Oper Res 119(2):427–439

Epstein R, Morales R, Serón J, Weintraub A (1999b) Use of OR systems in the Chilean forestry industry. Interfaces 29(1):7–29

Epstein R, Rönnqvist M, Weintraub A (2007) Chapter 20: Forest transportation. In: Weintraub A, Romero C, Bjørndal T, Epstein R (eds) Handbook of operations research in natural resources. Springer, New York, pp 391–403

Eriksson J, Rönnqvist M (2003) Proceedings of the 2nd Forest engineering conference, Växjö, Sweden, 12–15 May 2003. In: Decision support system/tools: transportation and route planning: Åkarweb – a web based planning system, pp 48–57

Flisberg P, Lidén B, Rönnqvist M (2009) A hybrid method based on linear programming and tabu search for routing of logging trucks. Comput Oper Res 36(4):1122–1144

Flisberg P, Lidén B, Rönnqvistm M, Selander J (2012) Route selection for best distances in road databases based on drivers' and customers' preferences. Can J For Res 42(6):1126–1140

Frisk M, Karlsson J, Rönnqvist M (2006) RoadOpt – a decision support system for road upgrading in forestry. Scand J For Res 21(7):5–15

Gronalt M, Hirsch P (2007) Log-truck scheduling with a tabu search strategy. In: Doerner KF, Gendreau M, Greistorfer P, Gutjahr WJ, Hartl RF, Reimann M (eds) Metaheuristics–progress in complex systems optimization. Springer, New York, pp 65–88

Henningsson M, Karlsson J, Rönnqvist M (2007) Optimization models for forest road upgrade planning. J Math Model Algorithms 6(1):3–23

Hirsch P (2011) Minimizing empty truck loads in round timber transport with tabu search strategies. Int J Inf Syst Supply Chain Manag 4(2):15–41

Linnainmaa S, Savola J, Jokinen O (1995) EPO: a knowledge based system for wood procurement management. In: Aikins JM, Shrobe H (eds) Proceedings of 7th conference on innovative applications of artificial intelligence, August 20–23, Montreal, Canada. AAAI Press, Menlo Park, pp 107–113

Marques AF, Rönnqvist M, D'Amours S, Weintraub A, Gonçalves J, Borges JG, Flisberg P (2012) Solving the raw materials reception problem using revenue management principles: an application to a Portuguese pulp mill, CIRRELT research paper 2012-29. CIRRELT, Montreal

McDonald T, Haridass K, Valenzuela J (2010) Mileage savings from optimization of coordinated trucking. In: 33rd annual COFE meeting, Auburn, 6–9 June

Murphy G (2003) Reducing trucks on the road through optimal route scheduling and shared log transport services. South J Appl For 27(3):198–205

Palmgren M, Rönnqvist M, Varbrand P (2003) A solution approach for log truck scheduling based on composite pricing and branch and bound. Int Trans Oper Res 10(5):433–447

Palmgren M, Rönnqvist M, Varbrand P (2004) A near-exact method to solve the log-truck scheduling problem. Int Trans Oper Res 11(4):447–464

Phillips R (2005) Pricing and revenue optimization. Stanford University Press, Stanford

Quante R, Meyr H, Fleischmann H (2009) Revenue management and demand fulfillment: matching applications, models, and software. OR Spectr 31(1):31–62
Rey PA, Muñoz JA, Weintraub A (2009) A column generation model for truck routing in the Chilean forest industry. Inf Syst Oper Res 47(3):215–221
Rönnqvist M (2012) OR challenges and experiences from solving industrial applications. Int Trans Oper Res 19(1–2):227–251
Rönnqvist M, Ryan D (1995) Solving truck despatch problems in real time. In: 31st annual conference of the operational research society of New Zealand, August 31–September 1, Wellington, New Zealand, pp 165–172
Rönnqvist M, Sahlin H, Carlsson D (1998) Operative planning and dispatching of forestry transportation, Research paper LiTH-MAT-R-1998-18. Linköping University, Sweden
Rummukainen T, Kinnari T, Laakso M (2009) Optimization of wood transportation. In: Paper making research symposium, Kuopio, Finland, 1–4 June
Talluri K, van Ryzin G (2004) The theory and practice of revenue management. Springer, New York
Weintraub A, Cholacky A (1991) A hierarchical approach to forest planning. For Sci 39(2): 439–460
Weintraub A, Epstein R, Morales R, Serón J, Traverso P (1996) A truck scheduling system improves efficiency in the forest industries. Interfaces 26(4):1–11
Weintraub A, Epstein R, Murphy G, Manley B (2000) The impact of environmental constraints on short term harvesting: use planning tools and mathematical models. Ann Oper Res 95:41–66
Xu H, Chen Z-L, Rajagopal S, Arunapuram S (2003) Solving a practical pickup and delivery problem. Transp Sci 37(3):347–364

Part II
Advanced Management Planning Topics

Chapter 9
Integrating Management Planning Levels with Decision Support Systems

José G. Borges, Alexandra S. Marques, Ljusk Ola Eriksson, and Manuel F. Marey

9.1 Introduction

The literature reports several approaches to classify computerized tools used to address forest management planning problems (e.g. Borges et al. 2003, 2014a; Gordon et al. 2005; Mower 1997; Nabuurs and Paivinen 1996; Rauscher et al. 2005; Reynolds et al. 2005, 2008; Schuster et al. 1993). Further, it provides several definitions of automated decision support systems (DSS) (e.g. Zahedi 1993; Mallach 1994; Turban and Aronson 2004). In this chapter we build from previous work (Reynolds et al. 2005, 2008) to define as DSS a computerized tool with a graphical user interface that includes a data management module, a growth and yield projection module and a solution module i.e. a module that provides guidance and support to define the timing and the location of industrial forest management options at various temporal scales. We thus understand DSS as a computer application that

J.G. Borges (✉)
Forest Research Centre, School of Agriculture, University of Lisbon, Tapada da Ajuda, 1349-017 Lisbon, Portugal
e-mail: joseborges@isa.ulisboa.pt

A.S. Marques
INESC Tec, Porto, Portugal
e-mail: alexandra.s.marques@inesctec.pt

L.O. Eriksson
Department of Forest Management and Geomatics, Swedish University of Agricultural Science, Uppsala, Sweden
e-mail: Ljusk.Ola.Eriksson@slu.se

M.F. Marey
Department of Agroforestry Engineering, Universidade de Santiago de Compostela, Lugo, Spain
e-mail: manuel.marey@usc.es

J.G. Borges et al. (eds.), *The Management of Industrial Forest Plantations: Theoretical Foundations and Applications*, Managing Forest Ecosystems 33,
DOI 10.1007/978-94-017-8899-1_9,

may encapsulate the information, the models and the methods described in earlier chapters to support strategic, tactical and operational industrial forests management scheduling.

Why such a computer application? The need for automation of strategic, tactical and operational decision processes drove the development of DSS applications to industrial forest management planning. The huge volume of data and information needed to support those processes called for computerized routines to address data processing efficiency requirements. Further efficiency as well as effectiveness requests triggered the adoption by forest managers of automated information systems that might provide timely retrieval and communication of information needed for developing their decision processes. Addressing decision-making effectiveness concerns further with models and methods described in Chaps. 6, 7 and 8 led forest managers to adopt interactive computerized modular structures to formulate and analyze plan proposals at various temporal scales. The usefulness of mathematical programming and simulation approaches as well as of heuristic techniques for industrial forest management scheduling depends on the possibility of its encapsulation in computerized modules within DSS.

In general, as Reynolds et al. (2005) pointed out, the innovation of DSS in forest management has been driven by a combination of factors that include advances in scientific understanding of forest systems and organizational needs for enhanced competitiveness. These two factors have played a key role in the development of DSS to address industrial forests management planning problems, namely by the vertically integrated forest industry. A recent review of forest management computerized tools used world-wide (Borges et al. 2014a) highlighted the role of the forest industry in developing and innovating DSS to address forest management planning problems in countries where industrial plantations are an important segment of the forest sector. The same review reported that the industry does take advantage of the integrated functionality of tools made available by DSS to enhance the effectiveness and the efficiency of industrial forest management planning decision processes, namely to address the integration of strategic, tactical and operational management planning levels.

Earlier computer-based modular structures to support industrial forest management scheduling often addressed the optimization of a single objective (e.g., net present value) under a context of timber flow regulation (Chap. 6). They often focused on stand-level decisions and on the improvement of strategic processes (Marques et al. 2011). The need to address other spatial and temporal scales led to the development of a wide array of new-generation DSS (e.g., Borges et al. 2003; Reynolds et al. 2008; Hetemaki and Nilsson 2005; Wikström et al. 2011). As Marques et al. (2011) pointed out, typically, these were multicomponent systems, with various combinations of models and optimization techniques, supported by database management systems and accessed by spatial and graphical user interfaces. More recently, the focus has been on the development of systems with integration, cooperation, and interoperability mechanisms required to support the interactions among the organizational structures involved in strategic, tactical and operational management scheduling (e.g. Ribeiro et al. 2005; Marques et al. 2010, 2011, 2012) as well as among all parties involved in the production and distribution segments of the wood supply chain (e.g., Weintraub and Romero 2006; D'Amours et al. 2008).

This chapter will discuss briefly the integrated functionality of modules prevalent in most industrial forest management DSS. Nevertheless, its primary focus will be on the architecture (i.e. the design) of DSS that may address the current challenges faced by industrial forest managers and the organizations where they operate. We will argue that DSS architecture approaches may be a critical success factor for the effective use of models and methods available to support and integrate strategic, tactical and operational industrial forests management scheduling as well as supply chain management. Borges et al. (2014a, b) emphasized that DSS architecture and design approaches characterized by a strong involvement of stakeholders are more prone to success. Stewart and Ray (2014) pointed out further that an increased focus on the process of developing DSS with end users is key to build trust and credibility for the DSS produced. We will build from the experience of industrial forest management DSS architectures reported by Ribeiro et al. (2005) and Marques et al. (2010, 2011, 2012) to highlight the key stages of this development process.

What You Will Learn in This Chapter

- The role of decision support systems in industrial forest management planning.
- The concept of architecture of a decision support system
- The stages involved in developing a system to support industrial forest management planning
- The integrated functionality of a forest management decision support system modular structure

9.2 Current Decision Support Systems

The review by Borges et al. (2014a) highlighted that the potential of DSS for integrating strategic, tactical and operational industrial forest management planning or for addressing supply chain management concerns is still seldom realized by currently available systems. Most DSS are used to address a single planning process or a single spatial scale. Moreover, often the development of DSS resulted from the cooperation of researchers and end users in the framework of research and outreach projects (Management Planning in Action 9.1). In fact, in several countries, researchers rank first than managers or consultants as users of DSS that address problems that have as single target the supply of timber Borges et al. (2014a).

DSS for addressing industrial forest management planning share key features of systems used in other sectors (Fig. 9.1) e.g. (i) a language system (LS) that enables users to communicate with and use the DSS (ii) a presentation system (PS) for displaying its outputs (iii) a knowledge system (KS) for storing all the input information and (iv) a problem processing system (PPS) (Burstein and Holsapple 2008; Borges et al. 2014a). As Eriksson et al. (2014) pointed out three of the main components (the LS, PS and KS) are representative systems:

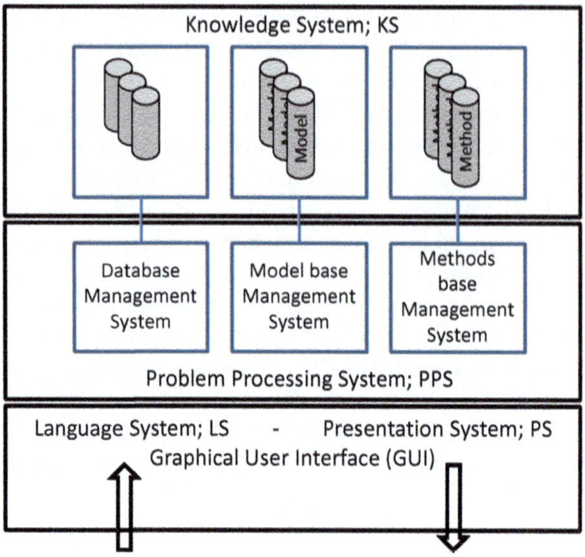

Fig. 9.1 Schematic diagram of the main components of a DSS (Eriksson et al. 2014)

the "communicative" system (LS) consists of all messages to the DSS from the user; the presentation system (PS) consists of all messages from the DSS to the user; and the knowledge system (KS) consists of all knowledge that is collected in the form of data or models held in the DSS. Eriksson et al. (2014) further characterized the components and functionality of each representative system. The knowledge system is sub-divided into three components – one that holds data (e.g. data on the forest concerned), one that holds models (e.g. models for predicting growth and yield) and one that holds methods (e.g. for calculating key statistics or a solver for optimizing a problem). The integrative part of the DSS is the problem processing system (PPS). The task of the PPS is, simply, to solve the problems the user has specified. In doing so the PPS must receive information coming from the LS, integrate the data, models and methods, and communicate the result to the PS.

The DSS used for industrial forest management planning reported by Borges et al. (2014a) are characterized by a wide range of formats and implementations of representative and processing systems. The range of temporal and spatial scales as well as of decision-making contexts addressed by the computerized tools helps explain this diversity. In general, the DSS that address the specificities of tactical and operational forest planning problems (as described in Chap. 8) rely in innovative quantitative approaches, making use of large amounts of information about the implementation of forest operations (e.g. use of available resources, production rates). In this case, the KS typically includes a geographical information system. Further it may include a multitude of sophisticated sensor technologies that collect and display the forest operations updates to facilitate close-to-real-time adaptive/reactive planning of forest harvesting and wood dispatching activities. Nevertheless, unlike DSS designed to address long-term strategic problems its model base does not include in most cases vegetation growth functionalities. This is consistent with the temporal horizon addressed by operational problems.

For example, when reporting the Chilean experience, Epstein et al. (2014) highlighted the development of operational planning applications involving the use of linear programming and heuristics to solve problems like the daily dispatch of trucks, road design, positioning of harvesting machinery in the field and logging. Several authors (e.g. Marques et al. 2014; Rodriguez and Nobre 2014; Keles and Baskent 2014; Epstein et al. 2014; Lamas et al. 2014; Kurttila et al. 2014) reported the use of optimization techniques (e.g. mathematical programming and heuristics) in DSS that address either tactical or strategic industrial forest management planning problems. Further, the model base of DSS designed to address strategic planning always includes either empirical or process-based growth and yield models. These models read the inventory data typically stored in relational databases and are used in conjunction with prescription writers to project stand-level prescriptions and corresponding timber outputs. Matrix writers take these prescriptions to generate the mathematical models used by the methods base. The latter is reported to include also simulation methods in the case of some strategic planning applications.

Management Planning in Action 9.1: The Heureka Decision Support System

Located mainly in the Norwegian spruce and Scots pine dominated boreal and hemiboreal zones and rich in forests (55 % of land area) Sweden is the largest roundwood producer in the European Union. The Swedish forest is intensively used for timber production although nature conservation and environmental considerations have increased during the last decades. In the early 2000s, a program for developing advanced decision support tools to address current management planning challenges was started at the Swedish University of Agriculture Sciences (SLU) (http://www.slu.se/en/). Initially funded by the university, the Heureka research programme was expanded in 2002 and 2005 when the research, the forest, and the environmental sectors joined as co-funders. The first version of the Heureka suite of software was released in 2009 (Heureka 2010). In the last phase, the research program was led by a board with representatives from, among others, SLU and the forestry and the environmental sectors. Reference groups for different user categories were also active.

The success of any forest management planning effort depends on the quality of temporal projections of biological growth of trees and stands. Thus, a basic idea for the Heureka system is to have a suite of models/systems targeting different users and problem areas built on a common kernel of models of tree and stand development. Three main models/systems are available in the Heureka suite: PlanWise for analysis of individual stands (interactive simulation), PlanWise for long term planning at large and small estates

(continued)

(continued)

(includes modules for strategic as well as tactical planning, both supported by optimization procedures) and RegWise for regional analyses (simulation). Besides models for trees and stand development the common kernel contains models for other ecosystem processes as well as economic and social models. The Heureka suite also includes a number of supportive software for, e.g., data import and multi criteria decision analysis (Wikström et al. 2011).

The system architecture chosen aims at a system easy to modify and update (Wikström et al. 2011). The architecture is multi-tier containing three main layers (Base, Domain, and Application layer). In the system, the forest is represented in a hierarchical manner, the *treatment unit* (TU, typically stands) as the bottom node. The TU contains *cells*, linked to *reference units* (real or simulated sample plots) in turn containing information on individual *trees*. Examples of nodes above the TU are estates and regions. The Heureka Decision Support System was selected for presentation in 2013 at a Workshop of the Community of Practice of Forest Management Decision Support Systems (http://forestdss.org/).

Borges et al. (2014b) pointed out that the information regarding the development of the DSS is scarce and is often limited to who developed the DSS and for what purpose. It does not describe the approach followed to design the DSS and it does not characterize the involvement of decision-makers and stakeholders in this architecture. Most development projects that were reported by Borges et al (2014a) focused on DSS to be used by researchers or consultants. This helps explain why many DSS are used by researchers rather than by forest managers. It points further to a potential problem, i.e. the development of several current DSS may not have addressed issues such as robustness or scalability that are key for effective and continuous use by industrial forest managers. The next section will focus on DSS architectural design approaches that may help circumvent system limitations and promote the use of models and methods available to support and integrate strategic, tactical and operational industrial forests management scheduling as well as supply chain management.

Management Planning in Action 9.2: The SADfLOR Decision Support System

SADfLOR started as a DSS developed under Portuguese Research and Technological Development projects coordinated by the Forest Research Center at the School of Agriculture in Lisbon (http://www.isa.ulisboa.pt/cef). Its aim was to demonstrate and apply the potential of current decision-support tools to enhance planning for forest management. Case studies included

(continued)

(continued)

several industrial plantations owned by the main Portuguese pulp and paper firms. Research and development teams included further the Forest Service, the Conservation Agency, forest landowner associations, and other local non-governmental organizations.

As described by Borges et al. (2003) and Reynolds et al. (2008) the prototype system architecture implemented a multiple-criteria spatial DSS. All potential end-users from participating institutions were involved. Key aspects in development were (a) database design and interaction; (b) linkage to growth and yield models; (c) interactive silviculture modeling; (d) design and linkage to a management model base (mathematical representations of scheduling problems in ecosystem management) and (e) techniques for reporting solutions (e.g. Falcão and Borges 2005). The modular structure of the system includes a management information system (MIS), a prescription simulator, a scenario-design and solution module, and a report writer.

The SADfLOR prototype was completed in 1999 and was used successfully in the context of several outreach efforts that targeted among others the forest industry. Namely it was adapted to assist with strategic management planning for the major pulp and paper industry groups (Grupo Portucel Soporcel – see Management Planning in Action 6.1 – Silvicaima and Celbi Stora Enso) in Portugal (e.g. Borges and Falcão 2000).

Addressing issues such as the strengthening of the involvement of the forest industry stakeholders in the development process, robustness and scalability led to the encapsulation of core SADfLOR functionalities by enterprise planning (EA) architectures (Sect. 9.3) in the context of several projects. The first such project focused on the development of an integrated planning system for the pulp and paper industry (Grupo Portucel Soporcel) (Ribeiro et al. 2005). Recently, an EA approach targeted the development of a SADfLOR toolbox to support management planning at a regional scale (Marques et al. 2013) and provided modules to address climate change scenarios in industrial plantations management scheduling (Chap. 10) and to facilitate adaptive management approaches (Rammer et al. 2014). The SADfLOR web-based toolbox was selected for presentation in 2013 at a Workshop of the Community of Practice of Forest Management Decision Support Systems (http://forestdss.org/)

9.3 Industrial Plantations Decision Support Systems Design

The DSS used by large scale forest owners and the industry for industrial forest management scheduling are often characterized by fragmented activity modeling approaches and independent single-activity information systems (Ronnqvist 2003).

As described in Sect. 9.2, the development of these DSS was frequently tailored to meet specific business requirements (e.g. long term planning or operational planning) within research and development projects that did not consider robustness and scalability for continuous use (Marques et al. 2012). Another shortcoming of the design process resulted from overlooking and underestimating the human dimension of information systems. Recent reviews of some of the more important and most recent developments of DSS in forest management, including examples from North and South America, Europe, Africa and Asia (e.g. Reynolds et al. 2008; Borges et al. 2014a; Packalen et al. 2013; Marques et al. 2014) have emphasized the need for a clear focus on the target users.

Addressing the human dimension of information systems, scalability and the frequent need for integrating decision processes may thus be key factors to successful information systems development and use. This will be in turn influential for an effective use of techniques described in earlier chapters to support strategic, tactical and operational decision processes. How to address each factor when developing a DSS will depend on the specificity of the industrial forest management planning context. Nevertheless, according to Borges et al. (2014b), these contexts are typically characterized by regional and forest-level problems that may involve several owners and yet are often managed by a single organization.

In this framework, Enterprise Architecture (EA) Spewak (1992) may provide an interesting platform for developing DSS suitable to most frequent industrial forest management planning contexts. EA emphasizes transparency, maintenance and scalability and is based upon stakeholders' involvement and proper testing. It thus addresses the key factors for successful development and use of DSS in those contexts. Moreover, it is an approach widely tested in several organizations in other sectors to address similar requirements – e.g. governmental organizations (e.g. Martin et al. 2004; Hjort-Madsen 2006), retail industry (Stecher 1993; Vasconcelos et al. 2003), furniture industry (Xu et al. 2007) and the defense electronics industry (Shunk et al. 2003).

Ribeiro et al. (2005) developed successfully an EA approach for specifying the Integrated Forest Management System for a major pulp and paper industry in Portugal (Management Planning in Action 9.3). Marques et al. (2011) extended this approach by including interactive workshops and other complementary participatory techniques (e.g., Kangas et al. 2008) to explicit and document stakeholders' knowledge, concerns, and requirements, thus enhancing their involvement in DSS design for industrial forest management planning. In fact, in both cases forest managers and practitioners as well as Information and Communication Technology (ICT) experts and decision makers' representatives were active members of the EA team, thus collaborating in all EA stages and validating all its outputs.

EA approaches encompass a series of interrelated stages to deduct the requirements of the future DSS from the description of the business processes of the organization. This is influential for ensuring the alignment between the ICT function and the business goals (Sousa et al. 2005). The first stage of an EA approach (Fig. 9.2) to developing a DSS for industrial forest management scheduling – Process Architecture (PA) – focuses on current and future business processes.

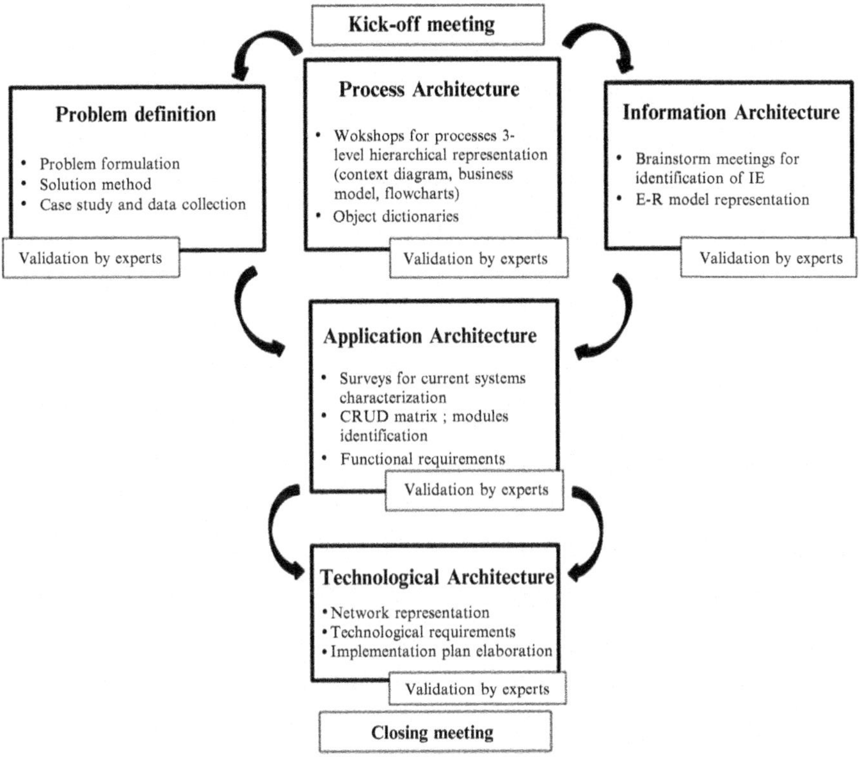

Fig. 9.2 Enterprise architecture components (Adapted from Marques 2012)

The second stage – Information Architecture (IA) – models the core business information, setting the basis for the future database structure. The third stage – Application Architecture (AA) – builds from the previous EA artifacts to identify the DSS modules and the respective functional requirements. The fourth EA stage – Technological Architecture (TA) – characterizes the technological requirements and the system development guidelines for the ICT developers.

9.3.1 Process Architecture

The first stage of EA — Process Architecture (PA) — focusses on modeling the processes to be supported by the DSS. It addresses the identification of the business processes, the characterization of the information needed and the information produced by each business activity and the identification of the actors (e.g. forest managers, ICT experts) responsible for each process as a whole as well as for conducting each activity within each process.

Business processes consist of sequences of interrelated activities that are performed by forest management actors to provide a specific output for an internal or external client. Each activity may be performed manually or may be fully automated within the DSS. The information required and produced by each activity within the business process is stored in the data system of the DSS.

In the case of DSSs for industrial forest management scheduling, the business processes encompass strategic tactical and operational (including transportation) decision processes undertaken by the forest managers, ICT experts and other actors.

This stage is key for the success of the EA approach as it lays the foundations for the development of its other components. It is a participated process involving all actors in the organization for which the DSS is being developed. Ribeiro et al. (2005) and Marques et al. (2011) reported approaches to support the participation of actors that included representatives from the organization administration board, forest planners, forest operational practitioners, ICT managers and forest certification experts. The representation of business processes is typically built over meeting room walls. Specifically, participants are asked to identify the process, actors, activities, and information (i.e. data flows) using different color Post-It notes. The results of the Post-It technique are easily perceived by all participants. The technique further facilitates the discussion and rearrangement of the business representations (Kangas et al. 2008).

A three-level hierarchical top-down structure is often used for the processes representation: the context diagram, the business model and the business processes levels (Marques 2012). The context diagram, as the top level representation, is an organization-centric diagram that presents the relations of the main organization with its customers, suppliers and other external actors For example, Marques et al. (2011) reported the context diagram of a forest management department in a vertically integrated pulp and paper industry, displaying the information exchanged with other departments of the industry as well as outside entities (Fig. 9.3). Often the organization concerns with timber supply are addressed by carrying out management planning activities in both self-owned and other industrial plantations. Moreover, harvesting and transportation operations may be contracted to outside firms. The organization-centric diagram is also very useful to address these contexts. The circular objects in the diagram represent internal and external entities/roles. These are displayed closer or farther from the center according to its business relevance. The connectors in the diagram highlight information flows among entities.

The intermediate level representation, i.e., the business model, identifies the company's business processes that handle both internal information flows (e.g. strategic, tactical and operational planning) and information flows with external entities (e.g. rental agreements, contracting of harvesting and transportation services) grouped according to its physical location (Fig. 9.4). This representation level provides an integrated vision of all ongoing processes to enhance the support to forest management scheduling and decision-making.

The flowcharts that are used as bottom level representations, detail the activities and information flows within each business process. Both activities and information

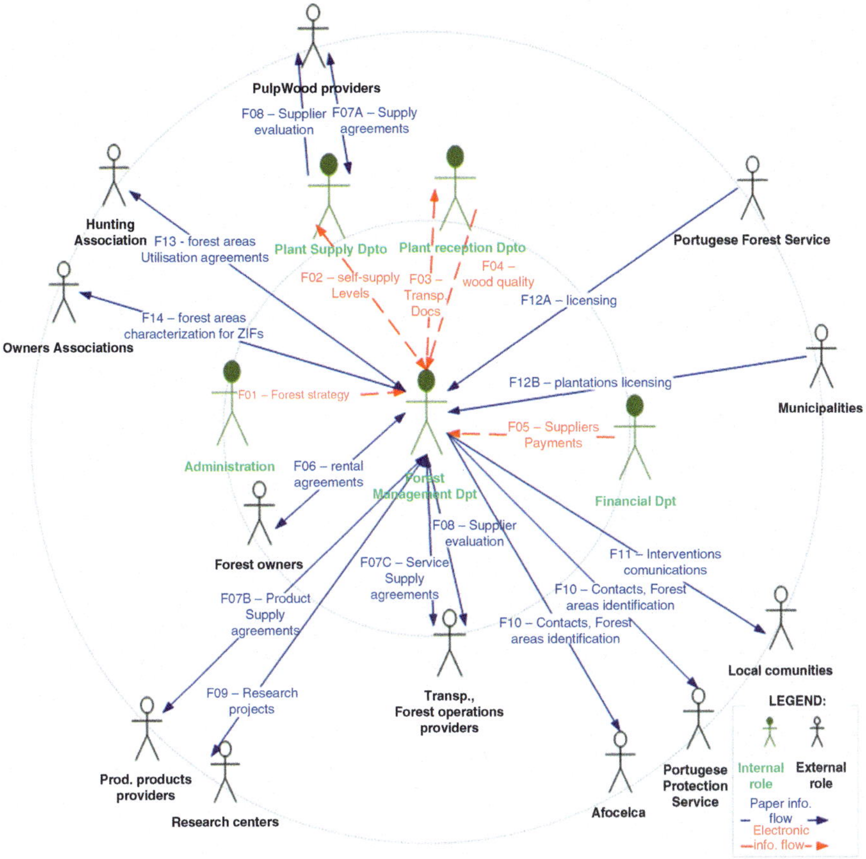

Fig. 9.3 Context diagram identifying the internal and external actors involved in Celbi Stora Enso pulpwood supply management as well as the information flows among them (Marques et al. 2011)

flows are sequenced according to its temporal execution and should follow the Business Processes Management Notation (BPMN). All activities and information flows are further defined in the corresponding dictionaries (Marques et al. 2012).

The first stage of the EA approach further inventories the computerized systems currently used by the landowner or the organization to support their current industrial forest management scheduling processes. This is influential for the diagnostic of their relevance to support the new business architecture as well as for the identification of its links to the new DSS. The PA also provides the basis for identifying the mathematical programming models and solution methods to include in the methods base of the DSS as it includes a comprehensive description of management planning goals and constraints.

BUSINESS PROCESSES

	FOREST PRODUCTION	WOOD LOGISTICS	P&P PRODUCTION	P&P LOGISTICS	P&P SALES
START. PLANNING	• Strategic forest planning • Forest areas registration • Land and wood acquilisitions and rentals	• Wood yard location, layout and capability • Routes and Transportation planning	• Plant location, layout and production capability • Pulb and paper demand and offer estimates	• Warehouses location, layout and capability • P&P distribution network design	• P&P market strategy (market selection, clients segmentation...) • New products and services • P&P pricing • Sales network design • P&P sales volumes estimates
TACT. PLANNING	• Tactical forest planning • Forest inventory planning • Forest roads management (maintanance, new roads) • Work-orders characterization	• Self-owned wood assortment and assignment • Wood deliveries agreements • Suppliers qualification • Equip. Management (trucks, harvesters...)	• Monthly qualified wood supply levels • Plant equipment management (maintanance, acquisitions) • Production factors supply management	• P&P plant stock management • Equip, Management • Transportation fleet management	• Client management (contracts) • Client assignment to sales points • P&P sales points, stock management
OPERAT. PLANNING	• Harvest scheduling • Forest sanity management • Forestry projects elaboration • Work orders adjudication • Logging routes definition	• Transp. scheduling • Crews scheduling • Wood reception at pulp plant • Trucks queue management	• Daily qualified wood supply levels • Daily plant equipment management • Production factors stock management • Working-shifts scheduling	• Internal warehouse management (Picking and packing)	• P&P marketing campains • Client management (P&P requests)
OPS&FIN. LFOLLOW-UP	• On-site forest operations follow-up • Financial follow-up; Wood production costs estimates. • Self-supply wood valorisat • Inventory follow-up	• Transportation follow-up • Wood yard availability follow-up • Wood quality evaluation • Suppliers evaluation	• Plant equip. functioning follow-up; Workers management (absences, alocation to equipments...) • P&P production levels and quality evaluation • P&P prod. costs estimates	• Transport follow-up • P&P stock availability follow-up • Suppliers evaluation	• Client requests reception and follow-up • P&P sales points stock availability follow-up • Clients satisf. Evaluation • Sales points evaluation

SUPPORT PROCESSES

• Human Resources management	• Environmental impact management	• Working Hygiene and Security Management	• Financial management	• IT management

Fig. 9.4 Pulp-paper supply chain process framework based on Celbi Stora Enso forest production and timber logistics business model (Marques et al. 2011)

9.3.2 Information Architecture

The EA second stage – Information Architecture (IA) – focusses on modelling the business information. It aims at aggregating the business information flows into high level information entities that may underlie the future database structure. Again this is a participated process that involves workshops and meetings that bring together forestry and ICT experts to identify the information needed by the organization to conduct industrial forest management scheduling. This is influential to strengthen the awareness about the value of information.

The IA evolves over three main steps (Ribeiro et al. 2005). The first step builds from the PA and analyses further and systematically the business processes' input/output information flows. The information entities involved in these flows are typically grouped to meet the information architecture target of a maximum of 25 Information Entities (IE) per project in order to facilitate the development of the Applications Architecture component (Marques et al. 2010).

Each IE is associated to an identifier and must be managed by a business process which is responsible for performing the respective acquisition, classification, quality control, presentation, distribution and assessment operations.

The second step involves the development of entity-relationship (E-R) diagrams to illustrate IA views from the perspective of business processes. Finally, each entity is characterized by its relevant attributes and an IE data dictionary is organized. For example, Marques et al. (2011) reported that the "stand" IE was characterized by biometric data collected in inventory plots, by type of ownership, and by the past record of forest operations as well as by the products it might provide (e.g. eucalyptus pulpwood and biomass residues). It was characterized further by an annual audit in order to plan forest operations required by the operational and tactical planning processes (e.g. harvest, regeneration, and transportation).

9.3.3 Application Architecture

The EA third stage – Application architecture (AA) – focusses on the identification of the DSS functional modules to support industrial forest management scheduling processes, as well as on the characterization of its requirements. It encompasses four steps (Ribeiro et al. 2005). Firstly, forest managers and practitioners and ICT experts meet in workshops to conduct a systematic analysis of the business processes to be supported by the DSS. This builds from tabular displays of business processes and IE in order to identify the entities manipulated by each activity. Then, the processes flowcharts are revisited in order to identify the relationships (Create, Read, Update, Delete) between each process and each IE to build a CRUD matrix (Marques et al. 2011).

Secondly, the cluster analysis of the resulting CRUD matrices (Fig. 9.5) groups business activities according to the IE they manipulate. In summary, clusters of functionally-related processes are built and the IE are also grouped until all Create, Update, Delete behaviors are displayed diagonally. Proper system consistency and business-ICT alignment rules are applied to identify the DSS functional modular components (Sousa and Pereira 2005). Accordingly, each process is to be supported by just one module and each IE is also to be fully manipulated by just one module (Marques et al. 2012). Interface IE may also be available for reading (R) by other modules.

Thirdly, workshop participants take both the business processes knowledge base and the results of cluster analysis to define the DSS modules to support each process and the way they should interface with each other. At the same time they identify the main data repositories needed. For example, Marques et al. (2011) reported the identification of seven pulpwood supply DSS modules where each module managed ("CRUD") an independent sub-set of IE and had specific functional requirements. The Forest Planning module encompassed further sub modules to support strategic, tactical and operational forest planning processes, a forest equipment management process, as well as a forest operations follow up process for updating biometrical data, events, and forest operations historical records.

Finally, each module is described in detail. This entails the definition of the set of functional requirements to manage its IE life cycle and to automate the

Activity (A) vs Entity (E)	E20 - Wood demand	E21 - Wood park capacity	E22 - Strategic information	E16 - Strategic parameters	E18 - Strategic plan	E23 - Tactical information	E17 - Tactical parameters	E25 - Tactical plan	E19 - Proposed tactical plan
A1 - Agregate strategic information	CR		C						
A2 - Organize strategic constraints				CRUD					
A6 - Reorganize strategic constraints				CRUD					
A3 - Strategic planning			R	R	C				
A4 - Strategic simulation validation					R				
A5 - Strategic simulation analysis					R				
A7 - Analise strategic information					R				
A8 - Organize strategic scenarios					RU				
A9 - Agregate tactical information	CR	CR				C			
A10 - Organize tactical constraints					R		CRUD		
A14 - Reorganize tactical constraints							CRUD		
A11 - Tactical planning						R	R		C
A12 - Tactical simulation validation									R
A13 - Tactical simulation analysis									R
A16 - Analise tactical information									R
A15 - Organize tactical scenarios									RU

Fig. 9.5 Fragment of a CRUD matrix to identify and cluster business processes activities according to the way they manipulate information entities (*C* create, *R* read, *U* update, *D* delete) (Ribeiro et al. 2005)

activities of the business processes supported by the module. It includes further the characterization of its data repositories, graphical user interfaces, and interfaces with other DSS modules as well as the profiling of its end users. Information is also provided about the impact of the proposed DSS modules on computerized tools currently being used to support industrial forest management scheduling. For example, Ribeiro et al. (2005) reported a strategic management planning module that included a sub module to extract the data needed for strategic planning, stored in an information system that was not part of the DSS, a sub module to generate strategic scenarios formulations and to search their optimal solutions and a sub module to compare and analyze solutions proposed for each scenario (Fig. 9.6).

9.3.4 *Technological Architecture*

The EA fourth stage – Technological architecture (TA) – focusses on the technological component of the DSS. It aims at characterizing the DSS technological requirements and producing the system development guidelines. It thus provides the technical foundation to support the BA, IA and AA. For that purpose, this stage encompasses a further detailed and systematic analysis of ICT systems being used

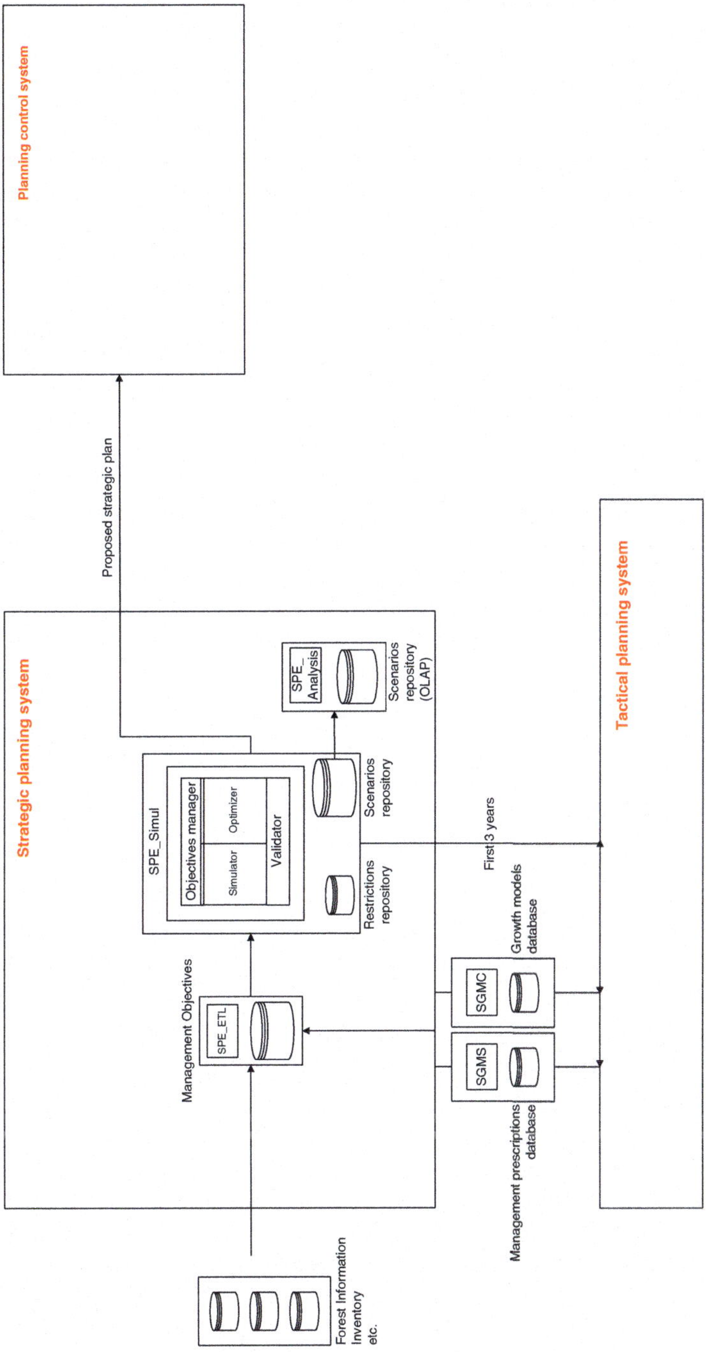

Fig. 9.6 Description of the strategic management planning module and of its interfaces with other DSS modules (Ribeiro et al. 2005)

by the industry or by the landowner. It is at this stage that a decision is made on whether each DSS module characterized by the AA needs to be developed from scratch or else it may be built by adapting ICT systems already available. The TA further inventories and selects relevant technological areas (e.g. network infrastructure, servers and workstations) and recommends technological solutions to support the development of each DSS module as well as its interfaces.

This EA stage is characterized by a strong involvement of the ICT experts in the organization and it is conducted according to general guidelines (e.g. the priority of business needs; the importance of re-using current technology, of using open standard software whenever possible, and of using ergonomic and intelligent human/machine interfaces). For example Marques et al. (2011) reported that the implementation of the industry local area network to support forest management scheduling should consider the adoption of Open Standards, the future expansion of traffic and servers, the adoption of a network management platform to detect and solve network failures, the collection and management of network maintenance data, the development of servers redundancy and other network failures tolerance mechanisms and centralized management of all network infrastructure.

The TA implementation plan provides a schedule as well as detailed guidelines for the development and deployment of all DSS modules to support the industrial forest strategic, tactical and operational management planning processes. Often a Change Management Project may run parallel to all development activities focusing on users training and system promotion across the entire organization (Marques et al. 2011).

Stakeholders are responsible for the validation of the Enterprise Architecture reports at the end of each stage. This may trigger new workshops and the update of the architecture. The EA final meeting is instrumental for suggesting initiatives to overcome potential obstacles to the architecture implementation and for planning the management of the EA initiative with the involvement of the end users.

Management Planning in Action 9.3: The Strategy for Integrating Planning Levels by a Vertically Integrated Pulp and Paper Industry in Portugal

The Portucel Soporcel Group (PSG) is one of the five largest European producers of uncoated woodfree paper (http://www.portucelsoporcel.com/). It is also the first European producer of Bleached Eucalyptus Kraft Pulp. It has a productive capacity that exceeds 1.5×10^6 tons of paper and 1.3×10^6 tons of pulp and has an annual turnover of 1.100×10^6 Euros. The PSG is the largest Portuguese forestry owner. It manages 1.5 % of the Portugal's land, 4.1 % of the country's forests and 15.6 % of its eucalyptus forest. This vertically integrated forest industry manages about 138×10^3 ha of property, of which the eucalyptus plantations occupy 76 %, and it is responsible for a diversity of

(continued)

(continued)

forestry assets spread over 172 local districts from North to South of Portugal. The production is carried out at 3 mills in Portugal with 1,950 employees.

In 2002, one of the Group's priorities was its strategy for the integration of planning levels. This framework prompted the development of a project aiming at the definition of architecture for an Integrated Forest Planning System that might effectively support strategic, tactical and operational planning as well as project and budget control business processes. The project was developed over a period of 3 months. The project team involved both consultants from technological firms (e.g. http://www.link.pt/) and the university (http://www.isa.ulisboa.pt/cef/) and actors in PSG involved in forest management planning. For example, all actors in PSG involved in the business participated in the development of the Business Architecture. Both the information technology (IT) and the forest resources management departments were permanently involved in the Enterprise Architecture planning process. In this project over 60 all-day meetings and workshops took place. The number of persons per meeting ranged from 7 to 10. Some people were present in all meetings (e.g. project manager, forest management planning director, IT director). Some people were present in the meetings when business issues related to their expertise were addressed (e.g. financial area, forest inventory, forest operations).

The Business Architecture provided a knowledge base about current forest planning at PSG. It further defined a model to integrate planning levels where strategic or long-term planning provided the framework for all other business processes. The Information Architecture provided the informational entities needed to support the PSG business processes and their relationships as well as a data dictionary. The Applications Architecture defined the applications to include in the DSS e.g. strategic, tactical and operational decision systems, project and budget monitoring systems, silviculture and growth and yield modelbase management systems and a planning control system. The recommendations of the Technological Architecture included technological solutions to each relevant technological area. The implementation/migration plan identified the impact of the new DSS on current IS in the organization. It presented business-oriented and technical-oriented implementation sequences. The former took into account on-going organizational system changes. It further listed expected benefits of the development of the DSS and its critical success factors. In summary, the EA process was crucial to an accurate understanding of how this forest industry works, enabling business changes to further integrate planning levels. For more details the reader is referred to Ribeiro et al. (2005).

9.4 Summary

This chapter discussed definitions of DSS and argued that the potential of models and methods described in earlier chapters to address industrial forest management planning may be fully realized if they are encapsulated in DSS. It highlighted the contribution of DSS to increase the efficiency and the effectiveness of industrial plantations management planning. A geo-referenced database DSS module is key to store and manage the inventory and topological data needed to address operational concerns in industrial plantations management. Addressing extended planning horizons requires projection capabilities that are made possible by automated simulators and prescription writers in a model base module in a DSS. The proposal of strategic, tactical and operational plans depends further on the automation of solution methods in the methods base module in a DSS. As Reynolds et al. (2005) pointed out the DSS approach has the potential to facilitate good decisions. It contributes to the efficiency of forest management by automating data management processes. Yet it puts emphasis on the improvement of the effectiveness of forest management by better representation of decision-making problems.

This chapter provided an overview of the experience of developing DSS to support industrial forest management planning and argued that DSS may help integrate strategic, tactical and operational management planning of industrial plantations and enhance supply chain management. The fulfillment of these goals depends on the architecture of DSS and the way it addresses the involvement of end users, the integration of planning processes and the robustness and scalability of the system.

An approach – Enterprise Architecture – was presented that may contribute to the success of the use of DSS in most industrial forest management planning contexts. It was shown how this approach might circumvent shortcomings perceived as critical obstacles to the use of DSS by forest managers.

9.5 Problems

1. Define decision support system and identify the modules needed to address industrial forest management planning.
2. How important is the integration of strategic, tactical and operational planning levels?
3. How may DSS contribute to this integration?
4. What are the critical factors for a successful use of DSS in industrial forest management planning.
5. A large-scale landowner hires you to develop a DSS that may help manage his industrial forest. Draft a proposal for the DSS development project.

Acknowledgements The development of this chapter was partially supported by the COST Action FORSYS Forest Management Decision Support Systems (FP0804), by the Erasmus Mundus Master Course 'Mediterranean Forestry and Natural Resources Management' (MEDfOR) and by funding from the European Union's Seventh Programme for research, technological development and demonstration under grant agreements: n° 282887 (INTEGRAL – Future-oriented integrated management of European forest landscapes) and n° PIRSES-GA-2010-269257 (ForEAdapt – Knowledge exchange between Europe and America on forest growth models and optimization for adaptive forestry).

References

Borges JG, Falcão A (2000) Simulação expedita de planos estratégicos para a área de eucaliptal da Stora Celbi, Celulose Beira Industrial, SA. Grupo de Economia e Planeamento em Recursos Florestais, Departamento de Engenharia Florestal, ISA, Lisboa, 490 p

Borges JG, Falcão A, Miragaia C, Marques P, Marques M (2003) A decision support system for forest resources management in Portugal. In: Arthaud GJ, Barrett TM (eds) System analysis in forest resources, vol 7, Managing forest ecosystems. Kluwer Academic, Dordrecht, pp 155–164

Borges JG, Nordström EM, Garcia-Gonzalo J, Hujala T, Trasobares A (eds) (2014a) Computer-based tools for supporting forest management. The experience and the expertise world-wide. SLU University Press (forthcoming)

Borges JG, Nordström EM, Hujala T, Garcia-Gonzalo J, Trasobares A, Rodriguez LCE (2014b) Addressing forest management planning problems with computerized tools. A synthesis of the experience world-wide. In: Borges JG, Nordström EM, Garcia-Gonzalo J, Hujala T, Trasobares A (eds) Computer-based tools for supporting forest management. The experience and the expertise world-wide. SLU University Press (forthcoming)

Burstein F, Holsapple C (eds) (2008) Handbook on decision support systems, International handbooks on information systems series, Handbook on decision support systems, 2. Springer, Berlin

D'Amours S, Ronnqvist M, Weintraub A (2008) Using operational research for supply chain planning in the forest products industry. Infor 46(4):265–281

Epstein R, Parada C, Weintraub A (2014) The design and use of forest decision support systems in Chile. In: Borges JG, Nordström EM, Garcia-Gonzalo J, Hujala T, Trasobares A (eds) Computer-based tools for supporting forest management. The experience and the expertise world-wide. SLU University Press (forthcoming)

Eriksson LO, Garcia-Gonzalo J, Trasobares A, Hujala T, Nordström EM, Borges JG (2014) Computerized decision support tools to address forest management planning problems: history and approach for assessing the state of art world-wide. In: Borges JG, Nordström EM, Garcia-Gonzalo J, Hujala T, Trasobares A (eds) Computer-based tools for supporting forest management. The experience and the expertise world-wide. SLU University Press (forthcoming)

Falcão A, Borges JG (2005) Designing decision support tools for Mediterranean forest ecosystems management: a case study in Portugal. Ann For Sci 62:751–760

Gordon SN, Johnson KN, Reynolds KM, Crist P, Brown N (2005) Decision support systems for forest biodiversity: evaluation of current systems and future needs. Final report, project A10, National Commission on Science and Sustainable Forestry, Washington, DC. URL: http://ncseonline.org/NCSSF/DSS/Documents/docs/NCSSF_A10_final_report_040128sg.doc

Hetemaki L, Nilsson S (eds) (2005) Information technology and the forest sector, IUFRO world series vol 18. International Union of Forest Research Organizations, Vienna, 235 p

Heureka (2010) The Heureka research programme. Final report for phase 2, October 2005 – September 2009. SLU. 119 pp

Hjort-Madsen K (2006) Enterprise architecture implementation and management: a case study on interoperability. In: HICSS 2006—39th Hawaii international conference on systems science, Kauai, p 71

Kangas A, Kangas J, Kurttila M (2008) Decision support for forest management. Springer, New York

Keles S, Baskent E (2014) The design and use of forest decision support systems in Turkey. In: Borges JG, Nordström EM, Garcia-Gonzalo J, Hujala T, Trasobares A (eds) Computer-based tools for supporting forest management. The experience and the expertise world-wide. SLU University Press (forthcoming)

Kurttila M, Heinonen T, Hujala T, Kangas A, Nuutinen T, Pukkala T, Pykäläinen J, Rasinmäki J, Tikkanen J (2014) The use of forest decision support systems in Finland. In: Borges JG, Nordström EM, Garcia-Gonzalo J, Hujala T, Trasobares A (eds) Computer-based tools for supporting forest management. The experience and the expertise world-wide. SLU University Press (forthcoming)

Lamas T, Nordstrom E-M, Sallnas O (2014) The design and use of forest decision support systems in Sweden. In: Borges JG, Nordström EM, Garcia-Gonzalo J, Hujala T, Trasobares A (eds) Computer-based tools for supporting forest management. The experience and the expertise world-wide. SLU University Press (forthcoming)

Mallach E (1994) Understanding decision support systems and expert systems. Irwin Inc., Chicago, 695 p

Marques AF (2012) New decision support tools for forest tactical and operational planning. PhD thesis, Instituto Superior de Agronomia, Universidade Técnica de Lisboa, 69 pp

Marques AF, Borges JG, Sousa P, Fonseca M, Garcia R, Batista G (2010) Applying enterprise architecture to the design of the integrated forest products supply chain management system. In: Varajão JE, Cruz-Cunha MM, Putnik GD, Trigo A (eds) Enterprise information systems international conference CENTERIS2010, Viana do Castelo, Portugal, October 20–22, 2010, Part II, Communications in computer and information science 110, Part 1. Berlin/Heidelberg, Springer, pp 32–40. doi:10.1007/978-3-642-16419-4_4

Marques A, Borges JG, Sousa P, Pinho AM (2011) An enterprise architecture approach to forest management decision support design. An application to pulpwood supply management in Portugal. Eur J For Res 30:935–948

Marques AF, Borges JG, Sousa PM, Fonseca M, Gonçalves J, Oliveira J (2012) An enterprise architecture approach for designing an integrated wood supply management system. In: Varajão J, Cruz-Cunha MJ, Trigo T (eds) Organizational integration of enterprise systems and resources: advancements and applications. IGI Global, Hershey, pp 1–22. doi:10.4018/978-1-4666-1764-3.ch001

Marques AF, Borges JG, Garcia-Gonzalo J, Lucas B, Melo I (2013) A participatory approach to design a toolbox to support forest management planning at regional level. For Syst 22:340–358

Marques A, Garcia-Gonzalo J, Borges JG (2014) Computer based tools for supporting forest management in Portugal. In: Borges JG, Nordström EM, Garcia-Gonzalo J, Hujala T, Trasobares A (eds) Computer-based tools for supporting forest management. The experience and the expertise world-wide. SLU University Press (forthcoming)

Martin N, Gregor S, Hart D (2004) Using a common architecture in Australian e-government: the case of smart service Queensland. In: Proceedings of the 6th international conference on electronic commerce, ACM, Delft, The Netherlands, pp 516–552

Mower HT (1997) Decision support systems for ecosystem management: an evaluation of existing systems, General technical report RM-GTR-296. USDA Forest Service, Fort Collins, Colorado, USA, 154 p

Nabuurs G, Paivinen R (1996) Large scale forestry scenario models – a compilation and review. EFI working paper 10, 174 p

Packalen T, Marques AF, Rasinmaki J, Ferreti F, Mounir F, Rodriguez LCE, Nobre S (2013) Review. A brief overview of forest management decision support systems (FMDSS) listed in the FORSYS wiki. For Syst 22(2):263–269. doi:10.5424/fs/2013222-03192

Rammer W, Vacik H, Palma J, Garcia-Gonzalo J, Borges JG, Lexer MJ (2014) A web-based ToolBox approach to support 1 adaptive forest management under climate change. Scand J For Res (forthcoming)

Rauscher HM, Reynolds KM (2005) Decision-support systems for forest management. Comput Electron Agric 49(1):1–5

Reynolds KM, Borges JG, Vacik H, Lexer MJ (2005) Information and communication technology in forest management and conservation. In: Hetemaki L, Nilsson S (eds) Information technology and the forest sector, vol 18, IUFRO world series. International Union of Forest Research Organizations, Vienna, pp 150–171

Reynolds KM, Twery M, Lexer MJ, Vacik H, Ray D, Shao G, Borges JG (2008) Decision support systems in natural resource management. In: Burstein F, Holsapple C (eds) Handbook on decision support systems, International handbooks on information systems series, Handbook on decision support system, 2. Springer, Berlin, pp 499–534

Ribeiro RP, Borges JG, Pereira CM, Sousa PM, Lé JP (2005) Designing an integrated forest planning system for the forest industry: an application in Portugal. In: Bevers M, Barrett TM (tech. comps.) Systems analysis in forest resources, General technical report PNW-GTR-656. Proceedings of the 2003 symposium, U.S. Department of Agriculture, Forest Service, Pacific Northwest Research Station, Stevenson/Portland, 7–9 October 2003, pp 89–96

Rodriguez LCE, Nobre S (2014) The use of decision support systems in Brasil. In: Borges JG, Nordström EM, Garcia-Gonzalo J, Hujala T, Trasobares A (eds) Computer-based tools for supporting forest management. The experience and the expertise world-wide. SLU University Press (forthcoming)

Rönnqvist M (2003) Optimization in forestry. Math Program Ser B 97:267–284

Shunk D, Kim J, Nam HY (2003) The application of an integrated enterprise modeling methodology—FIDO—to supply chain integration modeling. Comput Ind Eng 45:167–193

Schuster E, Leefers L, Thompson J (1993) A guide to computer-based analytical tools for implementing national forest plans, General technical report INT-296. USDA Forest Service, Ogden, Utah, USA, 269 p

Sousa P, Pereira C (2005) Enterprise architecture: business and IT alignment. In: Proceedings of the 2005 ACM symposium on applied computing, Santa Fe, New Mexico, pp 1344–1345

Sousa P, Pereira C, Alves Marques J (2005) Enterprise architecture alignment heuristics. Microsoft Archit J. http://msdn.microsoft.com/en-us/library/aa480042.aspx

Spewak S (1992) Enterprise architecture planning: developing a blueprint for data, applications and technology. Wiley-QED Publication, London

Stecher P (1993) Building business and application systems with theretail application architecture. IBM Syst J 32:278–306

Stewart A, Ray D (2014) Decision support systems for the sustainable management of forests in Britain. In: Borges JG, Nordström EM, Garcia-Gonzalo J, Hujala T, Trasobares A (eds) Computer-based tools for supporting forest management. The experience and the expertise world-wide. SLU University Press (forthcoming)

Turban E, Aronson J (2004) Decision support systems and intelligent systems, 7th edn. Prentice-Hall, Upper Saddle River, New Jersey, USA, 890 p

Vasconcelos A, Sousa P, Tribolet J (2003) Information system architectures. In: Proceedings of business excellence 2003, International conference on performance measures, benchmarking and best practices in new economy, Guimarães, June 2003

Weintraub A, Romero C (2006) Operations research models and the management of agricultural and forestry resources: a review and comparison. Interfaces 36:446–457

Wikström P, Edenius L, Elfving B, Eriksson LO, Lämås T, Sonesson J, Öhman K, Wallerman J, Waller C, Klintebäck F (2011) The Heureka forestry decision support system: an overview. Math Comput For Nat Res Sci (MCFNS) (in press)

Xu K, Liu L, Wu C (2007) A three-layered method for business processes discovery and its application in manufacturing industry. Comput Ind 58:265–278

Zahedi F (1993) Intelligent systems for business. Expert systems with neural networks. Wadsworth Publishing Company, Belmont, 658 p

Chapter 10
Addressing Risk in Forest Management Planning

Jordi Garcia-Gonzalo, Liliana Ferreira, and José G. Borges

10.1 Introduction

Risk and uncertainty have been used as equivalent terms. Samson et al. (2009) indicated that these two terms may be considered to be equivalent or separate concepts depending on the field and context in which they are used (e.g., engineering, operations research, finances). There are many definitions, but uncertainty may be defined as the lack of information (Rowe 1994). Uncertainty implies that in a certain situation a person does not dispose about information which quantitatively and qualitatively is appropriate to describe, prescribe or predict deterministically and numerically a system, its behavior or other characteristics Zimmermann (2000).

The word uncertainty has been equated with random variability, and it has been claimed that the classical probability theory will be sufficient for each situation (see Ferson and Ginzburg 1996; Zimmermann 2000). Risk has been defined as the expected loss due to a particular hazard for a given area and reference period (United Nations 1992). An expected loss is the product of the damage and its probability. And the damage is loss expressed in monetary terms. Example: The probability of a maritime pine stand to burn increases with stand density (Garcia-Gonzalo et al. 2012), whereas the damage is the decrease in the log price or the loss of revenues that results from non-salvageable timber. *Risk management* includes strategies and actions for reducing risk (Hollenstein 1997).

J. Garcia-Gonzalo (✉) • J.G. Borges
Forest Research Centre, School of Agriculture, University of Lisbon, Lisbon, Portugal
e-mail: Jordigarcia@isa.ulisboa.pt; joseborges@isa.ulisboa.pt

L. Ferreira
School of Technology and Management, Polytechnic Institute of Leiria, Campus 2, Morro do Lena-Alto do Vieiro, Centro de Investigacão Operacional, Leiria, Portugal
e-mail: liliana.ferreira@ipleiria.pt

J.G. Borges et al. (eds.), *The Management of Industrial Forest Plantations: Theoretical Foundations and Applications*, Managing Forest Ecosystems 33,
DOI 10.1007/978-94-017-8899-1_10, © Springer Science+Business Media Dordrecht 2014

Some authors have differentiated between endogenous and exogenous risk, for example, Thorsen and Hells (1998) divided the risk into exogenous or endogenous to the decision of the owner or manager. Endogenous would be when the forest manager or owner can control or modify the level of risk to which a stand is exposed through stand treatments (e.g. changing the stand structure through thinnings). Exogenous risk refers to factors that the forest owner cannot control, such as climate, features of surrounding. This nomenclature has been already adapted by many authors (e.g. González et al. 2005a, b; Garcia-Gonzalo et al. 2011a, b; Ferreira et al. 2012).

Under risk and uncertainty, the state of nature that would prevail is not known with certainty. Under risk, the probability of each state of nature occurring and, correspondingly, the probability distribution of consequences are known; otherwise, the decision is made under uncertainty (Kangas and Kangas 2004). The subject of risk analysis is a given system which includes valuable objects that might be lost or damaged. The system needs to be defined in terms of time and space and the seriousness of the expected hazard.

As forest planning is characterized by the long-term horizon of its outcomes; uncertainty and risk are thus closely related to the development of forest management plans. In addition, the consequences of a decision depend on many uncontrollable variables. These unknown factors determine the future state of nature, which in turn determine the outcome of the decision. Among these sources of uncertainty we find uncertainty related to the development of the trees, the performance of timber markets, the occurrence of catastrophic events (e.g. windthrows, wildfires, pests), or even to the preferences of the forest owner. On the other hand, as forest planning is based on inventory data, if this inventory is not accurate it becomes a source of risk.

Forest planning is ineffective if it ignores these sources of uncertainty and risk. The main reason is that projections of forest conditions and outcomes done during the planning process may be very different if they include risk. This would lead to overestimation of incomes or may lead to vulnerable forests to certain threats (e.g. fire, pests). Thus, optimizing forest management decisions without considering risk and uncertainty would lead to non-optimal decisions or even bad decisions (Pukkala 1998). For example, timber harvest in a forest area with a high wildfire risk would be much less than potential timber harvests if fire is ignored. This is especially important in industrial plantations where the decisions rely on forest productivity predictions. For example, in Portugal Eucalypt plantations covers around 23 % of the total forest area (AFN 2010) providing raw material for the pulp and paper industry. Wildfires are the most severe threat to eucalypt plantations and therefore, forest managers must contend with wildfire risk when developing and implementing forest management plans.

In this context, hazard and potential damage models that relate stand characteristics to the level of occurrence probability and the potential damage are a good tool to integrate risk in forest management (e.g. Schelhaas et al. 2003; González et al. 2006; Garcia-Gonzalo et al. 2012; Botequim et al. 2013; Marques et al. 2011a; Marques et al. 2012). These models allow computing the probability of outcomes and then they allow calculating expected values and they have included risk of sudden destruction in forest planning calculations (González et al. 2005a; Garcia-Gonzalo et al. 2012; Ferreira et al. 2011, 2012).

The variety of planning models is great, but applications of risk analysis are not very common in forest planning (Pasalodos-Tato et al. 2013). This chapter will show some examples on how to include uncertainty and risk in forest planning problems. For a more complete review the readers are referred to Yousefpour et al. (2012) and Pasalodos-Tato et al. (2013).

What You Will Learn in This Chapter

- Understand the concept of risk and uncertainty and understand their different sources.
- Understand how to integrate risk and uncertainty in forest planning.
- Understand when to use empirical or process based models depending on the source of uncertainty.
- Examples on the use a stand-level management optimizer that addresses risk and uncertainty.

10.2 Forest Growth Under Risk and Global Change Scenarios

Climate change may impact substantially the forest sector. Several studies point out to the warming of winters and to the increase of both the length of the dry season and the frequency of extreme events like forest fires in several parts of the world (Giorgi et al. 2004; Kjellström 2004; Moriondo et al. 2006; Christensen and Christensen 2007; Granier et al. 2007). This will impact growth and survival of plants as well as their geographical distribution and the composition of plant communities.

Several models are available, either empirical or process based to project forest growth and timber yields (Chaps. 3 and 4). Nevertheless, empirical models are not suitable for predicting the impact of climate change on the forest growth (Chap. 4). This is because they are based on the assumption that future conditions will stay the same as in the past. Thus, these models become inefficient of just not usable to support decision-making under environmental change scenarios. In the present context of climate change the process based models are becoming more popular in order to accommodate the effect of varying climate, even if the prediction ability of this type of models has not been as thoroughly validated as most growth and yield models have been (see Chap. 4 for more details).

Addressing climate change is a challenge to forest managers. This topic is gaining importance. For example, Nuutinen et al. (2006) used empirical tree-level models with transfer variables (Matala et al. 2005, 2006) estimated by the process-based model FinnFor (Kellomäki et al. 1993; Kellomäki and Väisänen et al. 1997) to include climate change in large-scale forest scenario analysis. More recently,

Garcia-Gonzalo et al. (2008) combined the use of Finnfor, a database, a wood products model and optimisation techniques to analyse forest management plans. The integration in DSSs of forest growth and yield models sensitive to environmental changes is gaining importance. As an example (Garcia-Gonzalo et al. 2014) have developed an innovative DSS (SADfLOR v ecc 1.0) to incorporate and test the use of a process-based model which is sensitive to environmental conditions, Glob3PG, as a vegetation growth projection tool. This DSS integrates four independent and compatible modules: (i) a management information module, (ii) a prescription generator module including a process-based model (GLOB3PG) allowing the users to generate and explore the outcomes of different forest prescriptions, (iii) a decision module to develop mathematical model which is solved, (iv) a reporting module to view and report results.

The use of this process-based model allows: (i) simulating the effect of intensive silviculture practices (i.e. initial stand density, stool selection), (ii) simulating growth under climate change and (iii) providing detailed stand structure information (diameter distribution, merchantable volumes to any top diameter).

In order for growth and yield models to be used effectively, they must be programmed and integrated within computer-based decision systems (Chap. 9). The DSSs that integrate forest growth and yield models sensitive to climate change may need specific information that may not be needed in classical systems. Typically the DSS will need much detailed information such soil information and climate data on their data component. Solutions proposed by a DSS are thus linked to specific climate change scenarios.

Tools to compare and analyse differences between management plans for each climate scenario are needed to generate as much information as possible for the decision maker. For example, one useful analysis is to check the performance of applying a management plan developed for current climate conditions but using a climate change scenario. The difference between this and the optimal plan for climate conditions would give opportunity cost of not adapting your management plan.

Nevertheless, empirical models may be used depending on the objectives of the management planning exercise and the source of uncertainty and risk. For example, if a management plan is mainly driven by economics (changes in wood price or tax rate) and no information or no importance is given to changes in environmental conditions, the decision between either empirical or process based model would be chosen based on the ability to predict accurately the economic returns as a consequence of the harvesting plan. However, when environmental aspects, e.g. nitrogen leaching, soil carbon storage, water recharge, etc., have to be considered in the management plan; it becomes important that a process based model should be used to provide the information needed for the solution space.

As explained in the introduction, risk may be linked to sudden destructive events (i.e. fire, windthrows, pest's attacks). In this case, models to predict the probability of occurrence of these events and the expected damage if the event occurs are needed. Combining these models with growth and yield models (see Sects. 10.3

and 10.4 for more details) during the simulations would generate risk scenarios in order to compute expected values.

What is an expected value? Expected value is a weighted average approach that involves multiplying each possible outcome in a situation with its probability to arrive at the expected outcome. Given a possible set of outputs from an hypothetical risky investment (X) with values $x_1, x_2, x_3, \ldots, x_n$, and respective probabilities of $p_1, p_2, p_3, \ldots p_n$, the expected value of X is given by the formula:

$$E(X) = x_1 p_1 + x_2 p_2 + x_3 p_3 + \cdots + x_n p_n. \quad (10.1)$$

Incorporating fire disturbance into forest management planning is a challenging topic because of the variability in the sources, the frequency, recurrence and severity of these events.

Management Planning in Action 10.1: A Participatory Approach to Design a Toolbox to Support Forest Management Planning Under Changing Climate at Regional Level

The Chamusca county is a rural and low population density municipality, extending over 74,599 ha in the Central Portugal. Forests extend over 51 % of the county territory where eucalypt and maritime pine plantations occupy over 62 % of the county forest area while cork and holm oak multi-functional forests occupy 25 % of this area. The remaining 3 % corresponds to protection areas. The forestland is predominantly private which significantly influences the management. Around 73 % of the area is managed by pulp and paper companies and a few large-scale non-industrial private forestland owners (NIPF) while the remaining area is managed by held by more than 2,200 NIPF which include small holdings with less than 1 ha. These stakeholders can act individually or grouped into forest associations and federations. In this area, ACHAR is the local forestland owners association (http://www.achar.pt/).

One of the concerns of the different stakeholders in the region is (e.g. ACHAR forest owners association) is the need to integrate models and tools to support management planning under global change scenarios. This framework prompted the development of a project aiming at the definition of architecture for a Forest Planning System that might effectively support strategic and tactical processes. The project team involved consultants from universities and actors in the Chamusca region involved in forest management planning. For example, different forest owners, involved in the business participated in the development of the Business Architecture. In this project some all-day meetings and workshops took place. The number of persons per meeting ranged from 3 to 7. This regional interaction network involves decision processes and information shared by 22 entities clustered into 13

(continued)

(continued)

stakeholders groups, including forestland companies and private forestland owners – acting individually or grouped into associations and federations -, national and regional offices of the forest authority, forest services providers, nongovernmental organizations and research centers.

The Business Architecture provided a knowledge base about current forest planning at Chamusca. The Information Architecture provided the informational entities needed to support the Chamusca business processes and their relationships as well as a data dictionary. The Applications Architecture defined the applications to include in the DSS e.g. strategic and tactical decision systems, silviculture and growth and yield model base management systems.

In this context the need for a process-based model calibrated for the main species growing in the region (i.e. maritime pine, eucalypt and cork oak is a priority). The use of a process based model is key to be able to make projections of forest growth under changing climatic conditions. These growth and yield simulators should be linked to a prescription writer that allows generating prescriptions to be tested and simulated. Thereafter a module to generate mathematical management models is needed in order to generate optimal management plans for the different climate scenarios simulated. Finally, a module to analyze results both numerically and graphically is needed.

10.3 Addressing Risk in Stand-Level Management Planning

In general, the objective of stand-level management planning is to determine the optimal combination of operations (e.g. thinnings, clear felling), the timing of entries, and the intensity and type of cuttings applied to a stand to best meet the management goals. This is done for each stand individually without considering neighboring stands. Different methods have been used to optimize stand management. Typically, this encompasses an automated search for the best combination of activities.

Classical planning models may be adapted in order to incorporate risk and uncertainty in the planning process. For example, in the Mediterranean area, if one observes the number of hectares burnt in forest plantations every year and the consequent loss of timber, the need to address fire risk in the management planning is evident. In fact, a catastrophic event may invalidate any forest prediction done based 100 % on growth and yield models without taking into account catastrophic events. However, incorporating stochastic (random) events in strategic planning problems is an interesting and difficult task. First of all, we may define what a stochastic event is. An event is defined as stochastic when we cannot predict with enough accuracy when and/or where this event will occur. This is the case of forest fires, windthrow events, insect outbreaks... The question decision makers may be more interested is how they should change or adapt their management in order to make their forests more resistant to these events. This calls for simultaneous optimization of rotation age, stand treatments and risk mitigation decisions (e.g. Reed 1987; Thorsen and Hells 1998; Amacher et al. 2005; González et al. 2005b, 2008; Pasalodos-Tato et al. 2009; Ferreira et al. 2011a, b; Garcia-Gonzalo et al. 2012).

Some former examples on the integration of risk in stand-level optimization are Kao (1982), Cauldfield (1988) and Valsta (1992). More recently, research on stand-level management scheduling models has addressed the need to treat risk as endogenous to stand management (Thorsen and Hells 1998; González et al. 2005a, b; Ferreira et al. 2012; Garcia-Gonzalo et al. 2012). This means that risk and damage depends on the management decisions that affect stand characteristics. In this context, recent research has focused on the development of planning oriented wildfire occurrence and damage models. These models use as predictors standard tree variables that are easy to obtain and that are under control of the forest manager (e.g. stand density may be changed through thinnings).

In order to develop these planning-oriented models inventory data, before and after the catastrophic event (e.g. fire, windthrow) is needed. Ideally, in addition to climatic data and physiographic data, one should include biometric data including variables (e.g. stand density) which may be controllable through management by the forest manager. This is especially important to be able to compute the effect of management operations on the level of risk and damage. Information of the event (e.g. fire location, perimeter of the fire, ignition point...) is also very important.

As an example, Marques et al. (2011b) and Garcia-Gonzalo et al. (2012) used logistic regression methods to develop models to predict the probability of fire occurrence and the expected damage if a fire occurs in even aged Maritime pine stands in Portugal. Thus, wildfire and inventory data analysis provided the models to estimate the probability of fire occurrence in a stand according to composition, vertical structure and biometric variables. This was instrumental to understand the effect of stand characteristics on wildfire occurrence probability. Wildfire and inventory data analysis further provided models to estimate the proportion of trees in a given stand that will die as a consequence of a wildfire event and to understand how stand-level variables impact mortality. This type of models thus provide information about the impact of management options on wildfire occurrence probability (González et al. 2005b; Marques et al. 2012; Garcia-Gonzalo et al. 2012; Botequim et al. 2013) and post-fire mortality (González et al. 2005b; Garcia-Gonzalo et al. 2011a; Marques et al. 2011b) in the main forest cover types in Portugal. Thus, in the framework of forest management planning, these models may be used to generate stand-level wildfire scenarios and the expected values associated to fire scenarios.

Projections by vegetation growth and yield models of conditions and outcomes of interest associated with each stand-level prescription under wildfire scenarios are thus made possible. In this context, including these equations in a simulation/optimization system would allow to examine likely effects of stand treatment programs on wildland fire effects on the stand and its expected net returns. Therefore, it allows for scheduling of fuel and harvesting treatments taking into account stand and/or landscape-level objectives (see Sects. 10.3 and 10.4).

10.3.1 Simulation

One of the most used techniques to introduce risk in stand-level management planning is the inclusion of random disturbance events (e.g. fire, windthrows) during the simulation of stand development. Then by repeating a considerable number of simulations with these random events an expected value of the outputs from the simulations can be computed. The typical example is Monte Carlo simulation.

Another method is stochastic simulation. For example, if we are considering fire, a value representing the historical probability of fire to occur during the horizon planning is used. Then, for each year of the simulation, a random number between 0 and 1 is draw, and if this number is higher than the historical probability the simulation assumes that fire is prompted and the stand is replaced. This process would be done for each simulation year for the whole horizon planning. Then, several repetitions of the whole simulation would be performed and an average would be obtained. In order to do this, analysis simulator is needed (including a growth and yield model and a fire probability model).

10.3.2 Stochastic Dynamic Programming and Markovian Processes

The use of DP concepts within a forestry framework have been addressed in Chap. 5 of this book. Specifically, the DP application to stand management scheduled has been introduced. In this chapter we introduce how the construction of a DP model has to be modified in order to address the development and use of stochastic dynamic programming to handle risk of sudden destruction of a stand (e.g. wildfire). The main difference with a typical dynamic programming problem is the inclusion of scenarios of fire occurrence and the expected mortality. The readers are referred to Ferreira et al. (2011 and 2012) for more details of the stochastic dynamic models to include fire.

An innovative approach to integrate forest and fire management planning in eucalypt stand has been developed by Ferreira et al. (2012). This method may be used by industrial forest owners to manage their eucalypt plantations. This would help them to select the sequence of management activities, including fuel treatment and timber harvests, that maximizes soil expectation value for a stand where wildfire risk is related to stand structure and fuel loads. In addition, they may estimate non-optimal prescriptions' opportunity costs in a wildfire risk context. This method has

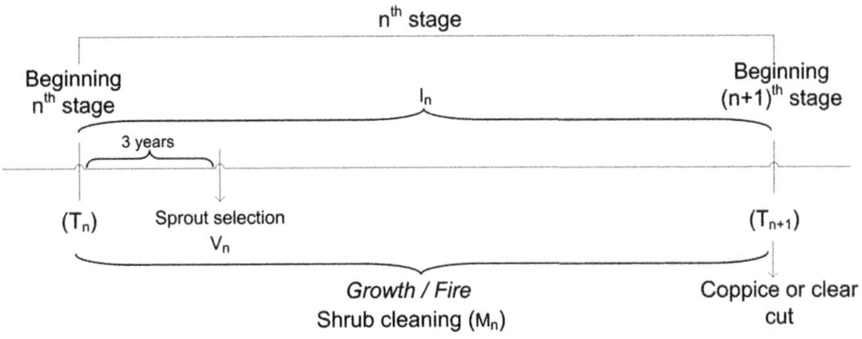

Fig. 10.1 Characterization of the nth stage, where n = 1, . . ., N identifies the stage, which is characterized by the number of harvests; (I_n, V_n, M_n) = set of variables that characterize the decision to be taken at the beginning of the nth stage; I_n is the number of years of the nth cycle, V_n corresponds to the average number of sprouts per stool after a stool thinning in the nth cycle and M_n represents the number of fuel treatments over the nth cycle. (T_n) = node that characterizes the state, at the beginning of the nth stage; corresponds to the number of years since the stand was planted until the harvest age at the end of (n − 1)th stage. This figure has been published in (Ferreira et al. 2012)

been recently implemented as a module within a decision support system described by Borges et al. (2003).

Ferreira et al. (2012) present a DP stochastic approach that aims at proposing coppice stand optimal management policies. This includes fuel treatments, stool thinning and cycle lengths and optimal number of cycles within a coppice system full rotation according to the stand state. The materials included a stand-level growth and yield model and models developed in WP4, i.e., the wildfire occurrence and damage models and the shrubs biomass accumulation model.

The DP formulation breaks the coppice stand management problem into stages that are characterized by the number of harvests (ie. the number of stages is determined by the maximum number of harvests over a whole coppice rotation). A maximum of 4 harvests are allowed, i.e., the optimal rotation may include 1, 2 or 3 coppice harvests and a clear-cut. The number of years since the stand was planted characterizes the state at each stage.

The arcs represent harvest decisions. In fact, each arc is associated with a vector (*In*, *Vn*, *Mn*) representing the whole set of management policies that may be implemented at the stage (Fig. 10.1): *In*, the number of years of the nth cycle, *Vn* the average number of sprouts per stool after a stool thinning in the nth cycle, *Mn* the number of fuel treatments over the nth cycle.

The backward recursive equations are used to solve the problem. The backward recursive method is the only solution approach that may provide insight about what path to follow (management policy to implement) in any given stand state. This is instrumental for integrating risk in management planning and for designing adaptive policies. Thus, the solution gives the optimal policy to apply for any possible state of the forest (Ferreira et al. 2012).

Scenarios are characterized by both the probability of catastrophe occurrence over time and the damage caused. A set of scenarios do not present mortality (i.e., there is no fire or the fire does not kill any trees), the second set includes all scenarios that present mortality therefore triggering regeneration. In this context, even if the decision maker plans a rotation length with I_n length, the actual length for a specific scenario j may be different (I_n^j). The use of the occurrence and damage scenarios allow incorporating wildfire risk into the stochastic model. The scenarios where developed using the models developed by Marques et al. (2011b) and Botequim et al. (2013). Wildfire occurrence probability increases with the number of trees per ha, the understory biomass, and the tree quadratic diameter. Whenever mortality occurs, the damage increases with the stand basal area.

The Eucalypt growth was estimated using the stand-level growth-and-yield model Globulus 3.0 developed by Tomé et al. (2006). This model allowed to determine the state of the stand at any stage. Understory growth was estimated according to a model developed by Botequim et al. (2009). According to this model, the understory biomass stock depends on its age and on the basal area of the eucalypt stand.

The SDP model is defined as follows:

$$\textit{Determine } Z = F_1(0) - CP \tag{10.2}$$

$$F_1(T_n) = \max_{\substack{I_n \in \Psi_n \\ V_n \in \Theta_n \\ M_n \in \Pi_n}} \{G_n(T_n, I_n, V_n, M_n), S_n(T_n)\}, \quad n = 1, \ldots, N \tag{10.3}$$

$$F_{N+1}(T_{N+1}) = S_{N+1}(T_{N+1}) \tag{10.4}$$

$$\begin{aligned} G_n(T_n, I_n, V_n, M_n) = & \sum_{j \in J^1} p^j(T_n, I_n, V_n, M_n)\left[B_n^j(T_n, I_n, V_n) - CV_n^j(T_n)\right. \\ & \left. - L_n^j(T_n, I_n, M_n) + F_{n+1}(T_n + I_n)\right] \\ & + \sum_{j \in J^2} p^j(T_n, I_n, V_n, M_n)\left[B_n^j(T_n, I_n, V_n) - CV_n^j(T_n)\right. \\ & \times \left. L_n^j(T_n, H^j, M_n) + S_{n+1}(T_n + H^j)\right], n \\ & = 1, \ldots, N \end{aligned} \tag{10.5}$$

$$T_{n+1}^j = T_n + I_n^j, \quad \textit{with } n = 1, \ldots, N, \quad j \in J^1 \cup J^2 \quad \textit{and} \quad T_1 = 0 \tag{10.6}$$

$$S_n(T_n) = \frac{F_1(0) + CR}{(1+i)^{T_n}}, \quad \textit{with } n = 2, \ldots, N+1 \quad \textit{and} \quad S_1(T_1) = 0 \tag{10.7}$$

where N is the maximum number of harvests over the coppice rotation; $n = 1, \ldots, N$ identifies the stage, characterized by the number of harvests (maximum 4); $N + 1$ identifies the end of stage N; T_n is the number of years since the stand was planted at the beginning of stage n, it corresponds to the value of state variables defining a DP network node, at the beginning of the nth stage; T_{N+1} identifies the number of years since the stand was planted at the end of stage N when the stand is harvested the Nth time; I_n and M_n are decisions regarding cycle length and fuel treatment scheduling over a cycle, respectively, with $n = 1, \ldots, N$; V_n is thinning option over a coppice cycle, with $n = 2, \ldots, N$; J^1 and J^2 are subsets of catastrophic scenarios that either do not or do force the stand to be regenerated, respectively; H^j is year when the scenario $j \in J^2$ occurs; Z is soil expectation value associated with the optimal management policy; $F_n(T_n)$ is the optimal value of network node T_n, at the beginning of the nth stage; G_n (T_n, I_n, V_n, M_n) is the expected value of network path out of node T_n if the management policy involving decisions I_n, V_n, and M_n is implemented at the beginning of the nth stage; $S_n(T_n)$ is the value of the bare land node that is connected to network node T_n, at the beginning of the nth stage; B_j^n (T_n, I_n, V_n) is the discounted financial return associated with the sale of wood after a harvest or a catastrophe, at the nth stage, under the jth scenario; $CV_n^j(T_n)$ is the discounted cost of a stool thinning option at the nth stage, under the jth scenario; L_n^j (T_n, I_n^j, M_n) is the discounted cost of a fuel treatment management scheduling option, at the nth stage, under the jth scenario; $p^j(T_n, I_n, V_n, M_n)$ is the probability of occurrence of the jth catastrophe scenario, during the nth stage (it depends on the number of years since planting and on the management policy implemented at the beginning of the nth stage); CR is the conversion cost at the end of each rotation; and CP is the plantation cost at the beginning of the first rotation.

In this model, Eq. 10.2 defines the management objective (i.e., Maximizing SEV) which is computed by the backward solution approach. An estimate of the SEV is needed to satisfy the boundary condition and initiate the solution process because the value of $F_1(0)$ is still unknown. Thus, an estimate of the bare land value at the end of the rotation is given to start the backward recursion process (value given by a function S). The SDP return function computes the discounted net return of a set of management policies over a full cycle. The recursive function encompasses two subfunctions: G and F. G considers catastrophic occurrence and probabilities of damage scenarios. It computes the total expected value of the net returns of management policies that may be implemented at each state T_n, at the beginning of the nth stage with the net return associated with the optimal management policy to be implemented at either the state T_{n+1}, if no complete stand destruction occurs, or with the estimate of the bare land value associated with the age when the wildfire occurs, which is provided by the function S. F selects the optimal path out of a node T_n, at the beginning of the nth stage. $F_n(T_n)$ thus identifies the optimal management policy when the stand is in state T_n. This policy encompasses a decision regarding whether to clearcut or to implement one further coppice cycle and decisions associated with this potential cycle (length, stool thinning and fuel treatment). At the end of the solution process, the optimal value $F_1(0)$ is compared with the estimate used. If these values are different, the iterative solution process continues using former $F_1(0)$ to re-estimate the land value (Hoganson et al. 2008). This method of successive approximations to the true SEV value can be proven to converge (Ferreira 2011).

Equations 10.3, 10.4 and 10.5 correspond to the DP recursive relations and determine the value of each node at the beginning of the nth stage, i.e., $F_n(T_n)$. The value of the F function (Eq. 10.3) depends on the value of the G function (Eq. 10.5), which is an expected value set by catastrophe scenarios probabilities, $p^j(T_n, I_n, V_n, M_n)$. Equations 10.4 provide the values of the F function at the end of the nth stage. Equations 10.6 correspond to the transition function and reflect the relationship between states of consecutive stages.

The number of years from the planting to the harvest of the stand at the end of stage n corresponds to the sum of the number of years from the planting to the harvest at the end of stage $n-1$ with the duration of the nth cycle.

In this study, for simplicity, it is assumed that only one catastrophe may occur over a cycle. H^j represents the year in the cycle when the catastrophe occurs. It is also assumed that if the catastrophe generates tree mortality, the stand must be regenerated. In addition, discounted return associated with the sale of live trees timber and also from salvaged timber is computed.

The DP return function includes the cost of stool thinning in coppice cycles over the rotation. This cost thus occurs only from the second cycle on. This value it is proportional to the number of sprouts thinned NV_n. This value is discounted $T_n + x$ years, x being the year in the cycle when the stool thinning takes place (only once over a coppice cycle). The DP return function also includes the cost of fuel treatments. Catastrophe occurrence will affect the number and timing of fuel treatments actually performed over the cycle. For example, if a fire occurs in year l of the nth cycle under scenario j, the understory biomass is destroyed, and thus there is no need to treat fuel.

Solving the SDP network using the backward recursion process provides information about the optimal management policy (cycle length, stool thinning and fuel treatment schedule) to implement in any situation.

10.3.3 Direct Search Methods

As explained, one of the objectives of stand-level management planning is to provide information to the land-owners or forest management to select the sequence of management activities (decision variables) that maximizes a specific objective. Optimization is an automated way of searching for the best set of decision variables. Since there are often several decision variables (e.g. thinning schedule, thinning intensities, rotation lengths . . .), methods for multidimensional optimization must be used.

Direct search methods are suitable for maximizing or minimizing functions that are non-smooth and non-differentiable (Bazaraa et al. 1993). One of these methods is the Hooke and Jeeves algorithm. The Hooke-Jeeves algorithm has been used in

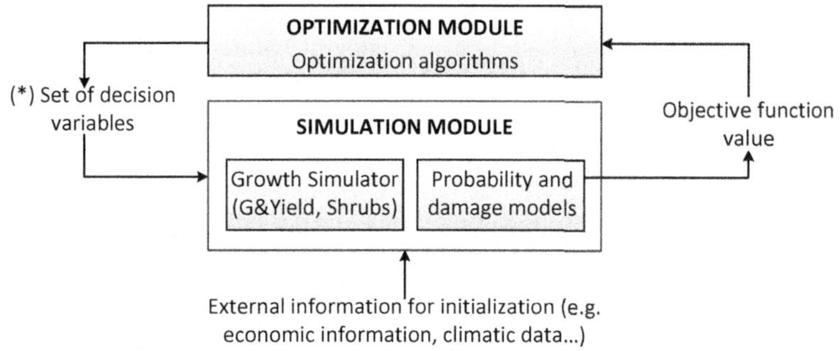

Fig. 10.2 Structure of the simulation optimization system. * an initial set of decision variables must be provided

several studies in forestry (e.g. Valsta 1990; Möykkynen et al. 2000; Miina and Pukkala 2000; Pukkala and Miina 2005; González et al. 2005b; Garcia-Gonzalo et al. 2011a).

This method uses two search modes: exploratory search in the direction of coordinate axes (one decision variable altered at a time) and pattern search in other directions (more than one decision variable altered simultaneously). See Baazara et al. (1993) for more details. Other methods so called population-based methods or evolutionary computation methods have also been used to solve complicated optimization problems. These methods include the differential evolution (Storn and Price 1997), particle swarm optimization (Kennedy and Eberhart 1995), and evolution strategy (Beyer and Schwefel 2002) and Nelder and Mead (1965). The difference between Hooke and Jeeves and population methods is that optimization works with a population of solutions instead of a single vector.

The drawback of these methods is that the solution reported is not necessarily the optimal due to the fact that the convergence of the methods to the global optimum is not guaranteed with objective functions which are neither convex nor differentiable (Miina 1998). An option to try to increase the probability of getting solutions that are near the global optimum, all optimizations may be repeated several times using different values to initialize the decision variables.

An example of the use of direct search methods in Maritime pine stand-level optimization in Portugal under risk of fire was presented by (Garcia-Gonzalo et al. 2011a). In this article the authors integrated stand growth and yield model (Falcão 1999), a shrub biomass model (Botequim et al. 2013), wildfire probability and damage models (Marques et al. 2012; Garcia-Gonzalo et al. 2011a) and non-linear optimization algorithms (e.g. Hook and Jeeves and populations methods) in a simulation-optimization system (Fig. 10.2). Even though this is a specific system developed originally by (Palahí and Pukkala 2003) and modified by Garcia-Gonzalo et al. (2011b) to include Portuguese models, this system is shown here to explain the general structure of an optimization-simulation system for stand-level management planning (Fig. 10.2).

The simulation optimization system has two components that are linked. The first component is the simulation system which integrates the growth and yield models (examples in Chap. 4) and the wildfire models. This module component projects vegetation growth and estimates the wildfire post-fire damage if a wildfire occurs. The probability of fire occurrence model (Marques et al. 2012) may be used to calculate the probability of a stand to burn. Then the post-fire mortality model may be used to estimate the proportion of trees in a given stand (i.e. the number of trees) that will die as a consequence of the wildfire event (Garcia-Gonzalo et al. 2011b). If a tree-level growth and yield model is available and we do have a complete list of trees and their respective biometric characteristics, another model may be used to estimate which trees from the stand will die (i.e., mortality models). In this case, the equation may be used to predict the probability of mortality of each tree in the stand and to build a list of all trees in the stand ordered according to this probability (trees with higher probability of mortality are ranked first in the list). The management planning model may then select the trees that will be assumed to die for planning purposes by going down the list and stopping when it reaches the number of trees that were estimated to die.

The second module incorporates and links the optimization methods to the stand simulator. All non-linear programming methods previously introduced in this section were included. In this system they were used to find the optimal management schedule, defined by a combination of decision variables, e.g. thinning (timing and intensity), fuel treatments (i.e. shrub cleanings) and rotation length.

In the example (Garcia-Gonzalo et al. 2011a) the maximized objective variable was the risk-adjusted soil expectation value (SEV). The risk of fire was defined as the product of fire probability and the damage caused by fire. Simulation module provides projections by vegetation growth and yield models of conditions and outcomes of interest associated with the decision variables (i.e. stand-level prescription) and the wildfire risk values in order to compute SEV. In Garcia-Gonzalo et al. 2011b for simplicity a constant probability of wildland fire was assumed.

The optimization starts with an initial set of decision variables and with the initialization of external data (e.g. prices of timber, costs of operations, interest rate, and depreciation of salvaged timber). The initial data will depend on the different models used in the simulation module. For example, process-based models may need description of soil and climatic data (see Chap. 4 for details). In addition, range values for each of the decision variables must be given (e.g. maximum and minimum basal area removal in thinnings, maximum and minimum rotation length . . .). The simulation module is run and the first value of the objective function is computed. Then the optimization algorithm selects a new set of decision variables and re-runs the simulation and computes the new objective function value. The process is repeated and the optimization stops depending on the stopping criteria defined in each algorithm.

The results of the HJ algorithm were compared with results obtained using population based methods. Population-based direct search methods work with a population of solutions instead of a single vector of decision variables (Pukkala 2009).

Stand management was optimized for different number of thinnings, shrub cleaning operations and risk scenarios. The solution provided the rotation age at which the stand should be clearcut if there is no fire (maximum rotation length). This was instrumental for private forest landowners to understand the range of management options available and its impact on economic, ecological and fire protection objectives.

10.4 Addressing Risk in Forest-Level Management Planning

Methods may incorporate risk and may provide information on the effects of forest management on that source of risk. For example, when incorporating wildfire risk provide information about the likely effects of spatial fuel treatment programs on wildfire behaviors and effects at the landscape scale.

10.4.1 Mathematical Programming

Simulation has been widely used in forestry. It may serve to test hypotheses or scenarios. However, simulation itself does not give answers on what is the most efficient way to manage the forest. On the contrary, optimization looks the best combination of management alternatives over the landscape in order to maximize/minimize certain objectives. This allows evaluate different plans of actions and it allows also to rank them according to certain criteria. In addition, it assess trade-offs between objectives. In planning problem at forest-level the activities related to multiple stands need to be simultaneously evaluated.

Generally, forest management models require the generation of mathematical programming matrices to describe the decision problem (e.g. Bettinger et al. 2009). In this context, prescription writer provide the coefficients for the resource capability modes and the policy models.

An example of the use of mathematical programming to cope with risk of wildfire in a maritime pine forest landscape in Portugal has been recently developed by Ferreira et al. (2014). In this research the wildfire occurrence and mortality models explained in previous section were also used to integrate wildfire risk in forested landscape management planning. In this example, a module to generate the stochastic model was included in the SADfLOR platform (Borges et al. 2003). The scheme to produce an optimized management plan including risk and uncertainty when using SADfLOR is shown in Fig. 10.3.

The forest management scheduling problem was designed as a linear programing Model I (Johnson and Sheurman 1977). This means that its solution encompasses the assignment of a prescription to each stand in the forest in order to achieve landscape-wide goals. The landscape-level model tries to reflect spatial and temporal relations between stand-level decisions. This model is spatially explicit.

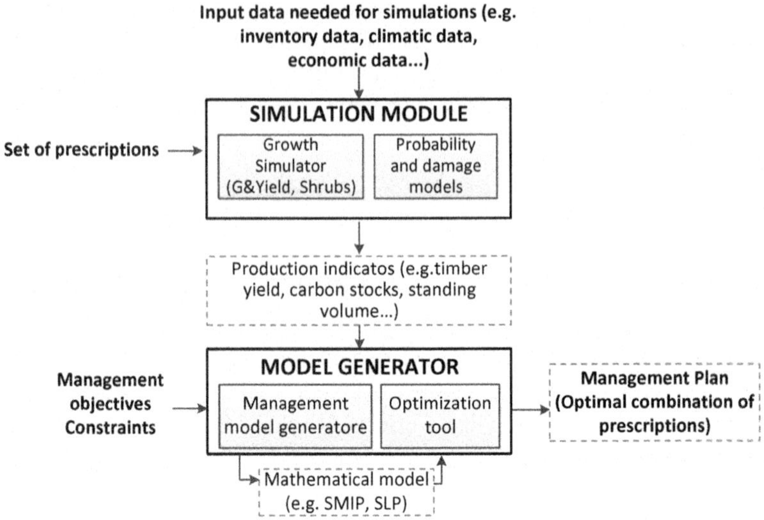

Fig. 10.3 Scheme for the generation of an optimal management plan including risk and uncertainty when using a DSS

This means that the model tries to recognize the impact of decisions made in a stand on the neighboring stands (e.g. the implementation of fuel treatments in a stand may help constraint the spread of fire to adjacent stands).

In this case, the information provided by the wildfire risk models presented in previous section was used to develop a stand-level wildfire resistance index. This index was adjusted to take into account the stand spatial context (e.g. information about the neighboring stands) and was incorporated into a landscape-wide wildfire resistance indicator. This information in addition to the growth and yield projections provided the information needed to develop a mixed integer programming approach to maximize the forest value while addressing timber even-flow and landscape wildfire resistance concerns.

The wildfire resistance indicator of any stand i was thus built to reflect the environmental and biometrical features (e.g. stand density, shrubs biomass) of both stand i and its adjacent stands (i.e., adjusted resistance indicator). The former contribute to the stand i specific resistance, while the latter are combined with the risk of wildfire spread to stand i to compute an adjusted resistance indicator. The specific resistance of stand i in period t is termed R_{it}: it reflects the proportion of trees in stand i that are prone to survive a wildfire occurring in planning period t, according to wildfire occurrence and post-fire mortality probability models (Garcia-Gonzalo et al. 2011a, b; Marques et al. 2012).

The adjusted resistance level of stand i in period t is termed RA_{it}. It incorporates the wildfire resistance levels of its adjacent stands and the risk of fire spread. It is defined as:

$$RA_{it} = R_{it} + (1 - w_i) \sum_{s \in \mathcal{V}\{i\}} \alpha_{is} (RA_{st} - R_{it}), \quad i \in \mathcal{I}, t \in \mathcal{T} \tag{10.8}$$

where

R_{it} = specific resistance of stand i in period t; $R_{it} \in [0,1]$;

$(1 - w_i)$ = weight to reflect the impact of neighboring stands on the wildfire resistance of stand i; $(1 - w_i) \in [0,1[$. w_i depends on parameters that reflect the size and shape of the stand i; we assume $w_i > 0$.

α_{is} = parameter reflecting the likelihood of a fire that occurs in stand s to spread to stand i; $0 \leq \alpha_{is} \leq 1$ and $\sum_{s \in \mathcal{V}\{i\}} \alpha_{is}$; where $\mathcal{V}\{i\}$ is the set of neighbors of stand i.

The difference $RA_{st} - R_{it}$ corresponds to the adjustment of the specific wildfire resistance of stand i. (R_{it}) might address the impact of its spatial context.

A stand specific wildfire resistance indicator might also be adjusted by considering the specific wildfire resistance of its neighbors rather than their adjusted wildfire resistance:

$$RA'_{it} = R_{it} + (1 - w_i) \sum_{s \in \mathcal{V}\{i\}} \alpha_{is} (R_{st} - R_{it}), i \in \mathcal{I}, t \in \mathcal{T}. \tag{10.9}$$

Nevertheless, in this case, the spatial context is interpreted more limitedly, as it is assumed that the wildfire resistance of a stand is impacted only by its adjacent stands rather than by a wider neighborhood. Thus, in this research, we assumed that a stand specific wildfire resistance indicator is to be adjusted according to the adjusted wildfire resistance of its neighbors s. In both cases (Eqs. 10.8 and 10.9) the wildfire resistance indicator value ranges from 0 and 1 (Ferreira 2011).

The approach to address wildfire risk concerns in forest management scheduling integrates the stand adjusted resistance index into a mixed integer programming (MIP) model that may be described as follows:

$$Max\ Z = \sum_{i=1}^{I} \sum_{k=1}^{K_i} c_{ik} A_i x_{ik} \tag{10.10}$$

$$\sum_{k=1}^{K_i} x_{ik} = 1, \quad i \in \mathcal{I} \tag{10.11}$$

$$\sum_{i=1}^{I} \sum_{k=1}^{K_i} v_{ikt} A_i x_{ik} = V_t, \quad t \in \mathcal{T} \tag{10.12}$$

$$\begin{aligned} V_t &\geq (1 - \lambda) V_{t-1}, \ t \in \mathcal{T} \setminus \{1\} \\ V_t &\leq (1 + \lambda) V_{t-1}, \ t \in \mathcal{T} \setminus \{1\} \end{aligned} \tag{10.13}$$

$$\sum_{i=1}^{I} A_i \sum_{k=1}^{K_i} F_{ik} x_{ik} \geq A_F f \tag{10.14}$$

$$R_{it} = \sum_{k=1}^{K_i} r_{ikt} x_{ik}, \quad i \in \mathcal{I}, t \in \mathcal{T} \tag{10.15}$$

$$RA_{it} = R_{it} + (1 - w_i) \sum_{s \in \mathcal{V}\{i\}} \alpha_{is} (RA_{st} - R_{it}), \quad i \in \mathcal{I}, t \in \mathcal{T} \tag{10.16}$$

$$\frac{A_i RA_{it}}{A_F} \geq RES_t, \quad t \in \mathcal{T} \tag{10.17}$$

$$x_{ik} \in \{0, 1\}, \quad i \in \mathcal{I}, k \in \mathcal{K}_i \tag{10.18}$$

Where i is the stand; k is a prescription that may be assigned to stand i; t is the planning period; $\mathcal{V}\{i\}$ is the set of stand i's neighbors;

Parameters:

c_{ik} = net present value, euros per hectare, resulting from the use of prescription k to manage stand i. It includes the value of ending inventory and it is adjusted to reflect the impact of wildfires;
A_i = area of stand i; in hectares;
A_F = total forest area, in hectares,
r_{ikt} = specific wildfire resistance of stand i in period t, if it is managed according to the prescription k;
RES_t = landscape-wide wildfire resistance in period t;
$(1-w_i)$ = as in Eq. 10.8;
α_{is} = as in Eq. 10.8;
v_{ikt} = volume per hectare harvested in period t if prescription k is assigned to stand i;
F_{ik} = age of stand i, at the end of the planning horizon, if prescription k is assigned to it;
f = minimum required average age of the forest, at the end of the planning horizon;
λ = maximum allowed variation of volume harvested in two consecutive periods;

Variables:

x_{ik} takes the value 1 if stand i is managed by prescription k or 0 otherwise.
R_{it} = specific resistance of stand i, in period t;
RA_{it} = adjusted wildfire resistance of stand i, in period t;
V_t = volume harvested in period t:

The objective function (Eq. 10.10) expresses the aim of maximizing the forest value. Equations 10.11 ensure that one and only one alternative is assigned to each stand. Equation 10.12 accounts the volume harvested in each planning period, while Eqs. 10.13 and 10.14 define the volume even-flow constraints. The latter state that the volume harvested in consecutive periods should not fluctuate more than λ. The concern about the state of the forest at the end of planning horizon is expressed by Eq. 10.14 that ensures that the average age of the ending inventory is greater or equal than a minimum value *f*.

Equations 10.15 compute the specific wildfire resistance of each stand in each period, according to the prescription assigned to it. Equations 10.16 compute the adjusted wildfire resistance of each stand, in each period as described in Eq. 10.8. In order to address wildfire risk concerns, harvest scheduling is constrained by setting a minimum level of forest-wide adjusted wildfire resistance, computed as the weighted average of its stands' adjusted wildfire resistance, where the weights correspond to the stand area (10.17).

The MIP model is deterministic and yet it may address wildfire risk by taking advantage of the information provided by the adjusted wildfire resistance. This indicator sets a protection level target and is used further to adjust the cost coefficients of the decision variables.

The value of the cost coefficient c_{ik} of a prescription *k* is typically computed by adding the net present value of its management options and the present value of ending inventory. In order to reflect the impact of wildfires, the former is adjusted so that it includes the present value of revenues resulting from the sale of salvaged burned wood (Sal_{ikt_int}):

$$\tilde{c}_{ik} = \sum_{t=1}^{T} (NPV_{iktint} + Sal_{iktint}) \tag{10.19}$$

Moreover, the net present value of other management options is also adjusted according to the probability of wildfire occurrence, the probability of mortality to occur after a wild and the proportion of dead trees if mortality indeed occurs. This information is encapsulated in the specific wildfire resistance indicators (r_{ikt}).

Management Planning in Action 10.2: A DSS for Enhanced Eucalypt Management Planning Under Climate Change

As a result of the workshops held in the participatory approach explained in Management Planning in Action 10.1 the necessity to integrate models and tools to support management planning, specially under global change scenarios where projections from the past may not be valid anymore was identified. In fact this was one of the big concerns of the forest owners association ACHAR (http://www.achar.pt/) and the individual forest owners and industrial plantation stakeholders. For this purpose a participatory approach (Enterprise Architecture methodology) to design a Forest Planning System was engaged.

(continued)

(continued)

The PA workshops with different stakeholders were instrumental for documenting the stakeholders' current decision processes as well as for identifying their concerns and expectations. The workshops also confirmed that the sophistication of the planning process tends to increase with the holding size. Both the industry and large-scale non industrial forest owners usually develop strategic (long term) plans at forest level and express their need for simulation/optimization systems to help search for the most profitable plan.

Based on the PA workshops information from industrial plantations stakeholders a new decision support system for Eucalypt plantations (SADfLOR v ecc 1.0) has been developed (Garcia-Gonzalo et al. 2014). This DSS integrates a process-based model calibrated for Portuguese conditions (Fontes et al. 2006) that is sensitive to changes in environmental conditions. It further integrates a prescription writer, an information system and an optimization module which includes exact and heuristic methods (e.g. Mixed Integer Programming models, Linear Programming and Simulated Annealing). This tool enables the decision maker to analyze impact of climate change on optimal forest plans. This is done by using different climate change scenarios for the simulations and optimizations. This system might effectively support strategic and tactical processes.

10.4.2 Heuristic Techniques

Heuristics have been acknowledged as a faster method when dealing with complex combinatorial methods at landscape level (Davis et al. 2001; Pukkala 2002). These methods can produce a good solution with shorter calculation time. However, these methods do not ensure the optimal solution.

In the case of heuristics, the scheme for the problem formulation and solving would be the same as in the Sect. 10.4.2 however the model generator module would no produce a exact mathematical formulation that would be solved by a solver or optimization tool. Instead, the information produced by the simulation module (prescription writer) would be used by a heuristic method for allocating prescriptions across the landscape taking into account multiple constraints and the objectives defined.

There are many different heuristics that have been used to integrate risk and uncertainty in forest planning. Pukkala and Kangas (1996) developed a method to integrate risk and attitude toward risk in forest planning combining scenario approach (to integrate risk in timber prices and in level of tree growth), a priority function (to compute the preferences and the attitude toward risk) and a heuristic optimization algorithm, to solve the problem. Meilby et al. (2001) included risk of wind in the objective function by means of scenario technique and then used simulated annealing to find optimal rotation length without even-flow cutting targets. Zeng et al. (2007) included risk of wind-throw minimization (and maximization) considering maximum cutting areas. Garcia-Gonzalo et al. (2008) included uncertainty in climatic conditions by means of scenarios of climate change and then used heuristics to find optimal management plans and to analyze the effect of not adapting management plans to climate change. In addition, Garcia-Gonzalo et al. (2014) developed and demonstrated the use of a DSS to develop management plans under uncertainty in climatic conditions. The last two papers used even-flow cutting constrains but did not include spatial constrains. Heinonen et al. (2009) also used risk of wind-throw as an objective variable in the optimization to analyze the effect of minimizing mean risk on the forest management.

More sophisticated approaches have been the integrated use of wildfire simulators into the DSS. For example, Kim et al. (2009) employed a wildfire simulator (FARSITE) and the heuristic great deluge algorithm to optimize the pattern of fuel management. A similar approach was followed by González-Olabarria and Pukkala (2011) but using a much simpler forest fire simulator. The risk of fire was integrated into the economic objective by incorporating potential wildfire losses in the expected net returns. Landscape metrics computing fire resistance were also included in problem formulations. The novelty of this last article is the use of recursive planning. This mean that within the optimization process the optimal management and fire risk were adjusted recursively to obtain an optimal management dependent on its associated fire risk.

As example of DSS used to incorporate risk and uncertainty, Graetz (2000) incorporates the use of a explicit fire model in the DSS (SafeD). This is a spatially explicit simulation/optimization system. It incorporates a stand prescription generator (Wedin 1999), the Forest Vegetation Simulator (FVS) as projection tool, and a spatially explicit fire model FARSITE (Finney 1999). It also includes a heuristic method of allocating activities across a landscape with multiple constraints.

10.5 Summary

As forest planning is characterized by the long-term horizon of its outcomes, uncertainty and risk are thus closely related to the development of forest management plans. Among these sources of uncertainty we find uncertainty related to the development of the trees, the performance of timber markets, the occurrence of catastrophic events (e.g. windthrows, wildfires, pests).

Forest planning is ineffective if it ignores these sources of uncertainty and risk. The main reason is that projections of forest conditions and outcomes done during the planning process may be very different if they include risk. Thus, optimizing forest management decisions without considering risk and uncertainty would lead to non-optimal decisions or even bad decisions (Pukkala 1998).

The variety of planning models is great, but applications of risk analysis are not very common in forest planning (Pasalodos-Tato et al. 2013). In this chapter we have introduced different techniques and models to incorporate risk and uncertainty into forest planning. We have focused on some novel models that show how to include risk of wildfire in forest planning problems.

This chapter may help the readers to understand the concept of risk and uncertainty and understand their different sources. Moreover, it may help to understand how to integrate risk and uncertainty in forest planning which models should be used (empirical vs process based models).

10.6 Problems

Propose a management model to incorporate risk of windthrow (or fire) in forest planning at landscape level.

Answer. Why do we need to use a process-based models when incorporating risk and uncertainty of climate change into management plans?

In Sect. 10.3.2 is presented an SDP model. Why is the backward recursive method used?

Acknowledgements The development of this chapter was partially supported by the projects PTDC/AGR-FOR/4526/2012 Models and Decision Support Systems for Adressing Risk and Uncertainty in Forest Planning (SADRI), PEst-OE/MAT/UI0152, PTDC/AGR-CFL/64146/2006 (Decision Support tools for integrating fire and forest management planning) and by the scholarship SFRH/BD/37172/2007, funded by the Portuguese Science Foundation (FCT). It was also financed by the ERDF – European Regional Development Fund through the COMPETE Programme (operational programme for competitiveness) within Project Flexible Design of Forest Fire Management Systems MIT/FSE/0064/2009, and it has received funding from the European Union's Seventh Programme for research, technological development and demonstration under grant agreements: (i) Nr 282887 INTEGRAL (Future-oriented integrated management of European forest landscapes), (ii) Nr PIRSES-GA-2010-269257 (ForEAdapt, FP7-PEOPLE-2010-IRSES).

References

Amacher G, Malik A, Haight R (2005) Not getting burned: the importance of fire prevention in forest management. Land Econ 81(2):284–302

Autoridade Florestal Nacional (AFN) (2010) Inventário Florestal Nacional 2005–2006 – Portugal Continental, (ed) Autoridade Florestal Nacional, Ministério da Agricultura do Desenvolvimento Rural e das Pescas, Lisbon, Portugal

Bazaraa MS, Sherali HD, Shetty CM (1993) Nonlinear programming. Theory and algorithms, 2nd edn. Wiley, Hoboken, 639 p. ISBN 0-471-55793-5

Bettinger P, Boston K, Siry J, Grebner DL (2009) Forest management and planning. Academic, New York

Beyer HG, Schwefel HP (2002) Evolution strategies: a comprehensive introduction. J Nat Comput 1(1):3–52

Borges JG, Falcão A, Miragaia C, Marques P, Marques M (2003) A decision support system for forest resources management in Portugal. In: Arthaud GJ, Barrett TM (eds) System analysis in forest resources. Proceedings of the eighth symposium, Snowmass Village, 27–30 Sept 2000. Kluwer, Dordrecht. Managing Forest Ecosystems 7:155–164

Botequim B, Borges P, Carreiras J, Oliveira MMO, Borges J (2009) Development of a shrub growth model in understory conditions (preliminary model). Tech Rep 7. FORCHANGE, Instituto Superior de Agronomia, Lisbon, Portugal

Botequim B, Garcia-Gonzalo J, Marques S, Borges JG, Tomé M, Oliveira MM, Tomé M (2013) Developing wildfire risk probability models for Eucalyptus globulus stands in Portugal. iForest 6:217–227. doi: 10.3832/ifor0821-006 (www.sisef.it/iforest/contents/?id=ifor0821-006)

Cauldfield JP (1988) A stochastic efficiency approach for determining the economic rotation of forest stand. For Sci 34:441–457

Christensen JH, Christensen OB (2007) A summary of the PRUDENCE model projections of changes in European climate by the end of this century. Clim Change 81:7–30

Davis LS, Johnson KN, Bettinger PS, Howard TE (2001) Forest management to sustain ecological, economic, and social values, 4th edn. McGraw-Hill, New York

Falcão A (1999) DUNAS—A growth model for the national forest of Leiria. In Proceedings of IUFRO workshop empirical and process-based models for forest tree and stand growth simulation, Oeiras, 21–27 Setembro 1997

Ferreira L (2011) Development of forest management models, under fire risk. PhD thesis, Universidade de Lisboa, Faculdade de Ciências, Lisbon, Portugal

Ferreira L, Constantino M, Borges JG (2011) Optimizing management of Pinus pinaster Ait stands, under risk of fire. An application in Portugal. Ann Oper Res. doi:10.1007/s10479-011-0845-z

Ferreira L, Constantino M, Borges JG (2011a) A stochastic approach to optimize Maritime pine (Pinus pinaster Ait.) stand management scheduling under fire risk. An application in Portugal. Ann Oper Res doi: 10.1007/s10479-011-0845-z (Published online January 2011)

Ferreira L, Constantino M, Borges JG, Garcia-Gonzalo J (2011b) A stochastic dynamic programming approach to optimize short-rotation coppice systems management scheduling. An application to eucalypt plantations under wild re risk in Portugal. Submitted manuscript

Ferreira L, Constantino M, Borges J, Garcia-Gonzalo J (2012) A stochastic dynamic programming approach to optimize short-rotation coppice systems management scheduling: an application to eucalypt plantations under wildfire risk in Portugal. For Sci 58:353–365

Ferreira L, Constantino M, Borges JG, Garcia-Gonzalo J (2014) A mixed integer programming approach to include wildfire risk in landscape level planning. For Sci (in print)

Ferson S, Ginzburg LR (1996) Different methods are needed to propagate ignorance and variability. Reliab Eng Syst Saf 54:133–144

Finney MA (1999) Mechanistic modelling of landscape fire patterns. In: Mladenoff DJ, Baker WL (eds) Spatial modelling of forest landscape change: approaches and applications. Cambridge University Press, Cambridge, UK, pp 186–209

Fontes L, Landsberg J, Tome J, Tome M, Pacheco CA, Soares P, Araujo C (2006) Calibration and testing of a generalized process-based model for use in Portuguese eucalyptus plantations. Can J For Res 36(12):3209–3221

Garcia-Gonzalo J, Jäger D, Lexer MJ, Peltola H, Briceño-Elizondo E, Kellomäki S (2008) Does climate change affect optimal planning solutions for multi-objective forest management? Allgemeine Forst und Jagdzeitung (AFJZ). Ger J Forest Res (GJFR) 179(5–6):78–95

Garcia-Gonzalo J, Pukkala T, Borges J (2011a) Integrating fire risk in stand management scheduling. An application to Maritime pine stands in Portugal. Ann Oper Res. doi:10.1007/s10479-011-0908-1

Garcia-Gonzalo J, Marques S, Borges JG, Botequim B, Oliveira MM, Tomé J, Tomé M (2011b) A three-step approach to post-fire mortality modelling in maritime pine (Pinus pinaster Ait.) stands for enhanced forest planning in Portugal. Forestry 84(2):197–206

Garcia-Gonzalo J, Zubizarreta-Gerendiain A, Ricardo A, Marques S, Botequim B, Borges JG, Oliveira MM, Tomé M, Pereira JMC (2012) Modelling wildfire risk in pure and mixed forest stands in Portugal. Allg Forst Jagdzg 183(11/12):238–248

Garcia-Gonzalo J, Borges JG, Palma JHN, Zuzibarreta-Gerendiain (2014) A decision support system for management planning of eucalyptus plantations facing climate change. Ann For Sci 71(2):187–199. http://link.springer.com/article/10.1007/s13595-013-0337-1

González JR, Palahí M, Trasobares T, Pukkala T (2005a) A fire probability model for forest stands in Catalonia (north-east Spain). Ann Forest Sci 63:169–176

González JR, Pukkala T, Palahí M (2005b) Optimising the management of Pinus sylvestris L. stand under risk of fire in Catalonia (north-east of Spain). Ann Forest Sci 62:493–501

González JR, Palahí M, Trasobares A, Pukkala T (2006) A fire probability model for forest stands in Catalonia (north-east Spain). Ann For Sci 63:169–176

González JR, Palahí M, Pukkala T, Trasobares A (2008) Optimising the management of Pinus nigra Arn. stands under endogenous risk of fire in Catalonia. Investig Agrar Sist Recurs For 17(1):10–17

González-Olabarria J, Pukkala T (2011) Integrating fire risk considerations in landscape-level forest planning. For Ecol Manage 261:278–287

Giorgi F, Bi X, Pal J (2004) Mean interannual and trends in a regional climate change experiment over Europe. II: Climate change scenarios (2071–2100). Clim Dyn 23:839–858

Graetz DH (2000) The SafeD Model: incorporating episodic disturbances and Heuristic programming into forest management planning for the Applegate River Watershed, Southwestern Oregon. M.S. thesis. Oregon State University, Corvallis, Oregon

Granier A, Reichstein M, Breda N, Janssens IA, Falge E, Ciais P, Grunwald T, Aubinet M, Berbigier P, Bernhofer C, Buchmann N, Facini O, Grassi G, Heinesch B, Ilvesniemi H, Keronen P, Knohl A, Kostner B, Lagergren F, Lindroth A, Longdoz B, Loustau D, Mateus J, Montagnani L, Nys C, Moors E, Papale D, Peiffer M, Pilegaard K, Pita G, Pumpanen J, Rambal S, Rebmann C, Rodrigues A, Seufert G, Tenhunen J, Vesala T, Wang Q (2007) Evidence for soil water control on carbon and water dynamics in European forests during the extremely dry year: 2003. Agr Forest Meteorol 143:123–145

Heinonen T, Pukkala T, Ikonen V-P, Peltola H, Venälaäinen A, Dupont S (2009) Integrating the risk of wind damage into forest planning. For Ecol Manag 258(7):1567–1577

Hoganson HM, Borges JG, Wei Y (2008) Coordinating management decisions of neighboring stands with dynamic programming. In Designing green landscapes, von Gadow K, Pukkala T (eds). Manag For Ecosyst 15:187–214

Hollenstein K (1997) Analyse, Bewertung und Management von Naturrisiken. Hochschulverlag AG der ETH, Zürich, 191 p

Johnson KN, Scheurman HL (1977) Techniques for prescribing optimal timber harvest and investment under different objectives – Discussion and synthesis. For Sci Monogr No 18. 31 pp

Kangas A, Kangas J (2004) Probability, possibility and evidence: approaches to consider risk and uncertainty in forestry decision analysis. Forest Policy Econ 6:169–188

Kao C (1982) Optimal stocking levels and rotation under risk. For Sci 28(4):711–719

Kellomäki S, Vaisanen H, Strandman H (1993) Finnfor: a model for calculating the response of boreal forest ecosystem to climate change. Joensuun Yliopisto, Metsä tieteellinen tiedekunta 6:3–120

Kellomäki S, Vaisanen H (1997) Modelling the dynamics of the forest ecosystem for climate change studies in the boreal conditions. Ecol Model 97:121–140

Kennedy J, Eberhart R (1995) Particle swarm optimization. In: Proceedings of the IEEE international conference on neural networks, Piscataway, pp 1942–1948

Kim Y-H, Bettinger P, Finney M (2009) Spatial optimization of the pattern of fuel management activities and subsequent effects on simulated wildfires. Eur J Oper Res 197:253–265

Kjellström E (2004) Recent and future signatures of climate change in Europe. Ambio 33:193–198

Marques S, Borges JG, Garcia-Gonzalo J, Moreira F, Carreiras JMB, Oliveira MM, Cantarinha A, Botequim B, Pereira JC (2011a) Characterization of wildfires in Portugal. Eur J For Res 130(5):775–784

Marques S, Garcia-Gonzalo J, Borges JG, Botequim B, Oliveira MM, Tomé J, Tomé M (2011b) Developing post-fire Eucalyptus globulus stand damage and tree mortality models for enhanced forest planning in Portugal. Silva Fenn 45(1):69–83

Marques S, Garcia-Gonzalo J, Botequim B, Ricardo A, Borges JG, Tomé M, Oliveira MM (2012) Assessing wildfire risk probability in Pinus pinaster Ait. stands in Portugal. For Syst 21(1):111–120

Matala J, Ojansuu JR, Peltola H, Sievanen R, Kellomäki S (2005) Introducing effects of temperature and CO_2 elevation on tree growth into a statistical growth and yield model. Ecol Model 161:95–116

Matala J, Ojansuu JR, Peltola H, Raitio H, Kellomäki S (2006) Modelling the response of tree growth to temperature and CO_2 elevation as related to the fertility and current temperature sum of a site. Ecol Model 199(1):39–52

Meilby H, Strange N, Thorsen BJ (2001) Optimal spatial harvest planning under risk of wind throw. For Ecol Manag 149:15–31

Miina J (1998) Preparation of management models using simulation and optimisation. In Pukkala T, Eerikäinen K (eds) Tree seedling production and management of plantation forests, University of Joensuu, Faculty of Forestry, Research Notes 68, pp 165–180

Miina J, Pukkala T (2000) Using numerical optimization for specifying individual-tree competition models. For Sci 46(2):277–283

Möykkynen T, Miina J, Pukkala T (2000) Optimising the management of a Picea abies stand under risk of butt rot. For Pathol 30:65–76

Moriondo M, Good P, Durao R, Bindi M, Giannakopoulos C, Corte-Real J (2006) Potential impact of climate change on fire risk in the Mediterranean area. Climate Res 31:85–95

Nelder JA, Mead R (1965) A simplex method for function minimization. Comput J 7:308–313

Nuutinen T, Matala J, Hirvela H, Harkonen K, Peltola H, Vaisanen H, Kellomaki S (2006) Regionally optimized forest management under changing climate. Clim Change 79(3–4): 315–333

Palahí M, Pukkala T (2003) Optimising the management of Scots pine (Pinus sylvestris L.) stands in Spain based on individual-tree models. Ann For Sci 60:105–114

Pasalodos-Tato M, Pukkala T, Castedo-Dorado F (2009) Models for the optimal management of Pinus radiata D. Don in Galicia (north–western Spain) under risk of fire. Allgemeine Forst und Jagdzeitung 180:238–249

Pasalodos-Tato M, Makinen A, Garcia-Gonzalo J, Borges JG, Lämas T, Eriksson O (2013) Assessing uncertainty and risk in forest planning and decision support systems: review of classical methods and introduction of new approaches. For Syst 22(2): 282–230. http://dx.doi.org/10.5424/fs/2013222-3063

Pukkala T (1998) Multiple risks in multi-objective forest planning: integration and importance. For Ecol Manag 111:265–284

Pukkala T (2002) Multi-objective forest planning: managing forest ecosystems. Kluwer Academic Publishers, Dordrecht

Pukkala T (2009) Population-based methods in the optimization of stand management. Silva Fenn 43(2):261–274

Pukkala T, Kangas J (1996) A method for integrating risk and attitude toward risk into forest planning. For Sci 42(2):198–205

Pukkala T, Miina J (2005) Optimising the management of a heterogeneous stand Silva Fennica 39(4):525–538

Rowe WD, 1994. Understanding uncertainty. Risk Analysis 14:743–750

Reed WJ (1987) Protecting a forest against fire: optimal protection patterns and harvest policies. Nat Resour Model 2:23–53

Samson S, Reneke JA, Wiecek MM (2009) A review of different perspectives on uncertainty and risk and an alternative modeling paradigm. Reliab Eng Syst Saf 94:558–567

Schelhaas M, Nabuurs G, Schuck A (2003) Natural disturbances in the European forests in the 19th and 20th centuries. Glob Chang Biol 9(11):1620–1633

Storn R, Price K (1997) Differential evolution – a simple and efficient heuristic for global optimization over continuous spaces. J Glob Optim 11:341–359

Thorsen BJ, Hells F (1998) Optimal stand management with endogenous risk of sudden destruction. For Ecol Manage 108:287–299

Tomé M, Soares P, Oliveira T (2006) O modelo GLOBULUS 30 Dados e equações Publicações GIMREF RC2/2006Universidade Técnica de Lisboa, Instituto Superior de Agronomia, Centro de Estudos Florestais, Lisboa

United Nations (1992) Internationally agreed glossary of basic terms related to disaster management. United Nations Department of Humanitarian Affairs, Geneva

Valsta LT (1990) A comparison of numerical methods for optimizing even aged stand management. Can J Forest Res 20:961–969

Valsta L (1992) A scenario approach to stochastic anticipatory optimization in stand management. For Sci 38(2):430–447

Wedin H (1999) Stand level prescription generator under multiple objectives. M.S. thesis, Oregon State University, Corvallis

Yousefpour R, Thorsen BJ, Jacobsen JB, Meilby H (2012) A review on decision-making under uncertainty and risk in forest management. Ann Forest Sci 69:1–15

Zeng H, Pukkala T, Peltola H (2007) The use of heuristic optimization in risk management of wind damage in forest planning. For Ecol Manage 241:189–199

Zimmermann H-J (2000) An application-oriented view of modeling uncertainty. Eur J Oper Res 122:190–198

Chapter 11
Integrating Nursery and Planting Activities

Silvana Ribeiro Nobre and Luiz C.E. Rodriguez

11.1 Introduction

Researchers in Brazilian industrial forest companies' generally determine a group of stands according to silvicultural prescriptions. Soil, climate and other site characteristics contribute to form homogeneous groups. Prescription in this context describes which clones, plantation densities, fertilization levels, irrigation regimes and other silvicultural practices are best suited for each group of stands. The forest companies must also establish clone distribution constraints to mitigate the disease risk that comes with having few clones in large continuous areas. Future production is the main objective of such efforts.

By contrast, forest managers in large forest companies make their decisions based on current industrial needs, which are subject to logistical constraints. In this chapter, we discuss a set of stands which were chosen to be harvested according to a long term harvesting plan as described in previous chapters of this book. An operational plan usually includes the sequence of the chosen stands subjected to logistics again and to current mill gate raw material demand. Economic, social and environmental issues are driving forces determining harvest scheduling.

The model described in this chapter links planting and harvesting; planting teams start their activities, a couple of months after harvesting, having in hand the best clone plants according to the prescription to establish a new forest.

The synchronization of nursery activities and silvicultural procedures requires a detailed plan, which accounts for all the components of a nursery and silviculture infra-structure.

S.R. Nobre (✉)
Department of Forest Management, Atrium Forest Consulting, São Paulo, Brazil
e-mail: silvana@atriumforest.com

L.C.E. Rodriguez
Department of Forest Sciences, University of São Paulo, São Paulo, Brazil
e-mail: lcer@usp.br

J.G. Borges et al. (eds.), *The Management of Industrial Forest Plantations: Theoretical Foundations and Applications*, Managing Forest Ecosystems 33,
DOI 10.1007/978-94-017-8899-1_11, © Springer Science+Business Media Dordrecht 2014

Linear programming techniques are used to model three processes together: nursery, planting and harvesting. The objective is to maximize future production (volume or expected pulp). The constraints are (i) nursery infra-structure, (ii) productivity of each nursery and planting activities, (iii) links among processes, (iv) planting team's productivity, (v) risks inside nursery regarding the amount of different clones in the production line, (vi) risks regarding geographic clone distribution. Decision variables are when clone-plant sets must start to be produced to meet field planting needs on time.

What You Will Learn in This Chapter

- What is a clone recommendation and prescription groups of stands.
- How one could organize the stands data to represent clone recommendation.
- How to link silviculture, nursery and harvesting processes.
- How a LP model could support daily decisions in nursery management.
- How to create a consistent planting plan to maximize future production.

11.2 Structure of Clone Recommendation

In the short term, forest managers must plan for planting immediately after clear cut operations. Beyond the usual planting schedule subject to operational constraints, a planting management plan should address industrial needs such as wood quality and security of supply (Sadic et al. 1998). Industrial forest units are optimized in terms of final production when dedicated to one specific market such as pulp, panels or charcoal. Productivity should be expressed by the final production (e.g. pulp per hectare).

In clonal plantations, specialized clones are developed to suit specific conditions in the field (Assis and Resende 2011). Forest researchers group the stands under similar soil and climate conditions, where each group is a technical recommendation unit (TRU). A proper prescription is developed to meet the needs of each TRU in terms of clone, fertilizing, planting density and a sequence of silviculture operations (Barros et al. 1986).

Usually, there is a list of clones that can be planted in a TRU, and they were chosen by researchers based on yield, adaptability and stability (Rosado 2012). A planting plan establishes when and which clones have to be planted in each stand subject to security and operational constraints of the process.

The structure of clone recommendation is illustrated in Fig. 11.1.

Figure 11.1 shows that Stands 1, 2 belong to a technical recommendation unit (TRU) number 1 while stands 3 and 4 belong to a TRU-2 and so forth. Managers must follow the fertilization x prescription, plant preferentially clone 1 in all stands of TRU-1.

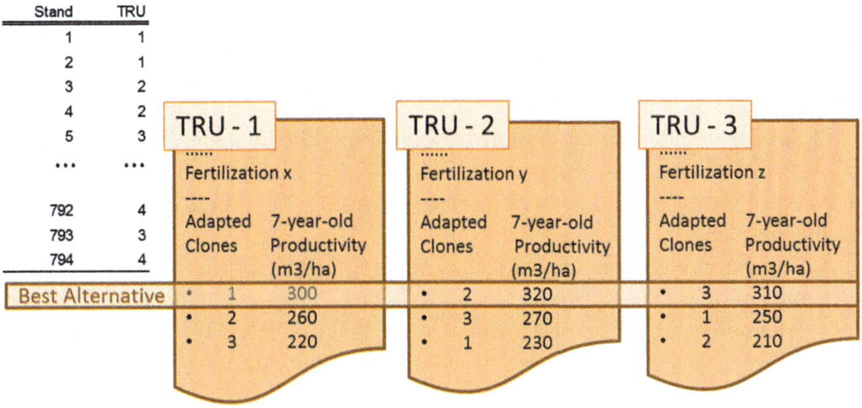

Fig. 11.1 Structure of clone recommendation

11.3 Future Production Optimization

To maximize the use of land, the future potential production should be maximized, and the model formulation is as follows:

11.3.1 [M 11.1] Maximize Future Production

$$Max\ Z = \sum_{i=1}^{I}\sum_{j=1}^{J} p_{ij} \ . \ x_{ij} \tag{11.1}$$

Subjected to

$$Total\ Area) \sum_{j=1}^{J} x_{ij} = X_i, \quad for \forall i \tag{11.2}$$

Where

Z probable future final product production;
p probable productivity of clone ***j*** when planted in stands ***i***;
x area of stand ***i*** where clone ***j*** is planted;
X total area of stand ***i***;
I number of stands in the plantation plan.

Model **[M 11.1]** establishes that managers should use the best clone in the total area of each stand. In the following sections, plantation process will be detailed, and further constraints will be defined to complete the model formulation that addresses the real plantation model.

If there were no other constraints but planting in the total area, managers would achieve the maximum potential production of the land; each stand would be planted with the clone that best suits its environmental needs. Table 11.2 shows an example of this problem.

In a scenario where the annual plan is planting 330 ha distributed in 5 stands, for example, all the 5 stands would belong to 3 TRU. Each recommendation unit has three alternative adapted clones (Roberds and Bishir 1997). Maximizing the future production means planting clone 1 in stands 1 and 2, clone 2 in stands 3 and 4, and clone 3 in stand 5.

However, the number of clones to be used in plantations is a fundamental problem that should be addressed by the planting management plan because of the forest health. The risk of catastrophic pests and diseases increases when the number of clones decreases. Roberds and Bishir (1997), using risk analyses, calculated that 30–40 unrelated clones provide protection against catastrophic failures in clone plantation.

A planting management plan should help managers to find an optimal monoclonal mosaic adapted to environment characteristics and subjected to operational constraints. The area of each clone can be limited to characterize the forest health constraint, and a new version of a model formulation is presented in **[M 11.2]**.

11.3.2 [M 11.2] Adding Forest Health Security Constraints

$$Max\ Z = \sum_{i=1}^{I} \sum_{j=1}^{J} p_{ij} \quad . \quad x_{ij} \tag{11.1}$$

Subjected to

$$\left.Total\ Area\right) \sum_{j=1}^{J} x_{ij} = X_i, \quad for \forall i \tag{11.2}$$

$$\left.Forest\ Health\right) \sum_{i=1}^{I} x_{ij} <= C_j, \quad for\ \forall j \tag{11.3}$$

Where

x_{ij} area of stand i, where clone j is planted;
C_j maximum Area of each Clone j;
J number of clones to be planted in the horizon;

Management Planning in Action 11.1: The Size of a Plantation Problem in Brazil

According to ABRAF – Brazilian Association of Forest Plantation Producers, in 2012, area planted in Eucalyptus and pine in Brazil reached 6.66 million hectares (ABRAF 2013). The segments of forest-based industry associated to planted forests are mainly pulp and paper, panels, charcoal and sawn wood and plywood.

The forest-based industry is spread from south to north in 15 states; they organize themselves in production units from 500 to 250,000 ha. Thirty five percent of those units have to manage 10,000–50,000 ha of plantation. Figure 11.1 shows the distribution of forest industry units in Brazil.

If a 35,000 ha unit is taken as an example, a large operational planning problem is generated to determine the production plan of a nursery that must produce the seedlings to feed the annual planting program. The following table presents the calculation of the problem size (Table 11.1 and Fig. 11.2).

Table 11.1 Problem size and complexity

Industry forest unit	35,000	ha
Productivity	45	m^3/ha.year
Clear cut	6	Years old
Annual replanting area	5,833	ha
Spacing (2.5 × 3 m)	6	m^2
Trees per area	1,667	Trees/ha
Operational efficiency	95 %	
Seedlings need	10,208,333	Seedlings/year
Typical Brazilian stand	20	ha
No. stands to plant in a year	292	Stands
Technical recommendation units	4	Units

(continued)

(continued)

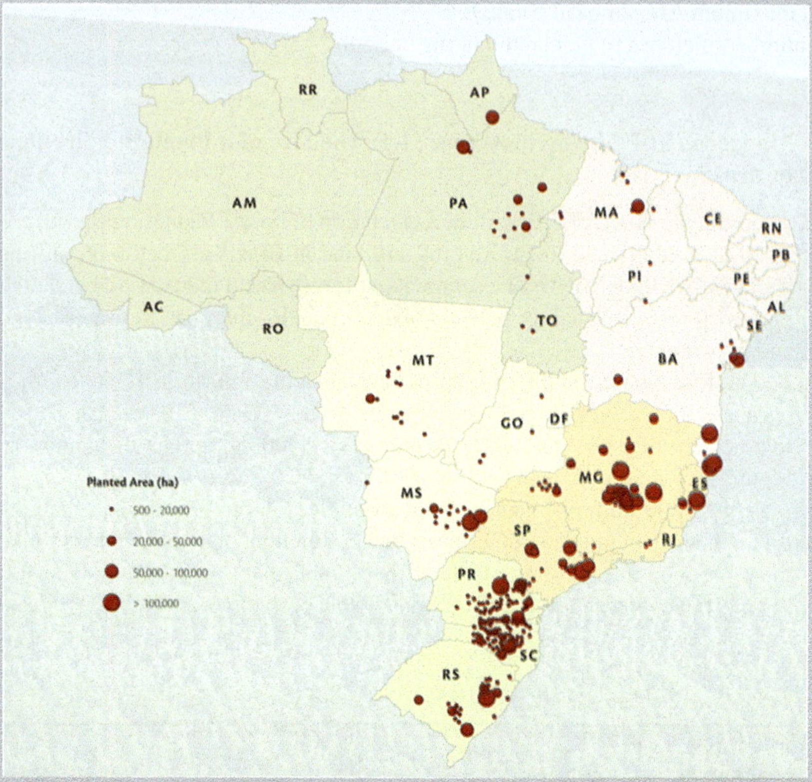

Fig. 11.2 Forest industry units' distribution (Source: ABRAF (2013))

In 1 year, on 10 % of the area we could have planted the 1st clone, but we planted the 2nd one

10 %	Area	>>>	583	ha
10 %	Production gap	>>>	30	m^3/ha
Production difference		>>>	17,500	m^3
Minimum Wood value		>>>	25	US$/$m^3$
Production Difference Value		>>>	437,500	US$

If the model considers the second constraint presented in **[M 11.2]**, the result would be slightly different; managers would no longer achieve the maximum production of the land. In the example shown in Table 11.2, only 100 ha of each clone could be planted; a proper result is presented in Table 11.3.

Table 11.2 Model 11.1 example – maximization of future production

Technical recommendation unit	TRU1						TRU2						TRU3			
Stands	St1			St2			St3			St4			St5			
Stand area	0	0	60	0	0	70	0	0	80	0	0	50	0	0	70	**330**
Clone	CL1	CL2	CL3	CL1	CL2	CL3	CL1	CL2	CL3	CL1	CL2	CL3	CL1	CL2	CL3	
Productivity (m^3/ha)	300	260	220	300	260	220	230	320	270	230	320	270	250	210	310	
Sum: Stand-clone area	**60**	0	0	**70**	0	0	0	**80**	0	0	**50**	0	0	0	**70**	
Sum: Stand area	0	0	**60**	0	0	**70**	0	0	**80**	0	0	**50**	0	0	**70**	**330**
Potential production (m^3)	**18,000**	0	0	**21,000**	0	0	0	**25,600**	0	0	**16,000**	0	0	0	**21,700**	**102,300**

Table 11.3 Model 11.2 example – forest health security constraint

Sum: Stand-clone area	**50**	0	**10**	**70**	0	0	0	**80**	0	0	**40**	**10**	0	0	**70**	
Sum: Stand area	0	0	**60**	0	0	**70**	0	0	**80**	0	0	**50**	0	0	**70**	**330**
Potential production (m^3)	**15,000**	0	**2,200**	**21,000**	0	0	0	**25,600**	0	0	**12,800**	**2,700**	0	0	**21,700**	**101,000**

11.4 Modeling Processes Together: Nursery, Planting and Harvesting

Planting management plan depends on a short-term harvest schedule. Planting operation happens after soil preparation; in fact, it must be a couple of months after harvesting. It means the right quantity of the right clone seedling has to be ready at the right moment because it cannot be stored. The clone seedling takes from 3 to 4 months to be ready after mini-cuttings collection from mini-stumps (Fig. 11.3) and cuttings planting (Fig. 11.4).

Furthermore, seedlings planting teams have to follow the plan according to their capacity constraint; they are usually geographically distributed and they should work in a limited region due to social issues and social law regulations.

Therefore, a highly synchronize short-term plan is required to meet industrial needs of wood quality in the future. There is not enough time to long planning process, when a short-term harvesting schedule is updated. The nursery schedule must be updated immediately to give time for seedlings growth.

However, there is a common complaint that a short-term harvesting plan changes the sequence of harvesting many times a year, and it is essentially relevant to the nursery mini-cutting collection and cuttings planting schedule.

Using the BPM – Business Process Management approach, nursery, planting and harvesting must be seen as part of a process to be integrated using IT (information technology) resources. Figure 11.5 illustrates the synchronization of the planning

Fig. 11.3 Collecting mini-cuttings from mini-stumps

Fig. 11.4 Cuttings planting

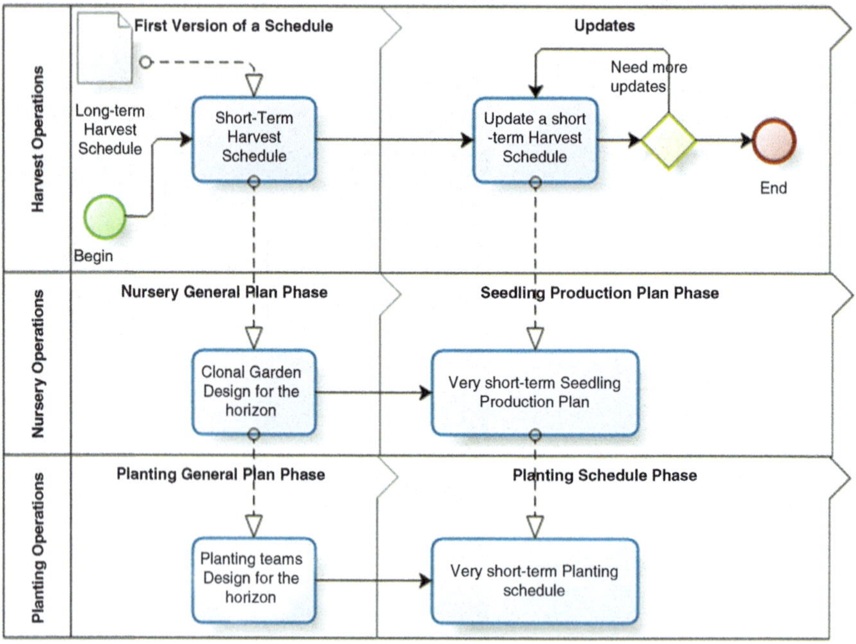

Fig. 11.5 Planning process synchronization

process. BPM can be defined as all efforts to improve fundamental activities of company's operations (Trkman 2010). It is possible to use the model described in this chapter if it is associated to other IT resources to create just-in-time plans.

11.4.1 Harvest Schedule Link

Besides the decision on which clone should be planted, the next step is to create the planting schedule subjected to the capacity of the teams prepared to do the job during the planning horizon. The time dimension should be added in the model; decision variable $\boldsymbol{x_{ij}}$ receives the dimension *period k.*

11.4.1.1 [M 11.3] Adding Time Dimension

$$Max\ Z = \sum_{i=1}^{I}\sum_{j=1}^{J}\sum_{k=1}^{K} p_{ij} \ . \ x_{ijk} \tag{11.4}$$

Subjected to

$$Total\ Area)\ \sum_{j=1}^{J}\sum_{k=1}^{K} x_{ijk} = X_i, \quad for \forall i \tag{11.5}$$

$$Forest\ Health)\ \sum_{i=1}^{I}\sum_{k=1}^{K} x_{ijk} <= C_j, \quad for\ \forall j \tag{11.6}$$

$$Planting)\ \sum_{i=1}^{I}\sum_{j=1}^{J} x_{ijk} <= F_k, \quad for \forall k \tag{11.7}$$

Where:

$\boldsymbol{x}$ area planted in a period $\boldsymbol{k}$, in stand $\boldsymbol{i}$, planted with a clone $\boldsymbol{j}$;
$\boldsymbol{k}$ period;
$\boldsymbol{K}$ number of periods in the horizon;
$\boldsymbol{F_k}$ field planting capacity of the team prepared to the horizon in a period $\boldsymbol{k}$.

If the manager needs to deal with many teams and their capacity varies because they work in different geographic regions or for any other reason, the dimension ***team*** should be created and the constraint (11.7) *Planting* must be slightly different:

11.4.1.2 [M 11.4] Adding Detailed Planning Constraint

$$Planting)\ \sum_{i=1}^{I}\sum_{j=1}^{J}\sum_{m=1}^{M} x_{ijkp} \leq F_{km}, \quad for \forall k\ and \forall m \tag{11.8}$$

Table 11.4 Harvest link constraint

		Technical recommendation unit														
		TRU1						TRU2						TRU3		
		Stand area (ha)														
		St1			St2			St3			St4			St5		
				60			70			80			50			70
		Stands Available to be planted (Harvest link constraint)														
Periods	1	**1**	**1**	**1**	0	0	0	0	0	0	0	0	0	0	0	0
	2	**1**	**1**	**1**	0	0	0	0	0	0	0	0	0	0	0	0
	3	**1**	**1**	**1**	0	0	0	0	0	0	0	0	0	**1**	**1**	**1**
	4	**1**	**1**	**1**	0	0	0	**1**	**1**	**1**	0	0	0	**1**	**1**	**1**
	5	0	0	0	0	0	0	**1**	**1**	**1**	0	0	0	**1**	**1**	**1**
	6	0	0	0	0	0	0	**1**	**1**	**1**	0	0	0	**1**	**1**	**1**
	7	0	0	0	0	0	0	**1**	**1**	**1**	**1**	**1**	**1**	**1**	**1**	**1**
	8	0	0	0	0	0	0	**1**	**1**	**1**	**1**	**1**	**1**	0	0	0
	9	0	0	0	**1**	**1**	**1**	0	0	0	**1**	**1**	**1**	0	0	0
	10	0	0	0	**1**	**1**	**1**	0	0	0	**1**	**1**	**1**	0	0	0
	11	0	0	0	**1**	**1**	**1**	0	0	0	0	0	0	0	0	0
	12	0	0	0	**1**	**1**	**1**	0	0	0	0	0	0	0	0	0

Where:

$\boldsymbol{F_{km}}$ field planting capacity of each different team $\boldsymbol{m}$ in a period $\boldsymbol{k}$

According to Stage (2003); the more connected to reality, the more the model is going to be used in a daily basis. In particular this case, if the model does not consider all constraints and details of the process, it cannot be used because there is no time for adjustments.

The Table 11.4 illustrates the link between the short-term harvest schedule and the planting schedule. This link can be represented by a yes-or-no variable saying when the stand should be planted, as shown in Fig. 11.6.

The planting management model needs a new constraint to represent the link between harvest schedule and planting schedule. Forest managers will need a matrix generation tool to build the set of constraints based on harvest schedule.

11.4.1.3 [M 11.5] Adding Harvest Link Constraint

$$\left.\textit{Harvest Link}\right) \sum\nolimits_{i=1}^{I} \sum\nolimits_{j=1}^{J} x_{ijk} = 0, \textit{ for some stands } i \textit{ in some periods } k \tag{11.9}$$

Table 11.5 shows an example of how these periods can be implemented adding these new dimensions into a MS Excel® spreadsheet presented in the last section. Consider having to plan 12 periods for the planting team and that they can plant only 30 ha per period.

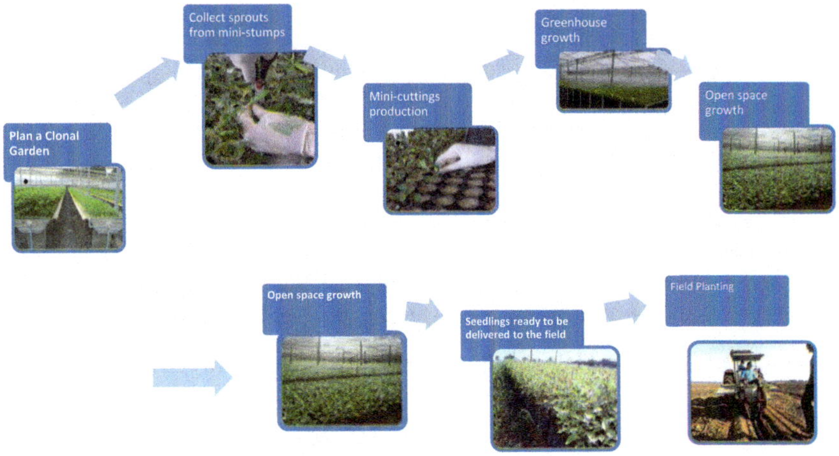

Fig. 11.6 Nursery and field planting process

Table 11.5 Decision variables per period

	Stand Area (ha)																	
	St1			St2			St3			St4			St5					
			60			70			80			50			70			
															330			
	Clone Productivity (m3 / ha)																	
	CL1	CL2	CL3	CL1	CL2	CL3	CL1	CL2	CL3	CL1	CL2	CL3	CL1	CL2	CL3			
	300	260	220	300	260	220	230	320	270	230	320	270	250	210	310			
	Sum: Stand-Clone Area (ha)																	
	60	0	0	70	0	0	0	50	30	0	50	0	0	0	70			
	Decision variables per Period															Planting (ha)		Maximum
1	30	0	0	0	0	0	0	0	0	0	0	0	0	0	0	30	<=	30
2	30	0	0	0	0	0	0	0	0	0	0	0	0	0	0	30	<=	30
3	0	0	0	0	0	0	0	0	0	0	0	0	0	0	30	30	<=	30
4	0	0	0	0	0	0	0	0	20	0	0	0	0	0	10	30	<=	30
5	0	0	0	0	0	0	0	0	0	0	0	0	0	0	30	30	<=	30
6	0	0	0	0	0	0	0	20	10	0	0	0	0	0	0	30	<=	30
7	0	0	0	0	0	0	0	30	0	0	0	0	0	0	0	30	<=	30
8	0	0	0	0	0	0	0	0	0	0	30	0	0	0	0	30	<=	30
9	0	0	0	10	0	0	0	0	0	0	20	0	0	0	0	30	<=	30
10	0	0	0	30	0	0	0	0	0	0	0	0	0	0	0	30	<=	30
11	0	0	0	30	0	0	0	0	0	0	0	0	0	0	0	30	<=	30
12	0	0	0	0	0	0	0	0	0	0	0	0	0	0	0	0	<=	30
																330	=	330

11.4.2 Seedling Production Schedule Link

Further improvements in planting management model integrate the nursery seedling production to the process. The nursery requires mini-cutting collection and cutting production plans to start seedling production. Creating a schedule for a nursery means establishing when and how many cuttings of each clone must be planted per day.

To calculate how many cuttings we need to plant, we should consider at least the following variables:

(a) **E** efficiency of planting or field survival is not 100 %. Some seedlings die after and before planting. We should consider **E%** – the efficiency of planting;
(b) **S** spacing in trees per hectare.
(c) **R** rooting inside the greenhouse. Many cuttings die before being a ready-to-plant seedling (Brondani and Wendling 2012). Each clone behaves in a different way in a nursery environment (Oskarsson and Brynleyfsdottir 2009).

We can use the following equation to calculate the required cuttings to be planted:

$$T_{jk-n} = \sum\nolimits_{i=1}^{I} \frac{x_{ijk} \cdot s_i}{E_i .\ R_j} \qquad (11.10)$$

Where:

T_{jk-n} is the required cuttings to be planted during the period k – n; that should occur n periods before planting clone j, in period k in all stands;
x_{ijk} area with a clone j, in the stand i, in period k;
S_i spacing that must be used to plant in stand i;
E_i field efficiency expected for the stand i;
R_j mini-cuttings rooting rate of clone j.

The required cuttings must be subjected to the nursery infrastructure. Usually, it includes the efficiency of personnel, the shed size, water supply, space and other limits. The maximum production per period can express the constraint related to cuttings production (Tables 11.6 and 11.7).

Using the Eq. (11.10), we can build a new constraint in the model to represent the seedling production link.

11.4.2.1 [M 11.6] Adding Cutting Capacity Constraint

$$\textit{Cutting Capacity}\Big) \sum\nolimits_{j=1}^{J} \sum\nolimits_{i=1}^{I} \frac{S_i .\ x_{ijk}}{E_i . R_j} \leq T'_{k-n} \quad for \forall\, j \ and \forall k \qquad (11.11)$$

Model **[M 11.7]** can integrate harvesting schedule, field planting schedule and cuttings plantation schedule. Decision variables can be used to generate a schedule itself.

Table 11.6 Example of clone cuttings requirements integrated to field planting and harvesting schedule

Technical recommendation unit	TRU1						TRU2						TRU3			
Stands																
	St1			St2			St3			St4			St5			
Replant efficiency			3 %			3 %			3 %			3 %			3 %	
Spacing (trees/ha)			1,667			1,667			1,111			1,111			1,515	
Stand area			60			70			80			50			70	330
Clones																Ha
	CL1	CL2	CL3	CL1	CL2	CL3	CL1	CL2	CL3	CL1	CL2	CL3	CL1	CL2	CL3	
Nursery efficiency (seedlings/mini-cuttings)	65 %	70 %	72 %	*65 %*	*70 %*	*72 %*	*65 %*	*70 %*	*72 %*	*65 %*	*70 %*	*72 %*	*65 %*	*70 %*	*72 %*	
Productivity	300	260	220	300	260	220	230	320	270	230	320	270	250	210	310	M3/ha
Stand-clone area	**35**	**8**	**17**	**56**	**7**	**7**	0	**79**	**1**	0	**36**	**14**	**9**	0	**61**	
Stand area			**60**			**70**			**80**			**50**			**70**	330
Seedling need	60,071	13,835	29,114	95,343	12,337	12,510	0	90,375	1,172	0	40,945	16,271	14,801	0	94,430	481,206
Potential production	10,496	2,095	3,730	16,659	1,868	1,603	0	25,272	276	0	11,450	3,839	2,371	0	18,760	98,420

Table 11.7 Example of clone cuttings requirements integrated to field planting and harvesting schedule

	Seedling requirements per clone					
	CL1	CL2	CL3	Sd Need		
1	16,250	13,835	21,425	51,510		
2	15,234	*0*	7,689	22,923		
3	22,544	*0*	26,325	48,869		
4	6,043	16,935	18,228	41,206		
5	*0*	25,445	12,116	37,561		
6	*0*	21,843	17,028	38,871		
7	14,801	8,271	20,733	43,806		
8	*0*	33,158	1,172	34,330		
9	23,843	9,345	9,094	42,282		
10	16,250	16,323	7,177	39,750		
11	16,250	12,337	*0*	28,587		
12	39,000	*0*	12,510	51,510		
	Mini-cuttings requirements per clone					
	CL1	CL2	CL3	**Mct need**		**Mct team constraint**
−2	25,000	19,764	10,679	55,443	<=	60,000
−1	23,437	0	36,563	60,000	<=	60,000
0	34,683	0	25,317	60,000	<=	60,000
1	9,297	24,193	16,827	50,317	<=	60,000
2	0	36,350	23,650	60,000	<=	60,000
3	0	31,204	28,796	60,000	<=	60,000
4	22,771	11,816	1,627	36,215	<=	60,000
5	0	47,369	12,631	60,000	<=	60,000
6	36,682	13,350	9,968	60,000	<=	60,000
7	25,000	23,318	0	48,318	<=	60,000
8	25,000	17,625	17,375	60,000	<=	60,000
9	60,000	0	0	60,000	<=	60,000

11.4.2.2 [M 11.7] Complete Integrated Model

$$Max\ Z = \sum_{i=1}^{I}\sum_{j=1}^{J}\sum_{k=1}^{K} p_{ij} \quad . \quad x_{ijk} \tag{11.4}$$

Subjected to

$$Total\ Area\Big)\ \sum_{j=1}^{J}\sum_{k=1}^{K} x_{ijk} = X_i, \quad for \forall i \tag{11.5}$$

$$Forest\ Health\Big)\ \sum_{i=1}^{I}\sum_{k=1}^{K} x_{ijk} \leq C_j, \quad for\ \forall j \tag{11.6}$$

$$Planting\Big)\ \sum_{i=1}^{I}\sum_{j=1}^{J} x_{ijk} \leq F_k, \quad for \forall k \tag{11.7}$$

$$Harvest\ Link)\ \sum_{i=1}^{I}\sum_{j=1}^{J} x_{ijk} = 0,\ for\ some\ stands\ i\ some\ periods\ k \quad (11.9)$$

$$Cutting\ Capacity)\ \sum_{j=1}^{J}\sum_{i=1}^{I} \frac{S_i .\ x_{ijk}}{E_i . R_j} <= T'_{k-n} \quad for \forall j\ and \forall k \quad (11.11)$$

Where:

Z probable future final product production
i stand;
j clone;
k period;
p_{ij} probable productivity of clone ***i*** when planted in stands ***j***
x_{ijk} area with a clone ***j***, in the stand ***i***, in period ***k;***
X_i total area of stand ***i***;
I number of stands in the plantation plan;
C_j maximum area to be planted of each clone ***j***;
J number of clones to be planted in the horizon;
F_k field planting capacity of the team prepared to the horizon in a period ***k;***
K number of periods in the horizon;
S_i spacing that must be used to plant in stand ***i;***
E_i field efficiency expected for the stand ***i;***
R_j cutting rooting rate of clone ***j***.
T'_{k-n} maximum amount of cuttings to be planted during the period k – n; that should occur n periods before planting all clone ***j***, in period ***k*** in all stands i.

11.5 Nursery Infra-structure Constraints

A process-oriented forest management should be able to bring together (i) the ecosystem processes view and (ii) the economic view (Fürst et al. 2007). The model presented in this chapter intends to bring together harvesting, planting and seedling production in a unique model to optimize the land use according to industrial objectives. Furthermore, the objective must be subjected to all natural and biological aspects of the entire process.

According to Fürst et al. (2007), the success of sustainable forest management depends on the ability to adapt management strategies to natural dynamics. A planting management plan, in order to meet the maximum production of the land, should still consider the biological infra-structure of a nursery: (i) the number of mini-stumps per clone in the clonal garden and (ii) the size of greenhouses.

11.5.1 Mini-stumps per Clone in the Clonal Garden

To have enough mini-cuttings of each clone, nurseries should have enough mini-stumps in clonal garden of each clone needed in the period. Actually, preparing a nursery to meet planting needs means considering the clonal garden design. Nursery managers must know in advance the planting plan of the next season to prepare the nursery in terms of how many mini-stumps they will need per clone (Brondani and Wendling 2012). Depending on the species, they must know the clones with more than a year in advance.

Therefore, one of the most relevant constraints of a nursery production is the amount of mini-stumps in the clonal garden. From this garden, sprouts will be collected periodically and transformed into mini-cuttings; to complete the planting model, we must consider two production constraints:

- The number of mini-stumps of each clone the nursery has in production;
- The speed the mini-stumps recover from the last collection.

Because of these constraints, nursery managers have, at least, two levels of planning:

- In the first level, after knowing a planting plan for a season, managers should design the clonal garden;
- In the second level, after knowing the very-short-term sequence of planting activities, they have to calculate the schedule of mini-cuttings planting.

Figure 11.5 shows the two levels of a nursery planning process. The mini-cuttings production schedule should be synchronized with the field planting schedule which, by its turn, must be synchronized with the short-term harvesting schedule. All of them must be subjected to the clonal garden infra-structure and should maximize the probable future production of the industry.

The productivity of a mini-stump is given by the number of sprouts that can be collected in a period to be transformed into mini-cuttings. The requirements of mini-stumps per clone per period can be written as follows:

$$D_{j\ k-n} = \frac{T_{j\ k-n}}{p_j} \tag{11.12}$$

Where:

$\boldsymbol{D_{jk-n}}$ mini-stumps requirements of the clone $\boldsymbol{j}$ in period $\boldsymbol{k-n}$;
$\boldsymbol{T_{jk-n}}$ mini-cuttings requirements of the clone $\boldsymbol{j}$ in period $\boldsymbol{k-n}$;
$\boldsymbol{d_j}$ how many sprouts can be collected from mini-stumps of the clone $\boldsymbol{j}$ in one period; a mini-cutting is generated from one sprout.

The value d_j is a biological parameter that depends on the clone; each clone has its own speed of regeneration from sprouts collection.

Now, the value of $T_{j\,k-n}$ in Eq. (11.10) can be substituted into the Eq. (11.11) to obtain the mini-stumps requirements as a function of primary values:

$$D_{j\ k-n} = \sum_{i=1}^{I} \frac{S_i\ .\ x_{ijk}}{E_j\ .\ R_j\ .\ d_j} \tag{11.13}$$

This constraint can be written as follows:

11.5.1.1 [M 11.8] Adding a Clonal Garden Constraint

$$Mini - Stumps\,\text{supp}\,ort\Big)\ \sum_{i=1}^{I} \frac{S_i\ .\ x_{ijk}}{E_i\ .\ R_i\ .\ d_j} \leq D_{jk-n} \qquad for\forall j\ and\forall k \tag{11.14}$$

Where

D_{jk-n} is the maximum that mini-stumps of clone ***j*** can support during period ***k − n***; that should occur n periods before planting clone ***j***, in period ***k*** in all stands;
x_{ijk} area with a clone ***j***, in the stand ***i***, in period ***k;***
S_i spacing that must be used to plant in stand ***i;***
E_i field efficiency expected for the stand ***i;***
R_j cutting rooting rate of clone ***j***.
p_j how many sprouts can be collected from mini-stumps of the clone ***j*** during the period ***k − n***

Tables 11.8 and 11.9 illustrate this constraint.

11.5.2 Greenhouse Size Constraint

After mini-cuttings planting, cuttings must go to the greenhouse to root and grow. There is a clear constraint related to the size of greenhouse prepared to receive the cuttings. Again, each clone, according to its adaptability to the environment, will take more or less time to root and grow inside the greenhouse.

The size of the greenhouse for the season is one of the results of the first planning level. During the season, for each period, cuttings production should be subjected to the greenhouse size constraint.

The size of greenhouse can be expressed in terms of a number of mini-cuttings it can contain in a period. Equation (11.10) shows how to calculate mini-cuttings requirements. The number of mini-cuttings in greenhouse must be less than the maximum number of mini-cuttings the greenhouse can contain.

Table 11.8 Process parameters per clone and technical recommendation unit (TRU)

Technical recommendation unit	TRU1						TRU2						TRU3		
Stands	St1			St2			St3			St4			St5		
Replant efficiency			3 %			3 %			3 %			3 %			3 %
Spacing (trees/ha)			1,667			1,667			1,111			1,111			1,515
Stand area			60			70			80			50			70
Clones	CL1	CL2	CL3	CL1	CL2	CL3	CL1	CL2	CL3	CL1	CL2	CL3	CL1	CL2	CL3
Clonal Garden Productivity (mini-cuttings/mini-stumps/month)	5	6	7	5	6	7	5	6	7	5	6	7	5	6	7
Nursery Rooting (seedlings/mini-cuttings)	65 %	70 %	72 %	65 %	70 %	72 %	65 %	70 %	72 %	65 %	70 %	72 %	65 %	70 %	72 %
Productivity (m3/ha)	300	260	220	300	260	220	230	320	270	230	320	270	250	210	310

Table 11.9 Mini-stumps support

Mini-stumps requirements per clone				Mini-stumps support in clonal garden per clone		
CL1	CL2	CL3		CL1	CL2	CL3
5,000	3,294	1,780	<=	16,000	8,000	10,000
4,687	0	6,094	<=	16,000	8,000	10,000
6,937	0	4,220	<=	16,000	8,000	10,000
1,859	4,032	2,805	<=	16,000	8,000	10,000
0	6,058	3,942	<=	16,000	8,000	10,000
0	5,201	4,799	<=	16,000	8,000	10,000
4,554	1,969	271	<=	16,000	8,000	10,000
0	7,895	2,105	<=	16,000	8,000	10,000
7,336	2,225	1,661	<=	16,000	8,000	10,000
5,000	3,886	0	<=	16,000	8,000	10,000
5,000	2,937	2,896	<=	16,000	8,000	10,000
12,000	0	0	<=	16,000	8,000	10,000

The constraint can be expressed as follows:

$$Greenhouse\ size\Big)\sum_{i=1}^{I}\sum_{j=1}^{J}\sum_{k=n}^{K}\frac{S_i.\ x_{ijk}}{E_i.\ R_j} \leq G_{k-n}\ for \forall k \qquad (11.15)$$

Where

$\boldsymbol{G_{k-n}}$ Maximum amount of mini-cuttings the Greenhouse can contain during period $\boldsymbol{k-n}$;

$\boldsymbol{x_{ijk}}$ area with a clone $\boldsymbol{j}$, in the stand $\boldsymbol{i}$, in period $\boldsymbol{k}$;

$\boldsymbol{S_i}$ spacing that must be used to plant in stand $\boldsymbol{i}$;

$\boldsymbol{E_i}$ field efficiency expected for the stand $\boldsymbol{i}$;

$\boldsymbol{R_j}$ cutting rooting rate of clone $\boldsymbol{j}$.

$\boldsymbol{p_j}$ how many sprouts can be collected from mini-stumps of the clone $\boldsymbol{j}$ during the period $\boldsymbol{k-n}$

$\boldsymbol{n}$ how many periods mini-cuttings will remain inside the greenhouse

It is necessary to emphasize that the amount of mini-cuttings that are planted in a period k will be inside of a greenhouse for n periods. The model should take into account the sum of mini-cuttings produced in n periods, since the period k – n.

If the nursery is working with clones that spend three periods to grow inside the greenhouse, it must be calculated to contain three times the capacity of mini-cuttings planting infra-structure. It takes part of the first level of planning, when managers are preparing the nursery for the planting season.

Figure 11.6 shows the entire process of seedling production in nursery prepared to feed a clonal forest. The picture shows the infrastructure of a Eucalyptus clonal nursery. There are other parts of the nursery process that could generate other constraints: (i) open space size and (ii) delivery infra-structure such as trucks or other equipment. However, all of them can be expressed in terms of the amount of mini-cuttings or seedlings they can deal with. The constraints can be built like the greenhouse size constraint.

11.6 A Tool to Implement a Continuous Planning Process

A nursery process should be analyzed as a continuous process like a supply chain process. When the first set of mini-cuttings in a planting season is ready to start the rooting and growth process, the greenhouse is empty. However, it is only the first time. From the first time on, nursery plan should consider the mini-cuttings that already are inside the greenhouse (Bae and Seo 2007).

The first step of a continuous planning should be reading a status of greenhouse occupancy from a nursery control system (Davis 2003). The expected data should contain how many mini-cuttings there are inside the greenhouse per clones and age.

The second step of the continuous planning system is reading the status of a short-term harvesting schedule from any harvest control system. These two sets of data complete the input of the model described in the previous sections of this chapter.

Figure 11.7 illustrates a complete system that could be developed to embed a model and produce an integrated nursery and field planting schedule. Nursery planning should often update nursery parameters based on control nursery data.

According to Sadic et al. (1998) past information from nursery records and knowledge accumulated through personal experiences are crucial in a process of intense planning. In other words, this model cannot give the benefits to the managers if there is no control system with parameters connected to reality.

At least parameters used as model coefficients should be controlled and updated. They can be grouped by source:

1. Geographic Information System (GIS) parameters
   - (X_i) Stand areas
   - Technical recommendation units and stands relationship
   - (E_i) Field planting efficiency
2. Genetic research parameters
   - (p_{ij}) Probable productivity of each clone and stand
   - (C_j) Maximum Area allowed to be planted per clone
   - (S_i) Tree Spacing
3. Field Planting parameters
   - (F_k) Field planting team capacity per period

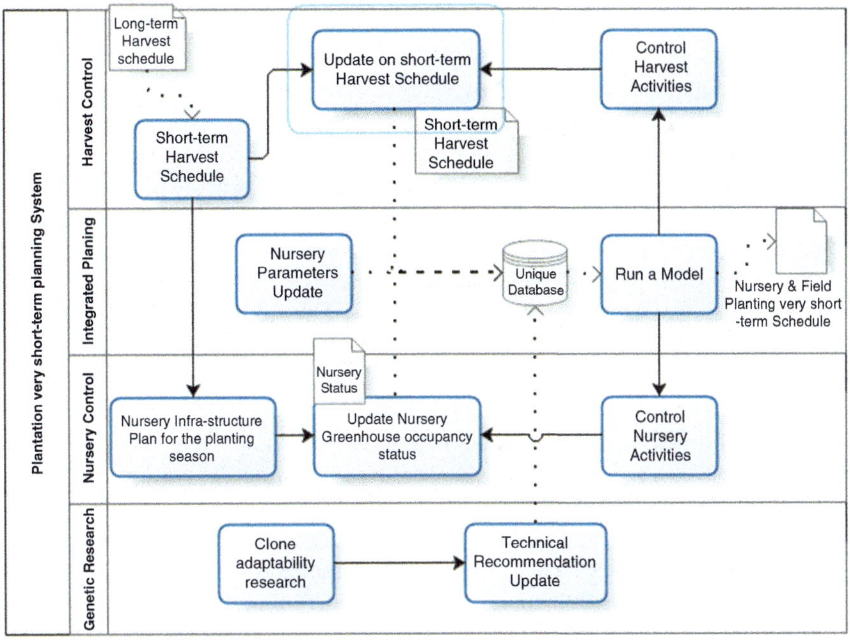

Fig. 11.7 Plantation very short-term planning system (Davis 2003)

4. Nursery parameters

- (T_{k-n}) Maximum mini-cuttings production per period
- (R_j) Rooting per clone
- (d_j) How many sprouts can be collected from mini-stumps of each clone per period
- (D_{jk-n}) Quantity of mini-stumps per clone in production in clonal garden during the horizon
- (G_{k-n}) Greenhouse capacity

The first step of an efficient system is accessing the last version of short-term harvest schedule and nursery inventory status. Then the system should perform the following steps: (i) read and interpret a set of listed parameters, (ii) generate a matrix, (iii) send to a solver, (iv) interpret a block of solver results, (v) calculate field planting and nursery schedule and (vi) produce reports.

According to Challa (2002), production model that integrates different disciplines to simulate and plan the whole production chain is a challenging development over the years. Also, putting a developed model into practice is still a challenge for forest managers.

Management Planning in Action 11.2

A large pulp company in Latin America developed a tool to implement the model described in this chapter. The aim of the project was to make effective nursery schedule to meet field planting. The industrial objective was maximizing the land use in terms of pulp yield per hectare.

After decades of investments, the R&D department had developed clones adapted to each micro region according to environmental factors. They know the future productivity of each clone if planted in each micro region in tons of pulp per hectare. However, before the project, they were not able to establish the right clone in the right stand.

Before the project, nursery managers complained that the short-term harvest schedule changed every week making nursery plans useless. It was common not having adapted clones to plant. However, it was a "push-button" system delivering a new plan for the nursery every month according to every small change in the harvest schedule.

The first step of the project was to make a map of nursery process, develop a small management system and determine parameters to support the model. The second step was convincing the stakeholders that the planning system should integrate three departments: harvesting, planting and nursery.

After the second year of operations, they could get the most impressive results they needed: future productivity had increased at least 2 m^3/ha.year in average MAI (mean annual increment), which represents more than US$10 million a year.

11.7 Summary

Productivity, site-fitness, micro daily planning of strategic genetic material distribution were discussed to determine how, when, where and which clone plants should be produced. This chapter covered a method and criteria used to allocate recommended clones to each technical recommendation unit during silvicultural activities.

This chapter describes a step-by-step nursery and planting model. Essential components of the optimization planning were included in the model, among which: (i) clone garden productivity; (ii) amount of mini-stumps available; (iii) mini-cuttings operational capacity; (iv) nursery infra-structure; (v) amount of time of each clone plant takes to grow; (vi) productivity of each clone in different technical recommendation units; (ix) number of periods to plant after harvesting.

11.8 Problems

1. Modify the proposed model to include differences in nursery productivity in summer and winter. Detailed suggestions, constraints and variables should be added.
2. Propose a MS Excel® spreadsheet to implement the model presented in this chapter.
3. Describe different objectives that can be used to represent future production if we have to optimize a pulp production.

References

ABRAF (2013) Year book statistical ABRAF. ABRAF, Brasília

Assis T, Resende M (2011) Genetic improvement of forest tree species. Crop Breed Appl Biotechnol 11 http://www.scielo.br/scielo.php?pid=S1984-70332011000500007&script=sci_arttext&tlng=es

Bae H, Seo Y (2007) BPM-based integration of supply chain process modeling, executing and monitoring. Int J Prod Res 45(11):2545–2566

Barros N, Novais R, Carmo DN (1986) Classificação nutricional de talhões florestais – descrição de uma metodologia. Revista Árvore 1:112–120

Brondani G, Wendling I (2012) Dynamics of adventitious rooting in mini-cuttings of Eucalyptus benthamii X eucalyptus dunni. Acta Scientiarum Agron 34(2):169–178

Challa H (2002) Crop models for greenhouse production systems. In: I. s. science (ed) Proceedings of the 4th international symposium on models for plant growth and control in greenhouses: modeling for the 21st century, Beltsville, MD, pp 47–53

Davis D (2003) The nursery management information system at J. Herbert Stone Nursery using MS Access. In: National proceedings: forest and conservation nursery associations 2002, Olympia, WA, pp 130–132

Fürst C, Vacik H, Lorz C, Makeschin F, Podrazky V, Janecek V (2007) Meeting the challenges of process-oriented forest management. For Ecol Manag 248:1–5

Oskarsson H, Brynleyfsdottir S (2009) The interaction of fertilization in nursery and field on survival, growth and frost heaving of birch and spruce. Iceland Agric Sci 22:59–68

Roberds J, Bishir J (1997) Risk analyses in clonal forestry. Can J For Res 27(3):425–432

Rosado A (2012) Seleção simultânea de clones de eucalipto de acordo com produtividade, estabilidade e adaptabilidade. Presquisa Agropecu Bras 47(7):964–971

Sadic R, Stock M, Wenny D, Robberecht R (1998) An expert system nursery management. AI Appl 12(1–3):41–50

Stage A (2003) How forest models are connected to reality: evaluation criteria for their use in a decision support. Can J For Res 33:410–421

Trkman P (2010) The critical success factors of business process management. Int J Inf Manag 10:125–134

Chapter 12
Dynamic Treatment Units in Forest Management Planning

Timo Pukkala, Petteri Packalén, and Tero Heinonen

12.1 Introduction

The concept "dynamic treatment unit" refers to such an approach in forest management planning, which does not consider stand compartments as permanent units of data collection, planning and management. The treatment units are created for the accomplishment of prescribed treatments, after which they are no longer required (Fig. 12.1). According to the forest management plan, the area within the dashed boundary should be thinned in 2020 (sub-areas 1 and 2), and the area within the thick solid boundary should be thinned in 2030 (sub-areas 2 and 3). The grey area (1 + 4) should be clear-cut in 2040. Each sub-area (1 + 2, 2 + 3, 1 + 4) only appears for the treatment. Between them there is just forest or trees, without any subdivision into compartments.

The interest towards using dynamic treatment units instead of fixed compartments is related to the ongoing changes in forest inventory. Inventory methods that are based on airborne laser scanning (ALS) are gaining popularity in several countries and in e.g. Finland and Norway ALS-inventory is already the main method in management-oriented forest inventory. The results of ALS-inventory can be calculated for any subdivision of the forest; the calculation units may be traditional stand compartments, micro stands formed by segmentation methods, or raster cells.

There are two approaches to carry out ALS based forest inventory. In the area-based approach, stand attributes are regressed against the height and density metrics of ALS echoes and a model is applied in a wall-to-wall manner using square cells or hexagons of approximately the same size as the sample plots. An alternative method is to detect individual trees and to predict tree level attributes from ALS data.

T. Pukkala (✉) • P. Packalén • T. Heinonen
University of Eastern Finland, Joensuu, Finland
e-mail: timo.pukkala@uef.fi

J.G. Borges et al. (eds.), *The Management of Industrial Forest Plantations: Theoretical Foundations and Applications*, Managing Forest Ecosystems 33,
DOI 10.1007/978-94-017-8899-1_12,

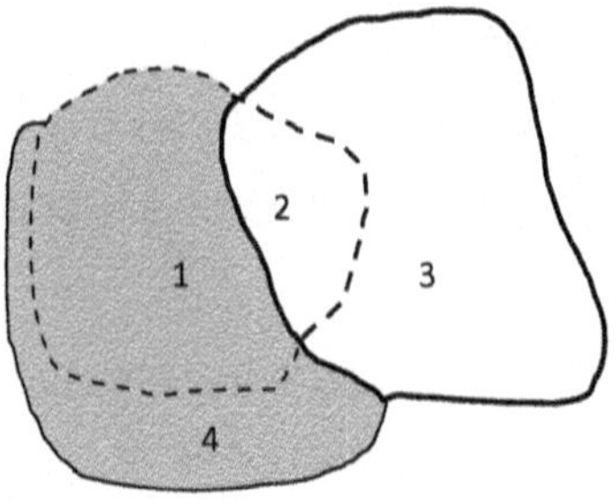

Fig. 12.1 Three dynamic treatment units: *dashed* (1 + 2), *thick solid* (2 + 3), and *grey* (1 + 4)

Detected trees can be aggregated into any spatial unit needed. Currently, most large scale ALS inventories use the area-based approach. Inventories that use space-borne data, such as Landsat or radar images, provide predictions for pixels corresponding to cells in ALS-inventory.

Although the dynamic units discussed in this chapter are called treatment units, dynamic spatial units can be formed also for forest features, not only for cuttings and other treatments. It is possible to aggregate raster cells that represent old-growth forest, certain habitat type, etc. The aim may be to reach a certain landscape structure in a particular year. In this case, the purpose of dynamic spatial units is to check that the landscape structure meets some ecological requirements. These kind of ecological spatial units are even more transient than the treatment units since they are only formed for making computational evaluations about landscape structure at certain moment. They are not used for management actions and they are never delineated in the forest. Of course they may continue existing in the forest but they may also start expanding or shrinking immediately after the evaluation year.

Spatial optimization is the most common method to form dynamic spatial units. The optimization problem is formulated in such a way that adjacent raster cells or micro stands having similar features or management prescriptions will be aggregated. However, optimization is not the only way. In Finland, UPM Kymmene has developed a so-called diffusion method to aggregate cutting operations. The forest is divided into small homogeneous segments (micro stands) by applying an automated segmentation algorithm to digital aerial photograph. Forest variables are interpreted for each micro stand. The algorithm calculates a harvest index for each segment, which describes the urgency or possibility of harvesting, based on stand age, stand density, tree size, etc. Displaying segments with harvesting index higher than a certain threshold usually shows a too fragmented harvest block map. The harvesting indices are therefore adjusted so that the index of a certain segment also depends on adjacent segments. Finally, a feasible harvest block proposal can be obtained (Fig. 12.2).

The main reason for using dynamic treatment units is that they may enable a more efficient utilization of forest resources than fixed compartments (Heinonen et al. 2007; Fig. 12.3). Fixed compartment boundaries can be regarded as constraints since the limits of treatments are forced to follow these boundaries. In many cases,

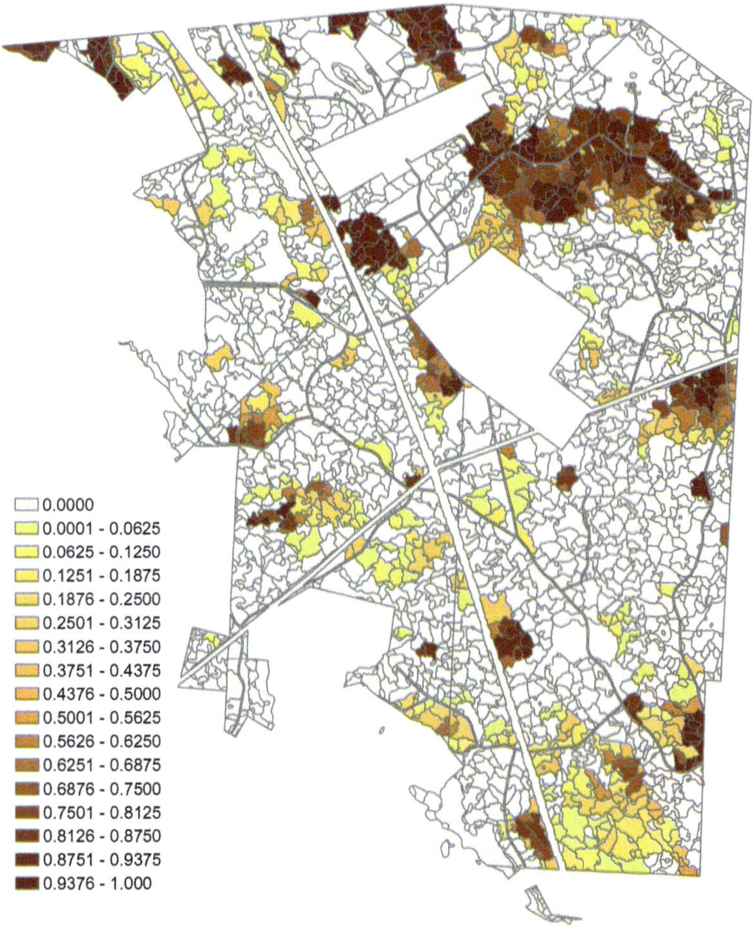

Fig. 12.2 The suitability for cutting (harvesting index) derived by the diffusion method of UPM Kymmene

this prevents optimal management since forest features do not necessarily follow compartment boundaries (Fig. 12.4). For example, if only a part of a compartment is financially mature, immediate cutting means too early cutting for a part of compartment, and postponed cutting results in too late cutting for a part of it. Sometimes it is optimal to divide also homogeneous compartments, for instance to exactly meet the constraints of a LP problem.

This chapter discusses methods that can be used to form dynamic treatment units by numerical optimization. The emphasis is on heuristic methods although integer programming could also be used. So-called decentralized heuristics are described, and results obtained with different methods are compared. Finally, some case study results from Finland and Brazil are reported.

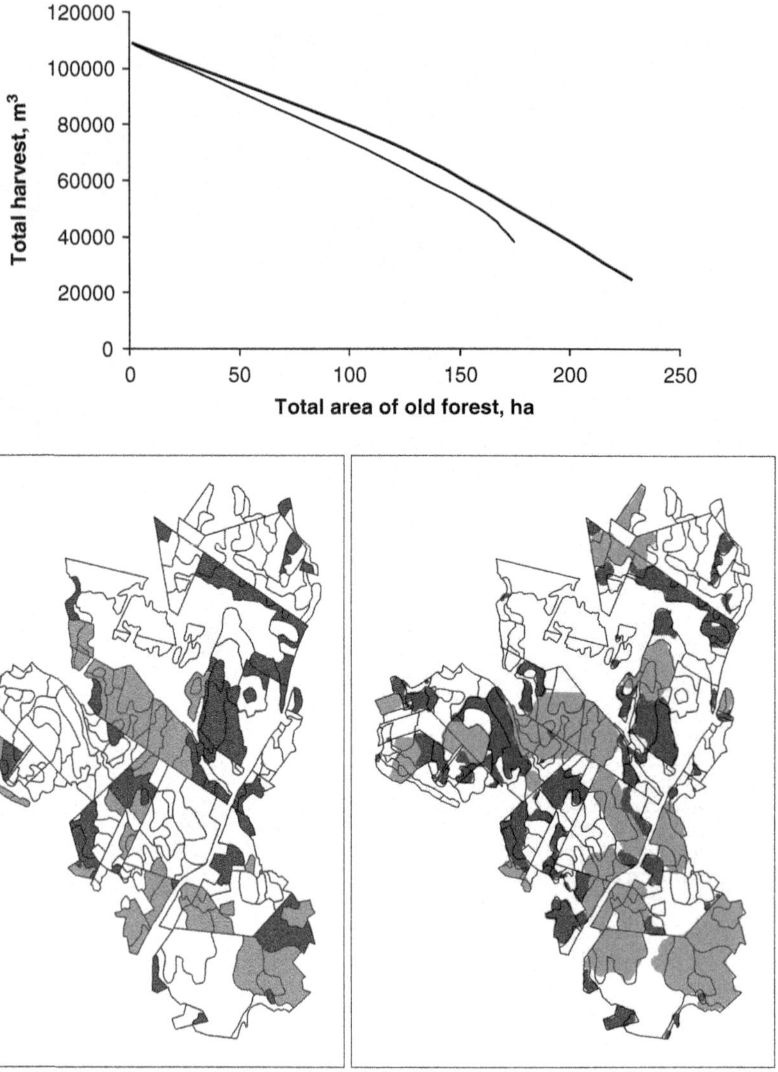

Fig. 12.3 *Top*: Trade-off curve between total harvest and the area of old forest when fixed compartments (*thin line*) or dynamic units (*thick line*) are used. *Bottom*: Cuttings and old forests in a management plan based on fixed compartments (*left*) and dynamic units created by spatial optimization (*right*). *Dark grey* indicates old-forest patches and *light gray* indicates harvest blocks

What You Will Learn in This Chapter

- The concept of dynamic treatment unit
- Utility function
- Utility-theoretic problem formulation
- Benefits of using dynamic treatment units

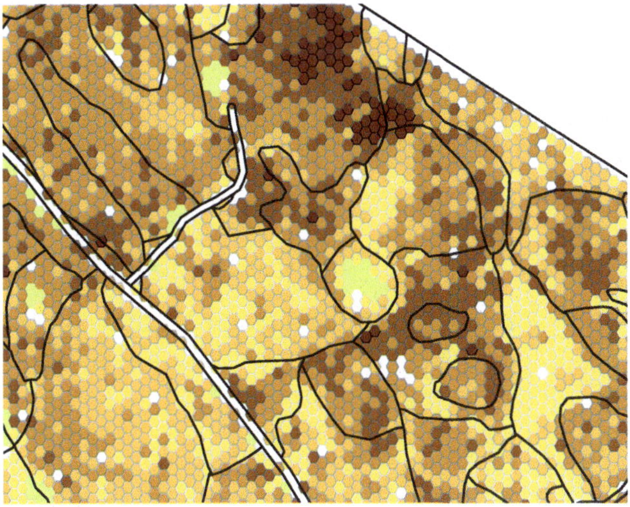

Fig. 12.4 Variation in growing-stock characteristics does not follow compartment boundaries (*solid lines*). The *brown tone* indicates standing volume interpreted for hexagons using ALS data. *Darker tone* implies higher volume

- The concepts of centralized and decentralized optimization
- The concept of cellular automaton
- Methods for decentralized optimization
- Two-stand (2-opt) move
- Spatial optimization

12.2 Problem Formulation

Forest planning problems can be formulated in several different ways. The reader is referred to the presentation of mathematical programming models and heuristic solution approaches in Chaps. 2 and 7. A problem formulation, which is common especially in Finland, is the utility function approach. The objective function and the whole planning problem may be formulated as follows:

$$\text{Max} \quad U = \sum_{k=1}^{K} w_k u_k\left(q_k\right)$$

Subject to

$$q_k = \mathrm{Q}_k\left(\mathbf{x}\right) \qquad k = 1, \ldots, K$$
$$\sum_{i=1}^{n_j} x_{ij} = 1 \qquad j = 1, \ldots, n$$
$$x_{ij} = \{0, 1\}$$

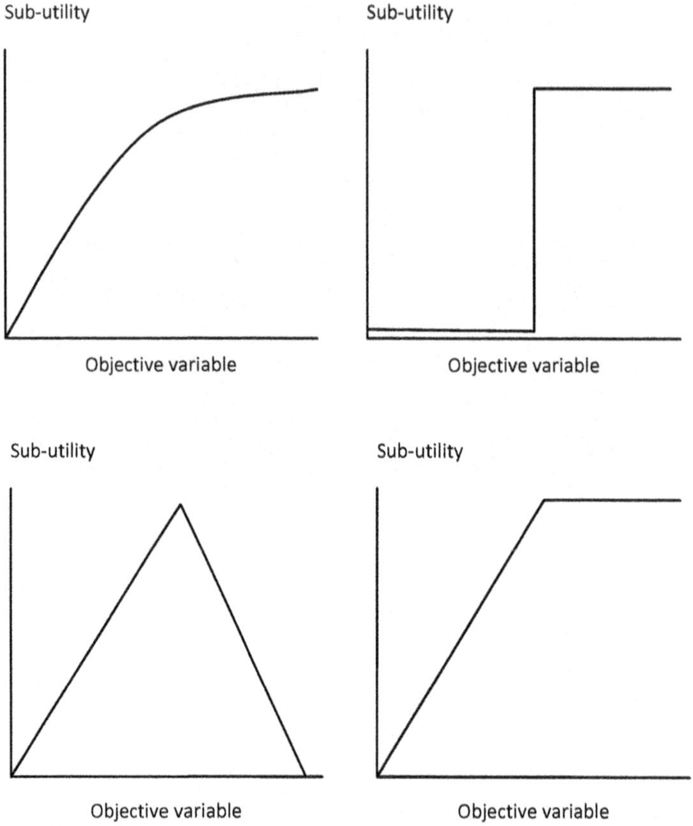

Fig. 12.5 Alternative sub-utility functions. *Top-left* describes decreasing marginal utility; *top-right* is a constraint-type function; *bottom-left* minimizes deviations from a target value; and *bottom-right* minimizes under-achievement

where **x** is a vector of 0–1 variables (x_{ij}) expressing whether ($x_{ij} = 1$) or not ($x_{ij} = 0$) treatment alternative i of stand j is included in the solution. There is a sub-utility function u_k for each objective variable, which removes the units of the original objective variable and scales it to range from zero to one. If the sum of the weights (w_k) is equal to one, as usual, the utility (U) will also range from zero to one. The sub-utility functions may have different forms (Fig. 12.5).

The utility function approaches may have constraints but there is seldom need for them since the sub-utility functions may be formulated in such a way that they guide the optimization process in the same way as constraints do. For example, the sub-utility function (top-right in Fig. 12.5) corresponds to a constraint, especially when used in a multiplicative utility function:

$$\text{Max} \quad U = \prod_{k=1}^{K} \left(u_k\left(q_k\right)\right)^{w_k}$$

In an additive utility function this sub-utility function (top-right in Fig. 12.5) determines whether the utility contribution of the objective variable is zero or one. If the weight of this variable (w_k) is high enough, the solution will have the corresponding variable at an acceptable level. If the principal objective variable, for instance net present value, has a sub-utility function similar to the top-left shape in Fig. 12.5 and the other functions are similar to the bottom-left shape, the formulation resembles the penalty function and goal programming approaches. The bottom-right shape corresponds to formulations in which only under-achievement is penalized.

12.3 Description of Possible Optimization Methods

Formulating the problem without using strict constraints may offer some computational advantages. The feasible region (which is now more conveniently called as "good region") may be approached also from outside, and it is possible to temporarily leave the good region. This makes it easier for the algorithm to get out from local optima and find new combinations of decision variables. The need of using strategic oscillation or temporary relaxation of constraints is minimal. There is no need to find an initial feasible solution since all solutions are feasible from the problem formulation point of view.

Another useful technique that facilitates search very near the border of the good or feasible region is that of using larger neighborhoods, which means making simultaneous changes in more than one stand or calculation unit. For example, a cutting prescription of stand 5 may be replaced by a cutting prescription in stand 6. These kinds of 2-stand (Heinonen and Pukkala 2004) or 2-opt (Bettinger et al. 1999) moves make it possible to alter the locations of cuttings without affecting much the total harvested volume. This is a benefit especially when optimization aims at aggregating cuttings or forest features.

The target of aggregating treatments or forest features can be incorporated in the problem formulation in several different ways. A simple way is to maximize, as one objective, the length or share of the boundary of such adjacent cells or compartments which are both cut in the same year or during the same period. This objective variable aggregates cuttings but it simultaneously maximizes the cutting area. When aiming at small and compact harvest blocks, it is possible to restrict the total cutting area or minimize the share of cut–uncut boundary. When there are requirements concerning the maximum area of harvest blocks or their size distribution, it is necessary to inspect which of the cut stands or pixels belong to the same harvest block. This is an additional computational burden, which may substantially slow down the optimization. In Finland, where managed forests are mosaics of stands of different age, site and species composition, and forested landscapes are interrupted by lakes and fields, there is seldom need to restrict the size of a single harvest block.

12.3.1 Decentralized Optimization

Standard meta heuristics developed for combinatorial optimization such as simulated annealing, tabu search, genetic algorithm and ant colony optimization, may be used to solve planning problems in which one aim is to aggregate treatments or forest features. However, when the calculation units are small and the area is large, optimization becomes easily very slow. This is because all objectives are global, i.e., they concern the whole area, which sometimes makes the calculation of their value more tedious. Moreover, the decision space is very large. The decision space consists of all possible combinations of treatment alternatives of stands. Even in a small case with 5 treatment alternatives per stand and 100 stands, there are as many as 7.88×10^{69} possible plans (1.15×10^{54} millions of plans for each person on earth).

Another approach is to decompose the optimization task into stand or cell level problems. This is called decentralized approach. In the above case, there are only 5 decision alternatives per stand and only 100 stands. If the objective is to maximize total harvest or wood production, only 5×100 evaluations need to be done, instead of 7.88×10^{69} evaluations if total enumeration is used in the former case. Since the objective variables are often more complicated than wood production or total harvest, and there may be constraints, special problem formulations and iterative search techniques are needed to guarantee that the stand or cell solutions are acceptable also at the forest level. Despite this, decentralized techniques may be the faster option with a high number of calculation units. Two of these decentralized approaches are discussed below, namely a cellular automaton approach (Heinonen and Pukkala 2007) and a spatial version of the reduced costs approach (Pukkala et al. 2009).

12.3.1.1 Cellular Automata

The cellular automaton approach originally proposed by Strange et al. (2001, 2002) and modified by Heinonen and Pukkala (2007) consists of two stages: local optimization and global optimization. Local optimization maximizes the objective function separately for each cell. The local function may include cell-level variables and neighborhood variables. For example, one may maximize net present value and the similarity of treatments of neighboring cells. At the forest level, this results in high net present value and aggregated treatments. However, forest level objectives such as certain total harvest or similar removals during successive time periods are not usually met. Therefore, the second stage is launched, in which the local objective function is augmented with a global objective function measuring the achievement of global objectives.

The algorithm begins with the selection of a random treatment schedule for each cell (or micro segment or stand). Then, mutation and innovation are applied to each cell with certain probabilities. A mutation is equal to replacing the current treatment schedule by a random schedule. An innovation means that the current treatment schedule is replaced by the best one simulated for the cell. The process of employing mutations and innovation is repeated for many iterations. For iteration t, the probabilities for innovation (P_I) and mutation (P_M) are as follows:

$$P_I = P_I^0(1 - t/T)^{\tau_I}$$

$$P_M = P_M^0(1 - t/T)^{\tau_M}$$

where $P_I{}^0$ is the initial probability of innovation, $P_M{}^0$, is the initial probability of mutation, T is the total number of iterations, and τ_I and τ_M are parameters (>0).

In the second stage, the cell level objective function (referred to as z_{local}) is replaced by the following function which includes a global component (z_{global}) for evaluating how well forest level targets (e.g. certain total removal) are met:

$$\text{Max} \quad z = \frac{A_{\text{cell}}}{A_{\text{forest}}} z_{\text{local}} + b z_{\text{global}}$$

A_{cell} is the area of the cell and A_{forest} is the surface area of the entire forest. Parameter b is the weight of the global objective function. The weight is gradually increased until the target levels of global objectives are met or they cannot be improved any more. In the second stage, all treatment schedules of all cells are systematically evaluated, and a better candidate always replaces the current treatment schedule. The global function z_{global} may be a utility function with sub-utility functions similar to the top-right, bottom-left, or bottom-right shapes of Fig. 12.5.

12.3.1.2 Spatial Version of the Reduced Cost Method

Another method for local decentralized optimization is a variant of the reduced cost method (Hoganson and Rose 1984) described in Pukkala et al. (2009). The cell-level objective function is the reduced cost derived from the linear programming theory

$$\text{Max} \quad RC_{ij} = c_{ij} - \sum_{l=1}^{L} a_{ijl} v_l$$

where RC_{ij} is the reduced cost of schedule i of cell j, c_{ij} is the amount of objective variable in this schedule, a_{ijl} is the value of constraining variable l produced by the schedule, and v_l is the dual price of constraint l. In the spatial application of the RC method, the single objective variable (c_{ij}) is replaced by a local priority function (P_{ij}), in which the priority may depend on neighborhood variables:

$$\text{Max} \quad RC_{ij} = P_{ij} - \sum_{l=1}^{L} a_{ijl} v_l = \sum_{k=1}^{K} w_k p_k \left(q_{ijk}\right) - \sum_{l=1}^{L} a_{ijl} v_l \qquad (12.1)$$

K is the number of local objective variables, w_k is the weight of objective k, p_k is a sub-priority function for local objective k, and q_{ijk} is the amount of objective k in treatment alternative i of cell j. Similarly to the CA method, a part of the local objectives may measure the similarity of the treatments or features of the cell with those of adjacent cells. Maximizing similarity leads to aggregation of treatments and features.

The dual prices of the global constraints are given initial values which may be equal to zero in the absence of better information. After completing an iteration, dual prices are updated by using the sub-gradient method. The constraints are divided into '≥' (falling short of the target is not allowed) and '≤' (exceeding the target is not allowed) types. An equality constraint (=) is replaced by two constraints: a ≤ and a ≥ constraint. The new dual price of a constraint ($v_{k,t+1}$) is obtained from:

$$v_{k,t+1} = v_{k,t} + Step_k\, Violation_{k,t}$$

$$Violation_{k,t} = \begin{cases} 0 \text{ if } q_{k,t} \le T_k \text{ for } \le \text{ constraint} \\ 0 \text{ if } q_{k,t} \ge T_k \text{ for } \ge \text{ constraint} \\ (q_k - T_k)/T_k \text{ if } q_{k,t} > T_k \text{ for } \le \text{ constraint} \\ (q_k - T_k)/T_k \text{ if } q_{k,t} < T_k \text{ for } \ge \text{ constraint} \end{cases}$$

where $q_{k,t}$ is the current amount of variable k in the whole forest, and T_k is the target value of variable k. Parameter $Step_k$ determines the rate of changing the dual price of constraint k. The whole optimization process with the RC method consists of the following steps:

1. Produce an initial solution
2. Set initial dual prices
3. Select the best schedule for every stand using Eq.12.1
4. Calculate the values of forest-level goals (constraining variables)
5. Update dual prices
6. Repeat Steps 3–5 until the forest level constraints are met

12.4 Comparison of Optimization Methods

The illustration and comparison of methods used to define dynamically the treatment units considered a 1,395-ha forest classified into micro segments formed by using automated segmentation of a digital aerial photograph. The stand characteristics of the micro segments were derived from field plots and ALS data by using the area-based interpretation approach. There were 3,136 micro segments with an average area of 0.44 ha (Fig. 12.6). The management plan was developed for two 10-year periods, and the maps show the cuttings of the second period. The aim was to maximize NPV, maximize Cut-Cut border (CC) during both periods, minimize Cut-Uncut border (CuC) and minimize the border between adjacent micro segments of which one is treated with final felling (RuR). Harvest incomes were calculated

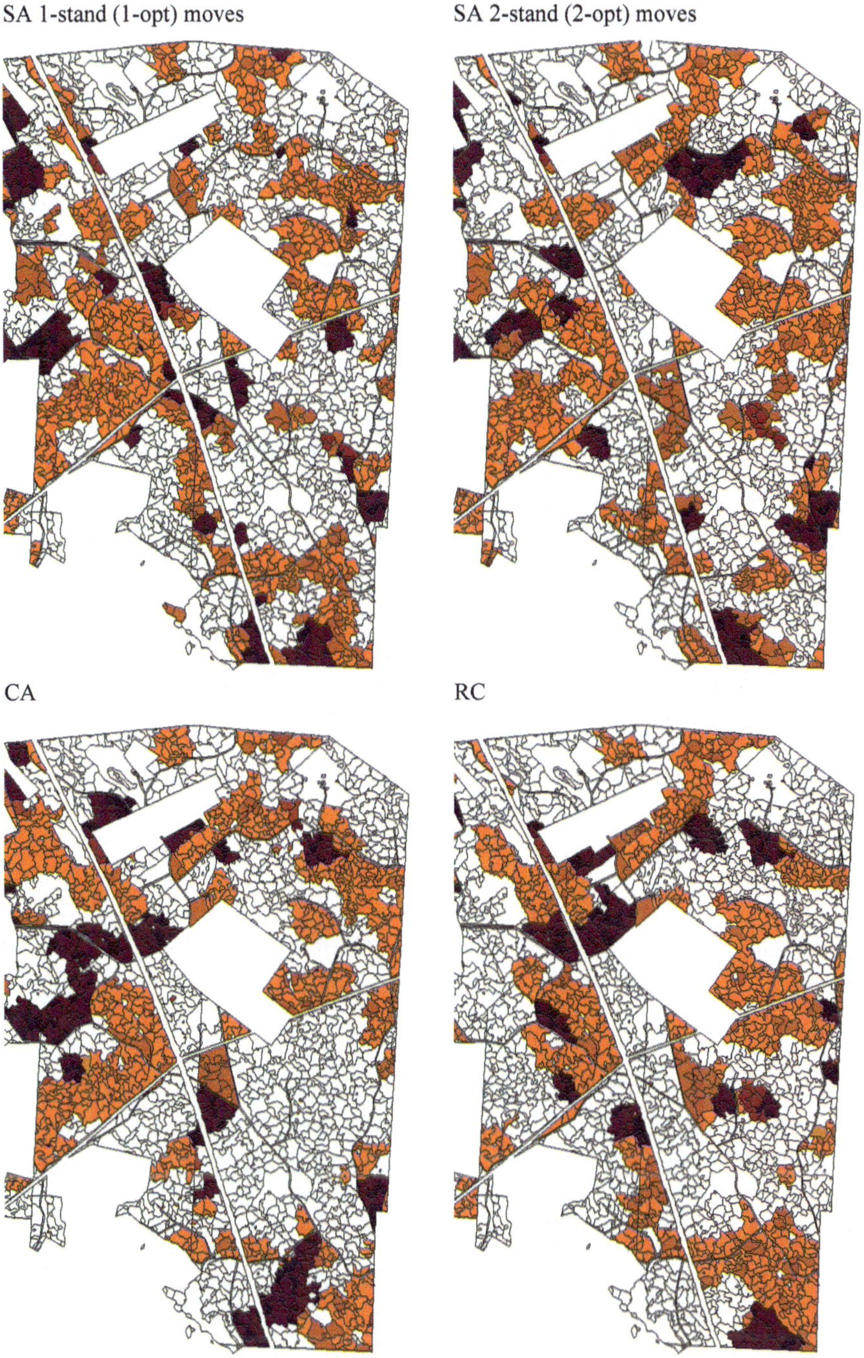

Fig. 12.6 Cutting prescriptions for the second 10-year period in solutions obtained with different optimization methods. *Purple tones* indicate regenerative cuttings and *brown tones* indicate thinnings

Table 12.1 Values of objective variables in the solutions obtained with different optimization methods

	SA 1-opt moves	SA 2-opt moves	CA	RC
NPV, maximized	8616437	**8634115**	*8500436*	8547434
CC, maximized	*0.3223*	0.3396	0.3231	**0.3524**
CuC, minimized	*0.1671*	0.1503	0.1588	**0.1226**
RuR, minimized	*0.0238*	**0.0212**	0.0425	0.0232
Time, s	167	187	**85**	*632*

The best value is in *boldface* and the worst value in *italics*

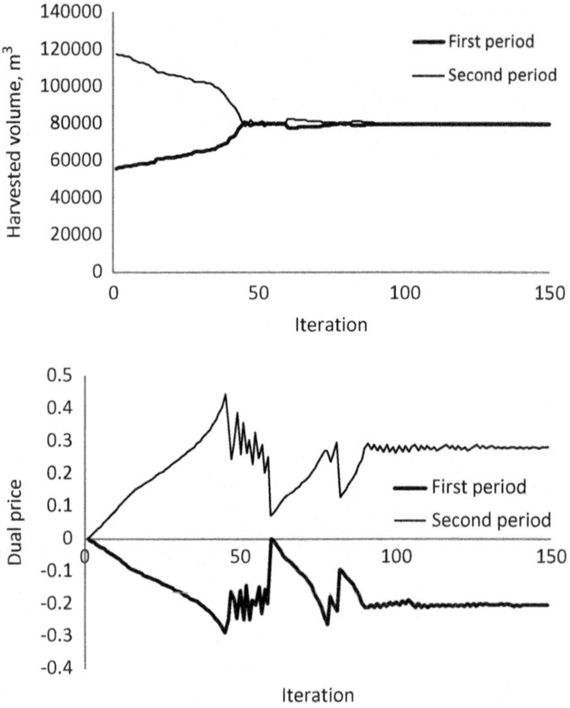

Fig. 12.7 Development of periodical cutting volumes (*top*) and dual prices for cutting volumes (*bottom*) during an RC optimization process in the forest shown in the maps of Fig. 12.6. The dual price of the first period is for the "Harvest ≥ 80,000" constraint and for the second period it is for the "Harvest ≤ 80,000" constraint. The other two dual prices (Harvest ≤ 80,000 during the first period and Harvest ≥ 80,000 during the second period) were very close to zero during the whole optimization process

with stumpage prices. Therefore, NPV did not depend of harvest block size. It describes the profitability of the treatment schedule provided that, in each harvest, the cell belongs to a large enough cutting block. The last objective (RuR) minimizes the length of the outer border of regeneration areas. The cutting target of both 10-year periods was 80,000 m^3. It had to be reached exactly, which means that both under- and over-achievement were minimized. Global optimization results obtained with simulated annealing (SA) are shown for comparison. Both 1-stand and 2-stand moves (1-opt and 2-opt moves) were used in SA.

The results show that CA was the fastest method and RC was the slowest. SA with 2-stand moves resulted in the highest NPV whereas RC was the best in terms of spatial objective variables (Table 12.1). SA with 2-stand moves was better than

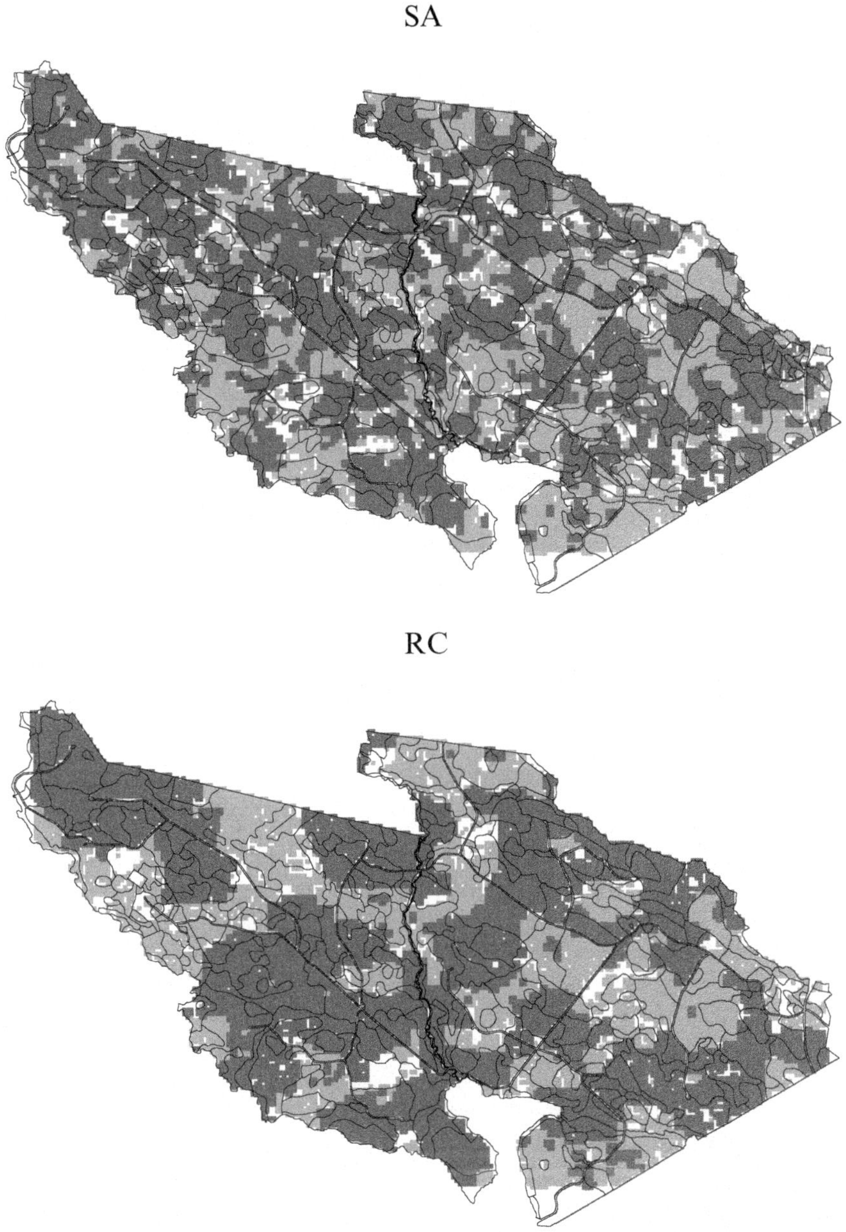

Fig. 12.8 Cutting areas (*dark grey*) of the third 20-year time period and old forests (*light grey*) at the end of the third time-period in solutions produced by simulated annealing (*SA*) and the reduced cost method (*RC*). The calculation units are pixels of Landsat 7 imagery

SA with 1-stand moves in terms of all four objective variables. The 10-year cutting targets were very closely reached in all solutions.

The speed of RC greatly depends on the initial guesses of dual prices and the step of changing the dual prices. In this exercise the initial guesses were not very good with a consequence that many iterations were required until the harvesting targets and dual prices converged to a constant value (Fig. 12.7).

In this comparison the calculation units were still large (4,400 m^2 on average) and their number was small (3,136). Increasing number of calculation units decreases the speed of optimization faster in centralized heuristics as compared to decentralized techniques.

Pukkala et al. (2009) compared SA, CA and RC in a case in which the calculation units were pixels of a Landsat 7 image. The stand characteristics were obtained by using the *k*-nn (*k* nearest neighbours) method (Tokola et al. 1996). Altogether 637 field plots were used in estimation. The study area consisted of about 18,000 pixels. NPV was maximized with periodical cutting targets while simultaneously aggregating both cuttings and old-forest patches. The parameters of the optimization methods were adjusted so that the running time was nearly the same in all methods.

NPV was 1 % higher in the SA solution, as compared to CA and RC, but the spatial objectives, i.e., cutting aggregations and old-forest aggregations were much better in CA and RC. RC found a clearly better aggregation than SA in the same computing time (Fig. 12.8). The landscape structure in the SA solution is much more fragmented than in the RC solution.

Management Planning in Action 12.1: Case Study Finland

Combined use of ALS-based inventory and dynamic treatment units was tested in a 440 ha forest holding in eastern Finland (Heinonen et al. 2008). The forest was rather heterogeneous and gradual changes in stand characteristics were common. The area-based approach was used to interpret the stand characteristics of calculation units. Hexagons of three different sizes (291, 541 or 1,182 m^2) were used as calculation units. The field data consisted of 447 circular plots 254 m^2 in size. Digital aerial photographs were used as additional source of data, to interpret the species compositions of the hexagons.

Variables calculated from the vertical distribution of the laser pulse data were used as predictors in models that predicted the stand characteristics required as inputs data in forest management planning. The interpreted spatial variables included species-specific stand basal areas, mean tree height, mean tree diameter, etc. The interpretation models were based on the field plot and laser pulse data of those hexagons in which field plots had been measured.

A forest planning software (Monsu) was used to simulate alternative treatment schedules for the hexagons for two 5-year periods. The reduced cost

(continued)

(continued)

method described above was used to maximize NPV while simultaneously aggregating cuttings. The cell-level objective function was

$$OF_{cell} = 0.05 \times (NPV/NPV_{max}) + 0.15 \times CC–0.60 \times CuC–0.20 \times RuR$$

where NPV is the net present value (€/ha) of the treatment schedule (including the net present value of ending inventory), NPV_{max} is the maximum NPV of all treatment schedules of all cells (€/ha), CC is the proportion of Cut–Cut border, of the total border length of the hexagon (proportion of the border of such adjacent cells which are both cut during the same 5-year period), CuC is the proportion of Cut–Uncut border, and RuR is the proportion of boundary length between such adjacent hexagons of which one is treated with final cut. NPV and CC were maximized whereas CuC and RuR were minimized. The most important objective was CuC, which means that much emphasis was given to getting compact harvest blocks with a short outer boundary.

Three forest level constraints were included to the optimization problem: the total harvest during the first and second 5-year period (H1, H2) had to be at least 15,000 m^3 and 2,000 m^3, respectively, and the ending volume after 10 years (V) had to be at least 50,000 m^3. Therefore the maximized cell level objective function was

$$RC = OF_{cell}–v_1H1–v_2H2 - v_3V$$

where v_1, v_2 and v_3 are the dual prices of forest level constraints. This function was maximized for every cell for many iterations. The dual prices of constraints were up-dated after every iteration on the basis of the deviation of the harvested or ending volume from its target value.

Cuttings do not follow the compartment boundaries that were drawn when the previous management plan was prepared (Fig. 12.9). The cuttings are more or less in the same places with both cell sizes (291 and 1,182 m^2) but the harvest blocks are more fragmented with the smaller size (Fig. 12.9). This is mainly because there are more cells which do not meet the cutting criteria when the cell size is small (stand basal area is too low for thinning, or tree size is too small for final felling), which means that cutting alternatives have not been simulated to those cells. This is not a problem within a thinning block, but individual uncut pixels within a clearfelling area are problematic for implementation. On the other hand, the smaller pixel size resulted in better objective function value (values of objective variables within the whole forest), suggesting that fine-grained partitioning of the forest facilitates a more efficient use of the forest resource.

(continued)

(continued)

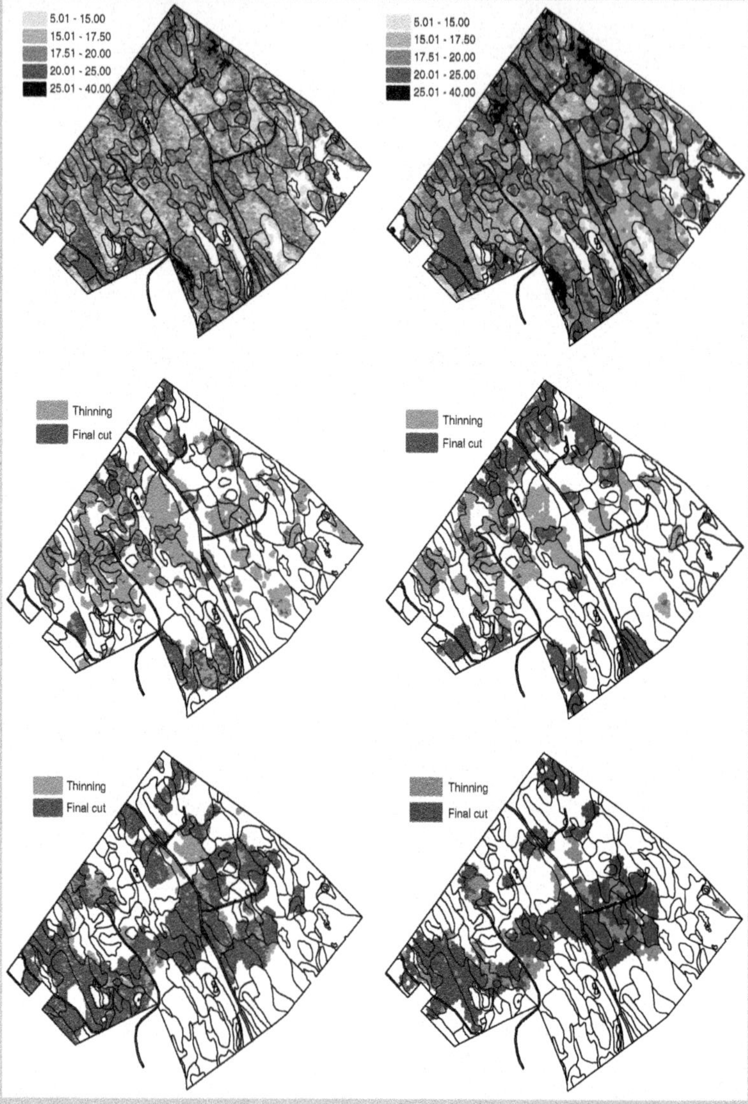

Fig. 12.9 Initial mean tree diameter interpreted for 291 m^2 (*left*) and 1,182 m^2 (*right*) hexagons. The maps in the *middle* show the thinnings (*light grey*) and final cuts (*dark grey*) during the first 5-year period and the maps at the *bottom* show the cuttings of the second 5-year period when the cell size is 291 m^2 (*left*) or 1,182 m^2 (*right*)

Management Planning in Action 12.2: Case Study Brazil

Another case study was conducted in a fast-growing eucalyptus pulpwood plantation in Bahia state, Brazil (Packalén et al. 2011). Since the compartments are large, it was hypothesized that there is variation in site productivity within the compartment with a consequence that the same rotation length must not be used within the entire compartment.

The field data consisted of 195 plots (531 m^2). The ALS data were collected using an Optech ALTM 3,100 laser scanning system. A digital terrain model (DTM) was generated from the ALS data using the ground point and the DTM was subtracted from the heights of the laser points to scale the ALS data to the ground level.

Linear mixed-effect modeling was used to fit models for dominant height (HD) and stand basal area (G) using variables calculated from ALS data as predictors (for instance percentiles of the height distribution of the echoes). In the used area-based approach, HD and G were interpreted for 300-m^2 hexagons. The future development of each cell was simulated for 10 years using different rotation lengths and a compatible growth and yield model. The site index of the hexagon was obtained from the known planting year and estimated dominant height. Figure 12.10 shows that there is considerable within-compartment variation in site index.

Fig. 12.10 Spatial variation in site index within a sub-area of eucalyptus pulpwood plantation in Bahia, Brazil. *Light tone* indicates good site

(continued)

(continued)

The local objective function maximized the ending growing stock volume and minimized the Cut–Uncut border. The total harvest of each year was constrained to at least 100,000 m^3. The reduced cost method was again used to solve the problems. Four different plans were produced, in which the weight of Cut–Uncut border varied from 0 (Plan 1) to 0.4 (Plan 4). The highest ending volume (712,000 m^3) with annual harvests of at least 100,000 m^3 was obtained without the aggregation target. When the weight of Cut–Uncut border was 0.2 (Plan 3 in Fig. 12.11), cuttings were already well aggregated, and the ending volume decreased only by 0.44 %. When original compartments were used, cuttings were only very slightly more aggregated than in Plan 3 but the ending volume was 4.95 % lower. The result suggests that the total production (harvested volume + ending volume) was higher when dynamic treatment unites were used, and this increase was not achieved at the cost or worsening size or shape of harvest blocks.

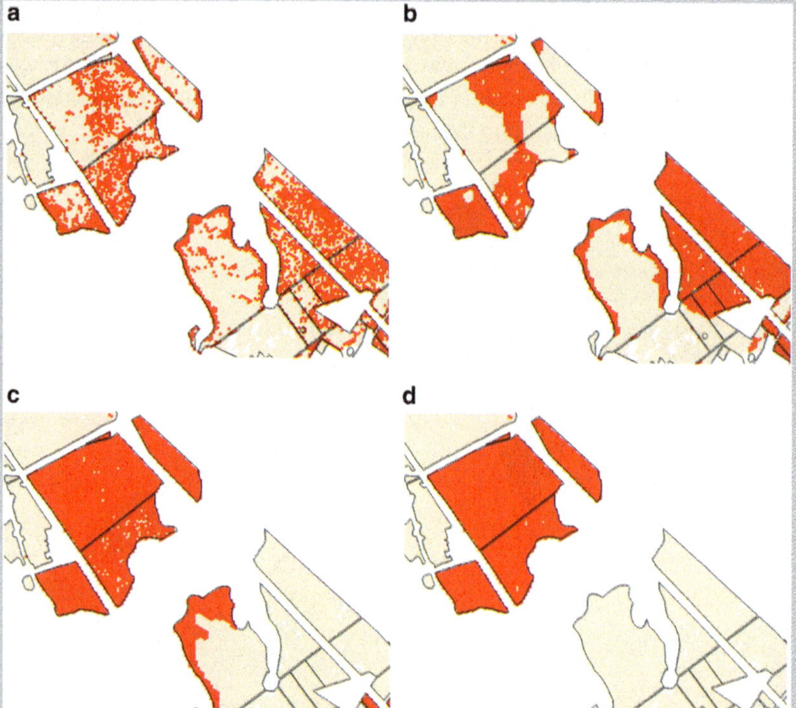

Fig. 12.11 Cuttings of the first year in a sub-area of a eucalyptus plantation in 4 management plan in which the weight of cutting aggregation increases from 0 (Plan 1) to 0.4 (Plan 4). (**a**) Plan 1 (Weight of CC = 0), (**b**) Plan 2 (Weight of CC = 0.1), (**c**) Plan 3 (Weight of CC = 0.2), (**d**) Plan 4 (Weight of CC = 0.4)

12.5 Discussion

The research and planning cases discussed in this chapter show that traditional fixed compartments are not necessary in forest management planning. Spatial optimization can be used to produce logical and practical treatment units. Some fine-tuning or filtering may sometimes be required to remove isolated one-to-few-cell treatment units, especially if the cell size is small and the forest is heterogeneous with much variation within short distance.

Dynamic treatment units improve the efficiency of forest use since the boundaries of financially mature forest can be placed in right places. Continuous habitat areas can be formed simultaneously with the formation of harvest blocks. Spatial optimization aiming at dynamic treatment units is compatible with the new forest inventory methods which are based on ALS. These methods are capable of producing inventory results for spatial units, which are too small as treatment units. However, the description of the forest is more detailed, allowing a more detailed management plan and a higher output of goods and services.

Standard meta heuristics can be used in optimization. They may be the best methods available if the quality of the solution is the only criterion. However, these methods may become too slow when the number of calculation units is very large. Since this is often the case in cell-based planning, alternative methods based on decentralized optimization have been developed and tested. Decentralized methods are fast in maximizing objectives which can be computed as sums or means of cell level values. They are also efficient in aggregating treatments or stands with similar features.

12.6 Problems

1. Describe the benefits of using dynamic treatment units instead of fixed stand compartments.
2. Handling of constraints in the utility-theoretic formulation of forest planning problems.
3. Advantages and disadvantages of decentralized heuristic optimization methods, as compared to centralized methods.

References

Bettinger P, Boston K, Sessions J (1999) Intensifying a heuristic forest harvest scheduling procedure with 2-opt decision choices. Can J For Res 29:1784–1792

Heinonen T, Pukkala T (2004) A comparison between one- and two-neighbourhoods in heuristic search with spatial forest management goals. Silva Fenn 38(3):319–332

Heinonen T, Pukkala T (2007) The use of cellular automaton approach in forest planning. Can J For Res 37:2188–2200

Heinonen T, Kurttila M, Pukkala T (2007) Possibilities to aggregate raster cells through spatial optimization in forest planning. Silva Fenn 41(1):89–103

Heinonen T, Leppänen V, Pukkala T, Gunia M (2008) Dynaamisiin käsittelykuvioihin perustuva metsäsuunnittelu. Metsät aikak 4(2008):281–293

Hoganson HM, Rose DW (1984) A simulation approach for optimal timber management scheduling. For Sci 30:220–238

Packalén P, Heinonen T, Pukkala T, Vauhkonen J, Maltamo M (2011) Dynamic treatment units in Eucalyptus plantations. For Sci 57(5):416–426

Pukkala T, Heinonen T, Kurttila M (2009) An application of the reduced cost approach to spatial forest planning. For Sci 55(1):13–22

Strange N, Meilby H, Bogetoft P (2001) Land use optimization using self-organizing algorithms. Nat Resour Model 14:541–574

Strange N, Meilby H, Thorsen JT (2002) Optimizing land use in afforestation areas using evolutionary self-organization. For Sci 48(3):543–555

Tokola T, Pitkänen J, Partinen S, Muinonen E (1996) Point accuracy of a non-parametric method in estimation of forest characteristics with different satellite materials. Int J Remote Sens 17(12):2333–2351

Chapter 13
Dealing with the Sustainability Issue for Industrial Plantation Management

Juan Carlos Giménez, Mercedes Bertomeu, Luis Diaz-Balteiro, and Carlos Romero

13.1 Introduction

In the field of Forestry, the idea of sustainability has always been assumed to be inherent to forest management practice. In fact, classical forest management textbooks have considered "environmental security" as a final goal of great importance to forest management, though it is recognized as being an ill-defined concept (Davis et al. 2001). Thus, even when goods and services demanded by societies were quite different from those today, maintaining the production capacity of renewable resources was already a dominant goal of the forestry profession, and forest programs have long specified prompt regeneration as a basic forest management constraint. However, assuming that goal has not been sufficient to ensure that all forest practices were sustainable.

Since the idea of sustainability was consolidated in the United Nations Conference on Environment and Development held in Rio de Janeiro in 1992, the problem of establishing an operational definition has gradually been tackled in all continents through regional processes, by which countries have attempted to fix a wide range of criteria and indicators consistent with several forest objectives.

Some parts of this chapter are based upon the paper by Giménez et al. (2013).

J.C. Giménez (✉) • M. Bertomeu
Grado en Ingeniería Forestal y del Medio Natural, Centro Universitario de Plasencia, University of Extremadura, Spain
e-mail: jcfernan@unex.es

L. Diaz-Balteiro
Departamento de Economía y Gestión Forestal, ETS Ingenieros de Montes, Technical University of Madrid, Madrid, Spain

C. Romero
ETS Ingenieros de Montes, Technical University of Madrid, Madrid, Spain

J.G. Borges et al. (eds.), *The Management of Industrial Forest Plantations: Theoretical Foundations and Applications*, Managing Forest Ecosystems 33,
DOI 10.1007/978-94-017-8899-1_13,

So, sustainability has been addressed from different viewpoints, though there is a general agreement on the need for articulating the idea from an initial common set of criteria and indicators. That is, the abstract concept must be formed by fixing a multidisciplinary list of criteria and indicators (Raison et al. 2001) encompassing several aspects. From an economic point of view and in relation with the maintenance of the utility provided by the aggregate capital, there is another approach related to the concepts of weak and strong sustainability.

In the case of forest plantations, sustainability has become more important due to two different aspects. On the one hand, forest managers and owners must have accurate measurements of sustainability as well as the capacity to predict the consequences of different management alternatives. On the other hand, intrinsic features of industrial plantations (single-species composition, exotic species, short rotations, intensive management prescriptions, etc.) have aroused suspicion in certain spheres, thus demanding some sort of sustainable management. Thus, while some years ago sustainability was just seen as being in relation to the productivity of plantations (Evans 1999) and the maintenance of the production capacity of a site in the long term (Watt et al. 2005), today it is considered that public services such as biodiversity preservation must also be considered.

Furthermore, this chapter has a clear connection with the chapter addressing forest certification (Chap. 15). However, its approach should not be viewed as a component of the process of forest certification (Schwarzbauer and Rametsteiner 2001; Eriksson et al. 2007), but as complementary to it. If the sustainability of forest plantations is to be compared, it is not enough to only focus on the area of the plantations certified. In fact, some authors (Higman et al. 2005, Ch. 8) describe some sustainability requirements that should be met by forest plantations, including standards from the FSC (Forest Stewardship Council certification) and some intergovernmental organizations, such as the International Tropical Timber Organization. Anyway, it is well known that not all plantations are certified. In addition, many countries have several competing certification systems. All these systems are unable to uniquely integrate in a single measurement the viewpoints of different stakeholders. In short, all these circumstances confirm the fact that certification systems are not the only way to prove forest plantations sustainability.

Many sustainability-related studies only focus on the analysis phase, describing and quantifying a previously defined set of indicators. Hence, this chapter aims to present a multicriteria methodology based on goal programming to resolve a weakness common to many studies: the lack of a method to aggregate the various sustainability indicators considered into a synthetic index. The methodology described provides a way to integrate different indicators and derive an overall ranking of management alternatives.

What You Will Learn in This Chapter

- Some notions about sustainability in industrial forest plantations
- Sustainability criteria and indicators
- Aggregating sustainability indicators by using a multicriteria approach
- Potentialities of quantifying sustainability of industrial plantations

13.2 Techniques Used to Measure Sustainability

Many studies have addressed sustainability from a prespecified set of indicators at different levels, from local to regional, and country levels. In the forestry field, several indicators have initially been set according to a small common number of criteria all over the world, thereby resulting in periodic measurements of them. Regardless of any shortcomings of these indicators, all these efforts have failed to respond to the immediate question of whether forest management in these countries is sustainable or not. As well stated by Kohl and Rametsteiner (2010, p. 60): "*this assessment depends on the precedence and valency attributed to the various aspects of sustainability*", from which the absence of a method for aggregating the various indicators can be derived.

In the case of forest plantations, some studies have defined sustainability incompletely, for example by only focusing on changes in certain soil properties and the production capacity of a site as a consequence of repeated rotations (Evans 2009). Other authors assess sustainability by taking into consideration various environmental indicators (Palmer et al. 2005; Jeffries et al. 2010), but do not include economic and social indicators.

Given the multidimensional nature of the concept (sustainability has environmental, economic, and social dimensions), several studies have attempted to portray it by means of multicriteria decision making (MCDM) techniques (see Diaz-Balteiro and Romero 2008). So, based on the classification of MCDM techniques in discrete and continuous ones, a brief review of the studies dealing with sustainability is summarized below, although only a few of them specifically deal with sustainability in industrial forest plantations.

Among the discrete techniques, one of those most used is the Analytic Hierarchy Process (AHP) developed by Saaty (1995). The AHP is a method of breaking down a decision making problem into a number of level hierarchies. The highest level is the overall objective of the problem, while the lowest includes the alternative plans. Next, the decision maker's priorities are derived by making paired comparisons on the relative importance of each element in terms of all the elements in the next higher level. Verbal judgments are then translated into numerical values and a set of overall priorities is derived. Examples of this technique can be seen in Mendoza and Prahbu (2000a, b), Mendoza and Dalton (2005), and Babaie-Kafaky et al. (2009). Vacik et al. (2007) used an extension of AHP, the ANP (Analytic Network Process), to characterize the sustainability of different forest management alternatives. A comparison between both methods can be seen in Wolfslehner et al. (2005), and Wolfslehner and Vacik (2008, 2011). Other authors suggested using qualitative MCDM techniques to integrate several sustainability criteria and indicators. Bousson (2001) applied the ELECTRE (*ELimination Et Choix Traduisant la REalité, ELimination and Choice Expressing REality*) method to choose the best management alternative regarding certain criteria, and Balana et al. (2010) compared several MCDM techniques to assess sustainability in community-owned forests in Ethiopia, while Huth et al. (2005) used the PROMETHEE method

(*Preference Ranking Organisation Method for Enrichment Evaluations*) for this purpose. Ducey and Larson (1999) implemented a hybrid methodology based on MCDM techniques and fuzzy theory to assess the sustainability of certain forest management alternatives. In another direction, Kangas et al. (1998) recommended the use of multi-attribute techniques to aggregate the indicators. Finally, Store (2009) combined these techniques with spatial analysis techniques.

In relation to the continuous MCDM problems, Diaz-Balteiro and Romero (2004a, b) proposed a methodology based on goal programming in order to conveniently aggregate the sustainability indicators. This methodology will be explained in detail in the following section. On the other hand, Maness and Farrell (2004) applied fuzzy multi-objective programming to aggregate the different criteria. Having arrived at this point, it is necessary to point out that the works cited in the last paragraphs are not linked to industrial forest plantations. That is to say, there are not many studies which analyze sustainability in plantations of this type using MDCM techniques, except for an occasional example relative to aspects connected with biofuels (Recchia et al. 2010).

The use of methodologies to aggregate various indicators has been well documented in other fields. Given the impossibility of presenting an exhaustive review, only some studies will be cited. Thus, Gómez-Limón and Sánchez-Fernández (2010) provided an aggregate measure of economic, social, and environmental indicators for farming systems in Palencia (Spain). López-Baldovín et al. (2006) used multiobjective programming to aggregate eight sustainability indicators in irrigation systems in the Guadalquivir River Basin. Van Calker et al. (2006) applied the Multiattribute Utility Theory to aggregate several indicators in a case study for dairy farms in The Netherlands. A similar approach can be seen in Blancas et al. (2010) to confront sustainability of coastal tourism in Andalusía (southern Spain) by using a goal programming approach. Based on multiobjective programming, Ruiz et al. (2011) addressed a sustainability problem in several municipalities in Andalusia (Spain). Finally, the Data Envelopment Analysis (DEA) was used by Munda (2005, 2006), Munda and Nardo (2009), Doukas et al. (2010), Murias et al. (2007), Zhou et al. (2007), and Hatefi and Torabi (2010) to aggregate several indicators.

13.3 Developing a Sustainability Index

13.3.1 Description of the Case Study

A neighbourhood community-owned forest called "Coroa" located in Northwestern Spain was used as a case study for the application of the forest management planning model. This forest is currently managed for pulpwood production by the Regional Forest Service through a management plan developed in 2004 by an external forest firm. Total land area covers 179.5 ha, 166.6 ha of which consists of

38 pure even-aged stands of *Eucalyptus globulus* Labill. and *Pinus radiata* D. Don. An isolated stand of *Pinus pinaster* Ait. (11.8 ha) and scattered clumps of *Castanea sativa* Miller and *Betula* spp. with a total area of 1.1 ha also exist in the area but were not considered in the analysis.

Most eucalypt stands (18 out of 28) were 14–16 years old plantations, while four stands were 7–8 years old, and six stands were 3–5 years old, covering a total area of 138.6 ha. Pine stand ages ranged from 7 to 25 years, and covered an area of 28 ha. All initial eucalypt stands were seedling stands except for the ones aged 3 years, which were at the second rotation interval.

The site index of stands was determined according to their ages as deduced from the years of planting and their dominant heights measured in the field survey. Next, timber volume predictions for eucalypt and Monterrey pine stands were derived from the growth and yield tables developed by Fernández López (1982) and Sánchez et al. (2003), respectively. In this respect, since the tables by Fernández López (1982) were only developed for the first two rotations, timber yields for the third rotation interval were assumed to be equal to those of the first rotation. The price of the eucalypt timber used was 30 and 32 €/m^3 for Monterrey pine (Asociación Forestal de Galicia 2007). Management costs included weed control (500 €/ha), site preparation and plantation costs (722.2 €/ha), fertilization (75 €/ha), stool thinning (400 €/ha), and stump clearing (900 €/ha). Finally, net present value of timber harvests was calculated using a 5 % discount rate.

13.3.1.1 Step I: Item Selection

Before describing the management alternatives and sustainability indicators considered, the following aspects of the mathematical model formulated to generate the management alternatives should be noted. First of all, since *Eucalyptus globulus* has the ability to sprout after harvest, stands created by planting can be coppiced in successive rotations. Thus, it was necessary to previously define the length of one full plantation cycle, that is, the number of rotation intervals or crops that will be harvested before stump removal and establishment of a new plantation, which has been called “the coppice problem” (Medema and Lyon 1985). Moreover, this length was fixed for future stands, i.e. for stands existing after the planning horizon, while the number of rotations could be smaller during the planning horizon. So, the model formulation took into consideration each stand rotation interval. The first rotation interval was the seedling rotation, in which the stand is made up of trees developed from planted seedlings. The second rotation interval was the first coppice rotation, where all standing trees have developed from stump sprouts. The third rotation interval was the second coppice rotation, and so on.

Second, it was also necessary to fix the length for each rotation interval, that is, the age at which trees would be harvested on each rotation interval. Thus, the rotation age could be different in each rotation interval, just in the same way that rotation ages vary with site quality. So, the age of a stand in a given rotation interval was also denoted in the model. Moreover, it was also assumed that the rotation age would be constant for each rotation interval after the planning horizon, while it could take several values during the planning horizon.

Thirdly, it should be noted that the ultimate goal of the model was to ensure the optimum constant timber yield in perpetuity after the planning horizon, namely, the long-term sustained yield (Davis et al. 2001). However, this goal was pursued in a different way to that used in the classical volume and area control strategies. Thereby, the volume control method ensures an even flow of timber during the planning horizon, i.e., in the near future, but at the sacrifice of not achieving the so-called fully regulated forest by the end of the planning horizon. On the other hand, the area control pursues this target by successively harvesting areas of equal size or productivity during the planning horizon, but at the cost of achieving a stable harvest during the planning horizon. As an alternative to both, the formulated model pursued the long-term sustained yield by forcing equal harvest volumes after the planning horizon. Hence, accounting rows were introduced to compute harvest volumes for one full plantation cycle after the planning horizon, assuming that only stands at final rotation age would be harvested in each year. Then, equalities between every two consecutive harvest volumes were added. In this manner, it was possible to obtain a steady-state structure without being restricted to the fully regulated forest achieved by the area control method, in which all rotation intervals and ages covered by the corresponding rotations coexist on equal areas at each site. In the steady-state structure achieved by the end of the planning horizon production would continue at a constant level, and the growing stock would remain unchanged in perpetuity. In this way, the planning horizon could be interpreted as being a conversion period, that is, as the time needed to convert the initial plantation into the steady-state structure.

The mathematical model corresponded to a strategic timber harvest scheduling problem, and the Model II formulation (Johnson and Scheurman 1977) was used to develop it. Regardless of the length of one full plantation cycle after the planning horizon, the model extended over two rotation intervals (length of the planning horizon). In addition, regarding eucalypt stands, two types of decision variables were defined to generate more flexible harvest schedules during the planning horizon. Thus, some decision variables represented the area harvested in a given stand and time period resulting in vegetative regeneration of the stand, while others represented the area harvested in a given stand and time period, and replaced by a new stand by planting after stump removal. Regarding Monterrey pine stands, decision variables were included to denote the area harvested in a given stand and time period, and planted with eucalypt, so that the pine stands were converted into eucalypt stands by the end of the planning horizon.

Moreover, in Model II, any hectare may pass through several decision variables before reaching the planning horizon. Thus, accounting equations were needed to link up the hectares from decision variables representing one generation to decision variables representing the next one, as well as equations to quantify the area of stands at the beginning of the planning horizon. That is, in order to monitor the evolution of the plantation area distribution during the planning horizon, new variables were defined for each stand to quantify the total area in a given rotation interval and age at the beginning of a given period according to the areas harvested in the preceding period. Thus, the model included constraints to force the total area harvested from each stand in each period to be less than, or equal to, the existing area. Furthermore, the decision variables were obliged to be greater than or equal to a prespecified minimum value should they be positive. Additionally, goals stating volume control during the planning horizon were included in the model to prevent as far as possible significant fluctuations over the planning horizon. Finally, accounting equations to compute harvest volumes and areas, and the net present value of harvests during the planning horizon were included, as well as nonnegativity constraints to prevent any of the decision variables from taking on negative values.

At this point, management alternatives were defined by considering alternative objective functions and different values for the number of rotation intervals in one full plantation cycle after the planning horizon and their corresponding rotation ages. With regard to timber harvests during the planning horizon, three objective functions were considered: (i) to maximize the net present value; (ii) to maximize the total harvest volume, and (iii) to maximize the net carbon captured. With respect to the number of rotation intervals, it was considered that one full plantation cycle (from planting to next planting) after the planning horizon could be alternatively comprised of two or three rotation intervals. In relation to final rotation age, since the eucalypt plantations in the case study are exclusively managed for pulpwood production, three alternative values were considered: 12, 15 and 18 years. In contrast, it was assumed that stands could be harvested within the intervals defined by the final rotation age ± 3 years during the planning horizon. For example, any stand could be harvested at any single age between 12 and 18 years in the planning horizon if the rotation age adopted for future stands (final rotation age) was 15 years. Consequently, 18 ($3 \times 2 \times 3$) management alternatives were defined.

Finally, assuming that forest sustainability involves economic, social, and environmental aspects, we took into consideration the following indicators and measured them for each management alternative during the planning horizon: (1) Net present value (NPV); (2) Total timber harvest volume (THV); (3) Net carbon captured (NCC); (4) Evenness of volumes harvested in consecutive time periods (DHV); (5) Harvest volume-stand growth ratio (RH/G), and (6) Average harvest area (AHA).

Net present value and total timber harvest volume are indicators of economic sustainability of management alternatives, since they measure profitability (in euros) of timber harvests and total harvest in physical terms (cubic meters), respectively. Therefore, both indicators are of the type "the more, the better". With respect to net carbon captured over the planning horizon, it is an indicator of environmental sustainability of the type "the more, the better", and it was derived from standing volumes and coefficients based on the density and carbon content of wood. The evenness of harvest volumes during the planning horizon simultaneously involves economic, social, and ecological considerations. So, at large forest scales, an even flow of timber guarantees the persistence of forest industries and specialized labour by providing regular employment. In addition, annual regeneration subsequent to harvests ensures continuous existence of several stand ages over the forest area.

Alternatives	
12R1NPV	rotation 12 years. One coppice rotation. Maximize Net present value
12R1THV	rotation 12 years. One coppice rotation. Maximize total timber harvest volume
12R1CC	rotation 12 years. One coppice rotation. Maximize Net carbon captured
12R2NPV	rotation 12 years. Two coppice rotations. Maximize Net present value
12R2THV	rotation 12 years. Two coppice rotations. Maximize total timber harvest volume
12R2CC	rotation 12 years. Two coppice rotations. Maximize Net carbon captured
15R1NPV	rotation 15 years. One coppice rotation. Maximize Net present value
15R1THV	rotation 15 years. One coppice rotation. Maximize total timber harvest volume
15R1CC	rotation 15 years. One coppice rotation. Maximize Net carbon captured
15R2NPV	rotation 15 years. Two coppice rotations. Maximize Net present value
15R2THV	rotation 15 years. Two coppice rotations. Maximize total timber harvest volume
15R2CC	rotation 15 years. Two coppice rotations. Maximize Net carbon captured
18R1NPV	rotation 18 years. One coppice rotation. Maximize Net present value
18R1THV	rotation 18 years. One coppice rotation. Maximize total timber harvest volume
18R1CC	rotation 18 years. One coppice rotation. Maximize Net carbon captured
18R2NPV	rotation 18 years. Two coppice rotations. Maximize Net present value
18R2THV	rotation 18 years. Two coppice rotations. Maximize total timber harvest volume
18R2CC	rotation 18 years. Two coppice rotations. Maximize Net carbon captured
Indicators	
NPV	Net Present Value
THV	Total timber harvest volume
NCC	Net carbon captured
DHV	Evenness of volumes harvested in consecutive time periods
RH/G	Harvest volume-stand growth ratio
AHA	Average harvest area

Fig. 13.1 Management alternatives and indicators considered in the case study

However, in the case study, this indicator was observed as a surrogate of an even flow of cash-flow. It was measured through the sum of the negative and positive deviation variables of the goals related to the strict equalities between consecutive harvest volumes during the planning horizon. Hence, it was an indicator of the type "the less, the better". With regard to harvest volume-stand growth ratio, it was calculated with the expression THV/(THV + FSV − ISV), thus assuming that stand growth could be obtained by adding the difference between the final and initial standing volumes (FSV − ISV) to the total harvest volume (THV). It is an indicator of the type "the more, the better". Lastly, average harvest area involves social considerations, as it is related to landscape visual impact, and it is of the type "the less, the better". Finally, in Fig. 13.1 all the management alternatives and indicators have been included.

13.3.1.2 Step II: Measuring

Once both the sustainability indicators and the management alternatives have been defined, the following step consists of generating a matrix with the values of each of the indicators for each of the alternatives, here called the indicator matrix. We must

		MAX NPV	MAX THV	MAX CC
+	NPV	2,997,015.5	2,979,985.5	2,548,466.5
+	THV	6,492.9	6,521.2	4,977.9
+	NCC	14,612.2	14,588.2	18,276.2
-	DHV	3,510.1	3,599.6	2,399.4
-	RH/G	1.15	1.15	1.19
-	AHA	2.97	3.00	2.56

Fig. 13.2 Indicator matrix for the alternatives 12R1

remember that, following the methodology developed in a previous section, the values in the matrix cells are derived from the solutions of the timber harvest scheduling model described before. Logically, if another methodology is opted for establishing the management alternatives, the matrix cells would correspond to the values of the indicators under each alternative.

As an example, Fig. 13.2 shows the indicator matrix for the three alternatives for a final rotation of 12 years and one coppice rotation in any full plantation cycle after the planning horizon. In this figure, the symbol "+" refers to the indicators of the type "the more, the better", whereas the symbol "–" refers to the indicators of the type "the less, the better". It can be observed that the indicators' values obtained by maximizing the net present value are rather similar to those obtained by maximizing the total timber harvest volume along the planning horizon. However, the indicators took different values when the objective was to maximize the net carbon captured throughout the planning horizon.

If we repeat the process for the rest of the alternatives, the matrix shown in Fig. 13.3 is obtained. For an easier reading, the indicators occupy columns 2–7, whereas each line corresponds to one of the alternatives defined.

Furthermore, the values shown in Fig. 13.3 differ by several orders of magnitude and are measured in different units (€, m^3, tC, and ha). Therefore, to aggregate all these values, they need to be normalized. Following Romero and Rehman (2003), *if the absolute values for the achievement levels of the several objectives are significantly different, then the scalarisation or normalisation of the degrees of closeness is necessary to avoid solutions biased towards those objectives that can achieve larger values.* For this purpose, Diaz-Balteiro and Romero (2004a) proposed the following method of normalization:

$$\overline{R}_{ij} = 1 - \frac{R_j^* - R_{ij}}{R_j^* - R_{*j}} = \frac{R_{ij} - R_{*j}}{R_j^* - R_{*j}} \qquad i = 1, \ldots, m; j = 1, \ldots, n \quad (13.1)$$

where:

$\overline{R}_{ij}$ is the normalized value of the *j*-th indicator under the *i*-th alternative.

R_j^* is the optimal value of the *j*-th indicator (ideal value). Thus, this value represents a maximum value if the indicator is of the type "the more, the better", and a minimum value when the indicator is of the type "the less, the better".

	NPV	THV	NCC	DHV	RH/G	AHA
12R1NPV	2997015,5	6492,9	14612,2	3510,1	1,153	2,967
12R1THV	2979985,5	6521,2	14588,2	3599,6	1,154	2,999
12R1CC	2548466,5	4977,9	18276,2	2399,4	1,191	2,563
12R2NPV	3018216,7	6297,6	14682,9	3178,1	1,163	2,364
12R2THV	3006057,7	6390,3	14983,6	3078,7	1,162	2,575
12R2CC	2549305,7	4873,6	18368,4	2405,0	1,199	2,342
15R1NPV	3175656,6	7383,2	18981,5	3374,6	1,070	2,806
15R1THV	3157136,6	7408,0	18849,1	2361,2	1,069	2,351
15R1CC	2453024,6	4786,3	22579,2	1952,8	1,105	2,085
15R2NPV	3154199,5	7144,0	19150,2	2904,5	1,076	2,150
15R2THV	3135687,5	7205,6	19313,8	2191,4	1,075	1,658
15R2CC	2437766,5	4655,3	22424,1	2256,9	1,112	2,012
18R1NPV	2795772,9	6003,2	22751,6	3150,2	1,042	2,108
18R1THV	2776264,9	6039,5	23440,5	2843,1	1,040	1,847
18R1CC	2301963,9	4455,8	25878,5	2886,1	1,052	2,002
18R2NPV	2809052,2	5909,2	23177,2	2958,0	1,043	1,769
18R2THV	2768187,2	5910,1	24063,5	2669,9	1,044	1,587
18R2CC	2303739,2	4212,8	26003,2	2264,0	1,061	1,865

Fig. 13.3 Indicator matrix for the case study

R_{ij} is the value achieved by the *j*-th indicator under the *i*-th alternative.
R_{*j} is the worst value achieved by the *j*-th indicator (anti-ideal or nadir value).

Using this normalization system, the sustainability indicators have no dimension, so that they can be used to implement any aggregation system. Moreover, the $m \times n$ values (18×6 in the case study) are now bounded between 0 and 1. Hence, for this normalization system, the ideal vector is $\overline{R}^* = (1, \ldots, 1)$, and the anti-ideal vector is $\overline{R}^* = (0, \ldots, 0)$. Using this procedure the normalized matrix shown in Fig. 13.4 is obtained. In summary, with this normalization system the indicators do not have any dimension and all of them are bounded between 0 and 1.

Next, the same procedure is applied in order to normalize the aspiration levels (targets) set for the different indicators. These aspiration levels are exogenous and they are determined by means of the experience accumulated by the authors. Thus, an achievement of 70 % of the ideal value attained by each indicator was considered to be a satisfactory degree of sustainability.

13.3.1.3 Step III: Aggregating

Once the values R_{ij} and the targets have been normalized, the following weighted goal programming model is formulated (Ignizio 1976) in order to obtain the "most sustainable" management alternative:

$$Min \sum_{j=1}^{m} \alpha_j n_j \tag{13.2}$$

	NPV	THV	NCC	DHV	RH/G	AHA
1 12R1NPV	0.796	0.714	0.002	0.054	0.288	0.022
2 12R1THV	0.776	0.722	***0.000***	***0.000***	0.282	***0.000***
3 12R1CC	0.282	0.239	0.323	0.729	0.052	0.309
4 12R2NPV	0.820	0.652	0.008	0.256	0.224	0.450
5 12R2THV	0.806	0.681	0.035	0.316	0.231	0.300
6 12R2CC	0.283	0.207	0.331	0.725	***0.000***	0.465
7 15R1NPV	**1.000**	0.992	0.385	0.137	0.814	0.137
8 15R1THV	0.979	**1.000**	0.373	0.752	0.817	0.459
9 15R1CC	0.173	0.179	0.700	**1.000**	0.591	0.647
10 15R2NPV	0.975	0.917	0.400	0.422	0.777	0.601
11 15R2THV	0.954	0.937	0.414	0.855	0.781	0.950
12 15R2CC	0.155	0.138	0.686	0.815	0.545	0.699
13 18R1NPV	0.565	0.560	0.715	0.273	0.987	0.631
14 18R1THV	0.543	0.572	0.776	0.459	**1.000**	0.816
15 18R1CC	***0.000***	0.076	0.989	0.433	0.928	0.706
16 18R2NPV	0.580	0.531	0.752	0.390	0.980	0.871
17 18R2THV	0.534	0.531	0.830	0.565	0.974	**1.000**
18 18R2CC	0.002	***0.000***	**1.000**	0.811	0.867	0.803

Fig. 13.4 Normalized indicator matrix for the case study

subject to:

$$\sum_{i=1}^{n} \overline{R}_{ij} X_i + n_j - p_j = \overline{t}_j \qquad j = 1, \ldots, m \tag{13.3}$$

$$\sum_{i=1}^{n} X_i = 1 \tag{13.4}$$

$$X_i \in \{0, 1\}; \;\; n_j \geq 0; \;\; p_j \geq 0 \tag{13.5}$$

where α_j denotes the weight that the decision maker attaches to the j-th indicator, that is, the relative importance of indicator j with respect to the rest of indicators; $\overline{R}_{ij}$ is the normalized value of the j-th indicator under the i-th management alternative; $\overline{t}_j$ is the normalized value of the target corresponding to the j-th indicator; n_j and p_j are the negative and positive deviation variables, respectively; and X_i represents the decision variables.

Thus, for the j-th indicator, Eq. 13.3 expresses the sum of the contributions of all management alternatives to the achievement of the corresponding target, while Eqs. 13.4 and 13.5 indicate that only one alternative can be chosen, given that decision variables X_i are binary ($X_i = 1$ if the i-th management alternative is chosen, and $X_i = 0$ otherwise). Finally, Eq. 13.2 is the so-called achievement function.

It should be noted that due to the normalization implemented, the achievement of all the indicators is of the type "the more, the better". Thus, the deviation variables n_j, measuring the under-achievement with respect to the normalized targets $\bar{t}_j$, are the only unwanted deviation variables to be included in the achievement function.

From a computational viewpoint, Model (2)–(5) is an integer linear programming problem. Hence, by resorting to commercial software like LINGO (Lindo Systems, 2011), the model can be easily solved to obtain the "most sustainable" management alternative. Moreover, by removing the resulting alternative and solving the model again, the second "most sustainable" alternative can be obtained. In this way, the ranking of management alternatives in terms of sustainability is obtained.

It should be noted that Model (2)–(5) provides the management alternative with the maximum aggregate achievement of the different indicators considered. However, if one of the indicators performs very poorly, the solution might be of dubious interest. In this respect, if we wanted to find the alternative that provides the most balanced solution regarding the achievement of the different goals, the following MINMAX (Chebyshev) goal programming model can be formulated (Romero 1991):

$$MinD \tag{13.6}$$

subject to:

Equations 13.3, 13.4, 13.5

$$\alpha_j n_j - D \leq 0 \qquad j = 1, \ldots, m \tag{13.7}$$

where variable D represents the maximum deviation. MINMAX model (6–7) provides the "best" solution in terms of a balance between the achievements of the different indicators, but the aggregate performance can be very poor (see Tamiz et al. 1998 for technical details about the properties of MINMAX solutions). Such dilemmas between "aggregate" and "balanced" performance can be mitigated by combining both models in the following extended goal programming formulation (Romero 2001):

$$Min\,(1-\lambda)\,D + \lambda \sum_{j=1}^{m} \alpha_j n_j \tag{13.8}$$

subject to:

Equations 13.3, 13.4, 13.5, and 13.7

For $\lambda = 1$, we have model (2)–(5), which provides the solution with the "maximum aggregate achievement", and for $\lambda = 0$, we have model (6–7), thus providing the solution with the "most balanced achievement". For intermediate values of control parameter λ, we obtain compromises between these two solutions. In summary, to determine the "most sustainable" management alternative, and/or the

λ=1		λ=0	
alternative	achievement function value	alternative	achievement function value
15R2THV	0.286	**18R2THV**	0.167
18R2THV	0.471	**18R1THV**	0.241
18R1THV	0.526	**15R2THV**	0.029
15R1THV	0.568	**15R2NPV**	0.300
18R2NPV	0.599	**18R2NPV**	0.310
15R2NPV	0.677	**15R1THV**	0.327
18R1NPV	0.771	**18R1NPV**	0.427
15R1CC	1.209	**15R1CC**	0.521
15R2CC	1.276	**15R2CC**	0.562
18R2CC	1.398	**15R1NPV**	0.563
15R1NPV	1.442	**12R1CC**	0.648
18R1CC	1.591	**12R2THV**	0.665
12R2NPV	1.909	**12R1NPV**	0.690
12R2THV	1.936	**12R2CC**	0.692
12R2CC	2.214	**12R2NPV**	0.692
12R1CC	2.295	**12R1THV**	0.698
12R1NPV	2.433	**18R1CC**	0.700
12R1THV	2.518	**18R2CC**	0.700

Fig. 13.5 Results obtained for $\lambda = 1$ and $\lambda = 0$

"ranking" of management alternatives in terms of sustainability, the following general extended goal programming model can be proposed:

$$\begin{aligned} & Min\,(1-\lambda)\,D + \lambda \sum_{j=1}^{m} \alpha_j n_j \\ & \text{subject to :} \\ & \sum_{i=1}^{n} \overline{R}_{ij} X_i + n_j - p_j = \overline{t}_j \quad j = 1, \ldots, m \\ & \sum_{i=1}^{n} X_i = 1 \\ & \alpha_j n_j - D \leq 0 \quad j = 1, \ldots, m \\ & X_i \in \{0,1\};\; n_j \geq 0;\; p_j \geq 0; \lambda \in [0,1] \end{aligned} \tag{13.9}$$

Starting from the hypothesis that the same weight has been given to the different sustainability indicators defined, Fig. 13.5 shows the results from applying the two goal programming models initially proposed: the model which supplies the most efficient solution or the best aggregated average ($\lambda = 1$, Model 2–5) on one hand, and the model which provides the most balanced solution ($\lambda = 0$, Model 6–7), on the other.

ranking	λ=1	λ=0.8	λ=0.6	λ=0.4	λ=0.2	λ=0
1	15R2THV	15R2THV	15R2THV	15R2THV	15R2THV	18R2THV
2	18R2THV	18R2THV	18R2THV	18R2THV	18R2THV	18R1THV
3	18R1THV	18R1THV	18R1THV	18R1THV	18R2NPV	15R2THV
4	15R1THV	15R1THV	18R2NPV	18R2NPV	18R1THV	15R2NPV
5	18R2NPV	18R2NPV	15R1THV	15R1THV	15R1THV	18R2NPV
6	15R2NPV	15R2NPV	15R2NPV	15R2NPV	15R2NPV	15R1THV
7	18R1NPV	18R1NPV	18R1NPV	18R1NPV	18R1NPV	18R1NPV
8	15R1CC	15R1CC	15R1CC	15R1NPV	15R1NPV	15R1CC
9	15R2CC	15R2CC	15R1NPV	15R1CC	15R1CC	15R2CC
10	18R2CC	15R1NPV	15R2CC	15R2CC	15R2CC	15R1NPV
11	15R1NPV	18R2CC	18R2CC	18R2CC	12R1CC	12R1CC
12	18R1CC	18R1CC	18R1CC	18R1CC	12R2CC	12R2THV
13	12R2NPV	12R2NPV	12R2NPV	12R2THV	18R2CC	12R1NPV
14	12R2THV	12R2THV	12R2THV	12R2NPV	18R1CC	12R2CC
15	12R2CC	12R2CC	12R2CC	12R2CC	12R2THV	12R2NPV
16	12R1CC	12R1CC	12R1CC	12R1CC	12R2NPV	12R1THV
17	12R1NPV	12R1NPV	12R1NPV	12R1NPV	12R1NPV	18R1CC
18	12R1THV	12R1THV	12R1THV	12R1THV	12R1THV	18R2CC

Fig. 13.6 Results obtained from the extended goal programming model

If we observe the most efficient solution, it can be checked how the most sustainable alternative entails a rotation of 15 years, and the first four solutions corresponded to management alternatives in which the total timber harvest volume was maximized. Our attention is drawn to the fact that, for this solution, all the alternatives involving a rotation of 12 years are the least sustainable ones, regardless of the objective function or the number of coppice rotations. When looking for the most balanced solution ($\lambda = 0$), the most sustainable solution is obtained when the management alternative involves maximizing total timber harvest volume, an 18-year rotation and two coppice rotations. In addition, there are alternatives that provide the same solution in terms of sustainability, since they present the same achievement function value (for an easy reading, these solutions have been shaded in Fig. 13.5). Finally, it should be pointed out that the most sustainable management alternatives for both types of solutions incorporate two coppice rotations and that a 12-year rotation is never one of the best alternatives for the sustainability of these plantations.

By solving Model (9) for different values of parameter λ, the overall sustainability of the management alternatives considered was obtained as shown in Fig. 13.6. It is interesting to note that control parameter λ trades-off the maximum aggregate achievement (measured by Eq. 13.2) with the most balanced achievement (measured by variable D of Model 6–7).

It is observed how the results for the most sustainable alternative only varied for the most balanced solution ($\lambda = 0$). The most sustainable alternative was always the same for all values of parameter λ equal to or greater than 0.2. From a forest management viewpoint, it is shown how the most sustainable alternatives always involved optimizing total timber harvest volume and two coppice rotations in the full plantation cycle. However, it should be stressed that these results are conditioned by the indicators selected, as well as by the assumption by which the same weight was assigned to each indicator.

13.4 Some Extensions

This section will raise some possible extensions of the models defined, with the aim of extending the analysis made in new directions. First, the possibility of considering a greater interaction of the stakeholders in the decision making process will be studied. Next, a model which attempts to offer a causal-type explanation of the rankings obtained will be presented. Finally, an adaptation of the method presented to determine the most sustainable plantation from a set of plantations will be analyzed, as initially proposed in this benchmarking exercise.

13.4.1 Integrating Different Weights for Each Indicator

In the previous models (e.g. Model (8)), the decision maker (DM) may assign different weights to the indicators proposed in terms of their relative importance. These weights can be set in different ways, such as by a direct interaction between the DM and a panel of experts. When introducing the new weight system in the models, it must be borne in mind that they can cause modifications in the initial solution. As an example, and given that no exercise was performed for the case study to obtain another vector of weights, a sensitivity analysis was carried out by changing the weights attached to the indicators with the aim of showing the variations triggered in the previous solutions. For this purpose, the weight given to each indicator was doubled while applying the *ceteris paribus* condition to the weights of the other indicators. Figure 13.7 shows the results derived from this sensitivity analysis.

The results showed that the rankings did not undergo any significant variation, except for the case in which the weight, given to the net carbon captured (w_3) was doubled, for which the most sustainable management alternative was different to the rest of the solutions. In fact, in that case, the most sustainable alternative involved maximizing the total timber harvest volume, and a rotation of 18 years. Note that three out of the four most sustainable alternatives implied an 18-year rotation, which is actually quite logical, because in that case we were prioritizing the net carbon captured. Finally, the interaction with the DM can be of use when fixing a target for each of the goals formulated in the models (Eq. 13.3). This exercise could be extended if there were different stakeholders and each of them gave a different weight to each of the indicators proposed. This multiplicity of stakeholders in the underlying decision-making process implies that it is of paramount importance for

ranking	w1=...w6=1	2w1	2w2	2w3	2w4	2w5	2w6
1	15R2THV	15R2THV	15R2THV	18R2THV	15R2THV	15R2THV	15R2THV
2	18R2THV	15R1THV	15R1THV	18R1THV	15R1THV	18R2THV	18R2THV
3	18R1THV	18R2THV	18R2THV	15R2THV	18R2THV	18R1THV	18R1THV
4	15R1THV	15R2NPV	18R1THV	18R2NPV	18R1THV	15R1THV	18R1NPV
5	18R2NPV	18R1THV	15R1NPV	18R1NPV	18R2NPV	18R2NPV	15R2NPV
6	15R2NPV	18R2NPV	18R2NPV	15R1THV	15R2NPV	15R2NPV	15R1THV
7	18R1NPV	18R1NPV	18R1NPV	15R2NPV	18R1NPV	18R1NPV	18R1NPV
8	15R1CC	15R1NPV	15R1NPV	15R1CC	15R1CC	15R1CC	15R1CC
9	15R2CC	15R1CC	15R1CC	15R2CC	15R2CC	18R2CC	15R2CC
10	18R2CC	15R2CC	15R2CC	18R2CC	18R2CC	15R2CC	18R2CC
11	15R1NPV	12R2NPV	12R2THV	18R1CC	18R1CC	15R1NPV	18R1CC
12	18R1CC	12R2THV	12R2NPV	15R1NPV	15R1NPV	18R1CC	15R1NPV
13	12R2NPV	18R2CC	18R2CC	12R2CC	12R2CC	12R2NPV	12R2NPV
14	12R2THV	18R1CC	18R1CC	12R2THV	12R1CC	12R2THV	12R2THV
15	12R2CC	12R1NPV	12R1NPV	12R2NPV	12R2THV	12R1NPV	12R2CC
16	12R1CC	12R1THV	12R1THV	12R1CC	12R2NPV	12R2CC	12R1CC
17	12R1NPV	12R2CC	12R2CC	12R1THV	12R1NPV	12R1THV	12R1NPV
18	12R1THV	12R1CC	12R1CC	12R1NPV	12R1THV	12R1CC	12R1THV

2wi means indicator i has a weight twice than other indicators

Fig. 13.7 Results obtained for the L1 model ($\lambda = 1$) with different weights

the management of forest plantations to adopt tools from the group decision-making discipline. In Diaz-Balteiro and Romero (2008), a review of papers focusing on this issue has been included. A case where MCDM methods and group decision-making are merged using goal programming can be found in Diaz-Balteiro et al. (2009).

13.4.1.1 A Benchmarking Case

The methodology described in the previous sections presents an additional utility in a context like the one occupying us: the management of industrial forest plantations. Indeed, the same methodology can be applied, not for determining which management alternative is the most sustainable, but for selecting which plantation is the most sustainable from an initially defined set. For example, an industry that owns a group of plantations, or a habitual buyer of the timber product supplied by those plantations, could be interested in finding out which of them is the most sustainable one.

Benchmarking is a term used in spheres like the entrepreneurial and operational research and is associated with the idea of comparing the result of one unit to others. When we talk about a "unit" this can be a company, a branch, a project, etc. In the present context, it would be associated with the idea of seeing which plantation is the most sustainable one from a more or less homogeneous group of plantations.

The first step in applying this methodology would obviously be to proceed to the selection of the plantations to be studied, as well as the indicators that would be considered in the analysis. At this point, we should need to obtain the values of each indicator on each plantation. In this process, an undesired circumstance might occur: the lack of data regarding one or several indicators on one or several plantations. If this happens, those plantations where it was not necessary to obtain the values of all the indicators initially defined could be removed from the analysis. This might complicate the study since it could limit the results. For example, if a battery of fifteen indicators was available but there was no value for one indicator on a given plantation, the possibility of comparing this plantation with the remaining ones would be discarded. To remedy this drawback there are some statistical techniques which permit to calculate the lost data by means of the imputation of the mean of the observations, or through techniques based on multiple regression. Thus, it would not be necessary to reduce the size of the group of plantations.

Once this problem had been corrected, a similar methodology to that described in this chapter could then be applied. It would be necessary to define the targets for each indicator and proceed to normalize the values of the indicators before solving the models previously presented. The results obtained would provide us with a sustainability ranking of the different plantations according to the indicators selected, their weights, and the value of the control parameter λ considered to be appropriate.

13.4.1.2 Explaining the Results

Once the ranking of management alternatives (or the most sustainable plantation) in terms of sustainability has been obtained, then this type of classification can be the object of a causal explanation. To achieve this purpose, the composite indicators of sustainability for the management alternatives are considered endogenous variables, while a set of tentative exogenous explanatory variables are defined. The endogenous and exogenous variables are linked by resorting to econometric models.

In short, it is aimed to go one step further and obtain an explanation justifying the ranking previously achieved using variables which are independent from the indicators considered. By means of an econometric model it is possible to find out which explanatory variables justify the degree of sustainability previously obtained from the described models (e.g. Model 8). Note that when setting up the rankings, the latter proceed from solutions of successive goal programming models, and in each one a numerical result has been obtained for the achievement function. Logically, the results from management alternatives (following the example given in 6.2) or plantations presenting initial positions in the respective ranking, have lower achievement function values than those corresponding to final positions in the respective ranking.

The next step would be to compile a set of explanatory variables which were not related to the indicators defined. For instance, in the case presented in 6.2, variables like the type of property ownership, the integration of each plantation in a certain forestry certification system, the area of each plantation, the clonal or genetic material used, etc., could potentially be considered as variables. However, it would be desirable for these variables to be independent of each other. That is to say, it could be said that many explanatory variables may be correlated with each other. Therefore, some statistical test should be applied to ensure that there is no significant correlation between the explanatory variables considered. The last step would consist of applying an econometric model in order to find the variables that could best explain the sustainability of the units (management alternatives or plantations) which make up the ranking being studied. For example, in Voces et al. (2012), there is a case in which an econometric model (based on the Ordinary Least Squares method) is employed to attempt to explain the sustainability of the wood industry in Europe. The ranking is made up of 17 countries, measured by 14 indicators and 39 independent variables have been initially defined to enable the cited ranking to be explained.

13.4.2 Summary

At present, there is an increasing interest in incorporating the idea of sustainability in the management of forest plantations. With this purpose in mind, this chapter starts by presenting how the concept of sustainability has been addressed so far in the forest management literature. After this, a case study related with a harvest scheduling problem for *Eucalyptus globulus* in Spain is presented. From this case, a set of feasible management alternatives as well as a battery of sustainability indicators of different nature (economic, environmental, and social) are defined. With the help of a goal programming methodology, a ranking of the management alternatives is obtained by simultaneously considering all the sustainability indicators. This type of solution is improved with the help of an extended goal programming formulation. In this way, several rankings are obtained according to the level of aggregate achievement of the indicators, as well as the level of balanced achievement among them. Finally, some extensions based on the introduction of preferential weights are suggested.

13.4.3 Review Questions

1. *Define a list of sustainability indicators in the forest plantations you are studying*
2. *Describe how you can normalize the indicators*
3. *What is the meaning of the pay-off or indicators matrix?*
4. *Formulate an extended goal programming formulation in order to calculate the forest plantation more sustainable*
5. *Define, in your own words, what is the meaning and the purpose of the term "benchmarking"*

Acknowledgements This research has been funded by the Spanish Ministry of Economy and Competitiveness under project AGL2011-2585 and the Autonomous Government of Madrid. Thanks are given to Diana Badder and PRS for English editing.

References

Asociación Forestal de Galicia (2007) http://www.asociacionforestal.org/

Babaie-Kafaky S, Mataji A, Sani NA (2009) Ecological capability assessment for multiple-use inforest areas using GIS- based multiple criteria decision making approach. Am J Environ Sci 5:714–721

Balana B, Mathijs E, Muys B (2010) Assessing the sustainability of forest management: an application of multi-criteria decision analysis to community forests in northern Ethiopia. J Environ Manag 91:1294–1304

Blancas FJ, Caballero R, González M, Lozano-Oyola M, Pérez F (2010) Goal programmingsynthetic indicators: an application for sustainable tourism in Andalusian coastal counties. Ecol Econ 69:2158–2172

Bousson E (2001) Development of a multicriteria decision support system adapted to multipleuseforest management: application to forest management at the management unit level in SouthernBelgium. In: Franc A, Laroussinie O, Karjalainen T (eds) Criteria and indicators for sustainable forest management at the forest management unit level, vol 38, EFI proceedings., pp 151–164

Davis LS, Johnson KN, Bettinger PS, Howrd TE (2001) Forest management, 4th edn. McGraw-Hill, New York

Diaz-Balteiro L, Romero C (2004a) Sustainability of forest management plans: a discrete goal programming approach. J Environ Manag 71(4):349–357

Diaz-Balteiro L, Romero C (2004b) In search of a natural systems sustainability index. Ecol Econ 49:401–405

Diaz-Balteiro L, Romero C (2008) Making forestry decisions with multiple criteria: a review and an assessment. For Ecol Manag 255:3222–3241

Diaz-Balteiro L, González-Pachón J, Romero C (2009) Forest management with multiple criteria and multiple stakeholders: an application to two public forests in Spain. Scand J For Res 24: 87–93

Doukas H, Karakosta C, Psarras J (2010) Computing with words to assess the sustainability ofrenewable energy options. Expert Syst Appl 37:5491–5497

Ducey MJ, Larson BC (1999) A fuzzy set approach to the problem of sustainability. For Ecol Manag 115:29–40

Eriksson OL, Sallnäs O, Ståhl G (2007) Forest certification and Swedish wood supply. For Policy Econ 9:452–463

Evans J (1999) Sustainability of forest plantations – the evidence. Issues paper, Department for International Development

Evans J (2009) Planted forests: uses, impacts, and sustainability. CABI – FAO, Wallingford

Fernández López A (1982) Evaluación del crecimiento y productividad del monte bajo de Eucalyptusglobulus en rotaciones sucesivas. In: Madrigal A, Álvarez JG, Rodríguez R, Rojo A (1999) Tablas de Producción para los Montes Españoles. Fundación Conde del Valle de Salazar, Madrid, pp 175–181

Gimenez JC, Bertomeu M, Diaz-Balteiro L, Romero C (2013) Optimal harvest scheduling in Eucalyptus plantations under a sustainability perspective. For Ecol Manag 291:367–376

Gómez-Limón JA, Sánchez-Fernández G (2010) Empirical evaluation of agricultural sustainabilityusing composite indicators. Ecol Econ 69:1062–1075

Hatefi SM, Torabi SA (2010) A common weight MCDA-DEA approach to construct composite indicators. Ecol Econ 70:114–120

Higman S, Mayers J, Bass S, Judd N, Nussbaum R (2005) The sustainable forestry handbook, 2nd edn. Earthscan, London

Huth A, Drechsler M, Köhler P (2005) Using multicriteria decision analysis and a forest growth model to assess impacts of tree harvesting in Dipterocarp lowland rain forests. For Ecol Manag 207:215–232

Ignizio JP (1976) Goal programming and extensions. Lexington Books, Lexington

Jeffries SB, Wentworthb TR, Lee Allen H (2010) Long-term effects of establishment practices on plant communities across successive rotations in a loblolly pine (Pinus taeda) plantation. For Ecol Manag 260:1548–1556

Johnson KN, Scheurman HL (1977) Techniques for prescribing optimal timber harvest and investment under different objectives: discussion and synthesis, Forest Science Monographs, 18

Kangas J, Alho JM, Kolehmainen O, Mononen A (1998) Analyzing consistency of experts' judgements-case of assessing forest biodiversity. For Sci 44:610–617

Kohl M, Rametsteiner E (2010) The state of Europe's forests: 2007 – report of the fifth ministerial conference on the protection of forests in Europe for sustainable forest management in Europe. In: Spathelf P (ed) Sustainable forest management in a changing world: a European perspective. Springer, Dordrecht, pp 53–60

Lindo Systems (2011) Lingo v. 13.0. Chicago, IL

López-Baldovin MJ, Gutiérrez-Martin C, Berbel J (2006) Multicriteria and multiperiod programming for scenario analysis in Guadalquivir river irrigated farming. J Oper Res Soc 57:499–509

Maness T, Farrell R (2004) A multi-objective scenario evaluation model for sustainable forest management using criteria and indicators. Can J For Res 34:2004–2017

Medema EL, Lyon GW (1985) The determination of financial rotation ages for coppicing tree species. For Sci 31:398–404

Mendoza GA, Dalton WJ (2005) Multi-stakeholder assessment of forest sustainability: multi-criteria analysis and the case of the Ontario forest assessment system. For Chron 81:222–228

Mendoza GA, Prabhu R (2000a) Development of a methodology for selecting criteria and indicators of sustainable forest management: a case study on participatory assessment. Environ Manag 26:659–673

Mendoza GA, Prabhu R (2000b) Multiple criteria decision making approaches to assessing forest sustainability using criteria and indicators: a case study. For Ecol Manag 131:107–126

Munda G (2005) Measuring sustainability: a multi-criterion framework. Environ Dev Sustain 7:117–134

Munda G (2006) A NAIADE based approach for sustainability benchmarking. Int J Environ Technol Manag 6:65–78

Munda G, Nardo M (2009) Non-compensatory/non-linear composite indicators for ranking countries: a defensible setting. Appl Econ 41:1513–1523

Murias P, de Miguel C, Rodríguez D (2007) A composite indicator for university quality assessment: the case of Spanish higher education system. Soc Indic Res 89:129–146

Palmer DJ, Lowe DJ, Payn TW, Höck BK, McLay CDA, Kimberley MO (2005) Soil and foliarphosphorus as indicators of sustainability for Pinus radiata plantation forestry in New Zealand. For Ecol Manag 220:140–154

Raison RJ, Brown A, Flinn D (2001) Criteria and indicators for sustainable forest management. CABI, Wallingford

Recchia L, Cini E, Corsi S (2010) Multicriteria analysis to evaluate the energetic reuse of riparian vegetation. Appl Energy 87:310–319

Romero C (1991) Handbook of critical issues in goal programming. Pergamon Press, Oxford

Romero C (2001) Extended lexicographic goal programming: a unifying approach. Omega Int J Manag Sci 29:63–71

Romero C, Rehman T (2003) Multiple criteria analysis for agricultural decisions, 2nd edn. Elsevier, Amsterdam

Ruiz F, Cabello JM, Luque M (2011) An application of reference point techniques to the calculation of synthetic sustainability indicators. J Oper Res Soc 62:189–197

Saaty TL (1995) Decision making for leaders, 3rd edn. RWS Publications, Pittsburg

Sánchez F, Rodríguez R, Rojo A, Álvarez JG, López C, Gorgoso J, Castedo F (2003) Crecimiento y tablas de producción de *Pinus radiata* D. Don en Galicia. Investig Agrar Sist Recur For 12:65–83

Schwarzbauer P, Rametsteiner E (2001) The impact of SFM-certification on forest product markets in Western Europe: an analysis using a forest sector simulation model. For Policy Econ 2241–256
Store R (2009) Sustainable locating of different forest uses. Land Use Policy 26:610–618
Tamiz M, Jones D, Romero C (1998) Goal programming for decision making: an overview of the current-state-of-the-art. Eur J Oper Res 111:569–581
Vacik H, Wolfslehner B, Seidl R, Lexer MJ (2007) Integrating the DPSIR approach and the analytic network process for the assessment of forest management strategies. In: Reynolds KM, Thomson AJ, Köhl M, Shannon MA, Ray D, Rennolls K (eds) Sustainable forestry: from monitoring and modelling to knowledge management and policy science. CABI, Wallingford, pp 393–411
van Calker KJ, Berentsen PBM, Romero C, Giessen GWJ, Huirne RBM (2006) Development and applications of a multi-attribute sustainability function for a Dutch dairy farming system. Ecol Econ 57:640–658
Voces R, Diaz-Balteiro L, Romero C (2012) Characterisation and explanation of the sustainability of the european wood manufacturing industries: a quantitative approach. Expert Syst Appl 39:6618–6627
Watt MS, Coker G, Clinton PW, Davis MR, Parfitt R, Simcock R, Garrett L, Payn T, Richardson B, Dunningham A (2005) Defining sustainability of plantation forests through identification of site quality indicators influencing productivity—a national view for New Zealand. For Ecol Manag 216:51–63
Wolfslehner B, Vacik H (2008) Evaluating sustainable forest management strategies with the analytic network process in a pressure-state-response framework. J Environ Manag 88:1–10
Wolfslehner B, Vacik H (2011) Mapping indicator models: from intuitive problem structuring to quantified decision-making in sustainable forest management. Ecol Indic 11:274–283
Wolfslehner B, Vacik H, Lexer MJ (2005) Application of the analytic network process in multi-criteria analysis of sustainable forest management. For Ecol Manag 207:157–170
Zhou P, Ang BW, Poh KL (2007) A mathematical programming approach to constructing composite indicators. Ecol Econ 62:291–297

Part III
The Industrial Forestry Supply Chain

Chapter 14
Carbon Sequestration and Industrial Plantations

Bruce Manley

14.1 Introduction

Since 2000, carbon has emerged as a product of industrial plantations that complements wood production. The total forest carbon market increased from 0.3 $MtCO_2$-e in 2002 to 30.1 $MtCO_2$-e in 2010. The component of the market most connected to plantation forests, afforestation/reforestation increased from <0.1 $MtCO_2$-e in 2002 to 5.8 $MtCO_2$-e in 2010 (Diaz et al. 2011). The scale and scope of this market is expected to continue to expand. It provides opportunities for growers of industrial plantations.

Forest growers with the ability to receive units for carbon sequestration and trade these units in a carbon market need to recognise the complementary cashflow generated by carbon. In order to do so they must be able to measure and predict carbon sequestration by industrial plantations. In conjunction with inputs on carbon prices and additional costs, this information can be used to evaluate the implications of carbon trading opportunities on forest profitability and forest management.

What You Will Learn in This Chapter

- The international framework that provides the background for carbon trading.
- The scope of the international carbon market and different approaches to carbon accounting.
- How carbon stocks are measured and modelled.
- The potential impact of carbon trading on forest management including profitability, silviculture and optimum rotation age.
- Risk associated with carbon trading and strategies for managing risk.

B. Manley (✉)
School of Forestry, University of Canterbury, Christchurch, New Zealand
e-mail: bruce.manley@canterbury.ac.nz

J.G. Borges et al. (eds.), *The Management of Industrial Forest Plantations: Theoretical Foundations and Applications*, Managing Forest Ecosystems 33,
DOI 10.1007/978-94-017-8899-1_14,

14.2 International Framework

This section will provide background on the international "rules" for carbon accounting. It will include a discussion of relevant parts of the Kyoto Protocol and what they mean for national reporting of carbon stocks. It will also discuss where international negotiations are heading – for example, the inclusion of harvested wood products. The purpose of the section is to:

- provide an understanding of the overarching environment within which carbon trading schemes are developing and operating.
- motivate the need for carbon measurement and modelling (a) at the national level for carbon reporting and (b) at the business level for carbon trading.

Much of the material for this section has been extracted as verbatim quotes or paraphrased from the United Nations Framework Conventions on Climate Change website (http://unfccc.int).

14.2.1 UNFCCC

The United Nations Framework Convention on Climate Change, an international treaty, was developed at the United Nations Conference on Environment and Development (UNCED), the "Earth Summit" held in Rio de Janeiro in 1992. The goal was "to cooperatively consider what they could do to limit average global temperature increases and the resulting climate change, and to cope with whatever impacts were, by then, inevitable".[1]

Parties to the Convention must submit national reports on implementation of the Convention to the Conference of the Parties (COP); in particular information on emissions and removals of greenhouse gases (GHGs).

14.2.2 Kyoto Protocol

In 1995, countries launched negotiations to strengthen the global response to climate change, and, 2 years later, adopted the Kyoto Protocol. The Kyoto Protocol sets binding targets for 37 industrialized countries for reducing greenhouse gas (GHG) emissions. Targets are expressed as levels of allowed emissions, or "assigned amounts," over the 2008–2012 commitment period. The allowed emissions are expressed as "assigned amount units" (AAUs). The targets represent an average reduction of 5 % against 1990 levels over the first 5-year commitment period 2008–2012.

[1]http://unfccc.int/essential_background/items/6031.php

The Kyoto Protocol operates within the UNFCCC. "The major distinction between the Protocol and the Convention is that while the Convention encouraged industrialised countries to stabilize GHG emissions, the Protocol commits them to do so".[2]

Under the Kyoto Protocol, countries must meet their targets primarily through national measures. These measures include land use, land-use change and forestry (LULUCF) activities.

Under Article 3.3 of the Kyoto Protocol, Parties can meet their commitments using "net changes in GHG emissions by sources and removals by sinks through direct human-induced LULUCF activities, limited to *afforestation*, *reforestation* and *deforestation* that occurred since 1990."[3]

Under Article 3.4 of the Kyoto Protocol, "Parties may elect additional human-induced activities related to LULUCF specifically, forest management, cropland management, grazing land management and revegetation, to be included in their accounting of anthropogenic GHG emissions and removals for the first commitment period. Upon election, this decision by a Party is fixed for the first commitment period."[4]

When LULUCF activities under Articles 3.3 and 3.4 result in a net removal of GHGs, a party can issue removal units (RMUs) on the basis of these activities as part of meeting its commitment.

However, they can also meet their targets using three market-based mechanisms:

- Clean development mechanism (CDM) allows a country with a commitment under the Kyoto Protocol to implement an emission-reduction project in developing countries. CDM projects can earn saleable certified emission reduction (CER) credits.
- Joint implementation (JI) allows a country with a commitment under the Kyoto Protocol to earn emission reduction units (ERUs) from an emission-reduction or emission removal project in another country with a commitment.
- Emissions trading which allows countries that have emission units to spare to sell this excess capacity to countries that are over their targets. Countries can trade AAUs, CERS or ERUs. They can also trade removal units (RMUs) from LULUCF activities under Articles 3.3 and 3.4 that result in a net removal of GHGs.

---

[2]http://unfccc.int/kyoto_protocol/items/2830.php

[3]http://unfccc.int/methods_and_science/lulucf/items/4129.php

[4]http://unfccc.int/methods_and_science/lulucf/items/4129.php

Text Box

Reservoirs

A component or components of the climate system where a greenhouse gas or a precursor of a greenhouse gas is stored. Trees are "reservoirs" for carbon dioxide.

Sink

Any process, activity or mechanism which removes a greenhouse gas, an aerosol or a precursor of a greenhouse gas from the atmosphere. Forests and other vegetation are considered sinks because they remove carbon dioxide through photosynthesis.

Carbon Sequestration

The process of removing carbon from the atmosphere and depositing it in a reservoir.

Source

Any process, activity or mechanism that releases a greenhouse gas, an aerosol, or a precursor of a greenhouse gas or aerosol into the atmosphere.

Carbon Offset

Where a greenhouse gas emission is reduced, avoided, or sequestered to compensate for emissions occurring elsewhere.

Definitions derived from UNFCCC

Although the Kyoto Protocol is primarily concerned with obligations at a national-level it does potentially impact on industrial plantations. At a general level it has generated a greater awareness of the role of plantations in carbon sequestration. It also has had more specific impacts:

- Government policy. The Kyoto Protocol provides an incentive for individual countries (or collectives of countries) to develop emissions trading schemes or other responses to provide incentives to individual entities to respond in a way that best helps the country meet its Kyoto Protocol obligations. For example the New Zealand ETS (NZ ETS) largely mimics Kyoto Protocol conventions. It was developed to provide "... a way for the New Zealand economy to meet our Kyoto obligations at the least possible cost ..."[5]

[5]www.mfe.govt.nz/issues/climate/greenhouse-gas-emissions/net-position/index.html. Accessed 30 May 2010.

- Tradeable commodity. Some types of CERs can be traded in the European Union Emissions Trading Scheme (EU ETS) where they can be purchased by governments to meet their Kyoto Protocol obligations or private entities to meet their EU ETS obligations. The price of CERs on the EU ETS provides a market benchmark for Kyoto Protocol-based units which is relevant for any ETS in which they can be traded. For example, the New Zealand ETS (NZ ETS) allows emitters to use some types of CERs to meet their carbon obligations.

A caveat here is that forestry CERs are not eligible for trading in either the EU ETS or NZ ETS. To reflect the non-permanence of forestry CDM projects two special types of CERs are issued; the temporary CER (tCER) that expires at the end of the commitment period and the long-term CER (lCER) that expires at the end of a project's crediting period (either 30 years fixed term or 20 years twice-renewable). On expiry temporary and long-term units must be replaced with other units by the buyer.

There are ongoing negotiations about what international agreement will succeed the Kyoto Protocol. At COP17/CMP7 in Durban in 2011 there was agreement to conclude an "agreed outcome with legal force" by 2015 for adoption by 2020. At COP18/CMP8 in Doha the Kyoto Protocol was extended to run from 1 January 2013 until 2020.

At the technical meetings associated with COP (Conference of the Parties) there has been agreement for the inclusion of harvested wood products (HWP). The Kyoto Protocol has the convention of "instantaneous oxidation" of carbon at time of harvest; i.e., that carbon associated with logs extracted at harvest is immediately released to the atmosphere. The instant oxidation assumption is a modelling surrogate for the assumption that the global pool of HWP is currently neither increasing nor decreasing.

In the negotiations leading up to COP15 in Copenhagen, it was argued that harvested wood products should be included in conventions to be adopted for post-2012 international commitments. This reflects the view[6] that "Maintaining the current 'instant oxidation' approach for post-2012 would be a poor outcome in terms of encouraging longer life wood products, investment in forests and, especially, allowing sustainable timber production." COP17 still has instantaneous oxidation as the default rule but allows for an emissions-to-atmosphere approach. Provided that transparent and verifiable data exists, accounting should include harvested wood products.

[6]http://www.maf.govt.nz/climatechange/international/nz-lulucf-submission-to-unfccc.pdf

Management Planning in Action 14.1: The Plantar Sequestration and Biomass Use Project in Brazil

"The Plantar Group, the World Bank's Prototype Carbon Fund and the BioCarbon Fund, with the support of Rabobank International, have been able to implement a successful pioneering project in the State of Minas Gerais, whereby the CDM enables the company to overcome the financial and economic barriers to establish new biomass plantations. Started in 2001, this is one of the first carbon finance projects developed, before the Kyoto Protocol entered into force. Without carbon finance, such plantations are neither economically viable nor is it possible for small-scale pig iron producers to obtain financing. More importantly, there is no other financial incentive to set aside large areas of native cerrado forests.

The Plantar Project creates emission reductions by: (i) reforesting 11,600 ha of deforested and abandoned pasture lands with sustainably managed and independently-certified plantations and (ii) uses renewable charcoal – a solid biofuel from new and additional planted forests – instead of coal coke in pig iron production. As a result, the company has become the first in the world capable of producing iron totally based on the use of renewable charcoal, with several additional environmental and social benefits. Today, the Plantar Project has become a model in the Brazilian iron industry. It has provided the basis for the integration of the CDM with several public policies, at the state and federal levels, aimed at promoting sustainable development and climate change mitigation. Particularly, it is paving the way for the State of Minas Gerais to promote the adoption of this project model to help achieve the State's goal of increasing the use of renewable charcoal in these sectors and contribute to the larger goal of reducing greenhouse gas emissions.

Environmental and Social Benefits:

The project demonstrates how sustainably managed plantations can generate new carbon stock and substitute the use of fossil fuels. It also shows how the private sector can use previously deforested and abandoned lands to help take the pressures off native forests, thereby promoting the conservation of these forests and their unique biodiversity. The project is also securing high-quality employment in rural areas with few other employment opportunities and prioritizing the hiring of local and neighboring residents."

https://wbcarbonfinance.org/Router.cfm?Page=Projport&ProjID=9600 accessed 11 January 2013

14.3 Scope of International Market

The purpose of this section is to review the development of markets for carbon sequestered by plantations and provide an understanding of the different approaches to carbon accounting.

Diaz et al. (2011) provide a forest carbon typology for forest carbon offset projects:

- Afforestation/Reforestation (AR).
- Improved Forest Management (IFM).
- Reduced Emissions from Deforestation and Forest Degradation (REDD).
- Agro-forestry.

The AR and IFM types are equivalent to the Kyoto Protocol Article 3.3 (afforestation, reforestation and deforestation) and Article 3.4 (forest management).

For industrial plantations the most relevant type is Afforestation/Reforestation ("the establishment of forest on areas without forest cover, capturing additional carbon in new tree biomass and other carbon pools; emissions reductions occur primarily through additional sequestration" Diaz et al. 2011). Also applicable is Improved Forest Management ("Existing forest areas are managed to increase carbon storage and/or to reduce carbon losses from harvesting or other silvicultural treatments; emissions reductions may occur through additional sequestration and/or avoided emissions" Diaz et al. 2011).

Forest carbon offset projects can also be differentiated into those that operate within a compliance carbon market from those within a voluntary carbon market. Diaz et al. (2011) list a diverse set of reasons for buyers engaging in the voluntary carbon market:

- Offsetting individual or corporate GHG emissions;
- Retail sale of credits or bundled environmental products to individuals or companies;
- Supporting environmentally friendly projects for corporate social responsibility purposes;
- Interest in integrating the valuation of ecosystem services into environmental management systems;
- Building experience and fluency in carbon market dynamics in preparation for compliance programs;
- Investing directly into forest carbon projects for a return on investment;
- Purchasing voluntary offsets for end-use as a pre-compliance hedge against coming regulatory liabilities;
- Resale speculation on the future value of forest credits.

Compliance carbon markets include Kyoto Protocol CDM (Clean Development Mechanism) projects, the New Zealand Emissions Trading Scheme (NZ ETS) and the New South Wales Greenhouse Gas Abatement Scheme (NSW GGAS). These are all AR projects with a total volume of 2.6 $MtCO_2e$ in 2010 compared to

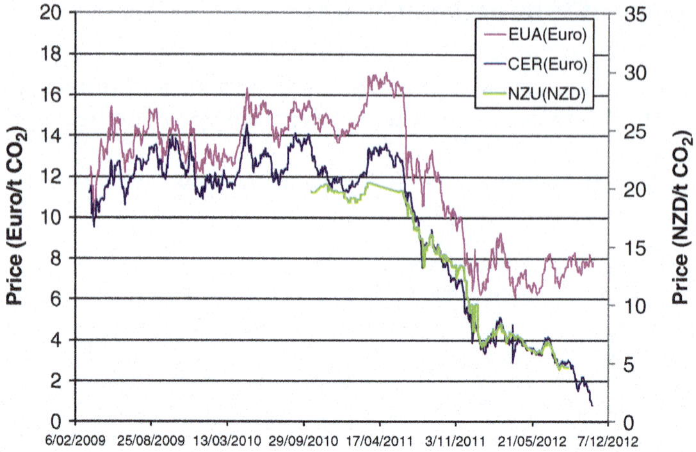

Fig. 14.1 Carbon prices for EUA (European Union Allowance), CER and NZU. Prices for EUA and CER are in Euro/t CO_2 (Source ECX) while the price for NZU is in NZD/t CO_2 (Source: Westpac)

27.6 $MtCO_2e$ for voluntary carbon markets. However the majority (71 %) of the voluntary market volume was from REDD projects with 13 % (3.5 $MtCO_2$-e) in each of the AR and IFM types (Diaz et al. 2011).

The supply estimated for 2011–2015 by project developers is 11.4 $MtCO_2$-e from AR projects and 7.3 $MtCO_2$-e from IFM projects in voluntary markets compared to 13.8 and 1.7 $MtCO_2$-e respectively in compliance markets (Diaz et al. 2011).

Carbon prices averaged US$5.60/t CO_2-e in 2010 in voluntary markets and $4.61/t CO_2-e in compliance markets. However there was a large differentiation in compliance market prices with CDM at $4.49 and the NZ ETS at $12.95 (Diaz et al. 2011). Carbon prices have shown a high level of volatility – Fig. 14.1 shows carbon prices for EUA, CER and NZU. Of the three units, the NZU is the only one that applies to carbon sequestration by forests – however the similarity in price between the CER and the NZU is apparent. Emitters in New Zealand can meet their obligations by purchasing CERs (but not forest-based tCERS or lCERs). Consequently there is linkage between the prices for NZUs and CERs. Factors that have caused price fluctuations in the EU ETS have also affected prices in the NZ ETS.

The compliance carbon markets are controlled by international, national or state regulations. The voluntary carbon market operates under a range of standards. The development of robust standards has been vital for the functioning of the voluntary market. Credible standards are necessary to ensure that credits from carbon offset projects bought by businesses and consumers can be trusted and have real environmental benefits. The dominant standard for the voluntary carbon market is VCS (Voluntary Carbon Standard).[7]

[7]http://www.v-c-s.org

Regulations or standards determine:

- How carbon is measured.
- The carbon credits or units received.
- Circumstances under which the units have to be surrendered.
- The credibility of the carbon units.
- How the carbon units can be traded.

Collectively these factors determine the benefits and risks associated with the carbon trading opportunity that the forest grower has.

14.4 Carbon Accounting Methods

Different approaches to carbon accounting (i.e., the issue and surrender of carbon units) include:

- Temporary units. Finite-term units with replacement. Units are issued for carbon sequestration but expire after a period of time. For example, the CDM issues two types of finite-term units, the temporary CER (tCER) that expires at the end of the commitment period and the long-term CER (lCER) that expires at the end of a project's crediting period (either 30 years fixed term or 20 years twice-renewable). On expiry temporary credits must be replaced with other units.
- Minimum carbon stock level. Units are issued for carbon sequestration but the forest grower must commit to maintaining sequestered carbon for a finite period. For example, the NSW GGAS carbon stocks have to be maintained for 100 years. The forest grower has no liability beyond 100 years.
- Stock change approach (Real-time accounting). Carbon stocks are periodically estimated. A forest grower receives units as carbon stocks increase but must surrender units as carbon stocks decrease, for example, after harvesting or after a catastrophic event (e.g., wind or fire). This is the approach adopted under the NZ ETS. It is also the general approach for California's greenhouse gas cap-and-trade program.[8]
- Long-term average approach (One-time stock change). The long-term average carbon stock (over some defined period) is estimated for the given species, site and silviculture (including rotation age). A forest grower receives units until the

[8]Forestry offsets under the California program include Reforestation, Improved Forest Management and Avoided Conversion projects. Each offset credit equals one metric ton of carbon dioxide and must be 'additional,' that is, over and above any reductions already required by law or regulation. Forest Projects must continue to monitor, verify and report offset project data for a period of 100 years following any offset credit issuance. A portion of offset credits issued to the forest offset project will be placed into a forest buffer account. If a significant disturbance occurs leading to an unintentional reversal that reduces the forest project's standing live carbon stocks below the forest project's baseline standing live carbon stocks then the forest buffer account is used. If reversal is intentional all credits received must be retired with, in the case of Improved Forest Management projects, the credits scaled up by a "compensation rate".

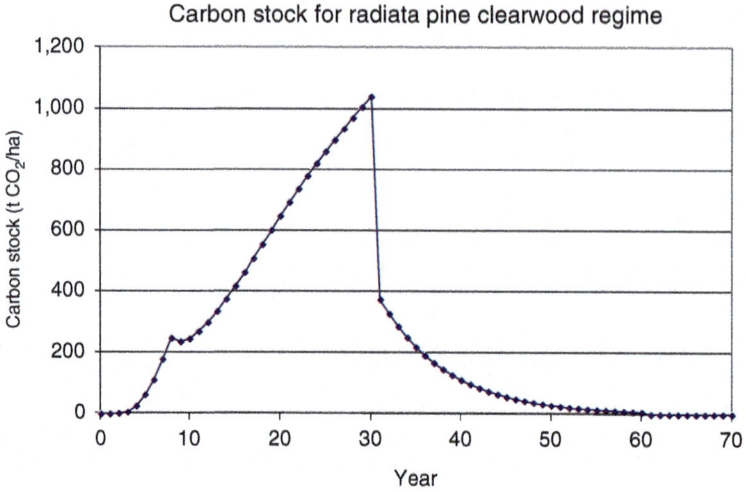

Fig. 14.2 Carbon stock profile for first rotation of a radiata pine clearwood example. It is assumed that there is instantaneous oxidation of carbon in logs that are harvested

long-term average is reached after which time no further units are received. The stand may be harvested without units having to be surrendered provided that the land is re-established in trees. This approach was originally described by Maclaren et al. (1995) and was proposed for Reforestation in the 2009 exposure draft of legislation for an Australian ETS (that was not enacted). It is the approach adopted by VCS in accounting for carbon for Afforestation, Reforestation and Revegetation projects.[9]

- Tonne-year accounting. Units are given each year for the total carbon stock. It was proposed by Moura-Costa and Wilson (2000) as a means of comparing the temporary sequestration of carbon with permanent sequestration or fossil fuel emission avoidance. An equivalence time is estimated and the inverse provides an equivalence factor. For example, for an equivalence time of 55 years the effect of sequestering 1 t of CO_2 for 1 year is deemed to be equivalent to 0.0182 (= 1/55) t of CO_2 that is permanently sequestered (or 0.0182 t of avoided emissions). This method does not require units to be surrendered.

A New Zealand plantation example is used to illustrate the different methods of carbon accounting. The example we will follow here is a radiata pine stand grown on an average New Zealand ex-farm site under a clearwood regime with planting of 1,000 stems/ha, pruning to 5.5 m in two lifts and thinning at age 8 years to 300 stems/ha. The carbon stock profile for a single 30 year rotation is shown in Fig. 14.2 while that for a succession of 30 year rotations is shown in Fig. 14.3. The change in carbon stock from year-to-year is given in Fig. 14.4.

[9]http://www.v-c-s.org/sites/v-c-s.org/files/VCS%20Guidance%2C%20Harvesting%20Examples.pdf

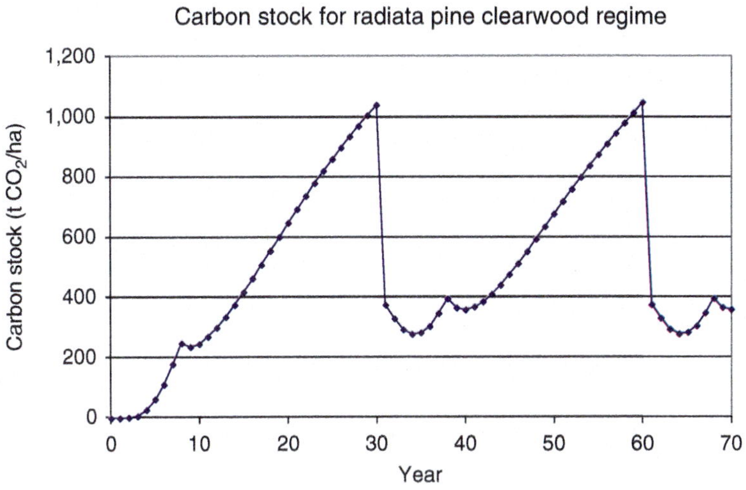

Fig. 14.3 Carbon stock profile for radiata pine clearwood example assuming that there is replanting. It is assumed that there is instantaneous oxidation of carbon in logs that are harvested. After the harvest of the first rotation the total carbon stocks are calculated as the sum of carbon stocks in residues from the first rotation plus the carbon stocks in the second rotation crop

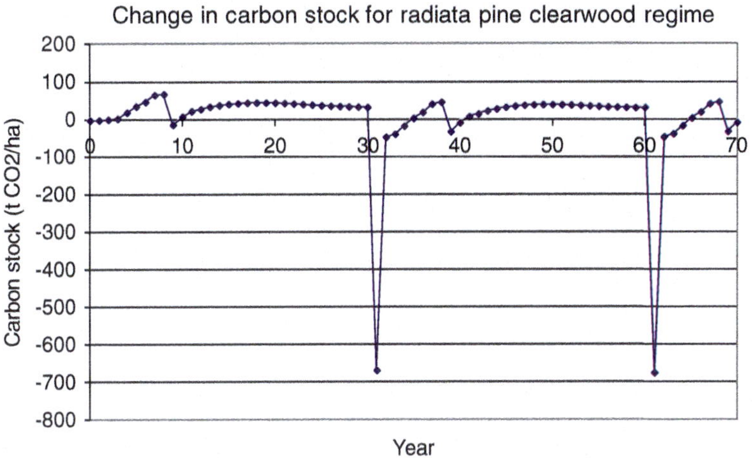

Fig. 14.4 Change in carbon stock for radiata pine clearwood example assuming that there is replanting

Figures 14.3 and 14.4 can be used to illustrate the opportunities for carbon trading under the different carbon accounting approaches:

- Temporary units. Assuming the same rules as apply to CDM projects temporary units could be received for either the first 5 years (or end of a commitment period if sooner) or 30 years following the initial 5 or 30 years of Fig. 14.4. On expiry temporary units would have to be replaced by the purchaser with other units.

- Minimum carbon stock level. The minimum carbon stock level of 279 t CO_2-e/ha occurs after 34 years (Fig. 14.3). Carbon units will be received annually (as shown in the initial part of Fig. 14.4) until 279 units have been received after 12 years. Thereafter no further carbon stocks are received. No carbon units have to be surrendered provided the minimum level is maintained.
- Carbon stock change. Figure 14.4 shows the carbon stocks that will be received or have to be surrendered.
- Long-term average approach. Suppose the long-term average is based on the asymptotic value for the example stand managed on a succession of 30 year rotations. The long-term average would be calculated as 577 t CO_2-e/ha. This value is reached after 19 years (Fig. 14.3). Consequently a total of 577 carbon units would be received following the profile of Fig. 14.4 for the first 19 years.
- Tonne-year accounting. Units would be received at the rate of 1/55 of the profile shown in Fig. 14.4; ie, at the rate of about 1 unit per year.

Note that these examples assume a single stand. In practice the area unit for carbon accounting might not be a single stand – rather the project or carbon accounting area might contain a number of stands at different ages. Carbon trading would be based on the combined profile.

14.5 Measurement and Modelling of Carbon Stocks

Carbon trading is based on changes in carbon stock over time. Rather than measuring carbon flux the standard approach is to compare measurements at two points in time. This section will describe approaches to measurement of carbon that use standard measurements such as tree diameter and height as well as basic wood density. It will also introduce models that predict the carbon stock development over time for stands of defined species, site and silviculture.

14.5.1 Pools

The pools that allow complete quantification of carbon stock changes for any category of land without double counting, are defined in IPCC (2006). A summary is given in Table 14.1.

Under the Kyoto Protocol, carbon that is removed off-site following thinning and harvesting is considered as an instant emission that year. The inclusion of harvested wood products would require the addition of an additional "harvested wood products" pool.

Table 14.1 Definition of carbon pools (Summary of Table 1.1 of IPCC 2006 – the limits of 2 mm and 10 cm are suggested values only)

Pool		Description
Biomass	Above ground biomass	All biomass of living vegetation above the soil including stems, stumps, branches, bark, seeds and foliage
	Below ground biomass	All biomass of live roots down to a 2 mm diameter
Dead organic matter	Dead wood	Includes all non-living woody biomass (larger than 10 cm) not contained in the litter, either standing, lying on the ground, or in the soil
	Litter	Includes all non-living woody biomass between 2 mm and 10 cm
Soils	Soil organic matter	Mineral soil carbon to a specified depth. Includes fine roots within the soil

14.5.2 Measurement of Carbon Stocks

Carbon stocks are estimated using relationships or models using basic tree measurements (diameter, height) as inputs. Different approaches are:

- Allometric relationships.
- Biomass expansion factors.
- Biomass allocation models.

14.5.2.1 Allometric Relationships

The term allometry refers to the relationship between the size of parts of an individual and the size of the whole individual. An example of a tree allometric relationship is:

$$\mathrm{agb} = \beta_0 + \beta_1 \left(\mathrm{d}^2\mathrm{h}\right)$$

where agb = tree above-ground biomass
d = tree diameter at breast height (dbh)
h = Tree height

The stand-level equivalent of this would be:

$$\mathrm{AGB} = \beta_0 + \beta_1 \,(\mathrm{GH})$$

where AGB = Stand above-ground biomass
G = Stand basal area
H = Stand height

Table 14.2 Ratio of below-ground biomass to above-ground biomass for example vegetation types

Vegetation type	Above-ground biomass (t/ha)	Ratio
Conifer plantation	<50	0.46
	50–150	0.32
	>150	0.23
Eucalypt plantation	<50	0.45
	50–150	0.35
	>150	0.20

IPCC (2003) Table 3a.1.8

These functions are derived using data from biomass studies. A wide range of forms exist (see for example Moore 2010). Other variables can be included; for example, wood density:

$$\text{agb} = \beta_0 + \beta_1 \left(\rho\ d^2 h\right)$$

where ρ = whole-stem wood density

Typically below-ground biomass is estimated as a proportion of above-ground biomass. For example, IPCC (2003) lists default values for different vegetation types. Table 14.2 gives the values for Conifer plantations and Eucalypt plantations.

It is more challenging to estimate carbon stock change for the other pools (Dead Organic Matter, Soils). Sometimes the default assumption is made that these pools are stable. For example in the NZ ETS, soil carbon is not included because changes in soil carbon are generally small and difficult to measure at reasonable cost (MAF 2008).

Carbon is calculated by multiplying biomass by the carbon fraction (0.5). Standard reporting and carbon trading uses units of CO_2 rather than C. As a final step, the units (tonnes) of CO_2 is calculated by scaling the tonnes of C by the ratio of atomic weights (=44/12).

14.5.2.2 Biomass Expansion Factors

These factors expand stem volume (or biomass) to account for other (above-ground) biomass components. For example:

$$\text{agb} = \text{BEF} * \text{v}$$

where v = stem volume

Sometimes density is made explicit:

$$\text{agb} = \text{BEF} * \rho\ \text{v}$$

14.5.2.3 Biomass Allocation Models

These models explicitly estimate the biomass in each pool. For example the C-change model uses stem volume production to predict the biomass and carbon content of the stem, crown, roots, forest floor and understorey of managed radiata pine stands. It integrates knowledge of growth partitioning, mortality and decay of tree components (Beets et al. 1999).

14.5.3 Modelling Changes in Carbon Stock

Growth models used in conjunction with allometric relationships, biomass expansion factors, or biomass allocation models can be used to forecast the change in carbon stocks for stands of defined species, site and silviculture. This enables financial analysis that includes carbon as well as traditional forestry with revenue coming from the sale of logs or timber.

14.6 Economic Optimisation of Plantations with Joint Production of Logs and Carbon

This section covers the incorporation of carbon into stand-level financial evaluation. It will illustrate how carbon trading provides a complementary cashflow profile to that from traditional forestry with the production of logs only – carbon cashflows (early revenues, costs after harvest) are the reverse of typical forestry cashflows (early costs, revenues after harvest). It will explore the consequence that this can have on forest profitability, choice of regime and optimum rotation age. The impact of including the HWP pool in carbon accounting is also evaluated. Initially the stock change approach to carbon accounting is considered. The effect of changing to the long-term average approach is then considered.

We will continue with the example of a radiata pine stand grown on an average New Zealand ex-farm site under a clearwood regime. Costs and revenues for this regime are given in Table 14.3.

The carbon stock profile was given in Fig. 14.3 and the carbon stock change in Fig. 14.4. Net cashflows for forestry and carbon are given in Fig. 14.5 with carbon accounting under the stock change approach and a constant carbon price of \$20/t CO_2-e and annual overhead costs for carbon of \$30/ha/year to cover the costs of measurement, auditing and administration.

A feature of Fig. 14.5 is how complementary the two different cashflow streams are. In most years the net revenue from carbon is sufficient to cover the costs of silviculture and forestry overheads. The revenue from harvest is sufficient to easily cover the cost of carbon units that have to be surrendered with the post-harvest reduction in carbon stocks (given the assumed log and carbon prices).

Table 14.3 Costs and revenues for example clearwood regime

Time (years)	Silvicultural cost ($/ha)	Annual overhead cost ($/ha)	Harvest revenue ($/ha)
0	1, 200	80	
1	100	80	
2	100	80	
3		80	
4		80	
5	700	80	
6		80	
7	600	80	
8	500	80	
9		80	
10		80	
11		80	
…		80	
29		80	
30			38,864

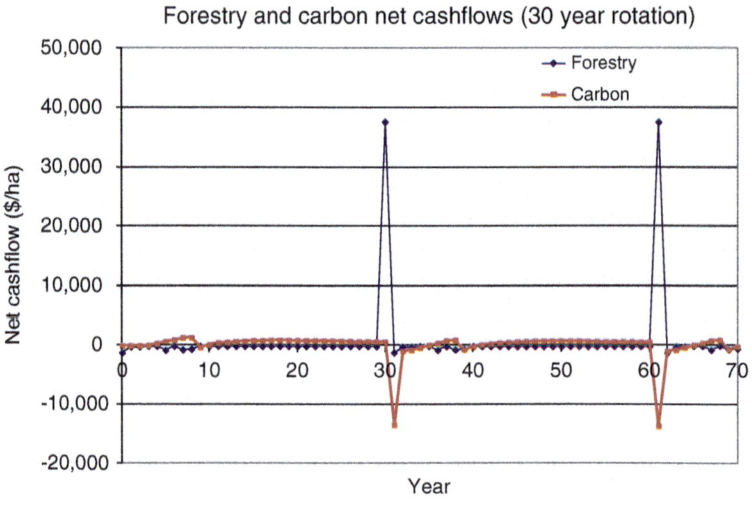

Fig. 14.5 Net cashflow for forestry and carbon for radiata pine clearwood example and a 30 year rotation

Figure 14.6 shows the LEV calculated, for a discount rate of 8 %, for a range of rotation ages for (i) forestry only and (ii) with carbon included at different prices. The inclusion of carbon gives a marked increase in LEV and also in the optimum rotation age. The higher the carbon price the greater the incentive to grow a stand on rather than harvest. At high carbon prices, harvest is substantially delayed. This has potential issues for wood supply and the wood-using industry.

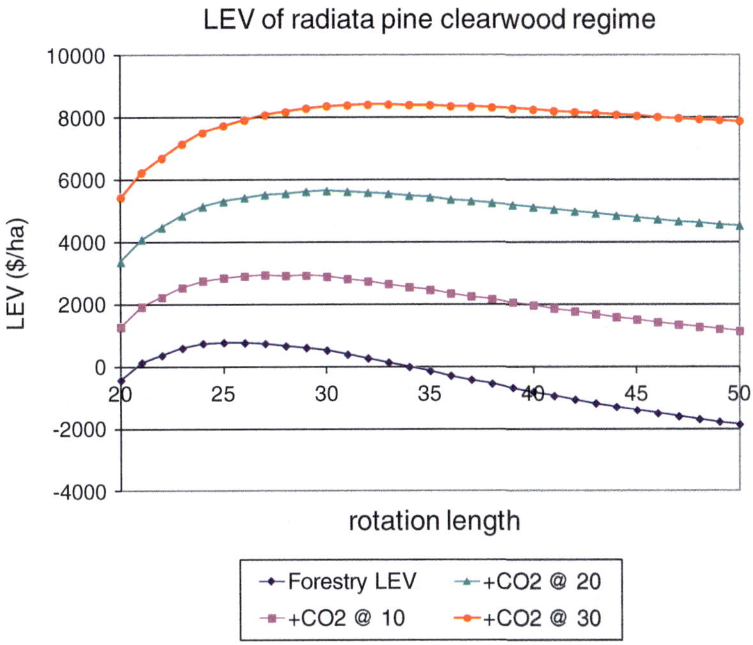

Fig. 14.6 LEV of radiata pine clearwood regime example when revenues come from (i) log sales only and (ii) log sales and carbon trading with carbon prices of \$10, \$20 and \$30/t CO_2-e

14.6.1 Choice of Silvicultural Regime

Consider an alternative silvicultural regime:

- No thin (plant 1,000 stems/ha, no thinning).

This regime is less intensive than the clearwood regime and has neither thinning nor pruning. The no thin regime has higher carbon stocks than the clearwood regime (Fig. 14.7).

Without carbon, the ranking of LEVs is Clearwood followed by No thin (Fig. 14.8). This ranking is reversed when carbon trading is included once the price of carbon exceeds about \$5/t CO_2-e. An increase in carbon price relative to log prices favours regimes that produce more biomass and hence sequester more carbon. Carbon trading also favours species that quickly increase biomass Maclaren et al. (2008).

14.6.2 Inclusion of Harvested Wood Products

IPCC (2003) provides default values for the half life in use for different HWP categories (Table 14.4). COP17/CMP7 in Durban aggregated the two panels categories and specified half-lives of 2 years for paper, 25 years for wood panels and 35 years for sawn wood.

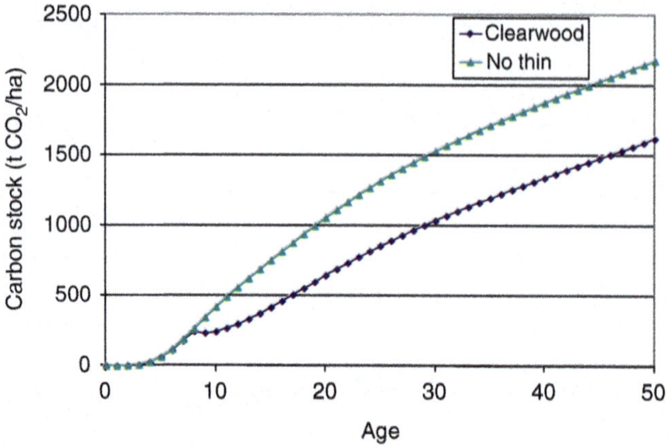

Fig. 14.7 Carbon stock by age for two different silvicultural regimes

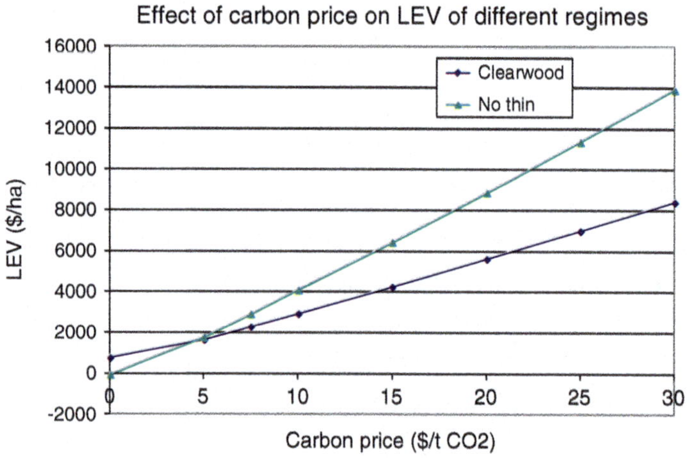

Fig. 14.8 Effect of carbon price on the LEV of two different regimes

Table 14.4 Half-lives for different HWP categories

HWP category	Half-life in use (years)
Saw wood	35
Veneer, plywood and structural panels	30
Non structural panels	20
Paper	2

IPCC (2003) Table 3a.1.3

Addition of the HWP pool to carbon accounting results in a different carbon stock profile (Fig. 14.9). The reduction immediately after harvest is diminished and the duration of carbon stocks is extended well beyond harvest age. The impact on LEV is not great (Fig. 14.10). Similarly the inclusion of HWP does not change optimum rotation lengths by any great extent – in the example it reduces optimum rotation

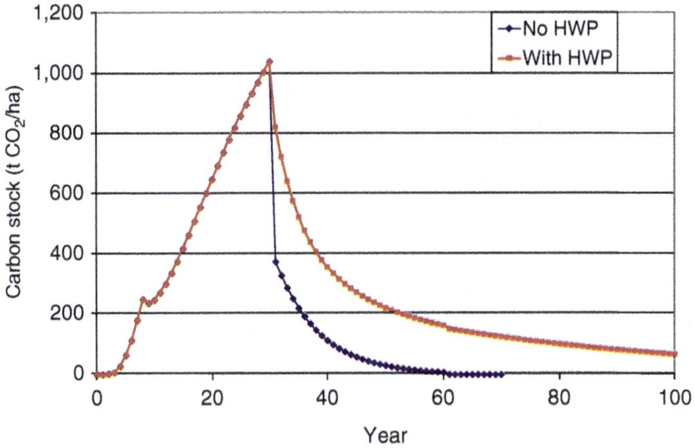

Fig. 14.9 Impact of the inclusion of HWP on carbon stock of a single rotation of the example radiata pine clearwood regime grown on a 30 year rotation. HWP are calculated using the half-lives in IPCC (2003)

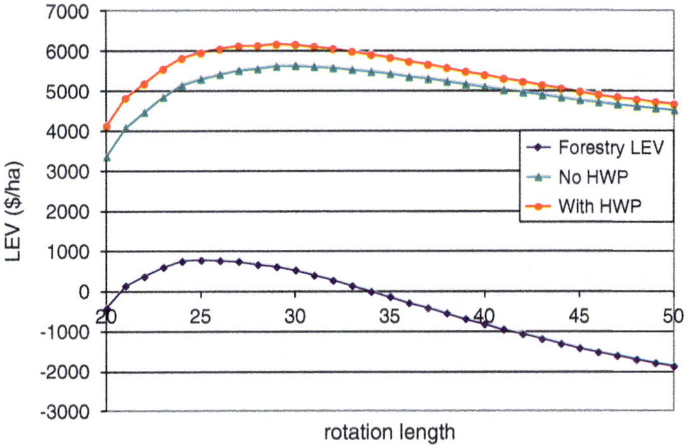

Fig. 14.10 Impact of the inclusion of HWP on the LEV of the example radiata pine clearwood regime grown on a 30 year rotation

age from 30 to 29 years. The greatest impact would be to reduce carbon price risk to the forest grower by (a) extending the period after harvesting over which units have to be surrendered; and (b) reducing the number of units that have to be surrendered, if harvesting is followed by replanting.

14.6.3 Impact of Carbon Accounting Approach

The example has been evaluated using the stock change approach. Adoption of the long-term average approach would have a modest impact on LEV (Fig. 14.11).

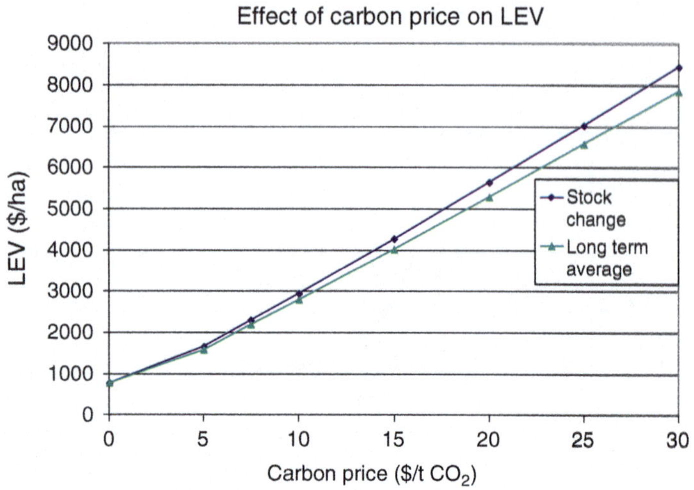

Fig. 14.11 Effect of carbon accounting approach on LEV at different rotation ages

However it removes an important component of risk that growers face under the stock change approach. Provided that they replant, growers do not have to surrender units after harvest. Consequently they are not exposed to the carbon price risk associated with post-harvest buy back of carbon units.

14.7 Strategies for Managing Risk

Key components of risk associated with carbon are that:

(i) An unexpected event (e.g. wind or fire) will require early surrender of units and create cashflow problems for the owner; i.e., physical risk that creates financial risk.
(ii) Carbon prices will be so high at the time of harvest that, under the stock change approach, owners will not be able to afford to harvest; ie, financial risk.

The first component might be dealt with in an ETS by not requiring the premature surrender of units provided that the tree crop is re-established. It can also be dealt with by the scheme administrator pooling this risk by retaining a proportion of units in a buffer account. Here we focus on the second component, carbon price risk, and discuss some of the strategies covered by Manley and Maclaren (2012).

14.7.1 Stand-Level Strategies

At the stand-level, carbon is not permanent. Harvesting causes a reduction in carbon stocks and, depending on the rules of the carbon accounting scheme, may trigger a

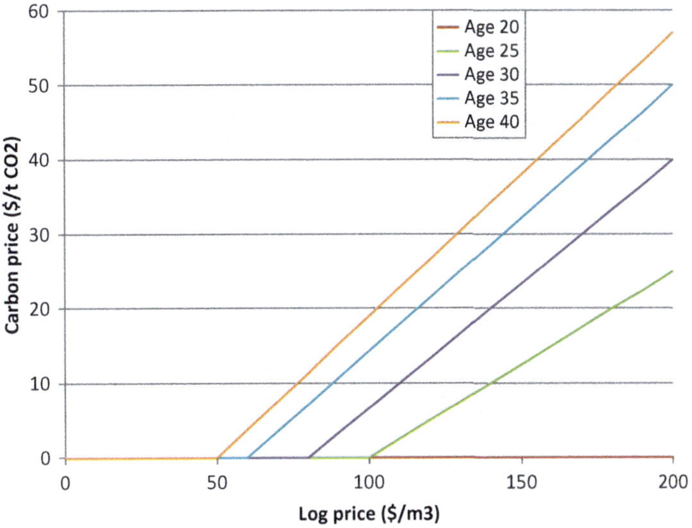

Fig. 14.12 Example of a reservation pricing strategy at example ages. Harvesting should occur when the combination of log and carbon prices is below the curve

liability. This is the case under the stock change approach. The need to surrender carbon units in the future at an unknown price creates risk. Stand-level strategies for managing this risk include delaying harvest, growing a valuable crop and trading only a proportion of carbon units received.

14.7.1.1 Option to Vary Harvest Age

There is a real option value associated with plantations because of the opportunity to vary harvest age in response to price – or never to harvest at all. Figure 14.12 gives an example of the reservation price strategy that could be adopted – it gives the combination of log and carbon prices at which harvesting should be carried out at different ages. If carbon prices stay high relative to carbon prices the grower will want to delay harvest – the focus of the business shifts from wood production to carbon sequestration via biomass accumulation.

14.7.1.2 Growing a Valuable Crop

One strategy to partially mitigate the carbon price risk is to grow a valuable crop; ie, to produce a crop that will give a stumpage revenue high enough to offset the cost of carbon units to be surrendered. Figure 14.13 shows that the radiata pine clearwood regime is relatively low risk in this regard – at a carbon price of \$20/t CO_2 stumpage revenues exceed the carbon unit surrender cost (ie, less than 100 %). Conversely the no thin regime is high risk.

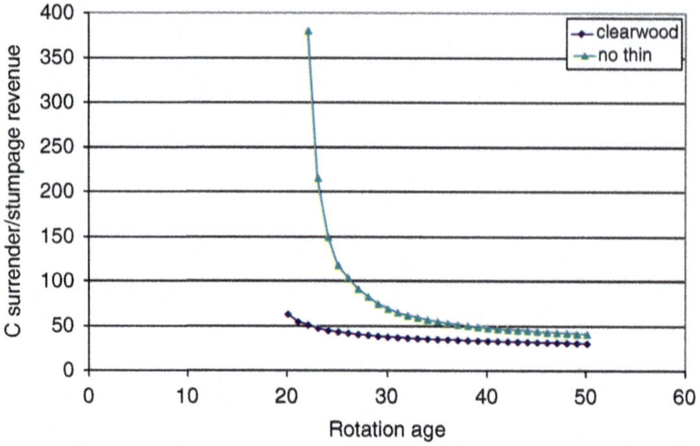

Fig. 14.13 Cost of carbon units to surrender at time of harvest as a percentage of stumpage revenue for different silvicultural regimes. Assumes a carbon price of \$20/t CO_2-e

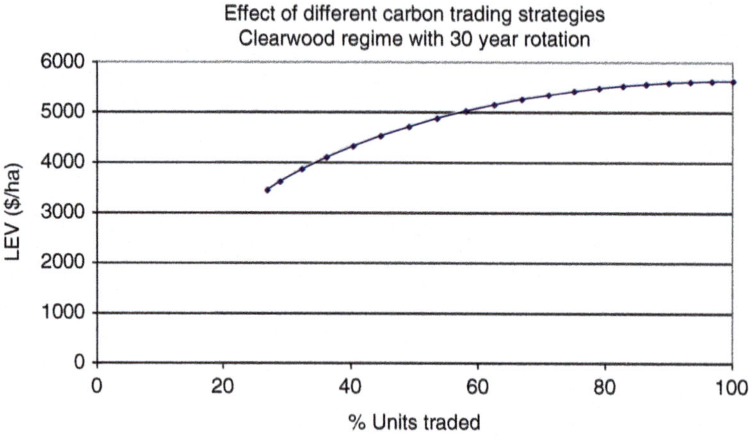

Fig. 14.14 Trade-off between percentage of carbon units received that are traded and LEV. Radiata pine clearwood regime with a 30 year rotation. Assumes a carbon price of \$20/t CO_2-e

14.7.1.3 Carbon Trading Strategy

Trading only a portion of the units received is another way to manage for carbon price risk. Rather than trading all units received some can be held and surrendered at the time of harvest.

So far it has been assumed that all units received are traded. Figure 14.14 shows the trade-off between the percentage of units traded and the LEV. The point at the bottom-left of the curve involves trading of only 279 units. This is low risk – no carbon units need to be surrendered because the level of carbon stock in the stand does not subsequently fall beneath 279 t CO_2–e/ha (provided that replanting occurs). However the opportunity cost is the reduction in LEV to $3,476/ha.

The curve is a form of risk-return frontier. Note how it is not linear. It is possible to achieve an LEV that is 95 % of the maximum LEV by trading only 71 % of units received. These units would be the first 71 % of units received – relating back to Fig. 14.3 they are the first 739 units received during the first 22 years. Units received from years 23 to 30 would not be traded.

14.7.2 Estate-Level Strategies

An important principle to understand is that while individual stands are not permanent, forest estates generally are. A forest estate of mixed age-classes (and mixed species) can be managed in a way that total carbon stocks never decline. As an extreme case, a normal forest could be developed by planting a constant area every year for a full rotation cycle – for example a 30 ha forest could be established over 30 years with 1 ha planted each year. When the first block of 1 ha is harvested the carbon emissions will be offset by the ongoing carbon sequestration in the remaining 29 blocks.

Consider four alternatives that include radiata pine grown under the clearwood regime. The alternatives differ in terms of when the land is planted and harvested in the first rotation (all have a rotation age of 30 years in the second and subsequent rotations):

1. Plant 30 ha this year and harvest at age 30 years.
2. Plant 30 ha this year and harvest 1 ha per year at ages 21 to 50.
3. Plant 15 ha this year and 15 ha in 15 years time. For both blocks harvest 1 ha per year between ages 25 and 39 years.
4. Plant 1 ha per year and harvest at age 30 years.

Strategy 1 is the stand-level strategy presented in Fig. 14.3 scaled up by 30 ha. The other three strategies are all alternative ways of developing a normal forest with a target rotation of 30 years. Each strategy provides a different carbon stock profile over time (Fig. 14.15). Note that it takes two rotations for a steady state of carbon to be reached – carbon stocks are higher during the second rotation as residual carbon from the first rotation is still present, particularly during early stages before it decays.

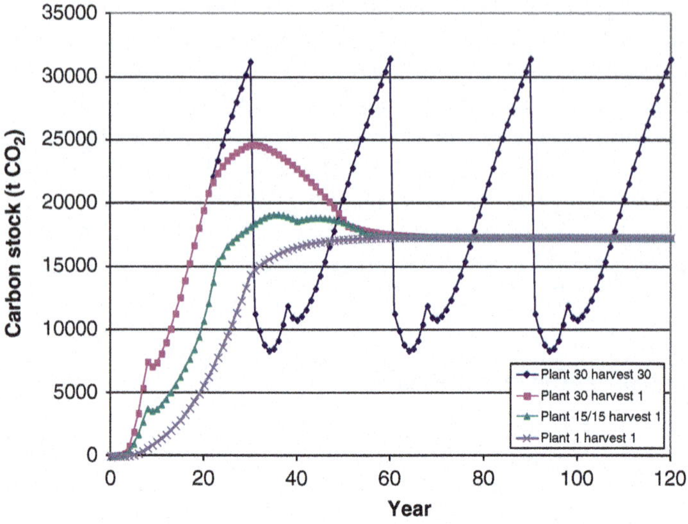

Fig. 14.15 Carbon stocks for four different strategies for example estate of 1,200 ha. Assumes clearwood regime

Management Planning in Action 14.2

City Forests have a 16,000 ha forest estate near Dunedin, New Zealand. Included in the estate is 4,000 ha of 'Kyoto-eligible' plantations that have been established since 1989 (Fig. 14.16).

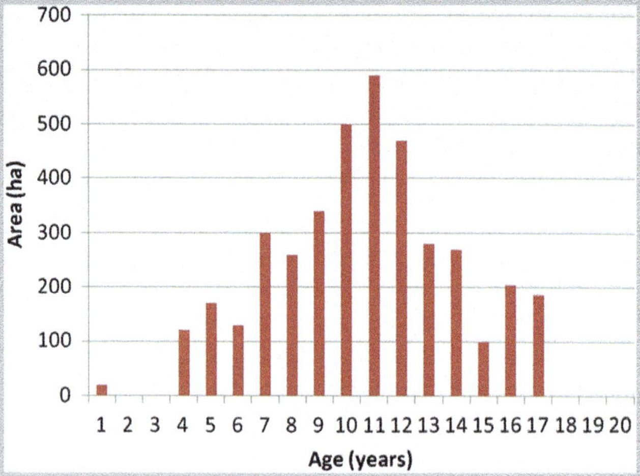

Fig. 14.16 Area by age-class of City Forests post-1989 plantings. Area is as at 2008 – the start of Kyoto Protocol commitment period 1. City Forest was able to earn NZUs for carbon sequestered since 1 January 2008

(continued)

(continued)

City Forests entered this estate into the NZ ETS in 2010 and have been claiming units annually. Since 2010 they have been selling NZUs to international energy companies with emissions obligations under the NZ ETS. They sold 11 million NZUs during 2010–2011 and a further six million NZUs during 2012–2013.

City Forests manage the 4,000 ha as a single estate for carbon trading. The strategy is to sell all units until the long-term minimum carbon stock level is reached. At this point they will hold the NZUs that they receive in order to use them to meet future surrender obligations. They will vary rotation age in order to increase the long-term minimum carbon stock level.

14.8 Summary

Although the Kyoto Protocol operates at a government-to-government level, it provides the framework within which markets for credits from forest carbon offset projects trading schemes are developing and operating. Forest carbon offset projects can be differentiated into those that operate within a compliance carbon market from those within a voluntary carbon market. While the compliance carbon markets are controlled by international, national or state regulations, the voluntary carbon market operates under a range of standards with the VCS having become the dominant standard. These regulations or standards determine the carbon accounting method to be used and how carbon is to be measured. Consequently they determine the benefits and risks associated with a forest grower's carbon trading opportunity.

Carbon trading has the potential to have a major impact on forest management. It provides a complementary cashflow to traditional forestry with early revenues offsetting initial forestry establishment and tending costs. The requirement, under the stock change approach to carbon accounting, to surrender units following harvesting creates a cost that is offset by harvest revenues.

The impact on forest profitability depends on carbon prices. With increasing carbon price:

- LEV increases;
- Optimum rotation age increases; and
- Less intensive silviculture with higher stockings is preferred.

An important risk associated with carbon is that carbon prices may be so high at the time of harvest that, under the stock change approach, owners will not be able to afford to harvest. Stand-level and estate-level strategies can be adopted to reduce this risk but with a reduction in profitability.

14.9 Web References

This topic is rapidly evolving with the ongoing development of international agreements and emissions trading schemes. Useful web references to check include:

United Nations Framework Convention on Climate Change unfccc.int/
World Bank Carbon Finance Web Site siteresources.worldbank.org/INTCARBON FINANCE/Resources/
NGO portals like: www.forestcarbonportal.com/

14.10 Problems

1. Kyoto Protocol rules have included instant oxidation of carbon in logs that are harvested. Why was this rule adopted in spite of it being unrealistic? What are advantages and disadvantages of incorporating the harvested wood products (HWP) pool into carbon accounting (under the stock change approach)? What are the potential impacts on a forest grower in terms of profitability and risk associated with carbon trading?
2. Develop a simple spreadsheet model to calculate the NPV (with a 7 % discount rate and end-of-year discounting) associated with carbon trading under three different carbon accounting approaches:

    (a) Stock change approach;
    (b) Minimum carbon stock level approach; and
    (c) Long-term average approach.

    Assume that carbon stocks increase at the rate of 30 t CO_2-e per year for a 30 year rotation. Following harvesting 690 of the 900 t CO_2-e are associated with harvested logs (for which instant oxidation is assumed). The balance of 210 t CO_2-e decays linearly at the rate of 10 % per year. Replanting occurs at the time of harvest.

    Assume a constant carbon price of \$10/t CO_2-e and a 150 year time horizon. Compare the three approaches in terms of NPV and risk.

3. An Emissions Trading Scheme (ETS) has the potential to have a large impact on forestry. It provides opportunities for forest growers but there is also risk.

    (a) Explain the potential impacts that an ETS could have on plantation forestry. Do this by contrasting (i) "traditional" forestry (i.e., forestry with revenues only from log sales) with (ii) "ETS forestry" (ie, forestry including cashflows from carbon trading).
    (b) What are the additional risks associated with ETS forestry?
    (c) Explain how this risk can be managed?

4. In the chapter on Greenhouse Effect in his book, 'Environmental Effects of Planted Forests', Piers Maclaren notes that "a common source of confusion arises

because of a failure to distinguish a ***stand*** of trees from a ***forest***". He is referring to the common perception that because of harvesting, forests do not provide long-term environmental benefits.

(a) Why does this confusion arise and why is this common perception not correct? As part of your answer you should use examples to differentiate between a stand and a forest estate in terms of the pattern of carbon stock development over time.

Concern has been expressed about the risk that a forester would face (under the stock change approach to carbon accounting) in taking carbon credits because of the need to surrender credits when a stand is harvested; credits could be worth so much at the time of harvest that the forester might not be able to afford to repurchase the credits necessary to harvest the stand.

(b) Suggest possible strategies the forester could apply to eliminate this risk; i.e., identify strategies at (i) the stand-level and (ii) the estate-level under which the forester could take and sell credits and never have to repurchase credits. (In answering this question ignore the risk of catastrophic loss caused by factors such as wind, fire, pests and disease.)

References

Beets PN, Robertson KA, Ford-Robertson JB, Gordon J, Maclaren JP (1999) Description and validation of C_Change: a model for simulating carbon content in managed Pinus radiata stands. N Z J For Sci 29(3):409–427

Diaz D, Hamilton K, Johnson E (2011) State of the forest carbon markets 2011. From canopy to currency. Ecosystem Marketplace, p 70. http://www.forest-trends.org/documents/files/doc_2963.pdf

IPCC (2003) Good practice guidance for land use, land-use change and forestry. National Greenhouse Gas Inventories Programme, Intergovernmental Panel on Climate Change (IPCC), IPCC/IGES, Hayama, Japan

IPCC (2006) IPCC guidelines for national greenhouse gas inventories, vol 4. Agriculture, forestry and other land use. Intergovernmental Panel on Climate Change (IPCC), IPCC/IGES, Hayama, Japan

Maclaren JP, Wakelin S, Te Morenga L (1995) Plantation forestry in New Zealand – concepts, methodologies and recent calculations, pp 216–237. In: Greenhouse abatement measures: no regrets action now. Proceedings of the 1995 invitation symposium Melbourne, 18–18 Oct 1995. Australian Academy of Technological Sciences and Engineering, p 305

Maclaren P, Manley B, Final Year School of Forestry Students (2008) Impact of the New Zealand emissions trading scheme on forest management. N Z J For 53(3):33–39

MAF (2008) A guide to forestry in the emissions trading scheme. Ministry of Agriculture and Forestry, Wellington

Manley B, Maclaren P (2012) Impact of carbon trading on forest management in New Zealand. For Policy Econ 24:35–40

Moore J (2010) Allometric equations to predict the total above-ground biomass of radiata pine trees. Ann For Sci 67(8):806

Moura-Costa P, Wilson C (2000) An equivalence factor between CO_2 avoided emissions and sequestration – description and application in forestry. Adapt Mitig Strateg Glob Change 5:51–60

Chapter 15
Certification of Industrial Plantations

Anna Tikina and John L. Innes

15.1 Overview of Forest Certification

Forest certification appeared in the mid-1990s as a response to concerns about the protection of biodiversity and about the livelihoods of forest-dependent communities (FSC n.d.). The concept of certifying and labelling responsible forest management, aimed at helping responsible companies to emerge or stay in the market (Perera and Vlosky 2006), subsequently served as a driver for the "certification race" (Cashore et al. 2003, 2004). Forest certification labels are intended to inform consumers about the specific characteristics of the product, i.e., that the product was produced according to the principles, criteria and indicators of sustainable forest management. The creation of competitive pressures against forest companies considered to be acting unsustainably or irresponsibly has been a major objective of forest certification (Fletcher et al. 2002).

Almost all large forest industry companies are engaged in forest certification: some hoped for higher prices for certified wood products (Chen et al. 2010; Aguilar and Vlosky 2007), some strove to be ahead of the competition (FPAC 2005), and some were concerned about demonstrating their quality of forest management (Vogt et al. 2000). Environmental non-governmental organizations (ENGOs) were the major proponents of the first standard – the Forest Stewardship Council (FSC) certification, which emerged in 1994.

The "race" to certify involved several factors. First, it did not take long before alternative standards were developed, thus resulting in competition between standards, although there has been remarkable convergence amongst the standards (Bernstein and Cashore 2004; Tollefson et al. 2008). Early examples of alternative standards included the Sustainable Forestry Initiative (SFI) in the U.S., the LEI

A. Tikina (✉) • J.L. Innes
Department of Forest Resources, University of British Columbia, Vancouver, BC, Canada
e-mail: anna.tikina@ubc.ca

J.G. Borges et al. (eds.), *The Management of Industrial Forest Plantations: Theoretical Foundations and Applications*, Managing Forest Ecosystems 33,
DOI 10.1007/978-94-017-8899-1_15,

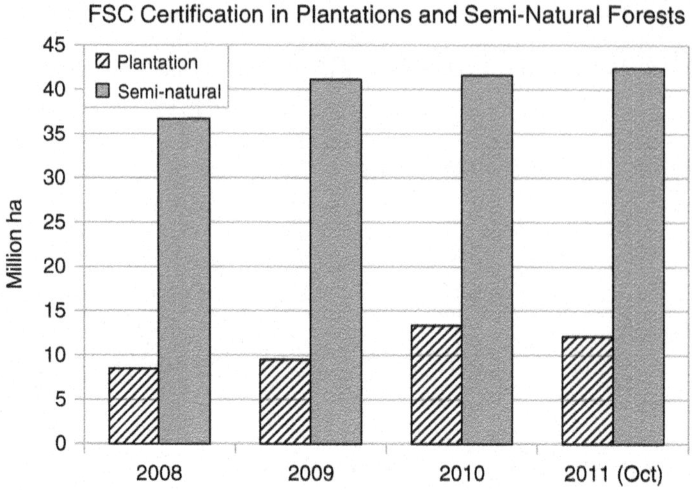

Fig. 15.1 Dynamics of the area of FSC-certified plantations and semi-natural forests (Source: FSC International, http://www.fsc.org/ppt_graphs.html?&no_cache=1&dlpath=facts_figures)

standard in Indonesia, and the Canadian Standards Association Z809 standard. In addition to the national standards, the PEFC (the Programme for Endorsement of Forest Certification) is an umbrella organization that covers a number of national forest certification initiatives, ensuring that they achieve a minimum of standard. Proponents of the FSC claim that its international FSC office also acts as an umbrella organization for national and regional standards, but the FSC standard also certifies forest management in many countries that do not have their own national or regional standards, using a generic FSC standard.

The area certified by a standard has become one of the most quoted measures in the competition among standards. This in itself is controversial as the number of certified hectares cannot be directly translated into the quality of forest management and its sustainability. Moreover, the figures are heavily skewed towards the very large forest licenses in Canada, where a single certification can cover millions of hectares. Outside the boreal forest, forest properties tend to be much smaller, significantly increasing the costs of certification per unit area. As a result, while certification has become widely adopted in the management of natural temperate and boreal forests, the area of tropical forests that has certified remains small (Leslie 2004; Tikina and Innes 2008). Since the mid-1990s, the certification standards have been working on this issue, and one part of the solution includes the possibility of certifying plantations.

The certification of plantations was not included in the original plans for a FSC standard. Principle 10, dealing with plantation management, was added to the standard in 1996, 2 years after the first standard appeared. Since then, the area of certified plantations has been growing at a fast pace, not only through FSC certification (Fig. 15.1; Karmann and Smith (2009) report that plantations occupied around 8 % of all FSC-certified forests in 2008), but also through national standards endorsed by the PEFC. Forest certification is often viewed as a viable mechanism for ensuring sustainable forest management in plantations (van Bodegom et al. 2008; Schulze et al. 2008; de Lima et al. 2009), but skepticism about the postulated positive effects of forest certification on rural livelihoods and the environment has also been voiced (Klooster 2010; Pokorny et al. 2012; Menne and Carrere 2007; World Rainforest Movement 2003). For example, Klooster (2010) argued that the strategies for managing and conserving natural forests as well as the engagement of community stakeholders were inadequately addressed by FSC in plantation certification. The requirements of forest certification standards can also add significant costs to plantations (Van Deusen et al. 2010) and thus need to be considered when determining the economics of establishing plantations. On the other hand, both the demand for timber and fibre and concern about the environmental and social impacts of timber and fibre production have been increasing. Forest certification is a mechanism that can provide some assurance on the legality and sustainability goals of plantation management.

15.1.1 Links with Other Chapters

Certification of industrial plantations is linked to the social and environmental considerations of plantation establishment that are described in Chap. 1. When properly applied, forest certification standards provide a major impetus to include social and environmental factors in forest management. Forest certification emerged as a market-based SFM tool, thus, the plantation economics and market conditions for the produced timber and fibre (Chap. 5) are closely connected to the adoption and impacts of forest certification.

Monitoring has been a requirement of all forest certification standards, as forest certification is about adhering to long-term sustainability goals. Without monitoring efforts and adaptations in management, the road to sustainability could be much longer and bumpier. The protection of forest biodiversity was one of the issues that drove the development of forest certification, and all standards pay significant attention to establishing biodiversity safeguards and requiring actions aimed at protection of biodiversity. The study if the landscape and genetic diversity is linked to biodiversity provisions of forest certification.

As each certification standard contains a set of criteria and indicators (C&I), the C&I sometimes include those used in Chap. 13 for calculating a sustainability index.

Certifying plantations has become a significant market issue, as the demand for sustainably produced fibre is growing (Chaps. 16 and 17). Although the demand for the forest-originated fibre has not matched the demand for, for example, agricultural products in the production of biofuels, the importance of forest biomass in the sustainable production of energy cannot be overlooked.

What You Will Learn in This Chapter

- The links between sustainable forest management and forest certification will be examined through an analysis of different criteria and indicators (C&I) used to assess whether a forest area is being managed sustainably. How C&I have been adopted in forest certification will also be discussed.
- The adaptations to the standards to allow the certification of industrial forest plantations will be presented through examples of a selection of the different major certification schemes, and also the international certification organizations (FSC and PEFC).
- The differences in the standards in relation to the certification of plantations will be emphasized.
- Procurement policies ensuring that wood is being derived from legal and sustainably managed sources will be analyzed. The implications of this for the development of certification will be examined.
- The chapter concludes with a summary of the main concepts. The prospective avenues of the development of certification will also be discussed.

15.2 Certification and Sustainable Forest Management

The United Nations Conference on Environment and Development in 1992 (Earth Summit in Rio) produced only non-binding Forestry Principles. The absence of a binding international forestry agreement led to the intensification of the development of national and regional criteria and indicators (C&I) to guide forest management (Castaneda 2000).

A criterion is defined as "a principle or standard that a thing is judged by" (Poulsen et al. 2001) while an indicator is a means for measuring forest conditions and tracking subsequent changes (Cubbage 2003).

The common themes in many C&I systems include protecting biodiversity, extent and change in forest resources, maintaining protective functions, productive functions and the health of forests, as well as the provision of social and economic benefits from the forest, and an appropriate legal and institutional framework. The UN Food and Agriculture Organization (FAO) defines sustainable forest management through adherence to these criteria (FAO 2008). Figure 15.2 shows major C&I processes around the globe.

The status and the progress towards C&I is often checked through the application of national forest certification systems. LEI, an Indonesian forest certification system, utilizes different sets of C&I for natural and plantation forestry. Box 15.1 provides an example of some C&I used for LEI Sustainable Plantation Forest Management System.

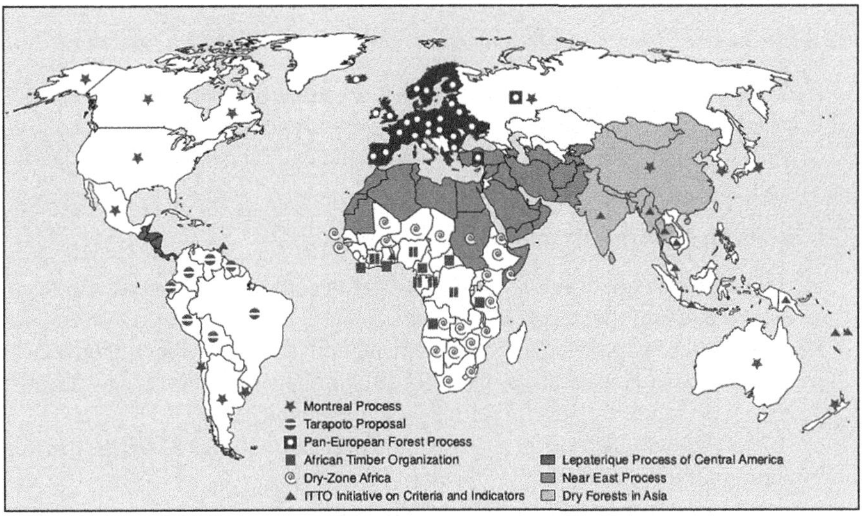

Fig. 15.2 SFM C&I processes (Source: Castaneda 2000)

Box 15.1: LEI (Indonesia) C&I for Plantation Forest Management

5.1 Production Function Sustainability
Criteria I: Resource Sustainability

P1.1 Land assurance as planted forest area.
P1.2 Forest fire management system . . .
. . . P1.9 Organization unit in forest management.

Criteria II: Forest Product Sustainability

P2.1 Forest disruption range.
P2.2 Seed availability . . .
. . . P2.7 The continuance and orderliness of funding for every activity aspects.

Criteria III: Business Sustainability

P3.1 Production area organization
P3.2 Efficiency on harvesting and utilization of planted forest products . . .
. . . P3.7 Forest stumpage asset enhancement.

5.2 Ecology/Environment Function Sustainability
Criteria I: Land and Water Quality Sustainability

E1.1 The percentage or actual breadth ratio of an appointed and well running conservation area, to its ideal breadth.

(continued)

(continued)

E1.2 The planning of effective production area structuring based on the appropriateness and capability of land and its water function continuance . . .
. . . E1.15 Active community involvement in the environmental-friendly land utilization pattern system.

Criteria II: Natural Diversity Sustainability

E2.1 The percentage of the actual breadth of the conservation area . . . to the ideal breadth of the conservation area.
E2.2 The area structuring of management unit based on the importance of flora/fauna conservation, plant forest stumpage protection, and forest resources that is very useful for the local community . . .
. . . E2.8 The existence of forest resources that can be utilized by the local community.

5.3 Social Economy and Cultural Function Sustainability
Criteria I: Community Access and Control Sustainability

S1.1 The certainty of the forest utilization area status.
S1.2 The certainty of access to the forest utilization by the community . . .
. . . S1.5 Management unit infrastructure utilization by the community.

Criteria II: Social and Cultural Integration Sustainability

S2.1 The management unit considers the social cultural impact to the community.
S2.2 Consequences towards utilization or damage of resources owned by local community . . .
. . . S2.9 Substantial management unit contributions to the economic development in the management unit area.

Criteria III: Labor Relation Sustainability

S3.1 Safety and health of workers.
S3.2 Labor involvement in contract arrangement . . .
. . . S3.7 Labor wage comply with local standard.

Source: LEI 2012. Sustainable Plantation Forest Management (SPFM) System. http://www.lei.or.id/plantation-forest-management-certification-system.

The C&I processes are aimed at balancing environmental, economic, social and cultural aspects of forest management. However, there is evidence that social and cultural aspects have attracted less attention from forest managers and other stakeholders (Gough et al. 2008). The difficulties in developing measurable C&I

for these aspects and the focus of the "green" movement on environmental aspects of forest management led to this bias. It is only recently that social and cultural concerns associated with of natural resource management have gained attention from the international policy, research and activist community. With the possible impacts that forest plantations could have on rural communities and livelihoods, monitoring of the social and cultural effects of forest management will remain a major task of forest certification.

15.3 Certification Schemes and Their Differences

Many forest certification standards have been developed over the past 20 years, involving multiple sets of C&I. For example, the Canadian Standards Association SFM standard (CSA Z809) was developed based on the C&I prepared by the Canadian Council of Forest Ministers, which closely resemble the international C&I of the Montreal Process. The original PEFC abbreviation stood for Pan-European Forest Certification, reflecting its origins in European certification practices and the Helsinki C&I. Later, PEFC became an international system for the endorsement of national certification standards in Europe, and was subsequently extended beyond Europe, reflecting the globalization of the wood products trade. This approach differs from FSC, which has a mix of a generic standard and national and regional standards based on that generic standard. A comparison of the FSC and PEFC positions on forest plantations shows the global extent of the two systems (Table 15.1). The development of national/regional standards by the FSC resulted in separate FSC standards for the management of natural forests and plantations, e.g., in Chile, where both standards were adopted in 2010.

Principle 10 was ratified and added into the FSC International Standard in 1996 (Box 15.2). The main requirements of the standards are the prohibition of the conversion of natural forests to plantations and a ban on the certification of plantations established after 1994. The cut-off date was introduced to the standard in 1999. The concern about certifying plantation forestry led in 2004 to a review of the sections dealing with plantations; this was completed 2 years later, but did not introduce any significant changes to the standard or the process. Recently, however, the FSC General Assembly produced Motion 18, which calls for the development of criteria and conditions by which plantations established after 1994 can still be FSC-certified (FSC General Assembly 2011). The critics of the motion see it as an opportunity for further deforestation and expansion of plantations at the cost of primary forests (Mongabay.com 2011). FSC, on the other hand, has emphasized the positive social and environmental effects of some plantations (e.g., in Latin America) (FSC General Assembly 2011). Although not likely to happen in the near future, the certification of FSC certification of plantations established after 1994 is being given significant consideration.

Table 15.1 Comparison of the FSC and PEFC and their approaches to plantation forestry

Global acceptance	
FSC (FSC 2011b)	Number of national/regional standards: 28 (in 20 countries) Number of certificates: 1,065 Forest Management; 21,535 Chain-of-Custody Area: 144 million hectares in 79 countries; Chain-of-Custody present in 107 countries
PEFC (PEFC 2011a)	Number of national standards: 31 (in 30 countries) Number of certificates: 517 Forest Management; 8,585 Chain-of-Custody Area: 240 million hectares in 27 countries; Chain-of-Custody present in 58 countries
Top countries with certified plantations	
FSC	United Kingdom, South Africa, Brazil reported by FSC (n.d.), closely followed in certified area by New Zealand (FSC 2011b)
PEFC	Possibly, USA (Southeast), Australia, Malaysia
Plantation definition	
FSC (FSC International n.d.).	Plantations are forest areas that lack most of the principal characteristics and key elements of native ecosystems, which result from the human activities of either planting, sowing or intensive silvicultural treatments Proposed new draft definition: A forest area established by planting or sowing with using either alien or native species, often with one or few species, regular spacing and even ages, and which lacks most of the principal characteristics and key elements of natural forests. Notes for exceptions explain, for example, how the definition applies to boreal and temperate forests (few species there do not constitute plantations).
PEFC (PEFC Council 2010)	Forest plantation/timber plantation/productive plantation: Forest or other wooded land of introduced species, and in some cases native species, established through planting or seeding mainly for production of wood or non-wood goods. Note 1: Includes all stands of introduced species established for production of wood or non-wood goods. Note 2: May include areas of native species characterized by few species, intensive land preparation (e.g., cultivation), straight tree lines and/or even-aged stands. Note 3: Application of the definition requires consideration of national forestry terminology and legal requirements.
Certifiable plantations cut-off date	
FSC	1994, with exceptions
PEFC	2011, with exceptions

(continued)

Table 15.1 (continued)

Area of certified plantations	
FSC (FSC International 2011b; Karmann and Smith 2009)	12 million hectares (approximately 8 % of total FSC-certified forests globally). 42 million hectares of semi-natural and mixed forests (approximately 29 % of total FSC-certified forests globally)
PEFC	No data (at least, no easily available data). Certificates are often given to "plantation and natural forests", and no summary on the global extent of PEFC-certified plantations has been provided. The absence of summaries may reflect the reluctance of PEFC to closely consider and report on PEFC-certified plantations.
Approaches to forest conversion	
FSC (FSC 1999)	6.10 Forest conversion to plantations or non-forest land uses shall not occur, except in circumstances where conversion: (a) entails a very limited portion of the forest management unit; and (b) does not occur on high conservation value forest areas; and (c) will enable clear, substantial, additional, secure, long term conservation benefits across the forest management unit.
PEFC (PEFC Council 2010)	5.1.11 Conversion of forests to other types of land use, including conversion of primary forests to forest plantations, shall not occur unless in justified circumstances where the conversion: (a) is in compliance with national and regional policy and legislation relevant for land use and forest management and is a result of national or regional land-use planning governed by a governmental or other official authority including consultation with materially and directly interested persons and organisations; and (b) entails a small proportion of forest type; and (c) does not have negative impacts on threatened (including vulnerable, rare or endangered) forest ecosystems, culturally and socially significant areas, important habitats of threatened species or other protected areas; and (d) makes a contribution to long-term conservation, economic, and social benefits.

The final clause of FSC Criterion 10.9, exempting some plantations that do not qualify for the two main requirements, has resulted in significant criticism: the frequent trading of forest land allows for the certification of plantations created after 1994, and these issues have not yet been resolved by the FSC. In addition, Criterion 6.10 of the FSC International Standard allows certification if the conversion to plantation occurs on a small area, outside of forests of high conservation value, and clearly brings substantial long-term benefits to the conservation of the remaining forests across the management unit (FSC 1999). This condition provides additional opportunities to certify plantations.

A recent draft of the FSC International Standard (FSC STD 01 001 V5 0 EN FSC Principles and Criteria; Forest Stewardship Council 2010) does not contain a Principle specifically devoted to forest plantations. Instead, the draft underlines the applicability of all Principles and Criteria to plantations. The new version of the standard is not currently finalized, but has drawn the attention of companies wishing to become certified and the international environmental community.

Box 15.2: FSC Principle 10: Plantations

Plantations shall be planned and managed in accordance with Principles and Criteria 1–9, and Principle 10 and its Criteria. While plantations can provide an array of social and economic benefits, and can contribute to satisfying the world's needs for forest products, they should complement the management of, reduce pressures on, and promote the restoration and conservation of natural forests.

10.1 The management objectives of the plantation, including natural forest conservation and restoration objectives, shall be explicitly stated in the management plan, and clearly demonstrated in the implementation of the plan.

10.2 The design and layout of plantations should promote the protection, restoration and conservation of natural forests, and not increase pressures on natural forests. Wildlife corridors, stream side zones and a mosaic of stands of different ages and rotation periods, shall be used in the layout of the plantation, consistent with the scale of the operation. The scale and layout of plantation blocks shall be consistent with the patterns of forest stands found within the natural landscape.

10.3 Diversity in the composition of plantations is preferred, so as to enhance economic, ecological and social stability. Such diversity may include the size and spatial distribution of management units within the landscape, number and genetic composition of species, age classes and structures.

10.4 The selection of species for planting shall be based on their overall suitability for the site and their appropriateness to the management objectives. In order to enhance the conservation of biological diversity, native species are preferred over exotic species in the establishment of plantations and the restoration of degraded ecosystems. Exotic species, which shall be used only when their performance is greater than that of native species, shall be carefully monitored to detect unusual mortality, disease, or insect outbreaks and adverse ecological impacts.

(continued)

(continued)

10.5 A proportion of the overall forest management area, appropriate to the scale of the plantation and to be determined in regional standards, shall be managed so as to restore the site to a natural forest cover.

10.6 Measures shall be taken to maintain or improve soil structure, fertility, and biological activity. The techniques and rate of harvesting, road and trail construction and maintenance, and the choice of species shall not result in long term soil degradation or adverse impacts on water quality, quantity or substantial deviation from stream course drainage patterns.

10.7 Measures shall be taken to prevent and minimize outbreaks of pests, diseases, fire and invasive plant introductions. Integrated pest management shall form an essential part of the management plan, with primary reliance on prevention and biological control methods rather than chemical pesticides and fertilizers. Plantation management should make every effort to move away from chemical pesticides and fertilizers, including their use in nurseries. The use of chemicals is also covered in Criteria 6.6 and 6.7.

10.8 Appropriate to the scale and diversity of the operation, monitoring of plantations shall include regular assessment of potential on-site and off-site ecological and social impacts, (e.g. natural regeneration, effects on water resources and soil fertility, and impacts on local welfare and social well-being), in addition to those elements addressed in principles 8, 6 and 4. No species should be planted on a large scale until local trials and/or experience have shown that they are ecologically well-adapted to the site, are not invasive, and do not have significant negative ecological impacts on other ecosystems. Special attention will be paid to social issues of land acquisition for plantations, especially the protection of local rights of ownership, use or access.

10.9 Plantations established in areas converted from natural forests after November 1994 normally shall not qualify for certification. Certification may be allowed in circumstances where sufficient evidence is submitted to the certification body that the manager/owner is not responsible directly or indirectly of such conversion.

Source: FSC (1996)
http://www.fsc.org/fileadmin/web-data/public/document_center/international_FSC_policies/standards/FSC_STD_01_001_V4_0_EN_FSC_Principles_and_Criteria.pdf

Besides plantations and natural forests, FSC certifies semi-natural and mixed forests. Although current and upcoming forest management assessments include these categories of forests (FSC International 2011a), the FSC International Standard does not define them. Other institutions stipulate the definitions, e.g., "Mixed natural forest and plantations include large areas certified as one block that contains both natural forests and plantations. Semi-natural areas are forests that have some elements of both natural forests and plantations" (WRI 2004). FAO quotes an old, now defunct, FSC-US web-page: Semi-natural forest is "a forest that has a different species composition from natural forests in the area" (FAO 2002). Smartwood (2004) defined semi-natural and mixed forests in Australia as "Forest areas where many of the principal characteristics and key elements of native ecosystems such as complexity, structure and diversity are present, as defined by the FSC P&C." A further complication is the introduction of the term 'planted forest', which is broader scope that most concepts of plantations (FAO 2006). The term includes industrial plantations, but it also includes semi-natural forests that have been regenerated by planting. The absence of clear definitions limits the ability to estimate the extent of FSC-certified plantations.

The analysis by Stupak et al. (2011) examines differences in the approaches to forest conversion. It underlines the variability among national standards in their requirements concerning conversion. The majority of the current national FSC standards contain more stringent requirements and add detail to the international criteria and indicators (Stupak et al. 2011).

Like the FSC, the PEFC did not specifically address plantations during the original development of their system and only referred to the plantation-related requirements of the International Tropical Timber Organization (ITTO) and the Pan-European Operational Level Guidelines for Sustainable Forest Management (PEOLG) (WRI 2007). The ITTO guidelines were prepared with the input of ENGOs, US agencies, trade associations, and the scientific community (ITTO 1993). The document delineated the prerequisites for the sustainable management of planted forests in the tropics, the principles and criteria of establishing plantations, and post-management activities (ITTO 1993). The PEOLG (MCPFE 1998) was based on the European SFM C&I developed at the Ministerial Conference on the Protection of Forests in Europe in Helsinki in 1993. The PEOLG has been criticized by non-governmental organizations – some ENGOs considered that the lack of objectives and sufficient detail in the document precluded it from becoming a forest certification standard (Fern 1997). Although PEOLG and Pan-European Criteria for Sustainable Forest Management still form the basis of PEFC requirements, in 2010 PEFC added specific requirements about the conversion of forests to plantations and plantation management (PEFC Council 2010). As with the FSC, one of the drivers for an increased focus on plantations was the discrepancy in certification area between the temperate and boreal regions vs. tropical regions (Leslie 2004; Auer 2012); and interest in the certification of plantations in the tropics has remained high (Whelan and Dwinnells 2010).

The changes in 2010 included a new document containing requirements for SFM standards to be endorsed by the PEFC (PEFC Council 2010). The document lists the conditions for natural and plantation forests (although the Appendix dealing with the certification of tropical natural forests is still in the "enquiry draft" condition (PEFC 2011b)). Besides general requirements that apply to plantations and to land conversion (the latter is shown in Table 15.1), the guidelines on how to interpret the requirements for plantation forests are described (PEFC Council 2010). In plantations, the application of certain planning requirements cannot be at the scale of a forest stand; they need to be applied at the bioregional scale of the whole forest management unit. This is particularly important if impacts on biodiversity are to be reduced (e.g., McShea et al. 2009). There are some requirements that need particular attention in plantation forests. They include nutrient removal from the site during timber harvesting and the assessment of impacts of "introduced species, provenances and varieties". The importance of buffer zones and set-aside zones around plantations is emphasized during the plantation establishment phase; these zones are aimed at preserving the environmental, ecological and cultural functions of forests. They provide protection for endangered species and important biotopes, and provide vertical and horizontal structure to forest stands, snags and downed logs, and natural regeneration. Another issue is the use of herbicides to control vegetation in plantations, which remains an on-going problem in Australia, New Zealand and the USA (Rolando et al. 2011; Van Deusen et al. 2010). Although the conditional phrases, such as "when economically feasible" and "where appropriate" may still weaken the rigour of the requirements, the amount of detail in the requirements has increased since the inception of the system.

The national standards accredited by PEFC rely on national C&I processes to conceptualize and define sustainable forest management (Lawes et al. 1999). These include the C&I developed by the Ministerial Conference on the Protection of Forests in Europe (MCPFE), the Montreal Process, ITTO C&I for tropical forests, ATO (African Timber Organisation) or ITTO for tropical African forests, the Lepaterique Process in Central America, C&I for dry forests in Africa and Asia, Near East Process, and the Tarapoto Proposal for the Amazon (see Grayson and Maynard (1997) for details of all these agreements). National standards developed according to the regional processes and endorsed by PEFC generally limit the possibilities of conversion of natural forests to other uses and provide more specific list of conditions when such conversion can happen (Stupak et al. 2011).

The plantations that have been certified have received much attention. The FSC certification of plantations have been analysed, with the results suggesting both support for forest certification (de Freitas 2003; de Lima et al. 2009) and concerns with the quality of certified forest management (World Rainforest Movement 2003; FSC-Watch 2008; Kröger and Nylund 2012). The following case study of certified plantations relates to Australia and an Australian company (Box 15.3).

Management Planning in Action 15.1: Australia – A Case Study of Certified Plantations

Two forest certification systems are available in Australia, the Australian Forestry Standard (AFS) and the FSC system. The AFS is based on ISO 14001 and the Montreal Process (http://www.forestrystandard.org.au/). The standard was endorsed by the PEFC in 2004 (McDermott et al. 2010). The

(continued)

(continued)

FSC Australian Standard for Forest Management has not been adopted yet, and the latest draft is dated 2008 (FSC Australia 2008). Uncertainties over the content of the national FSC standard in Australia has led to many companies choosing the AFS when certifying forest management, although more companies have chosen FSC for chain-of-custody certificates. AFS forest management certificate holders include a few large governmental companies (e.g., Forestry New South Wales with 2.5 million hectares of certified defined forest area, and the West Australia Forest Products Commission with 1.2 million hectares; both areas contain natural forest and plantations), with the remaining companies having much smaller areas of land.

Total certified area: AFS 100 million hectares; FSC 0.6 million hectares.
Number of Forest Management certificates: AFS 25; FSC 9
Number of Chain of Custody certificates: AFS 195 certificates; FSC 251 certificates
(PEFC update 2011); (FSC update October 2011)

The Australian government views establishment of plantations as a means of stimulating the economy (Plantations 2020 n.d.). Plantations occupy around two million hectares, split almost equally between hardwood and softwood plantations (Department of Agriculture, Fisheries and Forestry 2010). AFS defines plantations as "Stands of trees of either native or exotic species, created by the regular placement of cuttings, seedlings or seed selected for their wood-producing properties and managed intensively for the purposes of future timber harvesting". The AFS requirements on planning, establishing and managing plantations rely on compliance with the national and State legislation and regulations as well as protocols and principles (e.g., Department of Agriculture, Fisheries and Forestry 1995). The standard prohibits conversion of natural forest to plantations, except in specific cases. The cut-off date for certifiable plantations established by conversion is December 2006. The standard emphasizes the strong public concern about the conversion of natural forest and suggests that the conversion "should cease". However, the standard considers conversion permissible (when in compliance with other AFS requirements), when it provides opportunities for Indigenous people.

With the Australian FSC standard still being under development, the FSC certifiers currently use interim standards – the standards developed by their organizations and endorsed by FSC. For example, the Smartwood auditing company uses their own check-list of indicators to assess compliance with the FSC requirements, and the standard includes general FSC requirements towards plantations, such as forest management requirements, forest conversion conditions and the cut-off date for certifiable plantations.

(continued)

(continued)

The first company that was certified by FSC was Hancock Victorian Plantations (HVP) in 2004 (Hancock Timber Resource Group 2004). The company manages softwood and hardwood plantations. At the time, HVP managed about 173,000 ha of plantations (Smartwood 2004): around 16 % were hardwood plantations (*Eucalyptus* species) and the rest 74 % were softwood plantations of predominantly *Pinus radiata* (Tonkin 2006). Around 50,000 ha of native forest were managed for conservation purposes according to the company's policy to leave these forests in their natural state (AFS 2011). The latest FSC re-certification occurred in 2009, and HVP's forest management was found to be in compliance with the FSC Principles (Smartwood 2009). HVP lands are now dually certified to both AFS and FSC, and the company has maintained AFS certification on all forest land it obtained that had been AFS-certified prior to their purchase (Hancock Timber Resource Group 2010). While HVP and their certifiers have been severely criticized by some environmental groups for "greenwashing" (e.g., FSC Watch 2012), the dual certification indicates the company's intent to retain its market share in as many international markets and product niches as possible.

Recent analyses (Dare et al. 2011) indicate that certification of plantations in Australia has had a positive impact on community engagement practices by plantation owners, with the requirement for continual improvement helping to overcome some of the significant barriers to community engagement.

15.4 Certification and Procurement

The procurement of responsibly-produced timber has recently received attention because of legislative developments, such as the amendments to the Lacey Act of 2008 in the United States, and the adoption of the EU Timber Regulation and EU FLEGT Regulation of 2010. These documents require assurance that wood has been harvested from legal sources. Other procurement initiatives rely on forest certification and the corresponding standard requirements, e.g., Keurhout in the Netherlands (Keurhout n.d.). Several European countries, Canada, Japan and New Zealand are all examples of where national governments have developed procurement policies addressing the legality of wood products. According to Simula (2006), at least 10 % of all wood products are consumed by the public sector; thus, the requirements of this market segment is encouraging the forest industry to seek assurances over the legal origin of timber. With the rise of sustainability-oriented governmental procurement policies, forest certification is helping the process, as all standards require compliance with local laws. Although forest management and chain of custody certificates do not result in the requirements of, for example, the Lacey Act being waived, forest certification is reducing the level of scrutiny. Certified timber and fibre is considered a low risk of being illegally produced, and this may drive the wider adoption of forest certification. To further lessen the burden of certification for companies, the FSC aims to align its Controlled Wood standard with the requirements of the EU Timber Regulation, EU FLEGT process and the U.S. Lacey Act (FSC Canada 2011).

The PEFC-endorsed Sustainable Forestry Initiative (SFI) standard (developed in the USA) has placed considerable emphasis on the responsible procurement of timber. SFI standard requirements are applied not only to domestic timber production (e.g., plantations in the U.S. Southeast), but also to overseas forest tenures. The multinational companies that have operations in Latin America or elsewhere must abide by the SFI requirements. They must ensure that the timber they procure (this can be both purchased or produced timber) come from legal sources and that the legal requirements of the region of timber origin are met. In addition, there are special restrictions placed on the procurement of wood from certain areas, such biodiversity hotspots.

Green building standards and markets present another venue where forest certification can help with ensuring the use of legally procured timber. Green building standards such as LEED (US); ANSI National Green Building Standard (US), Green Globes (US and Canada), Built Green Canada, BREEAM (United Kingdom), the Green Building Council of Australia, and CASBEE (Japan) all encourage the use of certified timber in construction. While LEED only recognizes wood products certified by the FSC, all the other standards recognize timber certified by either FSC or PEFC-accredited standards.

The use of forest certification does not always assure full compliance to national and international regulations. The establishment of plantations is sometimes related to illegal "land-grabs" and may be associated with a lack of consideration of traditional forest uses and tenures (Molnar et al. 2011). Forest certification was intended to avoid the conflicts that can arise from confrontations between companies operating plantations and local communities with traditional tenures. The intent has not always translated into practice: claims of conflicts between companies with certified plantations and local communities have arisen in Brazil (Transnational Institute 2007) and Uganda (Grainger and Geary 2011). Such difficulties should not be allowed to detract from the potential benefits that well-planned afforestation projects could have in disadvantaged areas (see, for example, Landry and Chirwa 2011). Thus, although forest certification is positioned as a mechanism of assuring the legality of timber and fibre, the issues of certifier credibility and the legitimacy of particular standards need to be resolved.

15.5 Issues with Certifying Plantation Forestry

The possibility of certifying forest plantations has been a major issue affecting the perceived legitimacy of forest certification standards. Some standards, such as the SFI, have always accepted plantations, while other standards have not. Plantation policies are creating a challenge to the legitimacy of the FSC (Schepers 2009; Stupak et al. 2011). Both PEFC and FSC have been criticised over their provisions for certifying plantations. The controversy for many environmental activists lies in the difficulty of referring to plantations as sustainably managed forests when they are so different from natural forests. Plantations are also juxtaposed with community-based forestry and the expansion of plantation area can limit or preclude opportunities for establishing community forests (World Rainforest Movement 2003; Nebel et al. 2005). However, some communities are actively establishing plantations, and the issue is really one of land tenure legitimacy. A greater emphasis in standards on the long-term social sustainability of plantations could help resolve this and other social issues (Gouldin 2006).

Another problem arises from the loop-holes and insufficient oversight of regional certification bodies. Although the problem is applicable to all forests in areas with weak governance, not only certified plantations, cases where certificates have been revoked coincide with areas where plantation forestry is common (e.g., Brazil, Malaysia, Uruguay). Discrepancies in forest certification audits cloud the legitimacy of certification in Uganda (Grainger and Geary 2011), Brazil (Schulze et al. 2008), and Cameroon (Cerutti et al. 2011). This problem is often linked with the issue of governance and transparency: a corrupt governance system is often very difficult to overcome (Cerutti et al. 2011), and the forest certification entities have to operate within the rules of the system.

The regional interpretations of the international FSC standard have also generated criticism. The interim standards in a country are often only loosely based on the international standard, with the majority of the requirements being developed in a way to allow for less stringent regional management rules (Cerutti et al. 2011) during the period that certification is trying to establish a presence in a country.

The definition of plantations can also be viewed as a problem. The definitions used by FSC and PEFC contain many additional notes (Table 15.1), and each limits or extends the scope of what can be called a plantation. Collecting statistics on all forests that qualify for the plantation definition is difficult (e.g., semi-natural planted forests), but will be helpful in analyzing the effects of forest certification on markets, ecosystems and livelihoods.

15.6 Summary

Forest certification of industrial plantations has become a widespread practice around the globe. The demand for timber and fibre, combined with the allocation of increasing areas of primary forest to reserves, is creating demands for more intensive forestry practices. Recent changes in governmental procurement policies and European and U.S. import rules increasingly require that legality be demonstrated and that wood be sourced from responsibly managed forests. Forest certification ensures compliance with local regulations, and eases the "burden of proof", provided that there are no non-conformities noted in an individual company's certification. Two international systems – PEFC and FSC – endorse forest certification standards that strive to meet these requirements. Both systems address the certification and management of forest plantations, and contain details on what kind of plantations can be certified. These requirements include (a) the cut-off date for the establishment of plantations derived from the conversion of natural forests, (b) a prohibition (in most circumstances) on the creation of plantations through the conversion of natural forests, and (c) consideration of special management requirements. The regional or national FSC standards and national standards endorsed by PEFC take into account the country or regional specifics.

As a marketing mechanism, forest certification has gained legitimacy with governments, buyers' groups, industrial companies and their associations. A few environmental groups still actively oppose the certification of plantations. Their concerns are based on the claims of deforestation, the loss of biodiversity and the impacts on the livelihoods of forest-dependent communities, all of which have been connected with the expansion of forest plantations. Despite these concerns, the global acceptance of forest certification has been increasing. New standards are being developed and prepared for endorsement by the FSC and PEFC, and existing standards are refining their requirements to reflect developments in competing systems. It is very likely that forest certification will bring economic and social benefits to established and, possibly, future forest plantations.

15.7 Problems

A. What are the monetary and non-monetary benefits of forest certification for a forest plantation company?
B. How do governance issues within a region (country) influence forest certification in that region (country) and its acceptance by the international markets or ENGOs?
C. What steps does a company need to follow in order to get its plantations certified? (More advice on the matter can be found at http://www.fsc.org/5-steps-certification.html or http://www.pefc.org/certification-services/forest/advantages).

References

Aguilar FX, Vlosky RP (2007) Consumer willingness to pay price premiums for environmentally certified wood products in the U.S. For Policy Econ 9(8):1100–1112

Auer M (2012) Group forest certification for smallholders in Vietnam: An early test and future prospects. Hum Ecol 40:5–14

Australian Forestry Standard (2011) AFS Sustainable Forest Management certification register. http://www.forestrystandard.org.au/files/Registers/AS4708_Register_JUL11_comp.xls. Accessed 21 Nov 2011

Bernstein S, Cashore B (2004) Non-state global governance: is forest certification a legitimate alternative to a global forest convention? In: Kirton J, Trebilcock M (eds) Hard choices, soft law: voluntary standards in global trade, environment and social governance. Ashgate Press, Aldershot, pp 33–63

Cashore B, Auld G, Newsom D (2003) The United States' race to certify sustainable forestry: non-state environmental governance and the competition for policy-making authority. Bus Polit 5(3):219–259

Cashore B, Auld G, Newsom D (2004) Governing through markets: forest certification and the emergence of non-state authority. Yale University Press, New Haven

Castaneda F (2000) Criteria and indicators of sustainable forest management. UNASYLVA 203:34–40

Cerutti PO, Tacconi L, Nasi R, Lescuyer G (2011) Legal vs. certified timber: preliminary impacts of forest certification in Cameroon. For Policy Econ 13:184–190

Chen J, Innes JL, Tikina A (2010) Private cost-benefits of voluntary forest product certification. Int For Rev 12(1):1–12

Cubbage F (2003) Sustainable forest management, forest certification, tree improvement, and forest biotechnology. http://digital.library.okstate.edu/Forestry/sf27p006.pdf. Accessed 25 Jan 2012

Dare M, Schirmer J, Vanclay JF (2011) Does forest certification enhance community engagement in Australian plantation management? For Policy Econ 13(5):328–337

De Freitas A (2003) Brazil forest certification case study. Forest Trends, Washington, D.C. http://www.forest-trends.org/documents/files/doc_359.pdf. Accessed 21 Nov 2011

de Lima A, Keppe A, Maule F, Sparovek G, Alves M, Maule R (2009) Does certification make a difference? Impact assessment study on FSC/SAN certification in Brazil. Imaflora, Piracicaba, Brazil. http://www.imaflora.org/upload/repositorio/Does_certification_make_a_difference.pdf. Accessed 18 Nov 2011

Department of Agriculture, Fisheries and Forestry (1995) Forest practices related to wood production in plantations: national principles. http://www.daff.gov.au/__data/assets/pdf_file/0007/37609/principles_wood_production.pdf. Accessed 21 Nov 2011

Department of Agriculture, Fisheries and Forestry (2010) National plantation inventory Australia 2010. http://adl.brs.gov.au/data/warehouse/pe_brs90000004201/NPIupdate2010_20100525_ap14.pdf. Accessed 21 Nov 2011

FAO (2002) Semi-natural forests. http://www.fao.org/DOCREP/005/Y4171E/Y4171E47.htm. Accessed 25 Nov 2011

FAO (2006) Global planted forests thematic study. Results and analysis, vol 14, Planted forests and trees working paper. FAO, Rome

FAO (2008) What is sustainable forest management? http://www.fao.org/forestry/sfm/24447/en/. Accessed 21 Nov 2011

Fern (1997) Joint NGO declaration on the proposed Pan European Operational Level Guidelines on Sustainable Forest Management. http://www.fern.org/sites/fern.org/files/pubs/archive/peolg.html. Accessed 26 Nov 2011

Fletcher R, Rickenbach M, Hansen E (2002) Forest certification in North America, vol 1518, EC. Oregon State University, Extension Service, Corvallis

Forest Stewardship Council (2010) FSC Standard. FSC principles and criteria for forest stewardship. Supplemented by explanatory notes and rationales. Draft, http://vote.fsc.org/md.static/FSC-STD-01\T1\textendash001_V5\T1\textendash0_D5\T1\textendash0_EN_Explanatory_Notes+Rationales.pdf. Accessed 5 Mar 2012

FPAC (Forest Products Association of Canada) (2005) FPAC market acceptance. Customer briefing note. Forest certification. Forest Products Association of Canada, Ottawa

FSC (1999) FSC international standard. Forest Stewardship Council, Bonn

FSC Australia (2008) DRAFT 00 ("Straw Dog") of the FSC Australia National Standard – copy for public information. http://www.fscaustralia.org/sites/default/files/pdf/publicinfo%20draft%20july%207%2008.pdf. Accessed 21 Nov 2011

FSC Canada (2011) Review of company evaluation of FSC controlled wood (FSC-STD-40-005). http://www.fsccanada.org/news/CompanyEvaluationofFSCControlledWood.htm. Accessed 21 Nov 2011

FSC General Assembly (2011) Motions. http://motions.fsc.org/motions/. Accessed 12 Jan 2012

FSC International (n.d.) The FSC principles and criteria for responsible forest management. http://www.fsc.org/pc.html. Accessed 20 Nov 2011

FSC International (2011a) Upcoming forest management (FM) assessments. http://www.fsc.org/fmassessments.html. Accessed 25 Nov 2011

FSC International (2011b) Global FSC certificates: type and distribution. http://www.fsc.org/fileadmin/web-data/public/document_center/powerpoints_graphs/facts_figures/2011-09-15-Global-FSC-Certificates-EN.pdf. Accessed 25 Nov 2011

FSC Watch (2008) FSC dumps Asia Pulp and Paper – but who was to blame? http://www.fsc-watch.org/archives/2008/01/10/FSC_dumps_Asia_Pulp_. Accessed 30 Nov 2011

FSC Watch (2012) FSC certified company, Hancock, caught on film converting forest to plantations in the Strzezlicki Ranges, Australia. http://www.fsc-watch.org/archives/?policy~=Friends+of+the+Earth. Accessed 20 Jan 2011

Gough AD, Innes JL, Allen SD (2008) Development of common indicators of sustainable forest management. Ecol Indic 8(5):425–430

Gouldin CJ (2006) Forest Stewardship Council certification of industrial plantation forests. Allg Forst Jagdztg 177(3–4):42–47

Grainger M, Geary C (2011) The new forests company and its Uganda plantations. Oxfam International. p 20. http://www.oxfam.org/en/grow/policy/new-forests-company-and-its-uganda-plantations-oxfam-case-study. Accessed 23 Jan 2011

Grayson AJ, Maynard WB (eds) (1997) The world's forests – Rio + 5: international initiatives towards sustainable management. Commonwealth Forestry Association, Oxford

Hancock Timber Resource Group (2004) Hancock Victorian plantations receives certification for responsible social and environmental practices. http://www.htrg.com/news_02_11_04.htm. Accessed 15 Jan 2011

Hancock Timber Resource Group (2010) Hancock Queensland plantations awarded forestry plantations Queensland licence. http://www.htrg.com/news_05_17_10.htm. Accessed 25 Nov 2011

ITTO (1993) ITTO guidelines for the establishment and sustainable management of planted tropical forests. International Tropical Timber Organization, Yokohama. http://www.pefc.org/images/stories/documents/pefc-technical/ITTO-Planted.pdf. Accessed 24 Nov 2011

Karmann M, Smith A (2009) FSC reflected in scientific and professional literature. Literature study on the outcomes and impacts of FSC certification. http://www.fsc.org/fileadmin/web-data/public/document_center/publications/FSC_Policy_Series/Impacts_report_-_Karmann_2009.pdf. Accessed 13 Mar 2012

Klooster D (2010) Standardizing sustainable development? The Forest Stewardship Council's plantation policy review process as neoliberal environmental governance. Geoforum 41:117–129

Kröger M, Nylund J (2012) The conflict over Veracel pulpwood plantations in Brazil – application of ethical analysis. For Policy Econ 14(1):74–82

Landry J, Chirwa PW (2011) Analysis of the potential socio-economic impact of establishing plantation forestry on rural communities in Sanga district, Niassa province, Mozambique. Land Use Policy 28(3):542–551

Lawes MJ, Everard D, Eeley HAC (1999) Developing environmental criteria and indicators for sustainable plantation management: the South African perspective. S Afr J Sci 95(10):461–469

Leslie AD (2004) The impacts and mechanisms of certification. Int Rev 6(1):30–39

McDermott C, Cashore BW, Kanowski P (2010) Global environmental forest policies: an international comparison. Earthscan, Washington, DC

MCPFE (1998) Pan-European Operational Level Guidelines (PEOLG) for sustainable forest management. Ministerial conference on the protection of forests in Europe, Lisbon

McShea WJ, Stewart C, Peterson L, Erb P, Stuebing R, Giman B (2009) The importance of secondary forest blocks for terrestrial mammals with an *Acacia*/secondary forest matrix in Sarawak, Malaysia. Biol Conserv 14:3108–3119

Menne W, Carrere R (2007) Swaziland: the myth of sustainable timber plantations. World Rainforest Movement, Montevideo. www.wrm.org.uy/countries/Swaziland/Book_Swaziland.pdf. Accessed 14 Jan 2012

Molnar A, Barney K, DeVito M, Karsenty A, Elson D, Benavides M, Tipula P, Soria C, Shearman P, France M (2011) Large acquisition of rights on forest lands for tropical timber concessions and commercial wood plantations. Forest Trends, Washington, DC

Mongabay.com (2011) FSC mulls controversial motion to certify plantations responsible for recent deforestation. http://news.mongabay.com/2011/0624-hance_fsc_motion18.html. Accessed 24 Nov 2012

Nebel G, Quevedo L, Jacobsen J, Helles F (2005) Development and economic significance of forest certification: the case of FSC in Bolivia. For Policy Econ 7(2):175–186

PEFC (2011a) Fact and figures. http://www.pefc.org/about-pefc/who-we-are/facts-a-figures. Accessed 28 Nov 2011

PEFC (2011b) Appendix 2: additional interpretation and requirements for tropical natural forests. http://www.pefc.org/images/stories/documents/pefc-technical/Appendix_2_PEFC_ST_10032010__ED1_20110928.pdf. Accessed 21 Nov 2011

PEFC Council. (2010) Sustainable Forest Management – requirements. PEFC INTERNATIONAL STANDARD. Requirements for certification schemes. http://www.pefc.org/standards/technical-documentation/pefc-international-standards-2010/item/672. Accessed 12 Nov 2011

Perera P, Vlosky RP (2006) A history of forest certification. Louisiana Forest Products Development Center, Louisiana State University, Baton Rouge

Plantations 2020. N.d. Plantations for Australia: the 2020 vision. http://www.plantations2020.com.au/vision/index.html. Accessed 12 Jan 2012

Pokorny B, Johnson J, Medina G, Hoch L (2012) Market-based conservation of the Amazonian forests: revisiting win–win expectations. Geoforum 43(3):387–401

Poulsen J, Applegate G, Raymond D (2001) Linking C & I to a code of practice for industrial tropical tree plantations. Center for International Forestry Research, Jakarta

Rolando CA, Watt MS, Zabkiewicz JA (2011) The potential cost of environmental certification to vegetation management in plantation forests: a New Zealand case study. Can J For Res 41(5):986–993

Schepers DH (2009) Challenges to legitimacy at the Forest Stewardship Council. J Bus Ethics 92:279–290

Schulze M, Grogan J, Vidal E (2008) Forest certification in Amazonia: standards matter. Oryx 42:229–239

Simula M (2006) Public procurement policies for forest products and their impacts. Draft discussion paper. Food and Agriculture Organization of the United Nations, Rome, 8 Aug 2006

Smartwood (2004) Forest management public summary for Hancock Victorian Plantations Pty. Limited. http://www.rainforest-alliance.org/sites/default/files/site-documents/forestry/documents/hancock_victorian.pdf. Accessed 14 Jan 2012

Smartwood (2009) Forest management certification assessment report for Hancock Victoria Plantations Pty. Ltd. in Melbourne, Australia. http://info.fsc.org/servlet/servlet.FileDownload?retURL=%2Fape. Accessed 14 Jan 2012

Stupak I, Lattimore B, Titus BD, Smith CT (2011) Criteria and indicators for sustainable forest fuel production and harvesting: a review of current standards for sustainable forest management. Biomass Bioenerg 35(8):3287–3308

Tikina AV, Innes JL (2008) A framework for assessing the effectiveness of forest certification. Can J For Res 38(6):1357–1365

Tollefson C, Gale F, Haley D (2008) Setting the standard: certification, governance and the Forest Stewardship Council. UBC Press, Vancouver

Tonkin M (2006) Forest Stewardship Council Certification. Australian forest growers conference 2006. http://www.planningplantations.com.au/assets/pdfs/management/certification/FSC_HancocksVictorianPlantations_MTonkin.pdf. Accessed 12 Jan 2012

Transnational Institute (2007) Brazil: an overview of monoculture eucalyptus plantations. http://www.tni.org/article/brazil-overview-monoculture-eucalyptus-plantations. Accessed 28 Jan 2012

van Bodegom AJ, van den Berg J, van der Meer P (2008) Forest plantations for sustainable production in the tropics: key issues for decision-makers. The Netherlands: Wageningen University & Research Centre. http://www.cdi.wur.nl/NR/rdonlyres/A338310F-333B-4F7E-8358-0523CEE12A5E/134717/Rapport_Forestplantations_totaal_lowres_sec.pdf. Accessed 2 Nov 2011

Van Deusen PC, Wigley TB, Lucier AA (2010) Some indirect costs of forest certification. Forestry 83(4):389–394

Vogt K, Larson B, Gordon J, Vogt D, Fanzeres A (eds) (2000) Forest certification: roots, issues, challenges and benefits. CRC Press LLC, Boca Raton

Whelan T, Dwinnells E (2010) The role of certification in protecting the world's forests. In: Sethi SP (ed) Globalization and self-regulation: the crucial role that corporate codes play in global business. Macmillan, New York

World Rainforest Movement (2003) Certifying the uncertifiable: FSC certification of tree plantations in Thailand and Brazil. http://www.wrm.org.uy/actors/FSC/text.pdf. Accessed 26 Nov 2011

WRI (2004) Technical and source notes. http://earthtrends.wri.org/searchable_db/variablenotes.php?varid=311&theme=9. Accessed 10 Dec 2012

WRI (2007) Sustainable procurement of wood and paper products. http://www.sustainableforestprods.org/node/9. Accessed 26 Nov 2011

Chapter 16
Forestry Raw Materials Supply Chain Management

Kevin Boston

16.1 Introduction to Forestry Supply Chains

16.1.1 The Concept of Forestry Supply Chain

Since the end of World War II, trade restrictions have been liberalized (Wold et al. 2011). General Agreement on Tariffs and Trade (GATT) and regional trade agreements have removed many of the barriers to trade between nations. New agreements continue to be negotiated to further reduce tariffs between nations. These reductions in trade barriers allow countries to exploit their natural competitive advantages to produce goods for world markets. This has been especially true in the wood products industry with the rise of many wood producing regions around the world including Brazil, Chile and New Zealand; all have emphasized plantation forestry. The increased competition from a globalized wood products industry will require that increased efficiency from producers who seek to gain access to new markets or hold existing markets from new competitive pressures. Improved management of the forestry supply chain is a cornerstone for modern timber management to remain a viable enterprise.

The development and management of efficient supply chains have become one of the main methods used by all businesses to become and remain competitive in these globalized markets. Wal-Mart, one of the world leaders in retail sales (Rice 2010) has been described as having one of the most efficient supply chains. Customer demands have been increasing for not only lower prices, but improved quality of these goods with predictable and reliable delivery schedule. Large companies such as Wal-Mart's continuously improve their supply chain that have resulted

K. Boston (✉)
Department of Forest Engineering, Resources and Management, Oregon State University, Corvallis, OR, USA
e-mail: Kevin.Boston@oregonstate.edu

J.G. Borges et al. (eds.), *The Management of Industrial Forest Plantations: Theoretical Foundations and Applications*, Managing Forest Ecosystems 33,
DOI 10.1007/978-94-017-8899-1_16,

in increasingly efficient operations with reduced inventory costs and improved customer service (Chandran and Gupta 2003).

What You Will Learn in This Chapter

- The ability to define a supply chain;
- The ability to describe the components of the standard forestry supply chain;
- How does the supply chain integrate the strategic, tactical and operational planning to manage the supply chain,
- How change is managed in the supply chain,
- It will describe the limits of collaborative planning used in the supply chain

16.1.2 Defining the Supply Chain

For this chapter, supply chain management (SCM) will be considered in its broadest terms and includes manufacturing, raw materials procurement and sales functions as well as logistics into a network model that describes the entire product development from raw material to the final items. Supply chain management organizes the various actions, silviculture treatments, harvesting, transportation, and marketing, and combines them into an integrated system that supports all aspects of the forestry business. The result is a multi-commodity network that includes the various raw materials such as: logs, lumber, veneer products, wood chip, pulp, paper and biomass. Besides the physical products, the supply chain includes the information that flows across these networks. This includes information such as forecasted demand, orders, inventory levels, production, supply and yield data. The final flow on these networks is the financial transactions that include the invoices and payments. Thus, the arcs represent a multi-commodity bidirectional network model (Fig. 16.1).

16.1.3 The Uniqueness of the Forestry Supply Chain

Forest supply chains have many unique properties when compared with most other manufacturing processes and there is a need to study them separately. There is both a short and a long influence of time on the forestry supply chain. The long time frames are unique to forestry. Temperate forests have a rotation length requiring a minimum of 15 years with solid wood products having a rotation of approximately 25 years. In tropical forest, the rotation age may be reduced to 5 or 6 years on many plantations dedicated to pulp wood production. Coordinated planning among the consumers, suppliers and harvesting contractors is necessary to accommodate these long time-frames that dominate the forestry supply chain.

These long times limit the ability of the forest manager to intervene in the management of the supply chain. Raw materials sources are limited by a combination

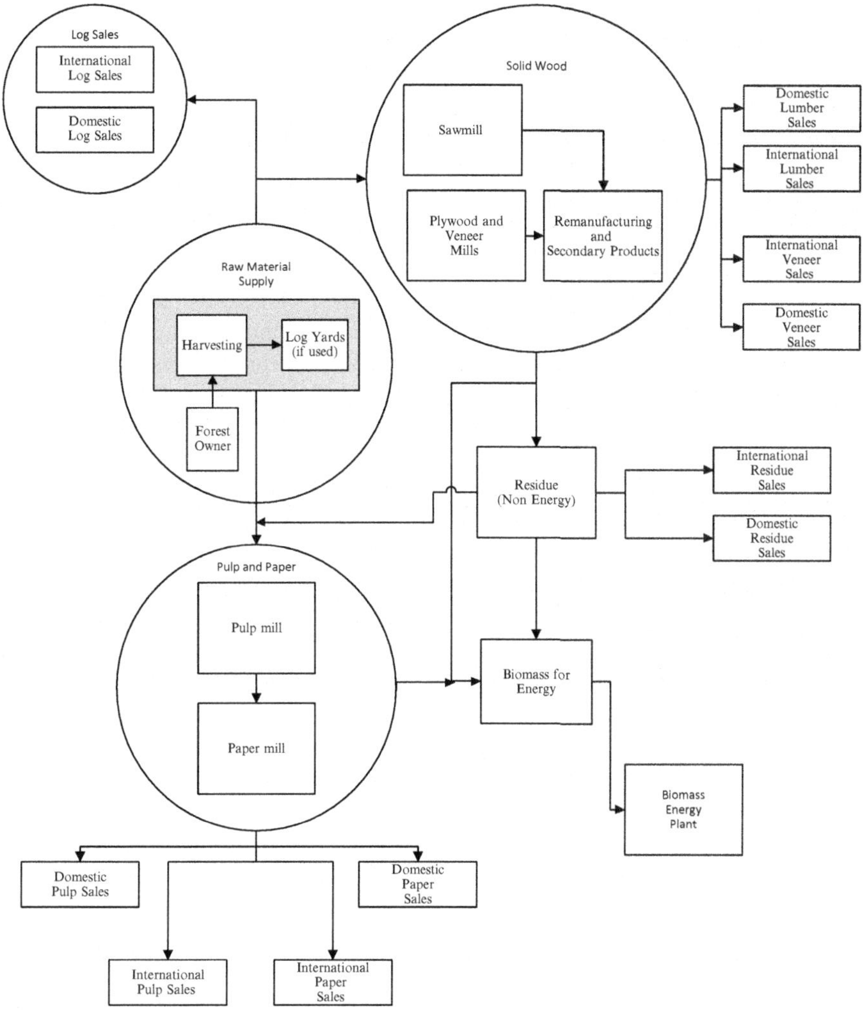

Fig. 16.1 The flow of materials through a forest supply chain

of the biological growth of the trees and the silvicultural decisions that are applied to the stands. Once the silvicultural regime is established, it is difficult to adjust to create new products to satisfy new demands. For example, a pruned saw log regime in temperate forests usually requires 6 to 10 years from the time of pruning to the time of the final harvest. This allows the tree to produce the maximum volume of clear-wood, wood free of knots, which is desired in the wood molding industry that demands higher value log. Thus, the long-time frame between the silvicultural decisions and the results adds significant uncertainty to the supply chain.

Besides the long-term consequence of time in managing the forest supply chain, there is often a short-term time frame that must be incorporated into the supply chain. Many species of pine, which are commonly used in plantation forestry, are susceptible to fungal staining, often from the genus, *Ceratocystis*. Although this does not generally attack the wood fibers, it can lower the value for the pulp and lumber by darkening the color to wood (Bruce et al. 2003). In warm wet weather, especially in the tropics or subtropical areas, sap-staining can occur within days of falling. Expensive chemical treatments, antisap-staining chemicals, can be applied to the logs to minimize these losses; another approach is to manage the supply chain to minimize the log inventory during these times when the damage from sap staining fungi is high. The result is a supply chain that resembles the fresh-produce markets, but without the ability to pick the fruit before it is ready so that it may ripen during transportation. Logs are exposed the moment the tree is severed from the stump. This becomes a significant problem in cable harvesting operations where each corridor must be felled prior to installing the lines for yarding. It is not uncommon for the last logs yarded from a corridor to be degraded before they reach the landing. Alternative practices will need to be developed to either avoid steep sites during the season when sap staining is most likely to occur or to evolve a staged-falling system that will have a higher cost to extract logs from these sites.

Most manufacturing items are an assembly industry; smaller parts or components are combined into larger items. A computer is a good example of this kind of industry; many components, which may be common to many other consumer goods, are combined into the larger items that are finally sold to the consumers. The forestry supply chain is a decoupling process; one that begins with a larger item and creates products by disassembling it into smaller products. Logs are produced from tree stem, and lumber is produced from logs that were produced from the larger tree stem. Residues from many of the solid wood manufacturing processes are inputs into other products such as paper, medium density fiberboard or used in energy and animal bedding products.

The result of a decoupling manufacturing system is that it will generate co-products or multiple products during the manufacturing process. Continuing with our pruned log example, the silvicultural regime is designed to maximize the volume of clear or knot-free wood in the pruned, bottom or butt log. The next two logs are industrial grade logs that have large limbs due to the wider spacing selected in the regime and the top log is a pulp. When the tree is felled, all three products are produced. Often, the market for these three logs can differ; the demand for pruned and pulp may be high but the demand for the industrial grade log is low. The supply chain manager will need to determine the value of tree when it is severed based on the current market conditions. The problem with co-products is best explained with an example. Assume there is a strong demand for the pruned and pulp logs; however, the remaining industrial logs have a low demand and correspondingly lower price. Thus, when the tree is felled to satisfy the pruned and pulp log demands, the industrial grade logs are produced but into a poor market. The options for the supply chain manager are to place the product into inventory and pay the holding costs and perhaps suffer the value loss from sap staining while waiting for the price for this log grade to improve, or sell the product into a lower value market. In this case, one may be able sell the logs a pulp log. This is often called down-grading.

Thus, the decision to harvest the tree will require that valuation of all the products that are produced at the time of felling.

The fourth reason that the forestry supply chain differs from others is the lack of comprehensive data used in the management and planning of supply chain. Many manufacturing systems have full tracking of all shipments. Six-sigma goals for data and product quality are common with complex enterprise planning system developed to exploit these rich data sets. Unfortunately, the data used in managing the forestry supply chain is much less precise for several reasons. One, is the supply data are from samples and not a census used in many other manufacturing sectors; therefore there are imprecision associated with sample data. Although these are plantation forests, they are biological systems and confidence intervals around the volume estimates can be large, typically plus or minus ten percent at one to two standard deviations. Thus, there is an imprecise estimate of the volume per hectare from the sample data.

Besides, the inventory data, logging production rates are notoriously difficult to predict. There are many variables such as: average tree size; others include the physical parameters of the harvesting unit such as the slope, soil strength and average logging distance and then there is variation found in the work force.

Summarizing the issues with forestry supply chains, it is complicated by the decoupling nature of the manufacturing methods used that produced co products that may not match customer demand. Often, it is a perishable product. The length of time required to produce trees further reduces the flexibility of the supply chain to adjust to different demands. To respond to these issues, most primary forestry supply chains are organized under two basic strategies. One is the cut-to-order and the other is cut to inventory.

16.1.4 How Are Supply Chains Deployed by Various Business Structures

16.1.4.1 The Vertically Integrated Form

Supply chains are the vehicle to implement a company's strategy. The business structure can influence how the supply chain is developed and managed. The business structure describes the relationship among the parties in the supply chain. Perhaps the simplest model is the vertically integrated company, one company that owns the forest and all of the manufacturing facilities. In addition, the corporation owns all of the harvesting and transportations equipment and the work accomplished exclusively by employees of that corporation. The benefits of this model are twofold; one is that there is a common objective for all of the actors, to maximize the profits of the overall corporation. They are able share all of the information regarding production, cost, market capacity and future development among all of the parties without the fear of violating anti-trust regulations that may exist between independent companies in many counties.

The second benefit of having all of the elements owned by a single supply chain is that the costs and the benefits are shared within the same company and it becomes much easier to implement changes to supply chain when one is must is only concerned of its net impact on the entire supply chain. For example, Hay and Dahl (1984) developed and deployed an integrated planning system to support the stump-to-product activities. They were able to show that proper allocation of logs to mills could, by encouraging the participants to think across their individual business boundaries, result in a $1.9 million dollar improvement when coordinated planning was used for the two facilities (Hay and Dahl 1984). The forestry, logging and hauling were all working for a single entity to improve the performance of the supply chain through the optimal allocation of the logs to the two mills.

16.1.4.2 The Multi-party Enterprises

The other example is when the forestry supply chain is composed of multiple firms. The forest land owner can be a timberland investment organization that only owns timberlands and no processing facilities. It sells stumpage to a wood dealer or direct to manufacturing sites, that hires the logging and hauling contractors. The logging contractor performs the logging and the initial merchandizing of material that is hauled to the various manufacturing facilities. In this simple example, there are four different parties in the supply chain that only considers the delivery of logs to the mill.

To illustrate how difficult it is to share the information, let's consider an example, a forest landowner has decided to improve his road system and has compacted and smoothed the road surface. He has born the cost for this activity with the anticipation that the hauling costs is reduced by increasing the travel speed, reducing fuel consumption, and lowering the tire wear on the vehicle with the elimination of the ruts. The purchasers may not recognize the benefit, or may not have the decision support tools to fully quantify the benefit and will most likely ignore it. Therefore, the likely result is that the hauling contractor will provide a bid based on distance or average road condition and will obtain the full benefit of the improved road conditions. There will be little return on the investment for the landowner. Thus, there is unlikely to be an incentive for the landowner to improve the road as the information that will be exchanged with the other parties may be insufficient for the recovery of the costs. The result of an informationally-starved system may have long-term consequences that limit investment and innovation as the rewards may not be forthcoming in a multiparty supply chain.

However, if we were the fully integrated firm, the reduction in hauling costs would flow directly to the single firm and the benefits would at a minimum be reflected in the fuel and maintenance cost savings for the logging trucks. Troncoso et al. (2011) using a Chilean example compared the value changes between an integrated and decoupled supply chains. The integrated firm has the goal to maximize the value for all of the products sold while the decoupled only is concerned with the logs sales Their results demonstrated that an integrated firm can have a 5 % increase in NPV with a 8.5 % increase in profits when compared with the from the decoupled industry that does use the planning information for the entire rotation than from piecemeal planning (Troncoso et al. 2011). This demonstrates the potential value for planning for the entire supply chain.

16.2 Forest Business Models

Due to the uniqueness of the forestry supply chain, the potential value lost due sap staining, the impact of the long planning horizon, use of imprecise planning data, companies have adopted one of two basic approaches to manage their business, cut-to-inventory or cut-to-order. Each of the systems was designed to minimize the impacts of the some of the difficulties found in the forestry supply chain.

16.2.1 Cut to Inventory or Cut to Stock Methods

The first strategy is the cut-to-inventory method. The system avoids much of the problems used to balance the inventory and production with market forecasts by decoupling the forest from the mill. This is often used in those areas where the forest is inaccessible due to weather restrictions such as the monsoonal rain, snow, or spring thaw and inventory levels must be established to allow the manufacturing facilities to continue to operate without access to the raw material for a portion of the year. The mills will develop its production plans based on the logs that are in inventory. The data used will most likely be obtained from scaled logs. This will provide them with the known quantity and quality of logs available to the mill to fill orders. For the logging contractors, they will need to maintain production levels to sustain the log yard inventory at appropriate levels allowing the manufacturing facility the ability to fully plan their operations. The second is to avoid destroying value in the logs through inappropriate bucking operations. Typically, this can be achieved by creating as many long logs as possible to allow for the maximum number of cross-cutting options in the future.

In the second strategy, cut-to-order, the goal is to have all logs harvested with an impending order from the customers when the tree is severed. It is often used in plantation forest in the Southern Hemisphere were the pine logs are a primary product (Penfold 2003) that is susceptible to sap staining. Management's goal with the cut-to-order system is to obtain the best match of the customer orders for logs with the supply in a manner to minimize the operating costs that includes the inventory costs. The system relies on data generated from the pre-harvest assessment completed prior to harvesting data and the logging production estimates to align the supply with the forecasted demand. Markets are developed for the projected supply and if the data remain imprecise, the markets may not be fulfilled or higher value logs used to fulfill orders for lower value logs. There is much greater risk of downgrading in the cut-to-order system as there is no inventory to buffer the production from the customer demand.

There are similarities in both business models; both methods rely on a detailed planning system to develop the strategic, tactical and operational forest plans to maximize the value for the supply chains. They must consider their supply and future markets and use detailed planning systems to align the markets and production. Finally, they must monitor their performance to develop opportunities for future supply chain improvements.

16.3 Components of the Supply Chain

16.3.1 Hierarchical Supply Chain Planning System

Planning is the dominate function of the supply chain management. It organizes the activities efficiently among the participants of the firm. At the simplest level, the goal of all supply chain planning is to match the supply, production capability, and demand in a seamless method. Frayret et al. (2007) describes how Canadian firms, through the adoption of a customer-centered environment, can improve the performance of the forest sector in Canada. Their method uses an agent-based approach to support the synchronization of the supply-chain operations (Frayret et al. 2007).

Others have suggested a hierarchical approach (Bettinger et al. 2009). This approach decomposes the problem into individual decisions that correspond with strategic, tactical and operational decision making that have traditionally been applied in forestry problems. The difference between hierarchical planning and the supply chain planning is the active incorporation of the other actors in the decision process (Fig. 16.2).

There are five main processes in the supply chain planning procedure. These are: demand management planning, supply planning, planning and scheduling, execution, and the final section are knowledge collection and reporting. These five functions each have a time frame associated with them that varies from long-term, strategic planning to the short-term, operational scheduling. For this example, the operational scheduling will be further divided into annual planning, a rolling 3-month or quarterly planning and weekly scheduling. These activities are commonly performed by most forest management organizations. Typically the strategic level of planning is for 2–3 rotations; while the tactical planning problem are generally one-half of a rotation and the operational plans are from the year to the week.

The system's goal is to align the supply with the demand using a best allocated harvesting and hauling capacity. This is achieved by continually refining the balancing of supply, demand and production capacity at the strategic level that uses coarser data to the formulation of weekly schedule that requires significantly

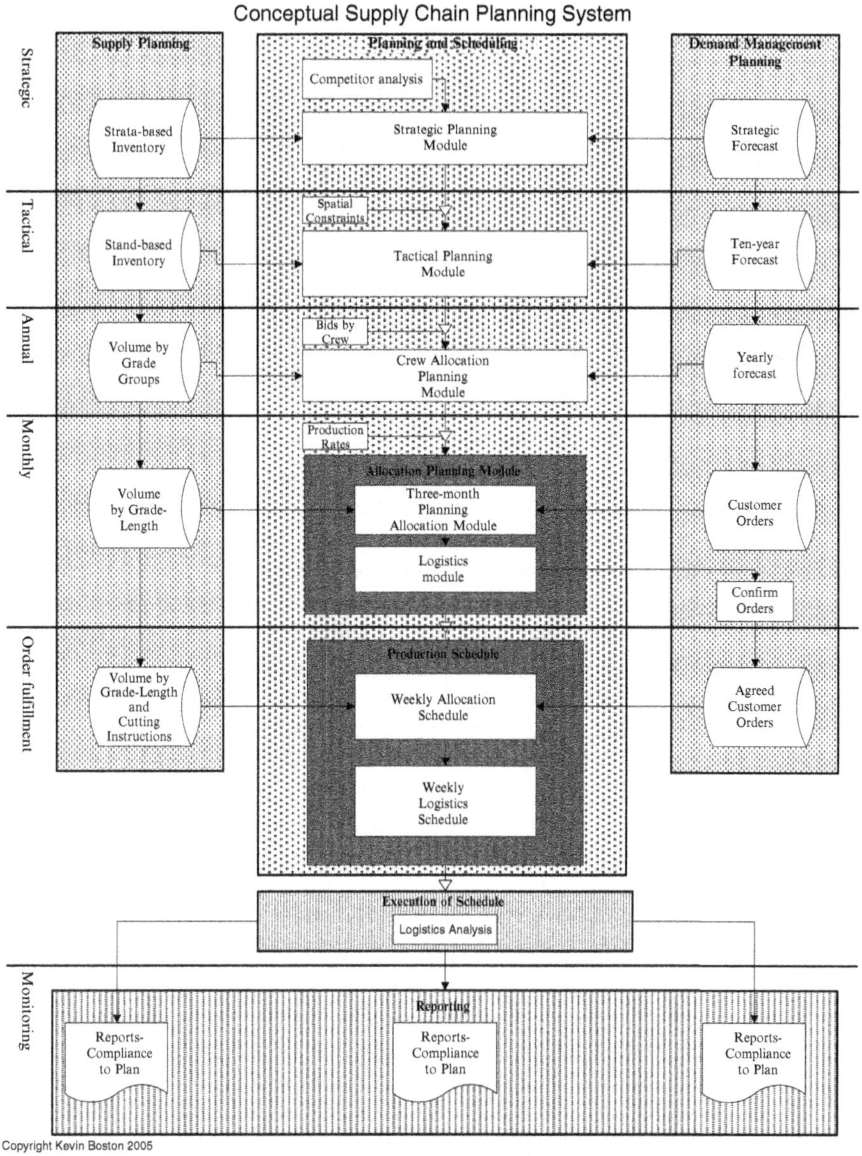

Fig. 16.2 A hierarchical supply chain management

refined data. This is process is not just an execution of a series of operations research models, but of the interactions among customers, contractors and forest managers to create the inputs and validating the results at each level in the hierarchy to achieve the desirable synchronization of the supply chains (Frayret et al. 2007).

16.3.2 Strategic Planning

The first level in the hierarchy is the strategic level. Its purpose is to develop the strategy that allows the firm to either develop or exploit competitive advantages through the execution of the supply chain.

The strategic plan is often produced for multiple rotations; therefore, the data used in the strategic level is typically coarse. The supply data is often limited to strata or croptype averages from sampled data. Yields for projected stands are typically computed for limited set silvicultural regimes using a growth and yield model. Thus, the volume is usually grouped into large grade-groups; for many plantation groups these include pruned, structural, appearance, industrial and pulp grades.

Demand management used in the strategic includes two types of analysis. One is the determination of customer importance. Importance can be defined using a variety of metrics. It may include those customers where there a long-term supply contractual relationship to deliver a specific quantity and quality of logs as compared to another firm that may only take a small volume of wood when prices are low, or it can based on the profitability of the customer.

The second type of analysis performed in the demand management is the price and volume forecasting. Generally, this is accomplished using econometric models that consider number macro-economic variables that influence the price for a given grade of lumber. Examples of these models include the Timber Assessment Market Model (TAMM) that produces estimates of supply by private and public land management that result in log demand and price for products at the regional levels (Adams 2007). Further analysis can be used to decompose the regional forecast into a local forecast that can be used at the mill working circle. Harvesting and transportation costs are often estimated through regionally developed models that account for much of the local conditions.

The goal of the strategic state is to determine if the quantity of wood offered for sale by the broad grade-groups are aligned with the demand from the customers and especially the key customers and harvesting and transportation capacity be able to deliver the material to these facilities at prices that warrant future or continual investment in the operations. Strategic analysis will determine the volume that is to be offered, the silvicultural prescriptions selected that allows the firm to be most successful. The results can usually be grouped into several possible groups. If there is a shortage of products from the fee timber land owners, then the strategic plan should evaluate the potential acquisition of new timberlands. If the desired returns are not achieved, new markets can be considered or disposal of tracks that have low economic returns or the analyst can create new or modify the silvicultural regimes to stands to reduce the economic expenses. Thus, the strategic planning is an important element in accounting for the long time encountered by in the forestry supply chains.

Linear programming (LP) is the most common operations research technique used to solve the strategic planning problem (see Chaps. 2 and 7). The goal of the LP objective function is to select among the all possible alternative the one that is most efficient based on the criteria used to compare the solutions. The most common criteria for evaluating the solutions is to maximize the discounted net revenue or total volume being produced. However, other measures can be developed and they may include measures of customer satisfaction, such as minimizing the deviation from key customers' forecasts. The individual firms will need to develop the objective functions that best correspond to the goals for their supply chain. The strategic objective is subject to physical constraints. These include factors that cannot be easily changed without major capital investment such as the purchase of new land or creating a new sawmill. These constraints are often considered fixed by the problem. The strategic objective is further constrained by the policy constraints. These constraints are more easily manipulated by the firm. Typically, they control the flow of goods or materials and can include even flow, which ignores market conditions, or it allows no more than a maximum variation in harvest, dollars spent, income between periods. This is useful to reduce the boom-bust cycles. It supports contractors by allowing some flexibility due to markets, but minimize the deviations in the work between period, or they can be used to model the minimum returns to investors or payments on a loan.

Strategic planning is more than just putting the data into a model and running it once. It is an iterative process where what-if scenarios are continually analyzed on each of the constraints to determine the impact of that constraint on the business. It can be used to value the importance of various customers and contractors to the company.

From the forest perspective, the result from the strategic plan is the harvested wood by grades group that will be supplied from the various forested tracts that will be delivered to each customer. It will specify the forest lands using a specific set of silvicultural regimes to satisfy customer demand. It will identify the timing, volume and grades that will either need to be procured as the fee lands are unable to produce this volume. If there is surplus volume, new markets can be sought or the products can be sold into existing markets but as lower value logs. For example, new export market, which can require time to develop, may be an alternative for surplus material that would be downgraded. The key from the strategic problem is to explore the alternatives for supply, demand and production capacity to develop a strategy for the maximum value, but that is grounded in the reality of the business.

The final aspect of strategic planning in the context of the supply chain is the role of senior management in understanding their supply chain. Senior management should have the role to craft the scenarios and reviewing the results; it is not to be accomplished by just analyst who report to management, but an active part of the company developing the strategies for its future operations.

16.3.3 Tactical Planning

Whereas the strategic plan has a focus on the long-term competitive plan and considers the entire forest estate. The tactical plan's goal is to implement this strategy, but include the spatial constraint that cannot be easily modeled at the strategic level. These plans are most often completed for smaller areas such as an individual catchment, or for an area of forest that enters a common point on a public road system at a forest gate or road shed. The planning horizons are generally between 10 and 25 years using annual periods, often a half of rotation in temperate plantation forestry.

A significant difference between the tactical and strategic plan is how the forest is characterized. At the strategic level, the planning units are usually strata or croptypes composed of multiple stands of common species and productivity units. While at the tactical level, the planning units are based on field-verified logging settings, where each polygon contains either the lowest cost harvesting feasible system. The tactical planning problems data inputs are the discrete harvest unit and the road segment; thus, the data is spatially explicit. The volume estimates are for individual logging units that may be composed of pieces multiple stands as harvest unit boundaries can often overlap stand boundaries. The data relies on either preharvest assessment or mid-rotation inventory. The data has a basis in field measurements and not from strata averages.

Besides the change in the characterization of the logging unit, the transportation network is part of the date used in the tactical forest plan. This simultaneous planning can result of in a 7 % increase in discount net revenue (Weintraub and Navon 1976). Often new plantation projects have a transportation system that was designed to facilitate establishment and may not support efficient harvesting. Significant gains may be available by developing a combined harvesting and transportation plan.

The final inputs usually are the proportional goals assign to that area established from the strategic plan. These include the harvest goals, volume, revenue and areas treated, that were determined from strategic analysis. The tactical forest plan is the spatial implementation of the strategic plan and new spatially restrictions are added to forest supply chain. This often includes the green-up constraint is a common element in many United States forest practices rules as well incorporated in several Australian states. Both the Forest Stewardship and Sustainable Forestry Initiatives have green-up constraints (Boston and Bettinger 2006). For example, the FSC standard for the Southern United States limits the size of an opening that can be created in a plantation forests to less than 31.5 ha (80 acres).

The problem formulation for the tactical forest planning model is a mixed-integer programming problem (see Chap. 8). These problems can quickly exceed the capacity of most commercially available problem solvers and plantation management organizations will likely need to develop customized solution techniques to solve these large problems. Traditional approaches, such as cutting plane algorithms and branch-and-bound techniques, can be used to solve small problems. However, for medium to large problems, there are a number of heuristic procedures have been developed and applied to solve these problems in the last 20 years. Bettinger and Chung (2004) provide an excellent review of the of spatial harvest scheduling systems that can be used to solve these problems. They are able to find solutions to these fast, but are often unable to determine the quality of the solutions.

However, many of the standard techniques such as simulated annealing and tabu search have history of finding solutions within 5 % of the optimal solutions when a comparison is completed (Bettinger et al. 2002).

These solutions techniques can be categorized into three broad classes. One is the gradient search categories such as tabu search. The second use stochastic techniques to solve these problems. Examples include simulated annealing and genetic algorithms.

The final categories are the hybrid algorithms that combine a variety of approaches to solve the heuristics (see Chaps. 7 and 8).

The result of the tactical planning is set of harvest units and road projects that are assigned to annual schedule and green-up constraints have been achieved. The limited ability of the forestry supply chain to respond to changes becomes more apparent with the solution from the tactical forest plan. Roads must be located, surveyed and built; this process can take a year in some areas. Once a tactical plan is implemented, it may take a year before some changes are available; the forestry supply chain is not very nimble.

16.3.4 Operational Planning

Operational planning can usually be divided into three levels, annual, monthly and quarterly. All levels of operational planning focus on the limited choices available during operational planning. The available harvest units are restricted by the results of the tactical plan. The other difference with operational plan is the demand emphasis has shifted from forecasted demand to customer orders. The solutions are centered on assigning orders to crews and units to best meet the customer demand.

16.3.4.1 Annual Planning

If the demand was even-throughout the year, the solution would be easy, determine the harvest capacity that meets the even-flow targets. However, the market rarely produces such a convenient demand level. Greene et al. (2004) demonstrated that poor planning was the third most common cause for unused logging capacity in the Southeastern United States. Thus, an effectively implemented supply chain can be used to improve the profitability and lower the cost of logging and hauling contractors by developing a work scheduling that allows these organizations to maximize their utilization through good supply chain scheduling. The annual plan results are twofold; one is to assign the harvesting crews to units and the other is determining the order in which the units are to be harvested. It supports the collaboration by allowing hauling and logging contractors to view the amount of work that is available for the year. As many corporations use a variety of key contractor or key supplier agreements, these groups can be given a priority for

Table 16.1 Comparison of profits based on variable utilization rates

Crew	Hours	Cost
A	2,000	65.07
B	1,400	80.39

assigning work to them to maximize their utilization and lower logging costs. For example, the cost of crew A working 2,000 h is compared with crew B working 1,400 in Table 16.1. Both crews have identical equipment, a feller-buncher, two skidders and a wheeled loader.

Thus, effective scheduling can increase the utilization of the logging work-force that can benefit both parties, the logger is able to have a stable work at a higher returns and the landowner may be able to achieve a lower cost per ton. Thus, the results of the annual plan are to determine the logging and trucking capacity need throughout the year to satisfy the customer demand.

The result of the annual plan is also a detail, bottom-up budget that has important tool for managing any forestry organization progress towards yearly targets. The budget establishes the benchmark for managing the business progress through the year.

16.3.4.2 Monthly and Weekly Scheduling

The second and third elements in operational plan are the monthly and weekly schedules. Both problems have the similar formulations and data requirements, orders for customers and the inventory and production level to estimate the supply that can be produced from the available unit pool. For the monthly problem, the first month's solution is usually finalized, orders are confirmed and crews are assigned to the next unit with subsequent months allowed limited changes. At the weekly level, there is usually certainty with the orders and under most circumstances; it is simply a confirmation of the weekly portion of the monthly order.

The supply is described as a result of applying a set of cutting instructions to stand of timber. These cutting instructions typically include a price for each length of log for a given log grade. The log grade is a composite of stem characteristics such as large and small–end diameter, branch size, knot size, degree of allowable sweep and defect. Thus, the yield file is dependent on the no only the presence of the product in the stand, but the relative importance of that log's value to other logs that can be produced. There are existing tools such as OPTICORT (Epstein et al. 1999) from Chile and MARVL (Manley et al. 1987) from New Zealand are examples of inventory tools that allow use of price-driven log stock tables.

The supply is limited to the current unit that are being harvested with the opportunity to create adjust is limited to changing the cutting instructions.

16.3.5 Execution Phase

The execution phase begins when the cutting instructions produced by the weekly plans are sent out to the harvesting crews. Trees are felled and bucked to satisfy customer orders and trucks will need to be assigned to haul the loads to the various mills.

The execution phase contains the logistics analysis that determines the trucking resources necessary to pick-up and deliver the wood on a daily basis. It requires a two-step procedure; the first step is the planning phase that uses the estimated production from the harvest crews along with the application of the cutting instruction to determine the rate at which full loads will be created. This is used to create a daily plan for trucks using vehicle routing solutions such as those describe in ASICAM (Weintraub et al. 1996) to route the trucks to best fulfill the orders. Trucks are assigned a route that pickups and delivers a load within the maximum length of time that a driver may be behind the wheel. In many jurisdictions, the limit is 10 h maximum. The model is constrained to have the driver return home at the end of the day.

The second component is a dispatch system that fine tunes the daily schedule by assigning the truck to landing using real-time data. The use of these logistical truck scheduling tools has resulted in significant improvements in the efficiency of the trucking resources by reducing the number of trucks needed while improving the utilization of the trucks (Weintraub et al. 1996). It is an excellent examples of how the benefits of supply chain planning can result in improvements for all of the participants by lowering the hauling costs paid by the forest enterprise, increasing the utilization for the remaining trucks that can increase the profit per truck, reducing the inventory on the landings that improves the safety for the harvesting crews, while improving the timeliness of the delivers for the customer.

16.3.6 Reporting and Data Modeling

To accomplish the goals of the hierarchical supply chain planning system requires a significant investment in data and information from a variety of sources. Inventory and growth and yield data will need to be collected and processed. Price and sales forecasts are generated from both analytical models and interactions with customers for all grades and species combinations. Logging crews will provide not only cost estimates but production estimates for each logging unit. Finally, truck and loading times, travel speed will need to be estimated to develop the logistical support for the execution phase of the supply chain. Much of this data is based on samples or estimates by the various groups. Therefore, one of the key elements of the reporting area is to collect these estimates.

While work is being performed, many transactions have occurred. Logs are produced and hauled to landings, invoices are submitted and received for the various producers and consumers, and finally payments are distributed. This data is used to not only create the necessary audit trail for these transactions, but it serves as the basis for continual improvement by allowing the supply chain managers the opportunity to compare the forecasted data with the actual data. Learning the causes for the difference can aid in the improvement in future forecasted data. Thus, growth and yield models, econometric sales and price forecasting models, logging production models can all be validated with this data.

The validation of existing models, the improved estimation of forecasts models for prices, volume, cost and revenues can significantly improve the nature of the decisions being made as the data used in the decision-making is continually improved. An additional benefit is that the firms involved in the supply chain will better understand the key drivers of the business and can focus its management attention on these areas.

16.4 The Interaction with Supply Chain

Despite the best efforts to plan, changes to supply chain will occur. Most of the time, these changes will be small. For example, if there is a surplus of volume being produced in one grade and a shortage being produced in another, then the action may be to adjust the cutting instructions. Thus, to manage the change in the supply, one can simply reverse the process working at the weekly, monthly or modifying the annual plan to meet the change. Dramatic changes, such as the Asian financial crisis in 1990s that had a significant changes for many Southern Hemisphere wood producers in the region may involve returning to the strategic and developing new strategies to account for the dramatic market changes.

However, in most circumstance those changes in production scheduling can be managed at the operational level. For example, if the overall production is too high, quotas that limit production can be applied to all or some of the crews to reduce production. However, the forest manager will need to understand that reducing the utilization of the logging crews will likely increase the logging and hauling costs. Those crews with high utilization incentives should be the last crews to place in quotas to prevent payment of penalties. If capacity remains too high, a new annual plan may need to be formulated.

If volume targets are not being achieved, there is an opportunity to increase the short-term work by having harvesting and hauling crews to work a 6-day work week. If volume targets remain unmet, the supply chain manager may need to resolve annual plan to add additional crews, or could possibly return to the tactical level to allow for units or crews with higher potential production to be the schedule the next operations. Unfortunately, these production systems are far from nimble and the full response to the changes may not be seen for several weeks.

16.5 Opportunities for Improvement

The issues with the supply chain involve three basic areas; one is collecting information. The goal is to collect better information that can be used to improve the supply chain's performance. The second area is tracking the progress of the products throughout the supply chain to provide the feedback that is necessary to support accurate financial transactions. This information has other uses; it can provide the data needed to improve the supply chain performance by creating feedback and data validation opportunities. The final area is development of the true collaborative processes with the business partners to understand the interactions among the supply chain elements. As many improvements in the supply chain require an investment by one party that will yield an improvement by another party in the supply chain. There needs to be an understanding of the relationships among the supply chain to better allocate the benefit and costs to encourage investments that make the overall supply chain more efficient.

16.5.1 Improved Characterization of the Resource

The use of LiDar to characterize the supply is being explored by numerous researchers. There are two platforms for this work; one is from an aerial platform that can be used to develop accurate estimates of tree count and average canopy heights to support volume estimation (Dubayah and Drake 2000). This information can be combined with information from traditional, ground-based forest inventory systems to either allow for lower cost or greater detailed inventory systems.

Terrestrial LiDar, images collected from the ground, has future potential in the plantation forests, as it can be used to improve the characterization of the individual tree (Mass et al. 2008). This can include a better determination of the volume of pruned section of tree or the volume in the internode length. The internode is the section of the stem between the branch whorls that if sufficient long enough it can result in high value peeling material without the cost of pruning. Additionally, terrestrial LiDar has the potential to characterize the sweep of the log, another variable that can impact the value of the logs that is difficult to accurately portray with current technology.

16.5.2 Improved Log Tracking

Historically, log branding and paint have been the most common technique for log tracking. However, it is does not yield much information and can be easily counterfeited. Often, a unique tracer element is added to the paint to improve the ownership detection. Today, bar codes are commonly attached to many logs;

especially those being shipped internationally, as the can provide more information that may include the size, location or date of felling. However, each tag currently must be scanned to obtain the information.

Some have suggested radio frequency tags; these allow for the storage of a great deal of information that can be read quickly. However, the cost of these tags when place on commercial plantation forestry operations can be excessive. The scale of large, commercial forest operations can result in perhaps ten million logs in a year and the cost even at pennies per tag can result in an annual cost in the millions. Another problem with both of these techniques is that managers of pulp and paper mills do not like and may often refuse to accept material that may contains plastic-coated paper or metal staples that can contaminate the raw materials.

Murphy and Franich (2004) have suggested that odors can be applied to logs and then detected with an electronic noise, similar to those used in airports that can detect remnants of explosives, be used to identify these smells unique to a log. Thus, the presence of a series of odors can be used to show ownership. The problem still remains that each log must be manually inspected to determine it origin.

There is similar work with micro-taggant paints that contain unique identification marks within the paint particle that can be detected with a microscope. These technologies may be useful to prevent or reduce log-theft by providing a unique identification of a log. Further work is needed to develop a log-tracking system that will support the supply chain. It should have the following characteristics, it can be scanned quickly while the logs remain on truck, not interfere with downstream manufacturing such as pulping or sawmilling as the marker should not contaminate or damage manufacturing process.

16.5.3 Collaborative Planning

The last area is the development of a collaborative understanding of the supply chain. This is especially important when the improvements require that one party make a capital investment that results in benefits to the other parties. For example, all of the members of the supply chain may realize that the transportation costs are too high and harm the competiveness of the entire supply chain. The forest landowner reconstructing roads to allow for vehicles with a larger payload or a higher travel speeds. The immediate benefit is to the transportation contractors as their costs are lower. If sufficient traffic travels that link, the fixed costs investment can easily be justified if it was a single entity, but it becomes more difficult when there are multiple parties in the supply chain. The supply chain participants will need to develop the trust in the data used to evaluate the options used to enhance the competitiveness of the supply chain. Welety and Becerra-Fernandez (2001) demonstrated how collaborative supply chain management was able to improve the productivity and customer service in the book publishing business by developing a collaborative approach to managing changes in the order cycle, similar work needs to be developed in the forest products industry. Having all parties understand how the changes will affect their cost, production and develop a willingness to collect and share the data to allow parties to recover their fair share of the benefits produced from the changes in the supply chain.

16.6 Summary

Efficient supply chain management is the key to a successful business. However, the forestry supply chain has a number of elements that make it unique that result in specialized applications. One is that forestry is primarily a decoupling business that creates co products that cannot use either the push, production-driven, or pull, customer demand-driven, approaches to determine the production levels as multiple products are created when a stem is bucked into logs or the log is milled into lumber. It requires a hybrid approach that uses a combination of the customer demand and production decision making when making decisions.

Time is a distinguishing factor of the forestry supply chain for two reasons. One is the length of time required to grow trees, even fast grown plantation forests can require 5 to 6 years to yield a product. Decisions must be made many years in advance of selling a product to market. Additionally, there are many features that limit the flexibility of the supply chain to change as time progressive due the seasonal nature of many forestry operations due to access limitations.

A hierarchical planning system that links the strategic, tactical and operational planning to best match the supply with customer demand and logging and hauling capacity is recommended as it incorporates the variable data quality found in the forestry supply chain. The quality of the data used in the hierarchical system supports the current practice of using progressively refined data.

Administering the supply chain results in the managing the deviations from the plans. There are only a few options, modifying the cutting instructions to switch production from one product to another, decrease production through the implementation of production quotas that will increase harvesting and transportation costs, or if possible contract with new crews to increase the hauling capacity, but this is often limited in the short-term.

The forest supply chain is a complex interaction among multiple parties. Improvements in data quality can reduce the large variation in the date use to manage the supply chain. There is significant work being accomplished using both ground and aerial –based LiDar to improve the characterization of the trees that can yield benefits was the products desired by the customers become more complex. The final area is collaborative supply chains to develop the trust among the participants to better share the costs and benefits that ultimately make the supply chain more competitive.

16.7 Questions

1. Describe the kinds of organizations that are found in your supply chain. Where do they get their information, resources and financial support from? What is the level of business sophistication they will have in your area? How it may limit their full participation in the supply chain management?
2. How does the perishability of a product impact the management of the supply chain? Describe the methods you would use to minimize the losses doe sapstaining if you were a harvesting contractor or a log truck dispatcher.
3. Time is a critical element in the forestry supply chain. Some products such as clear-wood require decisions be made 5–7 years in-advance of a customer order depending upon growth rates. Using the hierarchical planning system described the chapter, describes how you would incorporate the decision to prune a stands from the strategic and how would the coproduction of other wood products?
4. Greene et al. (2004) describes changes in market conditions to be the number cause for unused logging capacity in the southeastern United States. Describe how this is cost can be reduced using and efficient supply chain management techniques. Describe how you can use the principles of the supply chain planning system to motivate these contractors to participate in this management system.
5. Markets for forest products can change quickly at times – you are managing the supply chain – describe how you will respond to changes. In answering this problem, focus on what can be changed at the operation level and work backwards describing what can be easily changed at each level.
6. To improve the performance of supply chain requires the validation of forecasting results. Using your knowledge of forest harvesting, inventory and economics, develop an audit plan to improve the information used in the supply chain.
7. You manage a MDF plan that consumes a whole log and mill residue. If the demand for sawlogs has decreased, describe how this could potential increase or lower your raw material prices.

References

Adams D (2007) Solid wood-timber assessment market model (TAMM). In: Adams DM, Haynes RW (eds) Resource and market projects for forest policy development. Springer, New York, p 589

Bettinger P, Chung W (2004) The key literature of and trends, in forest-level management planning in North America, 1950–2001. Int For Rev 6(1):40–50

Bettinger P, Graetz D, Boston K, Sessions J, Chung W (2002) Eight heuristic planning techniques applied to three increasingly difficult wildlife-planning problems. Silva Fenn 36(2):561–584

Bettinger P, Boston K, Siry J, Grebner D (2009) Forest management and planning. Academic Press, New York, 331

Boston K, Bettinger P (2006) An economic and landscape evaluation of the green-up rules for California, Oregon, and Washington (USA). For Policy Econ 8:251–266

Bruce A, Stewart D, Verrall S, Wheatley RE (2003) Effect of volatiles from bacteria and yeast on the growth and pigmentation of sapstain fungi. Int Biodeterior Biodegrad 51:101–198

Chandran PM, Gupta V (2003) Wal-Mart's supply chain management practices. ICFIA Center for Management Research Case Study. Nagarjuna Hills, India

Dubayah RO, Drake JB (2000) Lidar remote sensing for forestry. J For 98(1):6, 44–46

Epstein R, Morals R, Serón J, Weintraub A (1999) The use of OR systems in Chilean Forest Industries. Interfaces 29(1):7–29

Frayret J-M, D'Amours S, Rpiseai A, Harvey S, Gaudreault J (2007) Agent-based supply-chain planning in the forest products industry. Int J Flex Manuf Sys 19:358–391

Greene WD, Mayo JH, deHoop CF, Egan AF (2004) Causes and costs of unused production capacity in the southern United States and Maine. For Prod J 54(5):29–37

Hay DA, Dahl PN (1984) Strategic and midterm planning of forest-to-product flows. Interfaces 14(5):33–43

Manley BR, Goulding CJ, Lawrence ME (1987) Stand assessment by log grades using MARVL. Forest Research Institute, New Zealand, Bulletin 132

Mass HG, Beinert A, Scheller S, Keane E (2008) Automatic forest inventory parameter determination from terrestrial laser scanner data. Int J Remote Sens 29(5):1579–1593

Murphy GE, Franich R (2004) Early experience with aroma tagging and electronic nose technology for log tracking. For Prod J 54(2):28–35

Penfold C (2003) Log making to market or to order. NZ Forest Industries, pp 20–23

Rice B (2010) Wal-Mart earns top spot in retail sales. Store Mag 8

Troncoso J, D'Amours S, Flisberg P, Rönnqvist M, Weintraub A (2011) A mixed integer programming model to evaluate integrated forest value chain – a case study in the Chilean Forest Industry. Interuniversity Research Centre on Enterprise Networks, Logistics and Transportation/Cirrelt, Montréal, p 28

Weintraub A, Navon D (1976) A Forest Management Planning Model Integrating Silvicultural and Transportation Activities. Manag Sci 22(12):1299–1309

Weintraub A, Epstein R, Morales R, Seron J, Traverso P (1996) A truck scheduling system improves efficiency in the forest industry. Interfaces 26(4):1–12

Welety B, Becerra-Fernandez I (2001) Managing trust and commitment in collaborative supply chain relationships. Commun ACM 44(6):67–73

Wold C, Gaines S, Block G (2011) Trade and the environment: law and policy. Carolina Academic Press, Durham, p 992

Chapter 17
Pulp and Paper Supply Chain Management

Sophie D'Amours, Marc-André Carle, and Mikael Rönnqvist

17.1 Introduction

The pulp and paper industry is important to the economies of several countries such as Sweden, Finland, Russia and Canada. In 2012, the Canadian forest products industry accounted for 1.9 % of gross domestic product and the $ 26.4 billion industry exports value and its 17 billion trade surplus is second only to Oil and Gas. Of the $ 26.4 billion export value, pulp and paper accounts for $ 17.3 billion. Moreover, pulp and paper industry is becoming increasingly important for some emerging economies like China, Brazil and Chile.

The importance of the industry also comes from the fact that it offers employment in many remote areas. The total direct employment of the pulp and paper sector in Canada during 2012 is 75,100. However, in this era of globalization of markets, relying on relatively abundant fibre supply is no longer a sufficient competitive advantage. New and emerging competitors offer products and services at lower cost. Consequently, the competitive position of enterprises is dependent on both operational excellence and innovation capability. High levels of customer service are proving to be a distinctive way to maintain and develop new market share.

Transforming a tree, or even a seed, into pulp and paper and then deliver it to a customer anywhere on the planet requires the intervention of several actors. These are responsible for: development and management of forests, logging, deployment of the forest road network, transport, trees, logs and residues, chipping, pulp

S. D'Amours (✉)
Department of Mechanical Engineering, Laval University, Laval, Canada
e-mail: Sophie.damours@gmc.ulaval.ca

M.-A. Carle
Department of Operations and Decision Systems, Laval University, Laval, Canada

M. Rönnqvist
FORAC, Université Laval, Québec, QC, Canada

J.G. Borges et al. (eds.), *The Management of Industrial Forest Plantations: Theoretical Foundations and Applications*, Managing Forest Ecosystems 33,
DOI 10.1007/978-94-017-8899-1_17,

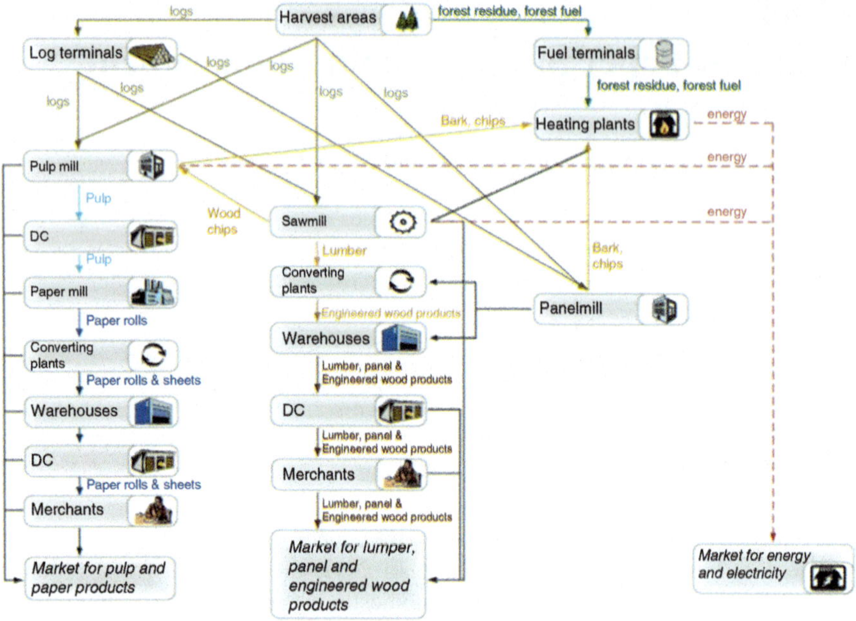

Fig. 17.1 The forest industry supply chain (From Carlsson et al. 2009)

making, paper making, paper conversion, distribution and delivery, recuperating and recycling. These actors are also responsible for managing the operations and use of resources to maximize their profits in respect of social and environmental constraints. All of these stakeholders as well as their interactions define the supply chain. Thus, the coordination of the use of resources (fiber, supply, equipment, assets, people, information and financial resources) creating value within the supply chain contributes significantly to improving the competitiveness of the chain. This coordination is achieved through information exchange, integrated or joint planning. Figure 17.1 illustrates the different actors of the forest products industry. It shows that the pulp and paper supply chain includes a great number of business units and interacts with the forest supply chain as well as the wood product supply chain in many different ways.

The objectives of planning in the supply chain is to capture new business opportunities through an optimum deployment of resources that improves corporate profitability by reducing operating costs and increasing revenues from satisfied customers. To do this, supply chain planning aims to synchronize operations and organize information sharing between business units of the chain in order to be more agile and to respond more efficiently to market signals and orders.

Companies cannot all aspire to the same level of performance because the markets they serve are different. These markets may be at different evolution stages, developing, expanding, fast-growing, maturing or declining. For example, companies operating in mature markets need to be excellent in terms of their operations and logistics. As a market grows, the qualification criteria become more numerous and demanding, companies must excel to differentiate themselves from the competition. For example, in the markets of the forest products industry to maturity, the price criterion is mostly a qualification since prices are set by the market. At this point, the ability of a company to differentiate itself in a sustainable manner thus depends on its distinctive competencies.

For example, the anticipation of supplies and their synchronization with the needs of plants contribute to increased levels of customer service. The service level is the percentage of shipments delivered on time according to the total number of deliveries. With this in mind, a reliable supply of wood, which allows the availability of the right timber at the right time, right place, can be a distinctive competence for differentiation of a business based on service levels.

Although there is a vast literature on SCM in general, due to the special characteristics of the pulp and paper industry there is a lack in this area. The purpose of this chapter is to provide an overview of the domain of supply chain planning in the pulp and paper supply chain and identify key methods and issues such as the important challenge of coordinating sales and operations as well as anticipating the transformation of the pulp and paper supply chain as the biorefinery supply chain is expanding as a new business model for many pulp and paper companies.

The paper markets can be divided into five main segments based on end-usage:

- Printing and writing (e.g. catalogues, copy paper, book paper and magazines)
- Newsprint (e.g. newspaper)
- Tissue (e.g. toilet or kitchen rolls, facial tissue)
- Container board (e.g. packaging boxes)
- Other paper and paperboard (e.g. paper boxes, paper bags, filters)

Capacity utilization rates in the P&P industry are generally very high (91 %). This is due to the fact that assets are extremely expensive often referred as "capital intensive" industry. Paper companies normally run continuously except for regular maintenance stops. This has an impact on the supply chain planning practices as one cannot run a pulp or paper mill on a stop and go mode.

Cellulose fibre normally accounts for more than 80 % of the weight content of the paper. The properties of the cellulose fibre itself are therefore crucial for the resulting properties of the paper. This constitutes a strong link backwards into the forest supply chain to the very origins of the fibre used in the paper-making. It is also very important that the pulping and paper-making processes do not destroy the properties needed further along in the chain.

The pulp and paper industry can be viewed as a large network of production units that gradually refines wood into consumer products (see Fig. 17.2). It is very rare that the entire refinement is made by a single company. The production network is linked to a procurement network which starts in the forest. This network may include several locations (wood yards or other storage points) where logs are just stored or transhipped before they go to production units. The production network is

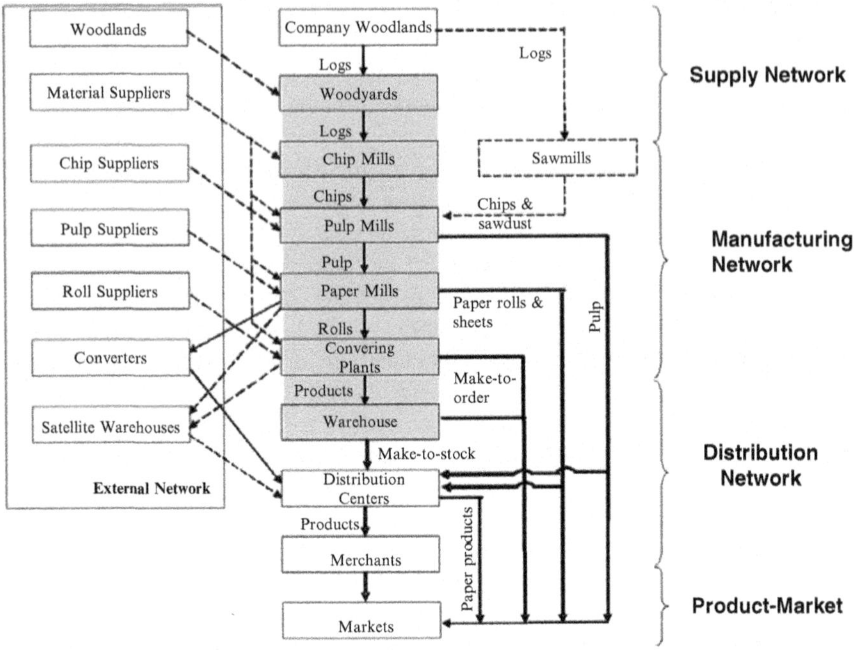

Fig. 17.2 The pulp and paper supply chain (From Martel et al. 2005)

also linked to a distribution network ending at merchants or retailers, who together with the final customers constitute the sales network. As a matter of fact there is actually a connection from the sales network back into the pulp and paper supply chain again.

There are four major processes in the pulp and paper supply chain; harvesting and transportation, pulp making, papermaking, and sales and distribution.

Harvesting is done by a set of harvest crews and the transportation by one or several transportation companies. The trees are cut and branches removed. Thereafter the tree is bucked (or cross-cut) into logs (with specific dimensions and quality). This process is typically done directly at harvest areas. Logs are transported from harvest areas directly to mills or through intermediate storage at terminals.

The overall harvest and transportation planning is often integrated. However, operational planning (e.g. routing of logging trucks) is done independently of, for instance, the bucking process. This issue has been discussed in previous chapters in this book.

The pulp making process converts pulp logs into chips. The chips (different species) are mixed following proprietary recipes with sawmill chips to produce the pulp desired. Chips are then boiled and washed, fibres are separated from lignin. Fibres are processed in a number of steps with chemicals, then bleached in order to produce fibres with certain brightness. This is a continuous process where the time

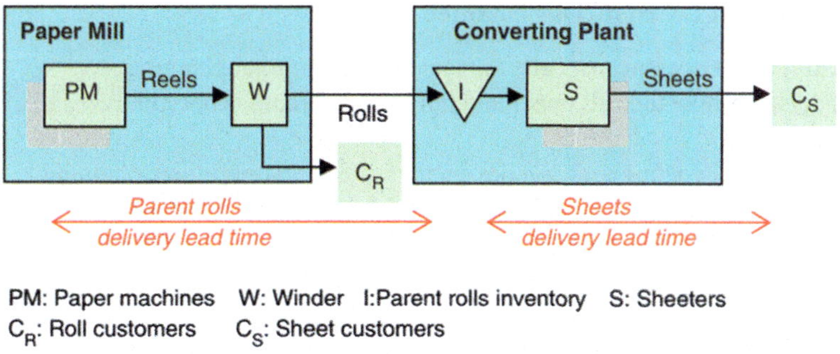

Fig. 17.3 Paper manufacturing process (From Chauhan et al. 2008)

from chipping to production of pulp is about 12 h. At integrated mills, the fibre is transported directly to the paper machines. In pulp mills, sheets of pulp are produced for further distribution to paper mills.

Paper making process involves feeding the paper machine (large mat) with pulp to form a paper web. By pressing the sheet the water is removed by force, then the web is dried by air and/or heat to produce the paper. At this stage the paper is uncoated. Coated paper is produced by fixing a thin layer of material in order to get matte, semi-matte/silk or gloss paper. Both uncoated and coated paper can have their surface polished through a process called calendering. Paper are produced on jumbo rolls, fixed width reels of a given paper grade, The reels are put on a winder and sliced into several rolls of smaller diameters and widths. The part of the reel not cut into rolls is trim loss which will be recuperated. Finally, the rolls are shipped to customers, or converted into cut-sheet finished products on a sheeter, which may also generate some trim loss. The rolls sheeted into finished products are known as parent rolls. Figure 17.3 illustrates the cutting process in paper making.

Pulp and paper are sold and distributed through a complex network of business units depending on the size of the producing company and the needs and type of the customers. For example, on one hand, large newsprints printers will deal directly with producers and one might even require that its paper be always produced on the same paper machine. On the order hand, fine printing paper (paper we use at home), will need to go through wholesalers, distributors, merchants, before end-users can buy. This implies that the producers may face different buying behaviors from its direct customers, going from contract-base approaches to spot-market buying. In long distribution supply chain, inventory management is a key issue as products need to be available and at the right location to respond quickly to the distribution chain.

In some sectors of the industry, like Tissue, all these steps can be found at a single location and the products are then directly packaged for distribution to the consumers through merchants and/or retailers. In other sectors, the products are delivered to printers or additional converting plants before entering the sales network.

What You Will Learn in This Chapter

- Distinguish between strategic, tactical and operational planning
- Understand how the full paper and pulp supply chain is coordinated and affected by its parts
- Understand in detail strategic design and sales and operations planning
- Understand how coordination of the chain can be organised
- Have knowledge to extend the pulp and paper SC into a biorefinery SC
- Have knowledge about open research topics that may inspire future work in this domain

The remaining chapter unfolds as follows. Section 17.2 describes key supply chain decisions in the pulp and paper industry. Section 17.3 focus on strategic design. Section 17.4 focus on sales and operations planning and Sect. 17.5 on cross chain coordination. The Sect. 17.6 provides a discussion on how the Pulp and paper SC can be extended into a biorefinery SC. It ends with a summary in Sect. 17.7, a few problems in Sect. 17.8 and literature references for further reading.

17.2 Key Supply Chain Decisions in the Pulp and Paper Industry

Supply chain planning in the forest products industry encompasses a wide range of decisions, from strategic design to tactical and operational planning which are driven by cost and customer service, integration, coordination and operational excellence. A reference used by many managers in this area is the SCOR Model Supply Chain Council (supply-chain.org/scor). The main challenges from a planning perspective is to ensure that decisions are not taken independently of each other and that the decisions system is coherently driven by corporate objectives and not by division nor by business units objectives.

A thorough understanding of markets and customers expectations is one of the cornerstones of integrated supply chain. The world-class companies use and exploit models of demand forecasting and anticipation for each product-market segment (defined as a specific set of customers who want the same product) volumes that will be required. They form cross-functional teams responsible for planning sales and operations. They work with their clients in order to better manage the demand chain. In general, these teams use statistical tools designed to forecast the needs and the capacity to meet the needs. In supply chain, demand management (information and demand forecasts) is a critical success factor for all stakeholders in the chain. For example, the sharing of real information at point of sale between the various stakeholders in the chain can reduce the negative impact of information distortion.

Integration refers to the organization of the supply chain, that is to say, the establishment of who does what and how. Generally, the company do not have full control of the chain, part of the activities being outsourced to others. However, it needs to master its own procurement decisions and choose wisely its suppliers and subcontractors. Clever managers can influence the selection of suppliers of suppliers with a clear vision of its business. They will also set the collaboration scheme. For example, they may call for suppliers to deliver their supply at the rate at which they will be consumed within the mill, under a just-in-time model. This reduces the work in progress and maximizes business agility and liquidity. The agility of the company is its ability to shift its production plans to take advantage of new business opportunities within a very short time.

This agility is also dependent on information sharing about supply chain flows and resources. In a supply chain, companies or business units are linked by flows. There are flows of goods, information flows and financial flows. Product flows include all transfers of finished goods, in process and raw materials flowing through the network to create value. Information flows include information between business units, for example information demand, supply, price, contract and transaction confirmations to be provided to governments. Financial flows include cash transfers between business units. They consist of receipts, refunds, miscellaneous payments, stumpage, etc.. Information on demand, supply and resources (assets, goods, financial and human resources) is key to the supply chain planning processes. Availability of high quality information is a necessary condition for good decision making. Benefits associated with a greater synchronization of transactions between companies or business units are well established. Information sharing do raise a number of challenges, one of which it related to the standardization of information exchange. In recent years, companies have worked on a voluntary basis in the definition of standards such as papiNet (a standard of the forest products industry) so that they can share a common representation, understanding and vocabulary of the supply chain components and decision processes in order among other things to achieve operational excellence.

Operational excellence refers to the optimal use of resources in order to achieve the strategic objectives. It is only achieved when the company manages its resources dynamically while keeping a high level of quality deliveries to the customers, constantly improving its processes and reducing as much as possible all sources of waste. In recent years, many companies have focused on the concepts of lean manufacturing to achieve operational excellence. The premise of these concepts is the synchronization of all actions according to meet demand in a timely way minimizing all sources of waste (trim, inventory, obsolescence, absence, quality based lost, handling and transportation, corrective maintenance, etc.). The lean manufacturing principles also had an impact of the definition of key performance indicators. For examples: service levels, velocity (cycle time/ time value), tack time, cash cycle time are KPI used by management. Service level is defined as the percentage of orders delivered in accordance to customer expectations in terms of delivery date, quantity, quality and location.

In order to achieve high delivery performance, value creation, agility and operational excellence, supply chain planning is required. The following subsections will discuss strategic decisions, tactical planning (more specially, sales and operations planning) and finally operational planning. To set the stage, a supply chain planning

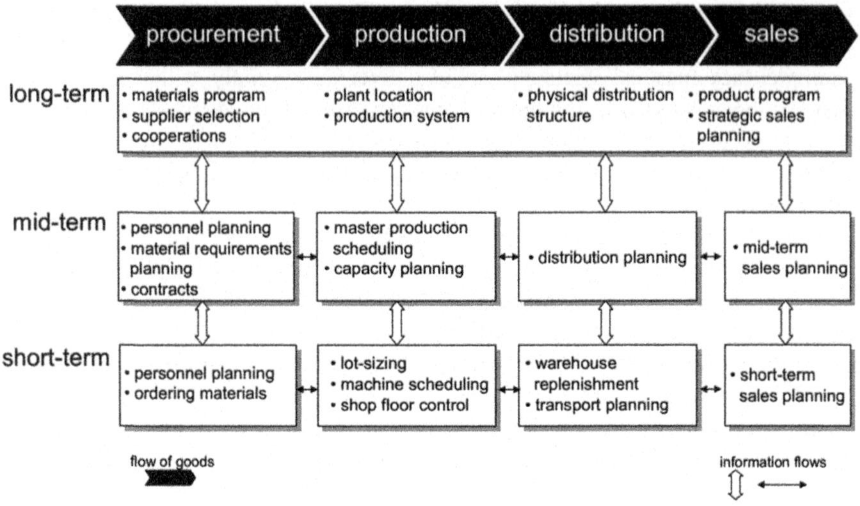

Fig. 17.4 Supply chain planning matrix (From Fleischmann et al. 2008)

matrix is introduced in Fig. 17.4. This figure is inspired by the work of Fleischmann et al. (2008). They have divided the supply chain planning into four main stages or processes. Procurement involves the operations directed towards providing the raw material and resources necessary for production. In the case of pulp and paper the most important raw material is wood. Production here referred as pulp and paper making is the next process in the chain. In this process the raw materials are converted into intermediary and/or finished products. Thereafter, distribution includes the logistics taking place to move the products either to companies further processing the product or to distribution centres, and finally to retailers. The sales process deals with all demand planning issues including customer or market selection, pricing strategy, forecasting and order promising policies.

17.2.1 Strategic Design

Strategic, or long-term, planning in the pulp and paper industry is indeed very long-term. For example, the rotation of forest growth can take more than 100 years, and a new pulp or paper mill is normally intended to last more than 30 years. Thus, strategic decision-making includes making choices related to forest management strategies, silviculture treatments, conservation areas, road construction, the opening/closing of mills, the location/acquisition or selling of new mills, process investments (e.g., machines, transportation equipment, information technology), product and market development, financial and operational disclosure, planning strategies (e.g., make-to-stock, make-to-order, cut-to-order) and inventory location (e.g., location of decoupling points and warehouses).

The planning approach chosen has a major impact on all investment decisions. For example, the capacity needs and the type of equipment required to support a make-to-stock strategy would be different than those needed to support a make-to-order strategy. Therefore, the planning approach defines important parameters with respect to the necessary technology, capacity, inventory levels and maximum distances to customers. Such decisions naturally involve evaluating how the investment will fit into the whole supply chain, including deciding which markets are available for the products based on anticipated market trends, how the distribution of the products should be carried out and at what cost, and how the production units should be supplied with the necessary wood fibers (i.e., wood or pulp). Other elements, such as energy supplies, might also be crucial.

The type of forest land tenure may also affect the way supply chain strategic decisions are made. Wood could come from public lands, private lands or both, with each type requiring different procurement programs. Other factors may also have to be considered. For example, governmental rules governing the amount of forest land to be set aside for biodiversity purposes, recreational use and/or carbon sequestration must be taken into account in any decision.

The literature about strategic supply chain planning provides a broad and rich examination of the domain and proposes a variety of different solution methods. However, very few articles focus on divergent alternative production processes or mixed demand behaviors (e.g., spot and contract-based demands). Specific methodological contributions are needed to remedy this lack. In addition, very little research in strategic supply chain planning pertains specifically to the forest products industry (FPI). The scarcity of research in this domain underlines the demand for knowledge about FPI implementation and the need to integrate the decisions related to forest management and forest operations into the other downstream supply chain planning decision-making processes.

A strategic supply chain planning strategy and process must also take into account the substantial amount of uncertainty inherent to any type of long-term planning. A company's vulnerability to different sources of risks and disruptions (supplier-, market- or technology based) depends on its corporate structure, financial situation and existing supply chain configuration and processes. Risk aversion (or lack thereof) also influences most strategic decision-making and should also be taken into account. However, the scientific literature is rather scarce on this topic, and more contributions are needed.

Decisions involved in strategic supply chain planning typically involve senior executives up to the CEO and are implemented across a company's entire value chain.

17.2.2 Tactical Planning (or S&OP)

Following strategic planning, the next level in the hierarchical planning structure is tactical or mid-term planning. Tactical planning is slightly different depending on whether a forest management problem or a production/distribution planning problem is being addressed. In forest management, hierarchical planning approaches are widely implemented as they permit the tactical planning problem to be initially addressed without taking spatial issues into account. Once this has been done, the problem is then tightly constrained spatially. While strategic forest management planning problems generally span 100 years, tactical planning problems are often reviewed annually over a 5-year planning period.

In planning problems dealing with production/distribution issues, tactical planning normally addresses the allocation rules that define which unit or group of units is responsible for executing the different supply chain activities or what resources or group of resources will be used. It also sets the rules in terms of production/distribution lead times, lot sizing and inventory policies. Tactical planning allows these two types of rules to be defined through a global analysis of the supply chain. Tactical planning also serves as a bridge between the long-term comprehensive strategic planning and the short-term detailed operational planning that has a direct influence on the actual operations in the chain (e.g., truck routing, production schedules). Tactical planning should also ensure that the subsequent operational planning conforms to the directives established during the strategic planning stage, even though the planning horizon is much shorter. Other typical tactical decisions concern allocating customers to mills and defining the necessary distribution capacity. The advanced planning required for distribution depends on the transportation mode. For example, ship and rail transportation typically need to be planned earlier than truck transportation.

Another important reason for tactical planning is tied to the seasonality of the supply chain, which increases the need for advance planning. Seasonality has a great influence on the procurement stage (i.e., the outbound flow of wood fiber from the forests). One reason for this seasonality is the shifting weather conditions throughout the year, which can make it impossible to transport logs/chips during certain periods due to a lack of carrying capacity on forest roads caused by the spring thaw, for example. In many areas of the world, seasonality also plays a role in harvesting operations. In the Nordic countries, for example, a relatively small proportion of the annual harvesting is done during the summer period (July–August). During this period, operations are instead focused on silvicultural management, including regeneration and cleaning activities. A large proportion of the wood is harvested during the winter when the ground is frozen, thus reducing the risk of damage while forwarding (or moving) the logs out of the forest. Seasonality can also affect the production stage (e.g., in Nordic countries, hardwood drying times can vary over the season) or the demand process (e.g., again in Nordic countries, most construction projects are not conducted during the winter period).

An example of a tactical planning task is production scheduling of pulp mills with regard to wood availability and distribution. The time horizon may vary in this planning between 6 and 12 months. It is crucial to account for the period of the year when wood availability is scarce in order to ensure a supply of certain assortments. In a chemical pulp mill, the cost of changing from one product to another is relatively high; therefore the number of product transitions is not that large (typically somewhere in the range of 6–24 per year). This makes it important to account for the scheduling of the production already at the tactical level. When it comes to paper, production transition costs are normally considerably smaller. This is why the scheduling sometimes is not explicitly accounted for at the tactical level. The purpose of this plan is to define guidelines on monthly levels for subsequent short-term planning. The wood supply department is given target volumes of different assortments to deliver to different mills the following month. This planning task stretches over several stages of the chain: procurement, production and distribution. There are also tactical planning tasks which are more restricted to one of the stages. In the procurement stage, for example, one planning task is to define catchment areas for the supply to different mills. A catchment area denotes the geographical area in which the wood from different harvesting locations is hauled to a certain mill. Often there is more than one mill requiring the same assortment. The problem in this case is normally to ensure that all mills are sufficiently supplied while the total transport cost is minimized.

In most companies, an important task is planning the annual budget for the following year. During this exercise, the company decides which products to offer to customers and in what quantities. In the process of elaborating these decisions, their implications for the whole supply chain (procurement, production and distribution) need to be evaluated and net profit maximization should be the aim. In that regard, Shapiro (2007) suggests that the tactical planning models be derived from the strategic planning models where the 0–1 variables related to the strategic decisions are fixed and the planning horizon is extended to a multi-period (multi-seasonal) horizon.

Management Planning in Action 17.1: Supplier Managed Inventory (SMI) at Södra Cell

Södra Cell is one of the largest pulp producers in the world with five pulp mills in Sweden and Norway. Its supply chain stretches from the forests where the wood is harvested, to the paper mill where different kinds of pulp may be mixed together to produce a certain paper product. The terminals, sea and land terminals, used for distributing pulp to customers in Europe are contracted from terminal operators. The demand for pulp products is mainly located across Europe including domestic demand in Sweden and Norway, see figure below. In the figure, terminals are located in the left and customers in the right part.

(continued)

(continued)

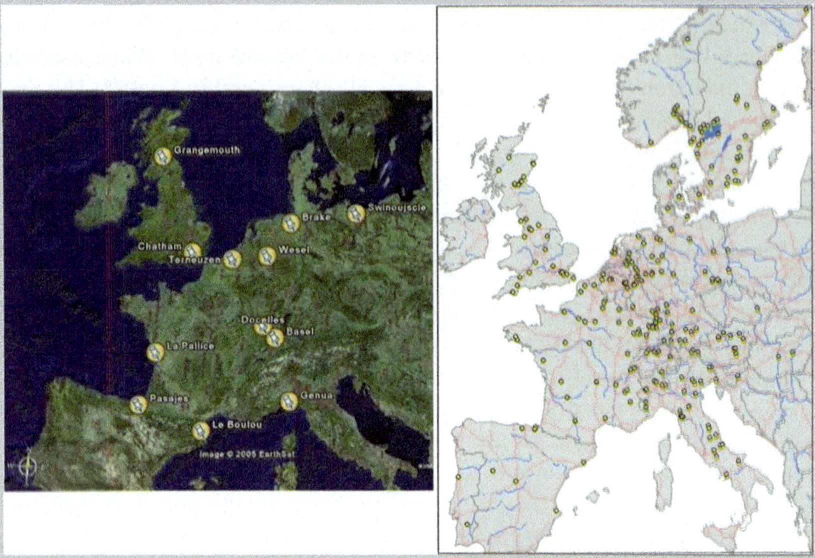

Södra Cell is dealing with roughly 150 customers in total. The distribution from pulp mills to customers use several transport modes including long term chartered vessels, train, barges, spot contracted ships, and trucks. Customers normally delivered through terminals can as an alternative also be delivered directly by trains or trucks, often to a higher cost however. Roughly 60 % of the delivered volume is transported by vessel and through terminals. These vessels trips may take up to 28 days to complete so planning ahead for the inventory is critical. Several levels of planning is conducted on strategical, tactical and operational planning. Distribution planning is done on an up to 3 month rotational planning horizon. Based on the anticipated deliveries to customers and the current production plan the distribution plan identifies the need for resupplying terminals. Direct deliveries from mills are planned in parallel with the routing of the long chartered vessels and ordering of spot vessels in order not to run out of pulp at the mill stock. This plan is more or less constantly reviewed with respect to e.g. actual production outcome and deliveries. The European pulp market used to be very traditional in that the sales term was CIF Northern European port where customers took over all responsibility for hauling the pulp to their paper mill. This is illustrated in the figure below. This often results in long lead times, often up to 60 days, between production at mills and the consumptions at customers.

(continued)

(continued)

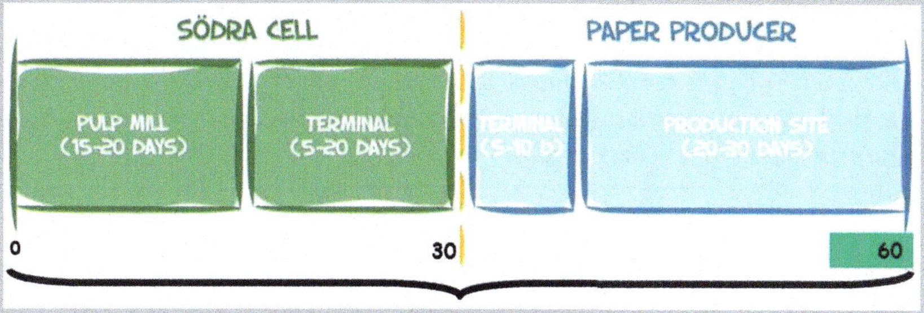

During the last 10 years this has totally changed to a situation where pulp suppliers takes care of the delivery to the final destination in the majority of cases. A concept that Södra Cell offered customers was Supplier Managed Inventory (SMI) where Södra Cell take full responsibility for all inventories including on customer's own site, see figure below.

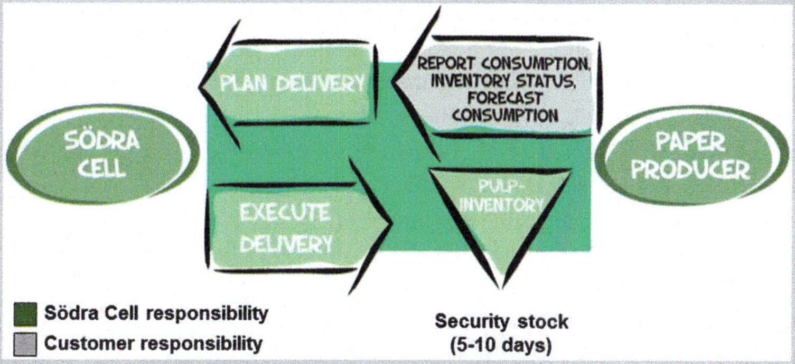

With SMI, Södra Cell as supplier, is responsible for having enough pulp available at the paper mill for the planned production. It is Södra Cell who does the delivery planning based on forecasted usage of pulp at the specific customer site. It is required that the customer updates the forecast every week for a number of weeks ahead. Based on a usage report filed every week, Södra Cell invoices the Customer for the consumed pulp. Södra Cell uses a strategic model for finding the optimal terminal structure (Gunnarsson et al. 2006), which partly defines the ability to offer good delivery service in terms of short notice deliveries etc. The introduction of SMI has several benefits for Södra Cell. It provides a more steady demand and hence it is possible to better make the production planning. Also, the contracts also provides the sales value which means that cash-flows then can be better forecasted. Södra Cell is today running about $^2/_3$ of their volume on SMI contracts and this

(continued)

(continued)

is believed to be a good level. A higher level gives less flexibility when the demand fluctuates too much among the SMI customers. One crucial issue is however to define the amount of pulp that should be kept in security stock at the terminals in order to cope with the uncertain demands This problem is further studied in Flisberg et al. (2012).

17.2.3 Operational Planning

The third level of planning is operational or short-term planning, which is the planning that precedes and determines real-world operations. For this reason, this planning process must adequately reflect the detailed reality in which the operations take place. The precise timing of operations is crucial. It is generally not enough to know the week or month that a certain action should take place; the time period must be defined in terms of days or hours. Operational planning is usually distributed among the different facilities, or units in the facilities, due to the enormous quantity of data that has to be manipulated at the operational level (e.g., number of Stock Keeping Units (SKU) and other specific resources).

The third level of planning is short-term or operative planning, which is the planning that precedes and decides real-world operative actions. Because of that, there are very high demands on this planning to adequately reflect in detail the reality in which the operations take place. The precise timing of operations is crucial. It is normally not adequate to know which week or month a certain action should be taken; it has to be defined in terms of days or hours. The operative planning is normally distributed to the different facilities or cells of the facilities because of the enormous quantity of data that needs to be manipulated at that level (e.g. SKU and specific resources).

One operative problem is the roll cutting problem in paper mills. Once the reel has been produced in the paper machine it must be cut into the rolls demanded by internal and external customers. The reel may be 5–10 m wide and 30 km long. The customer orders are for products that may be 0.5–1.0 m wide and 5 km long. The problem is deciding on cutting patterns, and the number of each, in order to satisfy customer order while minimizing the number of reels required during a given period of time.

Scheduling the different products moving through the manufacturing lines is also a typical operational planning problem, as is process control involving real-time operational planning decisions. Process control is particularly critical in the pulp and paper industry as the characteristics of the output products depend greatly on the precision of the chemical-fiber mix. Another type of operational planning problem deals with the problems related to transportation, specifically the routing and dispatching done at several points in the supply chain. For instance, it is necessary to route the trucks used for hauling wood from the forest to the mills or for shipping finished products from mills to customers or distribution centers.

Table 17.1 presents an illustrative summary of the strategic, tactical and operational planning decisions needed in the pulp and paper industry. This supply chain planning matrix was proposed by Carlsson et al. (2009) based on a series of case studies conducted with the Swedish pulp producer Södra Cell (Carlsson and Rönnqvist 2005).

The next sections will discuss challenging issues in decision making within the P&P supply chain. These relate to supply chain design, sales and operations planning, cross-chain coordination and finally, transformation of the pulp and paper business model toward a biorefinery business model.

17.3 Supply Chain Strategic Design in the Pulp and Paper Industry

Strategic planning and supply chain design deals with the physical and organizational structure of the pulp and paper production and distribution network. The decisions at this level concern the opening or closing of facilities such as pulp and/or paper mills, the technologies to implement in each mills, the acquisition of transportation capacity by mode (truck, rail, ships, etc.), the development of new products as well as determining what customer zone to serve from each facility.

Supply chain design is a complex endeavour in the pulp and paper industry for a number of reasons. As the industry is very capital-intensive, substantial long-term planning is critical to achieving sustainable competitive advantage. Furthermore, in order for the industry to be effective, the pulp and paper supply chain must be coordinated with other forest value chains which use different parts of logs and trees: forest products (lumber and engineered wood), panels, and bioenergy, among others. Finally, the pulp and paper has seen the emergence of major international companies through series of mergers and acquisitions. These large companies yield potential for more economies of scale at the cost of more complex planning.

Multiple views of the concept of supply chain design can be found in the literature, which can result in some confusion and ambiguity. Three are particularly relevant:

1. A single-mill view consisting of a single plant, its direct suppliers and customers (whether these are internal or external) and the product, financial and information flows between these units. This type of supply chain is found in Weigel et al. (2009);
2. A multi-site intra-organizational supply chain consisting of all the facilities owned or operated by a company, as well as product, financial and information flows between its facilities and those of its suppliers and customers. Despite the complexity of designing a supply chain with an integrated perspective, there are significant benefits in doing so Eksioglu et al. (2009). An example of such a supply chain is modeled in Gunnarsson et al. (2006);

Table 17.1 Supply chain planning matrix for the pulp and paper industry

	Supply	Pulp and paper making	Distribution	Sales
Strategic	Wood procurement strategy (private vs public land)	Location decisions	Warehouse location	Selection of markets (e.g., location, segment)
	Forest land acquisitions and harvesting contracts	Outsourcing decisions	Allocation of markets/ customers to warehouses	Customer segmentation
	Silvicultural regime and regeneration strategies	Technology and capacity investments	Logistics resource investments (e.g., warehouses, handling)	Product-solution portfolio
	Harvesting and transportation technology and capacity investment	Allocation of product families to facilities	Contracts with logistics providers	Pricing strategy
	Transportation and investment strategies (e.g., roads, construction, trucks, wagons, terminals, ships)	Order penetration point strategy Investments in information technology and planning systems (e.g., advance planning and scheduling technologies, ships)	Investments in information technology and planning systems (e.g., warehouse execution)	Service strategy Contracts Investments in information technology and planning systems (e.g., On-line tracking systems, CRM)
Tactical	Sourcing plan (log class planning)	Campaign duration	Warehouse management policies (e.g., dock management)	Aggregate demand planning per segment
	Aggregate harvesting planning	Product sequencing during the campaigns	Seasonal inventory target at DCs	Customer contracts
	Route definition and transshipment yard location and planning	Lot-sizing	Routing (Ship, train and truck)	Demand forecasting, safety stocks
	Allocation of harvesting and transportation equipment to cutting blocks	Outsourcing planning	3PL contracts	Available to promise aggregate need and planning

	Allocation of products/blocks to mills	Seasonal inventory target		Available to promise allocation rules (including rationing rules and substitution rules)
	Yard layout design	Parent roll assortment optimization		Allocation of products and customers to mills and DCs
	Log yard management policies	Temporary mill shutdowns		
Operational	Detailed log supply planning	Daily production plans for pulp mills/paper machines/winders/sheeters	Warehouse/DC inventory management	Available to promise consumption
	Forest to mill: daily carrier selection and routing	Mill to converter/DC/customer: daily carrier selection and routing	DC to customer: daily carrier selection and routing	Rationing
		Roll-cutting	Vehicle loads	Online ordering
		Process control		Customer inventory management and replenishment

Carlsson et al. (2009)
DC Distribution Center

3. An extended supply chain consisting of all the suppliers, manufacturers, distributors and customers participating in a value chain. Since several components of the value chain belong to different companies, achieving an integrated supply chain design is very difficult. Game theory concepts are often needed to model these supply chains.

Several modeling approaches can be used to formulate the supply chain network design problem. The following sections describe alternative approaches and their characteristics.

17.3.1 Distribution Network Design Models

The simplest supply chain design models available date back to the 1950s and focus on either facility location, distribution system design or allocation of customers to warehouses or facilities. When these models cover a single type of facility in the supply chain, (e.g. DCs), they are labeled as single-echelon network design problems; multi-echelon problems cover multiple facility types. Several variants of this model have been proposed in the literature. The objective function minimizes the sum of fixed facility opening/closing costs and variable production costs. Medium-sized models are relatively easy to solve using commercial solvers, while large-scale models may require more advanced techniques such as Benders' decomposition (Geoffrion and Graves 1974).

Despite their simplicity, location-allocation models have been seldom applied to the pulp and paper industry as their simple nature fail to capture several key elements of the P&P supply chain, namely divergent production processes and alternative product recipes. More advanced and industry-specific models and applications need to be developed. For example, Carlsson and Rönnqvist (2005) propose a model combining terminal location and ship routing for designing the kraft pulp distribution system of Södra Cell AB across Europe.

17.3.2 Production-Distribution Network Design Models

In addition to facility location and allocation, production-distribution network design models cover additional decision types. In pulping and paper manufacturing, different combinations of products and grades can be obtained from the inputs of each process. In such contexts, associated with each activity are a number of methods (recipes) that describe how inputs are transformed into outputs using different potential technologies (Philpott and Everett 2001). Typical decision variables are annual throughputs for a given facility using a specific recipe-technology combination. Weigel et al. (2009) proposed a design model for pulp and paper plants with multiple fibre suppliers, technologies and recipes, while Vila et al. (2006)

and M'Barek et al. (2010) proposes supply chain design models for divergent and process industries, respectively.

Another important inclusion is the addition of different capacity expansion options on facilities. A production *technology* is defined by the set of products it can manufacture as well as the recipes it can use. While some models use fixed site configurations, a more flexible approach is to consider alternative production technologies and/or capacity levels that can be installed into a facility. Some models go further and do not impose any *a priori* echelon structure and expect the optimization model to determine the best structure and mission for the facilities (Paquet et al. 2004). For example, in addition to increased capacity, a given pulp mill could be reconfigured so as to include capacity for paper production and thus becoming an integrated mill. These types of models are quite useful to asses bioenergy or co-generation technologies which can be used to transform by-products into energy sources for other production technologies. This approach is taken in Feng et al. (2013).

Although several applications consider a single period, usually covering a planning horizon of 1 year, multi-period and multi-season models also have been proposed. According to Martel (2005), multi-season models anticipate variations in demand and activity levels during a planning horizon (relevant for modeling harvesting activities), whereas multi-period models consider several supply chain design cycles over a long-term horizon. An integrated multi-season model for process industries is found in Vila et al. (2006), while a multi-period model is proposed in Everett et al. (2001).

Although several companies are vertically integrated and thus control all their supply chain from harvesting to product distribution, planning of the forest and manufacturing supply chains is typically done in a decoupled approach, without much regard for one another (Gunn 2009). Integrating forest management and supply chain design at the strategic level is one of the most promising areas for additional research. For example, Troncoso et al. (2011) compare the benefits of planning the industrial and forest value chains in a decoupled or integrated way; their results show a significant increase in NPV of the forest as well as profit increases of up to 8 % when forest and industrial planning is integrated.

17.3.3 Financial Considerations and Value Creation

Another crucial decision is whether to build a model that maximizes profitability or minimizes costs. In the context of supply chain design, these choices are often far from being equivalent. A cost minimization model assumes that a certain demand level for each product-market must be met, and seeks to produce and deliver these volumes at minimal cost. This approach is clearly well suited to problems where demand and product mix choices are fixed. An example of such is Gunnarsson et al. (2006), which proposes a model for selecting terminal locations combined with transport mode selection and ship routing in pulp distribution in Europe.

When considering all the supply chain, cost minimization does not necessary lead to increased profitability (Martel 2005). A profit-maximization model allows for additional considerations such as assessing which product mix would result in greater net income for the supply chain over a given year. This approach is actually used by Philpott and Everett (2001), Everett et al. (2001), who also include additional constraints on maximum capital expenditures per period. Weigel et al. (2009) presents an industrial case in which increased wood sorting activities lead to higher fiber quality input at the mill, which leads to increased profits, a conclusion that could not have been reached with a cost minimization model. Several metrics for value maximization have been proposed in the literature. While Everett et al. (2001) maximize discounted annual net earnings, other approaches use the net present value (NPV) (Feng et al. 2013) or economic value added (M'Barek et al. 2010) metrics.

As several pulp and paper supply chains operate at the multinational scale, several factors such as national taxation levels, transfer price options and regulations, environmental regulations, trade tariffs and currency exchange rates weigh heavily on a company's profitability. Adding these features however considerably increase the complexity of the supply chain design model. A detailed discussion on how to include international considerations into forest products supply chains when transfer prices are set can be found in Vila et al. (2006) and M'Barek et al. (2010). A bi-level optimization model to optimize transfer prices is presented in Vidal and Goetschalckx (2001), while Arntzen et al. (1995) discuss how to model duty avoidance strategies.

17.3.4 Uncertainty Modeling

The pulp and paper sector faces several sources of uncertainty: fibre quality and supply levels, product demand and prices, to name a few; international supply chains face an even higher number of risk factors, such as volatility of exchange rates. Even a supply chain design provided by the optimal solution of a deterministic optimization model may prove to be underperforming when supply or markets conditions change drastically. Weigel et al. (2009) conduct scenario analysis in which the impact of different fiber procurement costs and sales prices on profit is assessed. Another alternative is to generate scenarios beforehand and solve the model through a sample average approximation (SAA) based stochastic programming approach; an example of such approach is provided in Vila et al. (2007). Klibi and Martel (2011) propose a three-phase supply chain design methodology advocating the use of a mixed-integer model to generate a design and then evaluate this design through other tools such as simulations. When used on large, international supply chains, these methodologies usually result in very large mixed-integer programming models that are excessively difficult to solve. Alternate approaches such as Benders decomposition, progressive hedging or heuristics are often required.

17.4 Sales and Operations Planning in the Pulp and Paper Supply Chain

The key role of tactical planning in the pulp and paper supply chain is to balance supply with demand across the whole supply chain. Existing contributions will be reviewed, then we position this problem under the sales and operations planning (S&OP) paradigm.

17.4.1 Existing Tactical Planning Models

Several models have been proposed in the literature to solve tactical planning. Bredström et al. (2004) proposed a model covering both wood procurement, production planning of pulp mills and distribution up to aggregated demand sinks at Södra Cell. Both column generation and branch-and-bound techniques are used to solve the model.

Another key tactical decision relates to planning the production campaigns. This planning actually involves three interrelated decisions: determining the length of a campaign, the sequence of the products in the campaign, and finally deciding the lot sizes of each product in each campaign sequence. Bouchria et al. (2007) used a three-phase approach to tackle this problem.

Chauhan et al. (2008) deals with tactical demand fulfillment of sheeted paper in the paper industry. A sheet-to-order policy is used, in which parent rolls are produced to stock and sheeting operations are scheduled according to actual customer orders rather than demand forecasts. The authors propose a model to determine the assortment of parent rolls to keep in stock in order to minimize inventory and trim loss costs. The resulting nonlinear model was solved through two different methods, namely column generation and local search heuristics.

Another key problem arising at the tactical level is the coordinated wood procurement problem. This problem is actually comprised of several decision categories. Carlsson and Rönnqvist (2007) determine wood catchment areas (that is, an area for which specific wood assortments are delivered to certain mills) for different plants in a given region. These decisions affect the potential of efficient routing, including backhauling opportunities, and the model proposed by Carlsson and Rönnqvist (2007) accounts for those possibilities. The resulting model is solved through column generation. A similar wood procurement problem arising in a multi-company context was proposed by Beaudoin et al. (2007). The material flows in the supply chain are driven by both push and pull strategies and the models aims to maximize the value extracted from logs processed into the mills while preserving its freshness.

17.4.2 Sales and Operations Planning (S&OP)

There are significant advantages in planning wood procurement, production and sales centrally through the use of mathematical programming models. Sales and operations planning (S&OP) seems to be one of the most appropriate methods to achieve such coordination between procurement, higher service levels while cutting through costs. According to Feng et al. (2010), S&OP refers to an approach that integrates the cross-functional planning of sales, production, distribution, and procurement centrally. Ouhimmou et al. (2009) proposed an integrated tactical supply chain planning that was applied at Shermag inc., a wood furniture company located in Canada. The comparative study between the company's historical planning process and the proposed approach over a 2-year horizon showed that significant cost reductions of approximately 22 % could be achieved in comparison to a human-made decoupled planning approach.

S&OP can be further enhanced to incorporate revenue management principles, such as allocation of certain product volumes for specific classes of customers, in order to make sure that the least profitable customers are not serviced while the most profitable customers are not. To the best of our knowledge, no record of applying S&OP to the pulp and paper industry has been proposed in the literature. However, other studies have been performed in the lumber and oriented strand board (OSB) industries.

Feng et al. (2010) study the performance impact of three tactical planning systems: decoupled planning, partially and fully integrated S&OP. Partial integration refers to the integration of production and sales but with the procurement and distribution steps being decoupled, whereas all four steps are integrated in the fully integrated S&OP process. The authors propose mathematical programming models for each planning system and the performance of those is assessed through a rolling horizon-based simulation environment. While partial integration results in a profit increase of 3.5 % over the decoupled planning approach, the fully-integrated S&OP approach yields an additional 1 % profitability over the partial integration.

17.5 Cross Chains Coordination Within the Pulp and Paper Industry

Cross chains coordination occurs when different organizations decide to either drive decision or execution processes together by exchanging information or resources. Because supply chain are composed of different organizations, the collaborative processes supported by cross chains coordination play a critical role in achieving high supply chain performances. This may explain why, cross chains coordination is attracting high attention as it is viewed as a way to reduce logistics costs and to increase revenues. Examples of collaboration within the pulp and paper supply chain are numerous. For example, organizations have collaborate in regard for chips and logs supply, transportation, outsourcing of different production, converting and distribution stages.

Cross chains coordination may be set through contracts and supported by simple or sophisticated information exchange platforms (e.g. Internet based platforms). Third parties may be asked to coordinate the activities between the different organizations in order to set a fair and secure way to share the information and the resources.

These new collaborative approaches request the decision support systems to be able to deal with distributed and restricted information and to be able to propose collaborative solutions to all participants. This may imply that the "optimal solution" of integrating business decisions and activities is not feasible because some of the collaborating organization would not gain their fare share. In such cases, benefit sharing schemes are required. These are difficult to find, as they involve a good understanding of advanced economic and social theories. Although collaboration issues are tightly linked to any discussion on supply chains, it is only recently that OR has been used to evaluate the potential of collaboration for the forest products industry. Here are short descriptions of some of the work.

Given that many companies obtain their wood allocations from unevenly aged forests owned by the state, they often need to agree on a common in-forest harvesting plan. Beaudoin et al. (2007) addressed this problem proposing collaborative approaches to help the negotiation process converge on a profitable balanced solution. In their article, they first propose a planning approach to help each company establish its own optimal plan for several different scenarios. Then, they illustrate the value of collaboration for determining a final harvesting schedule.

The benefits of collaboration have also been explored in the context of transporting logs to mills. Often, many companies operate in different parts of the country, which provides opportunities for optimizing backhauling operations. This opportunity has been addressed in different parts of the world, using the specific wood allocation and trucking constraints found in each region. Frisk et al. (2010) (Sweden), Palander and Vaatainen (2005) (Finland), and Audy et al. (2012) (Canada) have all worked on different versions of this problem. They have also proposed models for sharing risks and benefits.

Finally, collaboration between paper mills and customers has been explored by Lehoux et al. (2009). Four different approaches to integration were simulated and optimized, starting with the traditional make-to-order, then moving toward continuous replenishment, vendor-managed inventory (VMI) and finally Collaborative Planning Forecasting and Replenishment (CPFR). Of all the tested scenarios, CPFR showed the greatest overall benefit. However, under certain economic conditions, customers may obtain a greater benefit from a continuous replenishment approach, while producers still obtain a greater benefit from the CPFR approach.

17.6 Extending the Pulp and Paper Supply Chain toward a Biorefinery Supply Chain

With the rapid development of bioenergy and biorefinery systems, an increasing number of publications have appeared in the area of bioenergy and biorefinery supply chain design. In bioenergy supply chain design, most of the contributions are focused on biomass supply chain optimization from harvesting areas to bioenergy plants with or without intermediate terminals determining plant locations, allocations, and capacities. In this vein, Kumar et al. (2003) analyzed the optimum plant size and power cost for power plants using three biomass fuels, agricultural residues (grain straw), whole boreal forest trees (logs, tops, and limbs), and forest harvesting residues (tops and limbs) respectively in western Canada (Alberta). Freppaz et al. (2004) developed a decision-support system (DSS) for forest biomass systems in an effort to find suitable plant locations and sizes as well as optimal supply areas within a region. Gunnarsson et al. (2004) presented an *MIP* model for the biomass supply chain problem which determined the optimum sourcing, transportation, and allocation of forest and sawmill residues to satisfy the demand at various known heating-plant locations in Sweden. Gronalt and Rauch (2007) proposed a simple stepwise heuristic approach for a regional woody biomass supply network design problem in which biomass was supplied from several forest areas to a number of energy plants in Austria. Kanzian et al. (2009) developed a deterministic *MIP* model for a regional woody biomass supply network from the forest to heating plants in Austria, including the optional use of intermediate terminals. Frombo et al. (2009) presented an *LP*-based strategic planning model to determine the optimal annual harvesting quantities and suitable power-plant capacities with different thermochemical conversion technologies at predetermined plant locations. Flisberg et al. (2012) develops a model to satisfy a number of heating plants with forest biomass for energy production. This model includes selection of transformation equipment at forest areas. Transportation can be done by multiple transportation modes including different truck types and train.

The integrated forest biorefinery (IFBR) paradigm also enable various business models, from the establishment of supply chains dedicated to bioenergy markets to the possibilities of integrating biorefinery activities to existing pulp-and-paper facilities and supply chains (Wising and Stuart 2006). Eksioglu et al. (2009) presented an *MIP* model for designing the biorefinery supply chain by determining the number, size, and location of the biorefineries needed to produce cellulosic ethanol using corn stover and woody biomass. Parker et al. (2010) evaluated the infrastructure requirements of hydrogen production from agricultural residues. A mixed-integer nonlinear programming model was developed to find the most efficient and economical configuration for the supply chain pathway.

In earlier studies, supply chain design problems have focused on the design of facility and distribution networks serving primary products within a single

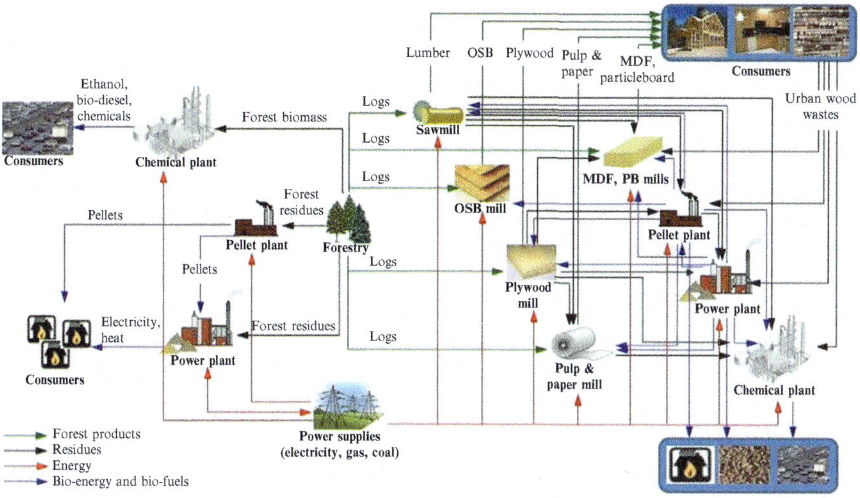

Fig. 17.5 Integrated bio-refinery and forest product supply chain framework

industrial environment. Taking biorefinery supply chain design as an example, the biorefineries are generally designed as standalone facilities with potential biomass supply sources and product markets (Eksioglu et al. 2009; Parker et al. 2010). A similar approach has been taken to bioenergy supply chain design problems (Kumar et al. 2003; Freppaz et al. 2004; Gunnarsson et al. 2004, Gronalt and Rauch 2007; Kanzian et al. 2009; Frombo et al. 2009). These supply chain design problems can be represented by the supply chain illustrated on the left-hand side of Fig. 17.5., while the IBRF supply chain is depicted on the right-hand side of Fig. 17.5. Because in biorefinery supply chain systems, performance depends strongly on the cost, consistency, and efficiency of the biomass supply, coordinating biomass supply and biorefinery manufacturing offers significant opportunities for effective biomass utilization, cost reduction, and value chain optimization. In this regard, techno-economic methods can and should be used to identify the most promising sets of biomass-product-process combinations to achieve maximum profitability (Ghezzaz and Stuart 2011). Machani et al. (2013) propose a multi-period supply chain design model in which biorefinery investment options are included in a pulp-and-paper supply chain. Feng et al. (2013) propose an integrated biorefinery supply chain design involving multiple facilities, production technologies as well as sources of biomass. Modeling the impacts on the entire supply chain is critical, as the rapid development of IFBR initiatives and facilities may create additional pressure on raw material supply volumes and prices for other components of the forest products industry that are used by bioenergy facilities and panel, pulp and paper mills.

17.7 Summary

This chapter proposed a description of the pulp and paper supply chain, with an emphasis on the special characteristics that make it different from most other contexts and industries. It described the key decisions and planning problems arising in managing the pulp and paper chain; these were classified according to the decision level (strategic, tactical or operational) as well as the scope of the supply chain covered by each problem. For each decision level, key contributions from the literature are reviewed and discussed. Where no specific contributions from the pulp-and-paper literature exist, we review more general papers from either the forest products or general divergent process industries. The fact that the literature on supply chain management in the pulp and paper industry is not abundant indicates a need for additional research. While there have been more papers integrating risk and uncertainty into account, there is still a lot of ground to cover in this area, especially at the strategic and tactical levels. While the sources of uncertainty affecting the pulp and paper supply chain are known, there have been very few methodological studies characterizing these uncertainties and exactly how they affect the chain. We expect to see more contributions to the field of pulp and paper supply chain management under uncertainty in the near future.

17.8 Problems

1. What would be key to consider when answering the following questions:
2. Should woody biomass and process residues be used for bioproducts or forest products manufacturing? What are the best tradeoffs and the optimal mix, and which market demands should be served?
3. What manufacturing facilities (e.g., power, pellets, biochemical/thermochemical plant) and how many of them should an organization use to satisfy its target market demand, and where should they be located?
4. Should different manufacturing facilities be built separately or partially or fully integrated?
5. Should these manufacturing facilities be built as greenfield plants or at an existing mill site, integrated with existing manufacturing activities such as a pulp and paper mill, sawmill, or wood composite mill to produce a new product mix using mill residues without additional transportation cost?
6. What technologies and at what capacities should be installed?
7. What products and in what quantities should each manufacturing facility be capable of producing?
8. Which raw material sources (or suppliers) should be used, and what should be the raw material allocations?

References

Arntzen B, Brown G, Harrison T, Trafton L (1995) Global supply chain management at digital equipment corporation. Interfaces 21(1):69–93

Audy J-F, D'Amours S, Rönnqvist M (2012) An empirical study on coalition formation and cost/savings allocation. Int J Prod Econ 136:13–27

Beaudoin D, LeBel L, Frayret J-M (2007) Tactical supply chain planning in the forest products industry through optimization and scenario-based analysis. Can J Forest Res 37(1):128–140

Bouchria H, Ouhimmou M, D'Amours S (2007) Lot sizing problem on a paper machine under cyclic production approach. Int J Prod Econ 105(2):318–328

Bredström D, Lundgren JT, Rönnqvist M, Carlsson D, Mason A (2004) Supply chain optimization in the pulp mill industry – IP models, column generation and novel constraint branches. Eur J Oper Res 156:2–22

Carlsson D, Rönnqvist M (2005) Supply chain management in forestry – case studies at Södra Cell AB. Eur J Oper Res 163:589–616

Carlsson D, Rönnqvist M (2007) Backhauling in forest transportation – models, methods and practical usage. Can J Forest Res 37:2612–2623

Carlsson D, D'Amours S, Martel A, Rönnqvist M (2009) Supply chain planning models in the pulp and paper industry. INFOR 47(3):167–183

Chauhan SS, Martel A, D'Amours S (2008) Roll assortment optimization in a paper mill: an integer programming approach. Comput Oper Res 35:614–627

Eksioglu SD, Acharya A, Leightley LE, Arora S (2009) Analyzing the design and management of biomass-to-biorefinery supply chain. Comput Indus Eng 57:1342–1352

Everett G, Aoude S, Philpott A (2001) Capital planning in the paper industry using COMPASS. In: Proceedings of the 36th conference of ORSNZ, University of Canterbury, Christchurch, New Zealand, 28th Nov–1st Dec 2001

Feng Y, D'Amours S, Beauregard R (2010) Simulation and performance evaluation of partially and fully integrated sales and operations planning. Int J Prod Res 48(19):5859–5883

Feng Y, D'Amours S, LeBel L, Nourelfath M (2013) Integrated forest biorefinery supply chain network design using mathematical programming approach. In: Stuart PR, El-Halwagi MM (eds) Integrated biorefineries: design, analysis and optimization. CRC Press, Boca Raton, pp 251–282

Fleischmann B, Meyr H, Wagner M (2008) Advanced planning. In: Stadtler H, Kilger C (eds) Supply chain management and advanced planning: concepts, models, software and case studies, 4th edn. Springer, Berlin, pp 81–106

Flisberg P, Frisk M, Rönnqvist M (2012) FuelOpt: a decision support system for forest fuel logistics. J Oper Res Soc 63:1600–1612

Freppaz D, Minciardi R, Robba M, Rovatti M, Sacile R, Taramasso A (2004) Optimizing forest biomass exploitation for energy supply at a regional level. Biomass Bioenergy 26(1):15–25

Frisk M, Göthe-Lundgren M, Jörnsten K, Rönnqvist M (2010) Cost allocation in collaborative forest transportation. Eur J Oper Res 205:448–458

Frombo F, Minciardi R, Robba M, Rosso F, Sacile R (2009) Planning woody biomass logistics for energy production: a strategic decision model. Biomass Bioenergy 33:372–383

Geoffrion A, Graves G (1974) Multicommodity distribution system design by benders decomposition. Manag Sci 20:822–844

Ghezzaz H, Stuart PR (2011) Biomass availability and process selection for an integrated forest biorefinery. Pulp Paper Canada 112(3):19–26

Gronalt M, Rauch P (2007) Designing a regional forest fuel supply network. Biomass and Bioenergy 31:393–402

Gunn E (2009) Some perspectives on strategic forest management models and the forest products supply chain. INFOR 47(3):261–272

Gunnarsson H, Rönnqvist M, Lundgren JT (2004) Supply chain modeling of forest fuel. Eur J Oper Res 158:103–123

Gunnarsson H, Rönnqvist M, Carlsson D (2006) A combined terminal location and ship routing problem. J Oper Res Soc 57:928–938
Kanzian C, Holzleitner F, Stampfer K, Ashton S (2009) Regional energy wood logistics – optimizing local fuel supply. Silva Fenn 43(1):113–128
Klibi W, Martel A (2011) Scenario-based supply chain network risk modeling. CIRRELT working papers, 2011-30
Kumar A, Cameron JB, Flynn PC (2003) Biomass power cost and optimum plant size in western Canada. Biomass Bioenergy 24:445–464
Lehoux N, D'Amours S, Langevin A (2009) Collaboration and decision models for a two-echelon supply chain: a case study in the pulp and paper industry. J Oper Logist 2(4):1–17
M'Barek W, Martel A, D'Amours S (2010) Designing multinational value-creating supply chain networks for the process industry. CIRRELT working papers, 2010-51
Machani M, Nourelfath M, D'Amours S (2013) A decision-support tool for evaluating the technical and economic potential of integrating bioenergy production within pulp and paper mills. CIRRELT working papers, 2013-07
Martel A (2005) The design of production-distribution networks: a mathematical programming approach. In: Geunes J, Pardalos PM (eds) Supply chain optimization. Springer, New York, pp 265–306
Martel A, M'Barek W, D'Amours S (2005) International factors in the design of multinational supply chains: the case of Canadian pulp and paper companies. CIRRELT working papers, 2005-AM-3
Ouhimmou M, D'Amours S, Beauregard R, Ait-Kadi D, Chauhan SS (2009) Optimization helps Shermag gain competitive edge. Interfaces 39(4):329–345
Palander T, Väätäinen J (2005) Impacts of interenterprise collaboration and backhauling on wood procurement in Finland. Scand J For Res 20(2):177–183
Paquet M, Martel A, Desaulniers G (2004) Including technology selection decisions in manufacturing network design models. Int J Comput Integr Manuf 17:117–125
Parker N, Fan Y, Ogden J (2010) From waste to hydrogen: an optimal design of energy production and distribution network. Transp Res Part E: Logist Transp Rev 46(4):534–545
Philpott A, Everett G (2001) Supply chain optimisation in the paper industry. Ann Oper Res 108(1):225–237
Shapiro JF (2007) Modeling the supply chain, 2nd edn. Thomson, Belmont
Troncoso J, D'Amours S, Flisberg P, Rönnqvist M, Weintraub A (2011) A mixed integer programming model to evaluate integrating strategies in the forest value chain – a case study in the Chilean forest industry. CIRRELT working papers, 2011-28
Vidal C, Goetschalckx M (2001) A global supply chain model with transfer pricing and transportation cost allocation. Eur J Oper Res 129:134–158
Vila D, Martel A, Beauregard R (2006) Designing logistics networks in divergent process industries: a methodology and its application to the lumber industry. Int J Prod Econ 102: 358–378
Vila D, Martel A, Beauregard R (2007) Taking market forces into account in the design of production-distribution networks: a positioning by anticipation approach. J Ind Manag Optim 3(1):29–50
Weigel G, D'Amours S, Martel A (2009) A modeling framework for maximizing value creation in pulp and paper mills. INFOR 47(3):247–260
Wising U, Stuart PR (2006) Identifying the Canadian forest biorefinery. Pulp Paper Canada 107(6):25–30

Chapter 18
Environmental Impact Assessment of Forest Operations and Pulp Manufacture

Sara González-García, Gumersindo Feijoo, María Teresa Moreira, Staffan Berg, and Roland Agnemo

18.1 Introduction

Nowadays, special attention is being paid on the environmental consequences derived from industrial sectors, specifically on wood products based sector, such as those relating to industrial emissions or to waste management. Thus, this chapter is focused on the identification of the environmental impacts derived throughout the life cycle of a specific wood based product, the dissolving pulp, as well as of the key environmental processes involved in its life cycle, from the production of the raw materials, that is the wood. Moreover, although the study was focused on this industrial wood product, the environmental consequences could be extended to other wood products.

18.1.1 Sustainable Development

The term sustainable development was defined by the Brundtland Commission to designate the development that "meets the needs of the present without compromising the ability of future generations to meet their own needs" (United Nations 1987). This definition was hazy although it shows the two principal features: the problem of the environmental degradation which is usually linked to an economic

---

S. González-García (✉) • G. Feijoo • M.T. Moreira
Department of Chemical Engineering, University of Santiago de Compostela, A Coruña, Spain
e-mail: sara.gonzalez@usc.es

S. Berg
The Forestry Research Institute of Sweden (Skogforsk), Uppsala Science Park, SE-751 83 Uppsala, Sweden

R. Agnemo
Domsjö Fabriker AB, S-891 86 Örnsköldsvik, Sweden

J.G. Borges et al. (eds.), *The Management of Industrial Forest Plantations: Theoretical Foundations and Applications*, Managing Forest Ecosystems 33,
DOI 10.1007/978-94-017-8899-1_18,

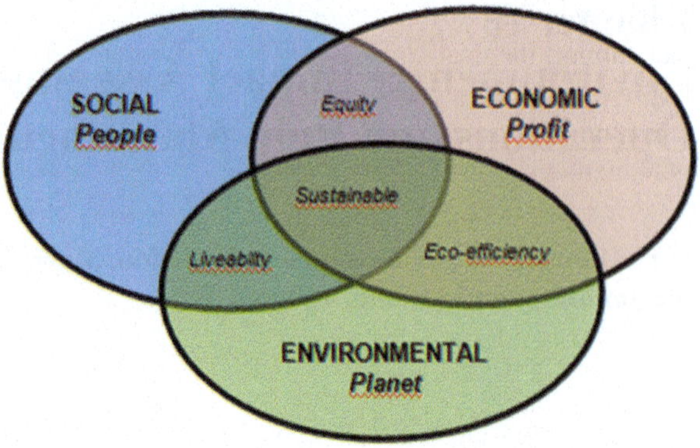

Fig. 18.1 Scheme of the three dimensions of sustainable development (Adapted from Adams 2006)

growth and the need for such growth to mitigate poverty (Adams 2006). Sustainable development is conceptually regarded as the intersection of three dimensions refers to environmental, economic and socio-political sustainability (Fig. 18.1).

The path towards sustainable development demands the modification of the current operational and environmental patterns. In this sense, it is necessary to pursue reductions in the consumption of materials and energy, as well as the mitigation of the corresponding environmental impacts (Feijoo et al. 2007a). Sustainability requires that human activity only uses nature's resources at a rate at which they can be replenished naturally. The long-term result of environmental degradation is the inability to sustain human life. Under this perspective, numerous environmental management tools have been developed with the aim of minimizing the environmental impacts linked to products, processes and services (Feijoo et al. 2007a).

18.1.2 Environmental Management Tools

Concepts such as responsible entrepreneurship and eco-efficiency reveal the current trend towards sustainable development when dealing with business. Thus, industries or industrial sectors should implement strategies that integrate the three pillars of sustainable development (Fig. 18.1). Environmental management tools have played, play and will play a major role in the establishment and confirmation of this trend (Baumann and Tillman 2004). They were developed to facilitate the enhancement of the environmental performance of industries and the integration of environmental, economic and social concerns.

Environmental management is focused on those activities that involve or may involve an impact on the environment. The advantages of performing environmental management include (Andersson 1998): (i) cost savings, (ii) legislative compliance, (iii) anticipation of future legislation, (iv) reduction of environmental risk, (v) fulfillment of supply chain requirements, (vi) improvement of relations with regulators, (vii) improvement of public image, (viii) increased market opportunities, and (ix) employee enthusiasm.

It is possible to find several types of environmental management tools. They can support the evaluation of environmental impacts (e.g. Life Cycle Assessment), improve product development (e.g. Eco-design) up to identify costs and benefits of environmental action (e.g. environmental accounts) (Bhamra 2004). Environmental management tools facilitate the execution of eco-efficiency strategies, life cycle thinking and environmental management systems (such as EMAS and ISO 14001) into the commercial network.

18.1.3 Life Cycle Assessment (LCA)

Environmental policy today focuses at the transition to sustainable production and consumption patterns. This is taking place in various ways and at various levels. Knowledge of the environmental impacts of production and consumption patterns is essential for improving the performance of industries and consumers. Integrated assessment of all environmental impacts from cradle-to-grave perspective is the base for achieving more sustainable products and services. One of the assessment tools widely used for this purpose is the Life Cycle Assessment –from now LCA (Guinée et al. 2001; Robert et al. 2002).

Although the concept of LCA evolved in the 1960s and there have been several efforts to develop LCA methodology since the 1970s, it has received much attention in environmental science fields since the 1990s. The Society of Environmental Toxicology and Chemistry (SETAC) has been involved in increasing the awareness and understanding of the concept of LCA. In the 1990s, SETAC in North America and the US Environmental Protection Agency (USEPA) sponsored workshops and several projects to develop and promote a consensus on a framework for conducting life cycle inventory analysis and impact assessment.

Similar efforts were undertaken by SETAC Europe, other international organizations (such as the International Organization for Standardization, ISO) and LCA practitioners worldwide. As a result of these efforts, agreement has been achieved on an overall LCA framework and a well-defined inventory methodology (ISO 14040 2006a).

LCA methodology has proved to be a valuable tool for documenting and analysing environmental considerations of systems (such as industrial plantations) that need to be part of the decision making process towards sustainability (Baumann and Tillman 2004). LCA offers a rational and comprehensive approach to environmental assessment encompassing all stages from a cradle to grave perspective, i.e., raw material extraction and processing; manufacture, transport and distribution; use; re-use, maintenance; recycling and final disposal (ISO 14040 2006a).

Development is considered as the principal area of application of LCA (Guinée et al. 2001) and it can be used to provide a picture as complete as possible of the interactions of an activity with the environment (e.g. when pulpwood is produced for further industrial uses), to identify major environmental impacts and the life-cycle stages or "hot-spots" contributing to these impacts, to compare environmental

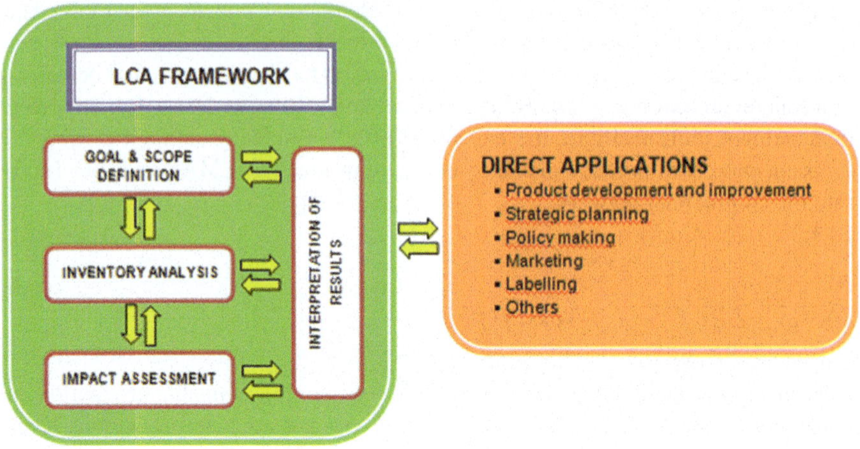

Fig. 18.2 LCA framework and applications (Adapted from ISO 14040 2006a)

impacts of alternative scenarios, to contribute to the understanding of the overall and interdependent nature of the environmental consequences of human activities and to provide decision makers with information on the environmental effects of these activities and to identify opportunities for environmental improvements. As presented in Fig. 18.2, four stages are distinguished for LCA studies (ISO 14040 2006a, b):

- Goal and scope definition.
- Inventory analysis.
- Impact assessment.
- Interpretation.

18.1.3.1 Goal and Scope Definition

This is the first stage of the LCA and probably the most important, since the elements defined here, such as purpose, scope, and main hypothesis are the key elements of the study (ISO 14040 2006a). Firstly, the goal is defined, as well as the reasons behind its implementation, the kind of decisions that will be made from the results obtained, and if these will be of internal use or external to inform the general public.

Secondly, the scope of the study must be described. This implies, among other elements, the definition of the system, its boundaries (conceptual, geographical and temporal), quality of the data used, the main hypothesis, as well as the limitations of the study.

A key issue in the scope is the definition of the functional unit (Baumann and Tillman 2004), which is the base of the calculations. It is often expressed in terms of amount of product, but it should really be related to the amount of product needed to perform a given function (Krozer and Vis 1998). For example, if it is assessed a pulp production system, the functional unit could be 1 t of bleached pulp and all the environmental flows should be referred to that unit. Thus, the adequate selection of a functional unit is of prime importance because different functional units could lead to different results for the same product systems (Kim and Dale 2006).

18.1.3.2 Inventory Analysis

The Life Cycle Inventory analysis (from now, LCI) is a methodology for estimating the consumption of resources and the qualities of waste flows and emissions caused by or otherwise attributable to the life cycle of the system under study (e.g. throughout the life cycle of the pulp production system) as defined in the scope (ISO 14040 2006a). This phase is the most effort intensive and time consuming compared to other phases in an LCA, mainly because of data collection. The processes within the life cycle and the associated material and energy flows as well as other exchanges are modelled to represent the system and its total inputs and outputs from and to the natural environment, respectively. It results in a system model and an inventory of environmental exchanges related to the functional unit. The activities of the LCI include (ISO 14040 2006a):

- Construction of the flowchart according to the system boundaries established in the goal and scope definition.
- Data collection for all the activities in the product system followed by the documentation of collected data.
- Calculation of the environmental loads (resource use and pollutant emissions) of the system in relation to the functional unit.

In order to make a representative and reproducible inventory, not only quality data is needed, but also a correct allocation of these data related to each of the subsystems to be evaluated (Feijoo et al. 2007b). A cause-effect relationship is established between raw material consumption/waste generation/system emissions and the activity or process which gives rise to the function being assessed (e.g. the production of pulp). In the case of systems accounting for only one product -monofunctional processes- (e.g. if only one type of pulp is produced in the production system under assessment), allocation is direct. However, problems emerge when dealing with overall input/output data for a system which produces more than one product -multifunctional processes- (e.g. if different types of pulp are produced simultaneously) or functions which influence more than one life cycle (e.g. open-loop recycling) (Feijoo et al. 2005). Multifunctional processes are those processes whose function requires the concurrence of more than one process. They include not only production processes which give rise to more than one product, but also waste treatment processes with more than one waste flow or energy generation (Ekvall and Finnveden 2001). In this type of systems, environmental burdens must be distributed among the different products or processes. With this purpose, inputs and outputs are allocated to the different products on the basis of procedures which must be clearly

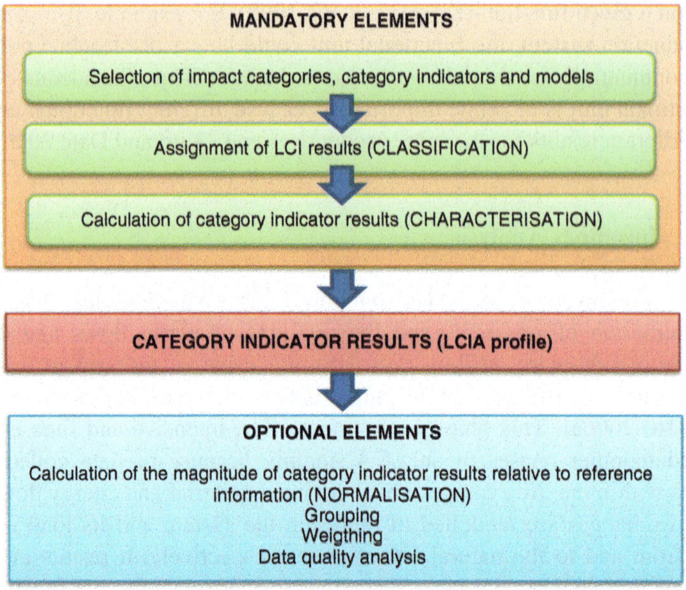

Fig. 18.3 Elements of LCIA as said by ISO 14040 (ISO 14040 2006a)

specified. Allocation procedures should capture the main features and relationships regarding inputs and outputs. The addition of the inputs and outputs allocated to one single unit process shall equal the addition of the inputs and outputs prior to allocation.

18.1.3.3 Impact Assessment

The Life Cycle Impact Assessment (LCIA) provides the indicators and the basis for analyzing the potential contributions of the resource extractions and wastes/emissions in an inventory to a number of potential impacts, within the framework of the goal and scope of the study. The result of the LCIA is an evaluation of a product life cycle (e.g. wood pulp production) on a functional unit basis (e.g. per tonne of pulp) in terms of several impacts categories (Baumann and Tillman 2004). Impact assessment in LCA generally consists of the following mandatory elements: classification, characterization, which converts LCI results to indicator results, and optional elements: normalization, grouping or weighting of the indicator results and data quality analysis techniques (mandatory in comparative assertions) (Fig. 18.3).

1. **Classification** is the process of assignment and initial aggregations of LCI data into common impact groups. For example, CO_2 and CH_4 emissions are classified in the category global warming potential.

2. **Characterization** is the assessment of the magnitude of potential impacts of each inventory flow into its corresponding environmental impact (e.g. modelling the potential impact of CO_2 and CH_4 on global warming potential). Characterization provides a way to directly compare the LCI results within each category. Characterization factors are commonly referred to as equivalency factors.
3. **Normalization** involves relating the environmental profile of the system to a broader data set or situation, e.g., relating the system's global warming potential to a country's global warming for one average year.
4. **Grouping** involves aggregating impact categories into one or more sets. Two possible procedures are followed: sorting of the category indicators on a nominal basis or ranking of the category indicators on an ordinal scale giving order or hierarchy.
5. **Weighting** is the phase where the environmental profile is reduced from a set of indicators to a single impact score by using weighting factors based on subjective value judgments.

There are a considerable number of impact assessment methodologies available to the LCA practitioner and, sometimes there is not one single obvious choice among them (Dreyer et al. 2003). In this study, the Dutch method Life Cycle Assessment – An Operational Guide to the ISO Standards (Guinée et al. 2001) has been selected, a midpoint methodology that has a widespread use in the LCA community (Dreyer et al. 2003). In particular, during the development of this research study, several updates have occurred and revised values of the characterization factors have been taken from the CML website (CML 2012). Several suggestions on complete sets of impact categories can be found in the LCA literature (Baumann and Tillman 2004). The selection of the categories studied in this study was based on the default list of impact categories reported by Guinèe et al. (2001). The unique exception was land use, which was not evaluated due to the lack of LCI data.

18.1.3.4 Interpretation of Results

This is the last stage of a LCA study, where the results obtained are presented in a straightforward way, presenting the critical sources of impacts and the options to reduce them. Interpretation involves a review of all the stages in the LCA process in order to check the consistency of the assumptions and the data quality in relation to the goal and scope of the study.

What You Will Learn in This Chapter

- the use of an environmental methodology for assessing the consumption of resources and the qualities of waste flows and emissions caused by or attributable to the life cycle of a production system
- the assessment from an environmental perspective of forest systems (pine and spruce) destined to wood production for industrial uses
- the identification of the most critical points with largest contributions to the environmental profile in the forest system

- the quantification of the environmental burdens derived from dissolving pulp production and the proposal of improvement alternatives for the environmental hot spots
- the results from this chapter should supply useful information to assist wood products industries in the aim of increasing their sustainability

18.2 Framework of Forest Operations in LCA Studies (State of Art)

Forest sector covers all the stakeholders with major interests in forestry, forest based materials and products. The forestry sector is well positioned to provide worldwide leadership in the practice of sustainable development. The forestry community is familiarized to a long-term perspective; it is reasonably knowledgeable about the response of forest ecosystems to natural and human disturbances; it is familiar with the sustained yield principle; and, in a few instances, it has attempted to practice a multiple and integrated use of forests (European Commission 2010).

Forest based products are biodegradable, reusable and recyclable and able to substitute products and energy from non-renewable sources. Furthermore, a great percentage of these products come from industrial plantations Environmental issues, together with energy and climate change are key priorities for forest-based industries and legislation helps the forest industry to become more eco-friendly. Forest biodiversity, Natura 2000 network, sustainable management and illegal logging are major concerns at the beginning of the chain (European Commission 2011). Within the forest sector those objectives include income and employment, social activities, healthy and productive forests, clean water and clean air. The three pillars are indispensable to support each other. The forest based sector needs to be economically strong so that part of the wealth generated can be used for forest conservation and other activities such as wastewater treatment. Moreover, it is needed environmental and social sustainability in order to be able to ensure the viability of the sector in the long run.

Sustainable development lies not with the natural changes that occur over time in undisturbed forests, but rather with the impact of human activity on the forest resource. While in the remote past, forest inhabitants across the globe utilized forest resources for subsistence with limited permanent impact on resources, the expansion of the agricultural frontier has led to significant and increasingly rapid permanent conversion of forest land to other uses (Holvoet and Muys 2004).

LCA of forestry and forest based products is relatively new and still developing which is already seen as an important tool to evaluate the environmental impact of forestry activities and forest products. It is evident that the utilization of wood as raw material needs an appropriate management as a key action to optimize the use of resources and to reduce the environmental impact associated. The Directive 2008/1/EC on integrated pollution prevention and control (IPPC) is very important for the forest based manufacturing processes (Official Journal of the European Union 2008).

Pulpwood from industrial plantations constitutes the main raw material of virgin paper pulp in developed countries and pulp manufacturing is the first non-food industrial utilization of plant biomass. Forest management practices considerably differ between countries and also may differ within the country. Forests face several environmental impacts derived from modern industrial society such as acidification, eutrophication and global warming as well as energy use, which must be considered when LCA methodology is applied to wooden pulp production.

So far, LCA has already been applied to evaluate the environmental impact of different wood products (Rex and Baumann 2007). Recent publications on LCA have studied wood products such as paper and pulp (Dias et al. 2007a; González-García et al. 2009a), boards (Rivela et al. 2006a, 2007; González-García et al. 2009b, 2011a), floor coverings (Petersen and Solberg 2004; Nebel et al. 2006), window frames (Richter and Gugerli 1996; Asif et al. 2002; Tarantini et al. 2011), walls (Werner 2001), furniture (Taylor and van Langenberg 2003), containers (González-García et al. 2011b), chairs (Fet et al. 2009), writing paper (Lopes et al. 2003) and packaging materials (Farreny et al. 2008) as well as the use of wood wastes in ephemeral architecture (Rivela et al. 2006b).

Moreover, numerous industries in the wood sector are being developed into wood based biorefineries producing a range of utilisable chemicals out of sidestreams and wastes. For example, black liquor (the residue from pulp production process) is an energy rich by-product. It also contains chemicals that are recovered to be reused in the pulping process. In addition, bioethanol obtained from fermentation or lignosulfonates from ultrafiltration of the black liquor can also be considered biorefinery by-products (González-García et al. 2011c).

Some of these studies have not analysed in detail the production of the feedstock or they have turned to databases. Therefore, its evaluation as well as the impact of the transport of round wood is needed and relevant as it could be a hot spot due to eutrophying related emissions. Only some papers can be found concerning the environmental effects of forest operations (silvicultural activities), nearly all them in Scandinavian countries (Aldentun 2002; Berg and Karjalainen 2003; Berg and Lindholm 2005; Lindholm and Berg 2005a, b; White et al. 2005; Michelsen 2007; Michelsen et al. 2008) where Nordic species such as Scots pine (Pinus sylvestris) and Norway spruce (Picea abies) have been evaluated in detail. Nowadays great efforts are being carried out in Iberian countries in order to have an environmental overview of the Iberian industrial plantations. This is the case of Portugal (Dias et al. 2007b) and Spain (González-García et al. 2009c, d), which analysed the productions associated to plantations of Maritime pine (Pinus pinaster) and Eucalyptus (E. globulus) from an LCA perspective. In addition, numerous studies were published in order to evaluate the material consumption in some forest operations (specifically logging) due to spare part utilization by harvesters and forwarders (Athanassiadis 2000; Athanassiadis et al. 2000) or to analyse the Swedish road transport (Eriksson et al. 1996) and Swedish wood distribution to specific wood based companies, specially pulp mills (González-García et al. 2009d).

18.3 LCA of Forest Operations in Pine and Spruce Industrial Plantations

This study comprises the production of the most important pulp wood species used nowadays in Swedish pulp mills: Norway spruce (Picea abies) and Scots pine (Pinus sylvestris) and the identification of the environmental loads associated with Swedish industrial plantations. Moreover, the pulpwood supply from the plantation landing to the pulp mill gate will be evaluated. Therefore, the main goal of this research work is to evaluate the environmental potential impacts linked to industrial wood production in Swedish plantations. To achieve this objective, inventory data from interviews and surveys from representative exploitations was collected. The identification of the activities with a significant environmental impact will make it possible to propose a framework for alternatives which leads to a better environmental performance.

18.3.1 Life Cycle Inventory

Field data were gathered by means of interviews, informal conversations and surveys of the forest operations carried out in Swedish industrial plantations located in Central and South Sweden (González-García et al. 2009c, d). It is important to make an observation here since large amounts of softwood from Baltic countries (Latvia, Estonia and Lithuania) are commonly processed in forest-based industries in Sweden. For this reason, the industrial plantations located in these countries were also considered but no significant differences in forest activities between both regions were assumed. Inventory data were completed with bibliographic resources when necessary. Forest management activities for the production of round softwood from field management to wood delivery at pulp mill gate were divided in three main subsystems (see Fig. 18.4): Silvicultural operations (S_i), Logging operations (S_{ii}) and Secondary hauling (S_{iii}).

Silviculture operations subsystem (S_i): the establishment of industrial plantation starts with site preparation that is, the eradication of unwanted vegetation and soil scarification by means of mechanised practices. Next, the regeneration takes place by manual planting and based fertilizer is applied in plantations located in Central Sweden. On the contrary, no fertilization is performed in South Sweden since soils are nitrogen rich. Some mechanized practices are carried out such as motor-manual cleaning and fertilization. Removal of unwanted vegetation in young stands is necessary to regulate tree species, composition, spatial distribution, growth and quality (González-García et al. 2009c).

Logging operations subsystem (S_{ii}): thinning, felling and extraction of wood are included in logging operations (González-García et al. 2009c). Around 40 % (immature stands) of stand is thinned to improve future stands development and to guarantee forest income (Eriksson 2006). Remaining 60 % of stand is felled.

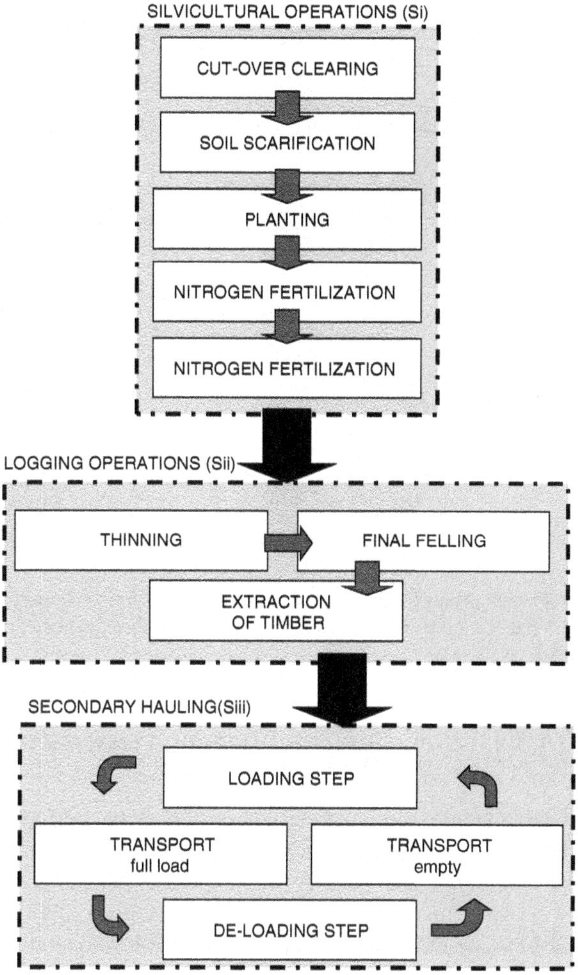

Fig. 18.4 Flowchart of the pulpwood production system under assessment

Both operations are carried out by means of a single-grip harvester. Commonly, standard Swedish stands are normally thinned 2–4 times during its growth cycle (Aldentun 2002). A forwarder is usually used in the extraction stage.

The inventory data for the South and Central Swedish industrial plantations corresponding to S_i and S_{ii} were extrapolated to Baltic countries (40 and 60 % of total instead of 30 and 45 % respectively).

Secondary hauling subsystem (S_{iii}): harvested pulpwood transportation from forest landing to the pulp mill gate is considered in this subsystem. Transport by road and sea was taken into account since 30 % of total wood processed comes from South Sweden, 45 % from Central Sweden and 25 % from Baltic countries (González-García et al. 2009c, d).

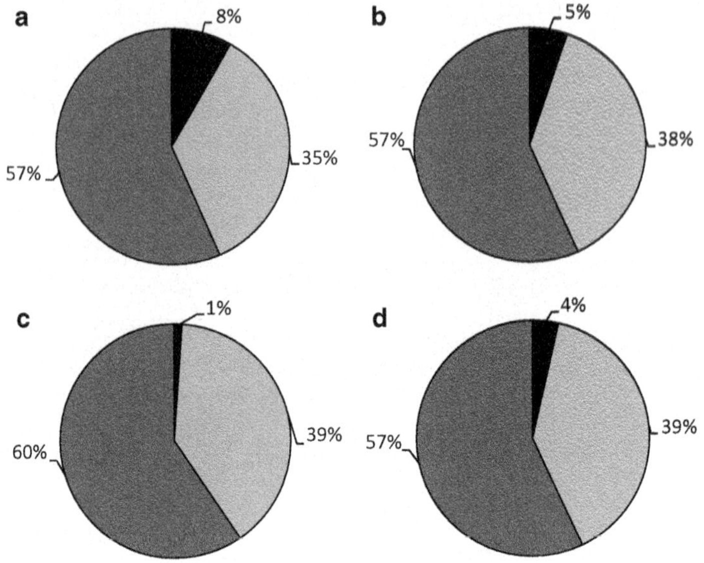

Fig. 18.5 Analysis of contributions per subsystem in impact categories under assessment. (**a**) Global warming, (**b**) Eutrophication, (**c**) Acidification, (**d**) Photochemical oxidants formation; (■Si ■Sii ■Siii)

18.3.2 Environmental Impact Assessment

Classification and characterization following ISO guidelines were applied to analyze the potential environmental impact of inputs (i.e. fertilizers and fossil fuel requirements) and outputs (diffuse and combustion emissions) from the LCI. LCA was performed using the CML 2 baseline 2000 V2.1 method (Guinée et al. 2001). The LCA software SimaPro 7.10 developed by PRé Consultants (2010) was used to perform the impact assessment stage. The impact categories considered under assessment were: Global Warming (GWP), Eutrophication (EP), Acidification (AP) and Photochemical Oxidants Formation (POP). According to the results (see Fig. 18.5), the contributions to all the impact categories under assessment were highly influenced by S_{iii} (more than 55 %) mainly due to the wood transport from Baltic countries and their corresponding high diesel consumption. Combustion emissions from fossil fuels in forest machinery significantly contributed to all impact categories under study (fossil CO_2, NO_x and sulphur emissions).

The low dose of fertilizer application supposed a small contribution to EP contrarily to other industrial plantations such as Eucalyptus in Iberian countries (González-García et al. 2009c). Wood transportation from landing to pulp mill uses large amounts of energy so special attention should be paid on this issue, especially when wood is delivered from Baltic countries (González-García et al. 2009c, d).

The environmental results are considerably affected by the mechanized degree, the effective work time and the nature of the stands in accordance with these results and the results obtained in other industrial plantations (González-García et al. 2009c, d). Additionally, weight and payload of freight lorries, which are commonly regulated, and where wide variations exist, can considerably influence the environmental profile. The use of more environmental friendly transport modules such as rail transport instead of lorry when wood is transported from South Sweden could allow important improvements in all impact categories under assessment mainly due to the lowest diesel requirements and thus, lowest combustion emissions, per functional unit (González-García et al. 2009d).

18.4 Environmental Assessment of Dissolving Cellulose Manufacture

The aim of this section was to evaluate the environmental potential impacts linked to Totally Chlorine Free (TCF) dissolving cellulose production using as main raw material the pulpwood assessed in the previous Sect. 18.3 (González-García et al. 2011c). The identification of the activities with a significant contribution to the environmental profile will make possible to propose a framework for alternatives which leads to a better environmental performance. The study under assessment (see Fig. 18.6) included all the industrial activities related to the production of dissolving cellulose in a Swedish biorefinery and corresponding co-products obtained in the mill: ethanol for chemical uses and lignosulfonates (González-García et al. 2011c). Thus, this study covers the whole cycle of the pulp mill from raw materials production (including industrial plantation activities – Sect. 18.3 of the current chapter) to the pulp mill gate (cradle-to-gate perspective).

The softwood logs directly delivered from the industrial plantations are debarked, chipped and fed to batch digesters to cook the cellulose adding the acid cooking (with HSO_3^- and SO_2). Subsequently, the resin content of the washed cellulose is reduced by means of an internal de-resination, where the resin is solubilized by alkali (González-García et al. 2011c). The following stage is the bleaching of cellulose in two stages in a completely close-loop bleaching sequence with a final peroxide bleaching step. The bleached cellulose (91–92 % ISO) is subsequently screened and dried. The black liquor from cooking and washing stages is evaporated to produce four streams. The first one is sent to the cogeneration plant to recover chemicals and to produce energy consumed in the process. The second stream is sent to the ethanol plant for fermentation where the hydrolyzed sugars are fermented to ethanol without enzymes. The third stream (52 % of solids) is sent to the lignosulfonate plant, where is dried and converted into fine dried powder (lignosulfonates). The fourth stream (remaining effluents) is sent to the wastewater treatment plant (WWTP) where under a first anaerobic and a second aerobic steps, biogas is produced and sludge used as soil conditioner (González-García et al. 2011c).

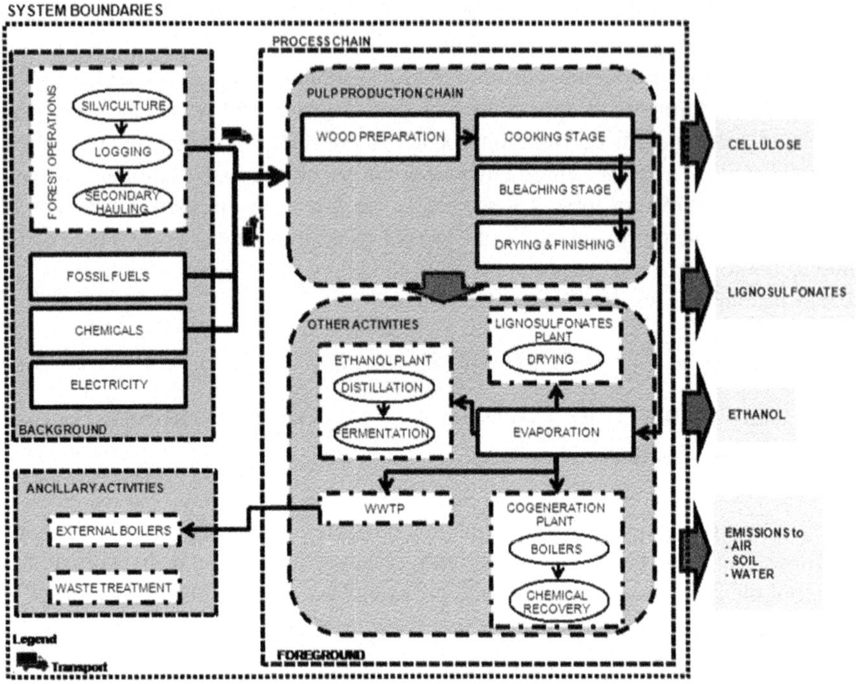

Fig. 18.6 Flowchart of the softwood based biorefinery under study

Ancillary activities correspond to the external boilers and general waste treatment. The biogas from the WWTP as well as resin and bark are burnt in external boilers to produce heat to supply the community and other small industries in the surroundings (González-García et al. 2011c).

18.4.1 Life Cycle Inventory

The pulp mill under study annually produces 210,000 t of dissolving cellulose (main product). Inventory data for the cellulose mill system were obtained from average annual data of on-site measurements (González-García et al. 2011c). These inventory data were completed, when necessary, with bibliographic resources and literature (González-García et al. 2011c).

18.4.2 Environmental Impact Assessment

In order to maintain the consistency with the previous section focused on industrial pulpwood production related operations, an identical methodology was used

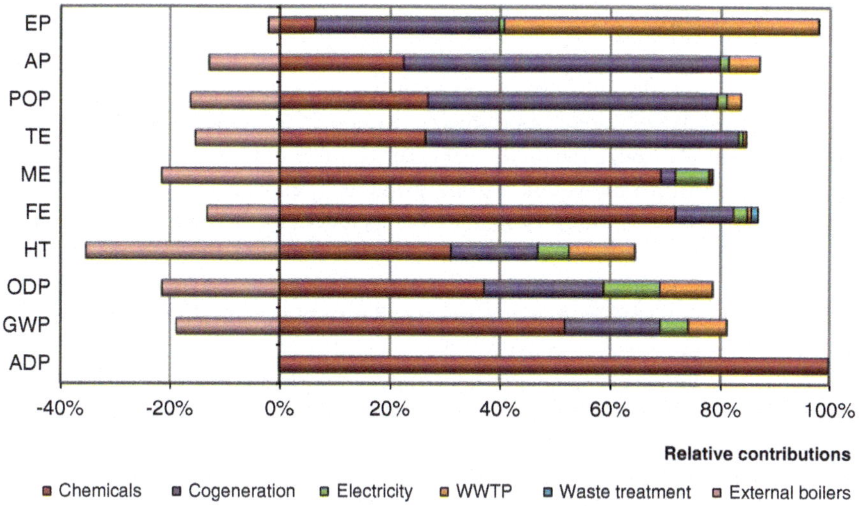

Fig. 18.7 Relative contributions from processes to each impact category

although all impact categories defined by the methodology were considered: abiotic depletion (ADP), global warming (GWP), ozone layer depletion (ODP), human toxicity (HT), fresh water aquatic ecotoxicity (FE), marine aquatic ecotoxicity (ME), terrestrial ecotoxicity (TE), photochemical oxidants formation (POP), acidification (AP) and eutrophication (EP), with the objective of presenting a comprehensive overview of the environmental performance of this industrial process.

According to the results, forest activities related to the production of the raw material played a minor role in the environmental profile. However, the production of chemicals consumed in the biorefinery activities (specifically in cooking and bleaching stages), the sludge treatment generated in the wastewater treatment plant (commonly as soil conditioner) and the on-site energy production system (cogeneration unit) were identified as key environmental factors. Specific actions associated to the reuse of wastes and improved gas treatment systems would improve the environmental profile of this production activity (see Fig. 18.7).

The production of chemical cellulose has been traditionally considered an important source of pollution due to its intensive energy consumption and the use of large amounts of chemicals, fuel and water. Nevertheless, this situation is changing due to the potential of wood biomass as a renewable source to produce value added products as well as energy and chemicals, minimizing environmental impacts, increasing sustainability and making this sector competitive and becoming less dependent on crude oil. In fact, this biorefinery not only produce cellulose but also other value co-products from waste streams such as lignosulfonates and ethanol (González-García et al. 2011c).

18.5 Summary

This chapter gives detailed information related to environmental concerns in the wood products based sector, and illustrates the role of LCA in addressing these concerns using Sweden case studies. This activity is multifunctional and comprises a wide range of operations that use the wood as a resource for producing goods. Few sectors have such multidisciplinary competences as this one. Wood industrial plantations face several environmental problems created by modern industrial society such as acidification, eutrophication and global warming as well as large amounts of energy are required in their related activities, which must be considered when LCA methodology is applied to wood based products.

Wood constitutes the main raw material of numerous products and cellulose manufacturing is the first non-food industrial utilization of plant biomass. Industrial plantations related operations were analysed in detail and several objectives were addressed in this chapter, which could be applied even to other wood based products:

(a) The study of the environmental loads associated with the wood plantations focused on spruce and pine industrial plantations (the main economically raw material for paper cellulose production).
(b) The study of the environmental loads associated with the production of spruce and pine TCF dissolving cellulose from a cradle-to-gate perspective in a biorefinery with the co-production of ethanol and lignosulfonates.

The conclusions from all these studies will enable to: (i) highlight the large influence of the geographical origin of wood processed not only in a pulp mill but also in any wood industry; (ii) provide useful information that can assist wood products industries in the aim of increasing their sustainability; (iii) identify the more environmental friendly pulpwood transport strategy and (iv) identify specific actions associated to reuse of wastes in the biorefinery as well as the introduction of superior gas treatment systems which could improve the environmental profile of this production process.

18.6 Problems

LCA offers a rational and comprehensive approach to environmental assessment of production systems as well as LCA studies can be performed in a whole range of different decision-making situations, ranging from internal use to public comparative use (Guinée et al. 2001). The nucleus characteristic of LCA is its holistic nature, which is its major strength and also its limitation since the broad scope of analysing the entire life cycle of products and processes can only be achieved at expenses of simplifying other aspects. Therefore, this is one of the problems of this environmental methodology.

All required inputs and emissions in many stages and operations of the life cycle are considered to be within the system boundaries. This includes not only inputs and emissions for production, distribution, use and disposal, but also indirect inputs and emissions (e.g. from the initial production of the energy used) regardless of when or where they occur. It shares some problems with other analysis techniques (e.g. system boundary selection), others are typical problems in LCA (i.e. allocation), importance of still others emerged in the last decades (i.e. spatial variation and site dependency) and all these problems are worthy of particular attention.

Acknowledgments This study was developed within the framework of the BIORENEW Integrated Project (Project reference: NMP2-CT-2006-026456) and has been partially financed by the Xunta de Galicia (Project Reference: GRC 2010/37). Dr. S. González-García would like to express her gratitude to the Spanish Ministry of Education for financial support (Grant No. AP2005-2374).

References

Adams WM (2006) The future of sustainability: re-thinking environment and development in the twenty-first century. Report of the IUCN renowned thinkers meeting, 19–31 January 2006

Aldentun Y (2002) Life cycle inventory of forest seedling production – from seed to regeneration site. J Clean Prod 10:47–55

Andersson I (1998) Environmental management tools for SMEs: a handbook. Environmental issues series. Luxembourg: Office for Official Publications of the European Communities; Copenhagen: European Environment Agency

Asif M, Davidson A, Muneer T (2002) Life cycle of window materials – a comparative assessment. Millenium Fellow School of Engineering, Napier University, Edinburgh

Athanassiadis D (2000) Energy consumption and exhaust emissions in mechanized timber harvesting operations in Sweden. Sci Total Environ 255:135–143

Athanassiadis D, Lidestav G, Wästerlund I (2000) Assessing material consumption due to spare part utilization by harvesters and forwarders. Int J For Eng 11:51–57

Baumann H, Tillman AM (2004) The Hitch Hilker's guide to LCA. An orientation in life cycle assessment methodology and application. Studentlitteratur, Lund. ISBN 9144023642

Berg S, Karjalainen T (2003) Comparison of greenhouse gas emissions from forest operations in Finland and Sweden. Forestry 76:271–284

Berg S, Lindholm EL (2005) Energy use and environmental impacts of forest operations in Sweden. J Clean Prod 13:33–42

Bhamra TA (2004) Eco-design: the search for development new strategies in product. Proc Inst Mech Eng B Int J Eng Sci 218(5):557–569

CML – Centre of Environmental Science (2012) Available from: http://www.leidenuniv.nl/interfac/cml/spp/projects/lca2/index.html. Accessed 23 Mar 2012

Dias AC, Arroja L, Capela I (2007a) Life cycle assessment of printing and writing paper produced in Portugal. Int J Life Cycle Assess 12:521–528

Dias AC, Arroja L, Capela I (2007b) Carbon dioxide emissions from forest operations in Portuguese eucalypt and maritime pine stands. Scand J For Res 22:422–432

Dreyer LC, Niemann AL, Hauschild M (2003) Comparison of three different LCIA methods: EDIP97, CML2001 and Eco-indicator 99. Int J Life Cycle Assess 8:191–200

Ekvall T, Finnveden G (2001) Allocation in ISO 14041 – a critical review. J Clean Prod 9:197–208

Eriksson E (2006) Thinning operations and their impact on biomass production in stands of Norway spruce and Scots pine. Biomass Bioenergy 30:848–854

Eriksson E, Blingue M, Lövgren G (1996) Life cycle assessment of the road transport sector. Sci Total Environ 189(190):69–76

European Commission (2010) Available from: http://ec.europa.eu/enterprise/sectors/wood-paper-printing/forestry/sustainable/index_en.htm. Accessed 18 July 2011

European Commission (2011) Available from: v http://ec.europa.eu/environment/nature/natura2000/index_en.htm. Accessed 19 July 2011

Farreny R, Gasol CM, Gabarrell X, Rieradevall J (2008) Life cycle assessment comparison among different reuse intensities for industrial wooden containers. Int J Life Cycle Assess 13:421–431

Feijoo G, Hospido A, Moreira MT (2005) Reglas de asignación en procesos multifuncionales. In: Clemente G, Sanjuán N, Vivancos JL (eds) Análisis de Ciclo de Vida: Aspectos metodológicos y casos prácticos. UPC Publisher, Valencia

Feijoo G, Hospido A, Moreira MT (2007a) Análisis de Ciclo de Vida (I): desarrollo sostenible y ACV. Ingeniería Química 443:153–161

Feijoo G, Hospido A, Gallego A, Rivela B, Moreira MT (2007b) Análisis de Ciclo de Vida (II): metodología y etapas. Ingeniería Química 444:114–125

Fet AM, Skaar C, Michelsen O (2009) Product category rules and environmental product declarations as tools to promote sustainable products: experiences from a case study of furniture production. Clean Technol Environ Policy 11:201–207

González-García S, Hospido A, Moreira MT, Romero J, Feijoo G (2009a) Environmental impact assessment of total chlorine free pulp from Eucalyptus globulus in Spain. J Clean Prod 17:1010–1016

González-García S, Moreira MT, Feijoo G (2009b) Environmental performance assessment of hardboard manufacture. Int J Life Cycle Assess 14:456–466

González-García S, Berg S, Feijoo G, Moreira MT (2009c) Environmental impacts of production and supply of pulpwood: Spanish and Swedish case studies. Int J Life Cycle Assess 14:340–353

González-García S, Berg S, Feijoo G, Moreira MT (2009d) Comparative environmental assessment of wood transport models. A case study of a Swedish Pulp Mill. Sci Total Environ 407:3530–3539

González-García S, Feijoo G, Heathcote C, Kandelbauer A, Moreira MT (2011a) Environmental assessment of Green hardboard production coupled with a Laccase activated system. J Clean Prod 19:445–453

González-García S, Silva FJ, Moreira MT, Castilla Pascual R, García-Lozano R, Gabarrell X, Rieradevall i Pons J, Feijoo G (2011b) Combined application of LCA and Eco-design for the sustainable procurement of wood boxes for wine bottles storage. Int J Life Cycle Assess 16:224–237

González-García S, Hospido A, Agnemo R, Svensson P, Selling E, Moreira MT, Feijoo G (2011c) Environmental life cycle assessment of a Swedish dissolving pulp mill integrated biorefinery. J Ind Ecol. doi:10.1111/j.1530-9290.2011.00354.x

Guinée JB, Gorrée M, Heijungs R, Huppes G, Kleijn R, de Koning A, van Oers L, Wegener A, Suh S, Udo de Haes HA (2001) Life cycle assessment. An operational guide to the ISO standards. Centre of Environmental Science, Leiden

Holvoet B, Muys B (2004) Sustainable forest management worldwide: a comparative assessment of standards. Int For Rev 6(2):99–122

ISO (2006a) ISO 14040 International standard, environmental management – life cycle assessment – principles and framework. International Organisation for Standardization, Geneva

ISO (2006b) ISO 14044 International standard, environmental management – life cycle assessment – requirements and guidelines. International Organisation for Standardisation, Geneva

Kim S, Dale BE (2006) Ethanol fuels: E10 or E85 – life cycle perspectives. Int J Life Cycle Assess 11:117–121

Krozer J, Vis JC (1998) How to get LCA in the right direction? J Clean Prod 6:53–61

Lindholm EL, Berg S (2005a) Energy requirement and environmental impact in timber transport. Scand J For Res 20:184–191

Lindholm EL, Berg S (2005b) Energy use in Swedish Forestry in 1972 and 1997. Int J For Eng 16:27–37

Lopes E, Dias A, Arroja L, Capela I, Pereira F (2003) Application of life cycle assessment to the Portuguese pulp and paper industry. J Clean Prod 11(1):51–59

Michelsen O (2007) Assessment of land use impact on biodiversity: proposal of a new methodology exemplified with forestry operations in Norway. Int J Life Cycle Assess 12:22–31

Michelsen O, Solli C, Strømman AH (2008) Environmental impact and added value in forestry operations in Norway. J Ind Ecol 12:69–81

Nebel B, Zimmer B, Wegener G (2006) Life cycle assessment of wood floor coverings. A representative study for the German flooring industry. Int J Life Cycle Assess 11:172–192

Official Journal of the European Union (2008) Directive 2008/1/EC of the European Parliament and of the Council of 15 January 2008 concerning integrated pollution prevention and control. Available from: http://eur-lex.europa.eu/LexUriServ/LexUriServ.do?uri=OJ:L:2008:024:0008:0029:EN:PDF. Accessed 19 Oct 2011

Petersen AK, Solberg B (2004) Greenhouse gas emissions and costs over the Life Cycle of Wood and alternative flooring materials. Climate Change 64:143–167

PRé Consultants (2010) Available from: http://www.pre.nl. Accessed 28 Oct 2010

Rex ELC, Baumann H (2007) Individual adaptation of industry LCA practice: results from two case studies in the Swedish Forest Products Industry. Int J Life Cycle Assess 12:266–271

Richter K, Gugerli H (1996) Wood and wood products in comparative life cycle assessment. Holz als Roh- und Werkstoff 54:225–231

Rivela B, Hospido A, Moreira MT, Feijoo G (2006a) Life cycle inventory of particleboard: a case study in the wood sector. Int J Life Cycle Assess 11:106–113

Rivela B, Moreira MT, Muñoz I, Rieradevall J, Feijoo G (2006b) Life cycle assessment of wood wastes: a case study of ephemeral architecture. Sci Total Environ 357:1–11

Rivela B, Moreira MT, Feijoo G (2007) Life cycle inventory of medium density fibreboard. Int J Life Cycle Assess 12:143–150

Robert KH, Schimdt-Bleek B, Aloisi de Larderel J, Basile G, Jansen JL, Kuehr R, Price Thomas P, Suzuki M, Hawken P, Wackernagel M (2002) Strategic sustainable development – selection, design and synergies of applied tools. J Clean Prod 10:197–214

Tarantini M, Dominici Loprieno A, Porta PL (2011) A life cycle approach to Green Public Procurement of building materials and elements: a case study on windows. Energy 36:2473–2482

Taylor J, van Langenberg K (2003) Review of the environmental impact of wood compared with alternative products used in the production of furniture. CSIRO Forestry and Forest Products Research and Development Corporation, Victoria

United Nations (1987) Report of the world commission on environment and development. General Assembly Resolution 42/187, 11 Dec 1987

Werner F (2001) Recycling of used wood – inclusion of end-of-life options in LCA. In: Jungmeier G (ed) Life cycle assessment of forestry and forest products; achievements of COST Action E9 working group 3 'end of life: recycling, disposal and energy generation'. Joanneum, Institute of Energy Research, Graz, pp 6/1–24

White MK, Gower ST, Ahl DE (2005) Life cycle inventories of round wood production in northern Wisconsin: inputs into an industrial forest carbon budget. For Ecol Manage 219:13–28

Index

J.G. Borges et al. (eds.), *The Management of Industrial Forest Plantations: Theoretical Foundations and Applications*, Managing Forest Ecosystems 33, DOI 10.1007/978-94-017-8899-1, © Springer Science+Business Media Dordrecht 2014

The manufacturer's authorised representative in the EU is Springer Nature Customer Service Centre GmbH, Europaplatz 3, 69115 Heidelberg, Germany. If you have any concerns regarding our products, please contact ProductSafety@springernature.com

Printed and bound by CPI Group (UK) Ltd, Croydon, CR0 4YY
30/09/2026
02225445-0015